CW01280409

MULTIPLE ACCESS PROTOCOLS FOR MOBILE COMMUNICATIONS

Multiple Access Protocols for Mobile Communications
GPRS, UMTS and Beyond

Alex Brand
Swisscom Mobile, Switzerland

Hamid Aghvami
King's College London, UK

JOHN WILEY & SONS, LTD

Copyright © 2002 by John Wiley & Sons, Ltd
Baffins Lane, Chichester,
West Sussex, PO19 1UD, England

National 01243 779777
International (+44) 1243 779777
e-mail (for orders and customer service enquiries): cs-books@wiley.co.uk
Visit our Home Page on http://www.wiley.co.uk or http://www.wiley.com

All Rights Reserved. No part of this publication may be reproduced, stored in a retrieval system, or transmitted, in any form or by any means, electronic, mechanical, photocopying, recording, scanning or otherwise, except under the terms of the Copyright Designs and Patents Act 1988 or under the terms of a licence issued by the Copyright Licensing Agency, 90 Tottenham Court Road, London, W1P 9HE, UK, without the permission in writing of the Publisher, with the exception of any material supplied specifically for the purpose of being entered and executed on a computer system, for exclusive use by the purchaser of the publication.

Neither the author(s) nor John Wiley & Sons, Ltd accept any responsibility or liability for loss or damage occasioned to any person or property through using the material, instructions, methods or ideas contained herein, or acting or refraining from acting as a result of such use. The author(s) and Publisher expressly disclaim all implied warranties, including merchantability of fitness for any particular purpose.

Designations used by companies to distinguish their products are often claimed as trademarks. In all instances where John Wiley & Sons, Ltd is aware of a claim, the product names appear in initial capital or capital letters. Readers, however, should contact the appropriate companies for more complete information regarding trademarks and registration.

Other Wiley Editorial Offices

John Wiley & Sons, Inc., 605 Third Avenue,
New York, NY 10158-0012, USA

WILEY-VCH Verlag GmbH
Pappelallee 3, D-69469 Weinheim, Germany

John Wiley & Sons Australia Ltd, 33 Park Road, Milton,
Queensland 4064, Australia

John Wiley & Sons (Canada) Ltd, 22 Worcester Road
Rexdale, Ontario, M9W 1L1, Canada

John Wiley & Sons (Asia) Pte Ltd, 2 Clementi Loop #02-01,
Jin Xing Distripark, Singapore 129 809

A catalogue record for this book is available from the British Library

British Library Cataloguing in Publication Data

Brand, Alex
    Multiple access protocols for mobile communications: GPRS, UMTS and beyond/
  Alex Brand, Hamid Aghvami
      p.cm.
    Includes bibliographical references and index.
    ISBN 0-471-49877-
    1. Global system for mobile communications. I. Aghvami, Hamid. II. Title.

TK5103.483 .B73 2001
621.382'12–dc21   2001055758
ISBN 0 471 49877 7
Typeset in 10/12pt Times by Laserwords Private Limited, Madras, India.
Printed and bound in Great Britain by Antony Rowe Ltd, Chippenham, Wiltshire.
This book is printed on acid-free paper responsibly manufactured from sustainable forestry, in which at least two trees are planted for each one used for paper production.

To Monica

CONTENTS

Preface		xv
Acknowledgements		xix
Abbreviations		xxi
Symbols		xxxi

1 Introduction — 1
  1.1 An Introduction to Cellular Communication Systems — 1
    1.1.1 The Cellular Concept — *1*
    1.1.2 Propagation Phenomena in Cellular Communications — *2*
    1.1.3 Basic Multiple Access Schemes — *3*
    1.1.4 Cell Clusters, Reuse Factor and Reuse Efficiency — *6*
    1.1.5 Types of Interference and Noise Affecting Communications — *6*
  1.2 The Emergence of the Internet and its Impact on Cellular Communications — 8
  1.3 The Importance of Multiple Access Protocols in Cellular Communications — 10
  1.4 A PRMA-based Protocol for Hybrid CDMA/TDMA — 12
    1.4.1 Why Combine CDMA and PRMA? — *12*
    1.4.2 Hybrid CDMA/TDMA Multiple Access Schemes — *14*
    1.4.3 Literature on Multiple Access Protocols for Packet CDMA — *15*
    1.4.4 Access Control in Combined CDMA/PRMA Protocols — *15*
    1.4.5 Summary — *21*

2 Cellular Mobile Communication Systems: From 1G to 4G — 23
  2.1 Advantages and Limitations of the Cellular Concept — 23
  2.2 1G and 2G Cellular Communication Systems — 25
    2.2.1 Analogue First Generation Cellular Systems — *25*
    2.2.2 Digital Second Generation Systems — *25*
  2.3 First 3G Systems — 27
    2.3.1 Requirements for 3G — *27*
    2.3.2 Evolution of 2G Systems towards 3G — *29*
    2.3.3 Worldwide 3G Standardisation Efforts — *31*
    2.3.4 The Third Generation Partnership Project (3GPP) — *32*
    2.3.5 The Universal Mobile Telecommunications System (UMTS) — *33*
    2.3.6 The Spectrum Situation for UMTS — *35*
    2.3.7 UTRA Modes vs UTRA Requirements — *36*

		2.3.8 3GPP2 and cdma2000	38
	2.4	Further Evolution of 3G	40
		2.4.1 Support of IP Multimedia Services through EGPRS and UMTS	40
		2.4.2 Improvements to cdma2000 1×RTT, UTRA FDD and TDD	41
		2.4.3 Additional UTRA Modes	42
	2.5	And 4G?	44
		2.5.1 From 1G to 3G	44
		2.5.2 Possible 4G Scenarios	44
		2.5.3 Wireless Local Area Network (WLAN) Standards	46
	2.6	Summary	48
3	**Multiple Access in Cellular Communication Systems**		**49**
	3.1	Multiple Access and the OSI Layers	49
	3.2	Basic Multiple Access Schemes	53
	3.3	Medium Access Control in 2G Cellular Systems	57
		3.3.1 Why Medium Access Control is Required	57
		3.3.2 Medium Access Control in GSM	58
	3.4	MAC Strategies for 2.5G Systems and Beyond	59
		3.4.1 On the Importance of Multiple Access Protocols	59
		3.4.2 Medium Access Control in CDMA	60
		3.4.3 Conflict-free or Contention-based Access?	62
	3.5	Review of Contention-based Multiple Access Protocols	63
		3.5.1 Random Access Protocols: ALOHA and S-ALOHA	64
		3.5.2 Increasing the Throughput with Splitting or Collision Resolution Algorithms	68
		3.5.3 Resource Auction Multiple Access	69
		3.5.4 Impact of Capture on Random Access Protocols	70
		3.5.5 Random Access with CDMA	72
		3.5.6 Protocols based on some Form of Channel Sensing	72
		3.5.7 Channel Sensing with CDMA	74
		3.5.8 A Case for Reservation ALOHA-based Protocols	75
	3.6	Packet Reservation Multiple Access: An R-ALOHA Protocol Supporting Real-time Traffic	76
		3.6.1 PRMA for Microcellular Communication Systems	76
		3.6.2 Description of 'Pure' PRMA	77
		3.6.3 Shortcomings of PRMA	79
		3.6.4 Proposed Modifications and Extensions to PRMA	81
		3.6.5 PRMA for Hybrid CDMA/TDMA	84
	3.7	MAC Requirements vs R-ALOHA Design Options	86
		3.7.1 3G Requirements Relevant for the MAC Layer	86
		3.7.2 Quality of Service Requirements and the MAC Layer	89
		3.7.3 A few R-ALOHA Design Options	92
		3.7.4 Suitable R-ALOHA Design Choices	94
	3.8	Summary and Scope of Further Investigations	96
4	**Multiple Access in GSM and (E)GPRS**		**99**
	4.1	Introduction	99
		4.1.1 The GSM System	99

		4.1.2 GSM Phases and Releases	*101*
		4.1.3 Scope of this Chapter	*104*
		4.1.4 Approach to the Description of the GSM Air Interface	*105*
4.2	Physical Channels in GSM		106
		4.2.1 GSM Carriers, Frequency Bands, and Modulation	*107*
		4.2.2 TDMA, the Basic Multiple Access Scheme — Frames, Time-slots and Bursts	*108*
		4.2.3 Slow Frequency Hopping and Interleaving	*111*
		4.2.4 Frame Structures: Hyperframe, Superframe and Multiframes	*115*
		4.2.5 Parameters describing the Physical Channel	*115*
4.3	Mapping of Logical Channels onto Physical Channels		115
		4.3.1 Traffic Channels	*115*
		4.3.2 Signalling and Control Channels	*116*
		4.3.3 Mapping of TCH and SACCH onto the 26-Multiframe	*120*
		4.3.4 Coding, Interleaving, and DTX for Voice on the TCH/F	*120*
		4.3.5 Coding and Interleaving on the SACCH	*124*
		4.3.6 The Broadcast Channel and the 51-Multiframe	*124*
4.4	The GSM RACH based on Slotted ALOHA		126
		4.4.1 Purpose of the RACH	*126*
		4.4.2 RACH Resources in GSM	*127*
		4.4.3 The Channel Request Message	*127*
		4.4.4 The RACH Algorithm	*128*
		4.4.5 Contention Resolution in GSM	*131*
		4.4.6 RACH Efficiency and Load Considerations	*132*
4.5	HSCSD and ECSD		134
		4.5.1 How to Increase Data-rates	*134*
		4.5.2 Basic Principles of HSCSD	*135*
		4.5.3 Handover in HSCSD	*136*
		4.5.4 HSCSD Multi-slot Configurations and MS Classes	*137*
		4.5.5 Enhanced Circuit-Switched Data (ECSD)	*139*
4.6	Resource Utilisation and Frequency Reuse		140
		4.6.1 When are Resources Used and for What?	*140*
		4.6.2 How to Assess Resource Utilisation	*143*
		4.6.3 Some Theoretical Considerations — The Erlang B Formula	*144*
		4.6.4 Resource Utilisation in Blocking-limited GSM	*145*
		4.6.5 Resource Utilisation in Interference-limited GSM	*152*
4.7	Introduction to GPRS		155
		4.7.1 The Purpose of GPRS: Support of Non-real-time Packet-data Services	*155*
		4.7.2 Air-Interface Proposals for GPRS	*157*
		4.7.3 Basic GPRS Principles	*158*
		4.7.4 GPRS System Architecture	*160*
		4.7.5 GPRS Protocol Stacks	*161*
		4.7.6 MS Classes	*163*
		4.7.7 Mobility Management and Session Management	*163*

4.8	GPRS Physical and Logical Channels	164
	4.8.1 The GPRS Logical Channels	164
	4.8.2 Mapping of Logical Channels onto Physical Channels	165
	4.8.3 Radio Resource Operating Modes	168
	4.8.4 The Half-Rate PDCH and Dual Transfer Mode	169
4.9	The GPRS Physical Layer	170
	4.9.1 Services offered and Functions performed by the Physical Link Layer	171
	4.9.2 The Radio Block Structure	171
	4.9.3 Channel Coding Schemes	171
	4.9.4 Theoretical GPRS Data-Rates	172
	4.9.5 'Real' GPRS Data-rates and Link Adaptation	175
	4.9.6 The Timing Advance Procedure	177
	4.9.7 Cell Reselection	179
	4.9.8 Power Control	179
4.10	The GPRS RLC/MAC	180
	4.10.1 Services offered and Functions performed by MAC and RLC	180
	4.10.2 The RLC Sub-layer	181
	4.10.3 Basic Features of the GPRS MAC	181
	4.10.4 Multiplexing Principles	183
	4.10.5 RLC/MAC Block Structure	184
	4.10.6 RLC/MAC Control Messages	185
	4.10.7 Mobile Originated Packet Transfer	188
	4.10.8 Mobile Terminated Packet Transfer	194
4.11	The GPRS Random Access Algorithm	197
	4.11.1 Why a New Random Access Scheme for GPRS?	197
	4.11.2 Stabilisation of the Random Access Algorithm	198
	4.11.3 Prioritisation at the Random Access	206
	4.11.4 The GPRS Random Access Algorithm	207
4.12	EGPRS	211
	4.12.1 EGPRS Coding Schemes and Link Quality Control	212
	4.12.2 Other EGPRS Additions and Issues	216
	4.12.3 EDGE Compact	218
	4.12.4 Further Evolution of GPRS	220

5 Models for the Physical Layer and for User Traffic Generation — **221**

5.1	How to Account for the Physical Layer?	221
	5.1.1 What to Account For and How?	221
	5.1.2 Using Approximations for Error Performance Assessment	222
	5.1.3 Modelling the UTRA TD/CDMA Physical Layer	223
	5.1.4 On Capture and Required Accuracy of Physical Layer Modelling	225
5.2	Accounting for MAI Generated by Random Codes	225
	5.2.1 On Gaussian Approximations for Error Performance Assessment	225
	5.2.2 The Standard Gaussian Approximation	227
	5.2.3 Deriving Packet Success Probabilities	228
	5.2.4 Importance of FEC Coding in CDMA	229

		5.2.5 Accounting for Intercell Interference	231
		5.2.6 Impact of Power Control Errors	236
	5.3	Perfect-collision Code-time-slot Model for TD/CDMA	237
		5.3.1 TD/CDMA as a Mode for the UMTS Terrestrial Radio Access	237
		5.3.2 The TD/CDMA Physical Layer Design Parameters	238
		5.3.3 In-Slot Protocols on TD/CDMA	240
	5.4	Accounting for both Code-collisions and MAI	241
	5.5	The Voice Traffic Model	242
		5.5.1 Choice of Model	242
		5.5.2 Description of the Chosen Source Model	244
		5.5.3 Model of Aggregate Voice Traffic	245
	5.6	Traffic Models for NRT Data	246
		5.6.1 Data Terminals	246
		5.6.2 The UMTS Web Browsing Model	247
		5.6.3 Proposed Email Model	250
		5.6.4 A Word on Traffic Asymmetry	252
		5.6.5 Random Data Traffic	253
	5.7	Some Considerations on Video Traffic Models	253
	5.8	Summary and some Notes on Terminology	255
6	**Multidimensional PRMA**		**257**
	6.1	A Word on Terminology	257
	6.2	Description of MD PRMA	258
		6.2.1 Some Fundamental Considerations and Assumptions	258
		6.2.2 The Channel Structure Considered	258
		6.2.3 Contention and Packet Dropping	259
		6.2.4 Accounting for Coding and Interleaving	261
		6.2.5 Duration of a Reservation Phase	261
		6.2.6 Downlink Signalling of Access Parameters and Acknowledgements	262
		6.2.7 Resource Allocation Strategies for Different Services	263
		6.2.8 Performance Measures for MD PRMA	263
	6.3	MD PRMA with Time-Division Duplexing	264
		6.3.1 Approaches to Time-Division Duplexing	264
		6.3.2 TDD with Alternating Uplink and Downlink Slots	266
		6.3.3 MD FRMA for TDD with a Single Switching-Point per Frame	266
	6.4	Load-based Access Control	267
		6.4.1 The Concept of Channel Access Functions	267
		6.4.2 Downlink Signalling with Load-based Access Control	269
		6.4.3 Load-based Access Control in MD PRMA vs Channel Load Sensing Protocol for Spread Slotted ALOHA	269
	6.5	Backlog-based Access Control	270
		6.5.1 Stabilisation of Slotted ALOHA with Ternary Feedback	270
		6.5.2 Pseudo-Bayesian Broadcast for Slotted ALOHA	270
		6.5.3 Bayesian Broadcast for Two-Carrier Slotted ALOHA	271

		6.5.4 *Bayesian Broadcast for MD PRMA with Orthogonal Code-Slots*	*272*
		6.5.5 *Accounting for Acknowledgement Delays*	*273*
		6.5.6 *Bayesian Broadcast for MD FRMA*	*274*
		6.5.7 *Estimation of the Arrival Rate*	*274*
		6.5.8 *Impact of MAI on Backlog Estimation*	*275*
	6.6	Combining Load- and Backlog-based Access Control	276
	6.7	Summary	276

7 MD PRMA with Load-based Access Control — 279

7.1 System Definition and Choice of Design Parameters — 279
    7.1.1 *System Definition and Simulation Approach* — *279*
    7.1.2 *Choice of Design Parameters* — *280*
7.2 The Random Access Protocol as a Benchmark — 281
    7.2.1 *Description of the Random Access Protocol* — *281*
    7.2.2 *Analysis of the Random Access Protocol* — *282*
    7.2.3 *Analysis vs Simulation Results* — *283*
    7.2.4 *On Multiplexing Efficiency with RAP* — *284*
7.3 Three More Benchmarks — 288
    7.3.1 *The Minimum-Variance Benchmark* — *288*
    7.3.2 *The 'Circuit-Switching' Benchmark* — *291*
    7.3.3 *Access Control based on Known Backlog* — *291*
7.4 Choosing Channel Access Functions — 293
    7.4.1 *The Heuristic Approach* — *293*
    7.4.2 *Semi-empirical Channel Access Functions* — *293*
7.5 On the Benefit of Channel Access Control — 298
    7.5.1 *Simulation Results vs Benchmarks* — *298*
    7.5.2 *Benefits of Fast Voice Activity Detection* — *301*
    7.5.3 *Interpretation of the Results and the 'Soft Capacity' Issue* — *302*
7.6 Impact of Power Control Errors and the Spreading Factor on Multiplexing Efficiency — 303
    7.6.1 *Impact of Power Control Errors on Access Control* — *303*
    7.6.2 *A Theoretical Study on the Impact of Power Control Errors and the Spreading Factor* — *305*
    7.6.3 *'Power Grouping': Another Way to Combat Power Control Errors?* — *308*
7.7 Summary — 308

8 MD PRMA on Code-Time-Slots — 311

8.1 System Definition and Simulation Approach — 311
    8.1.1 *System Definition and Choice of Design Parameters* — *311*
    8.1.2 *Simulation Approach, Traffic Parameters and Performance Measures* — *313*
    8.1.3 *Analysis of MD PRMA* — *313*
8.2 Comparison of PRMA, MD PRMA and RCMA Performances — 314
    8.2.1 *Simulation Results, No Interleaving* — *314*
    8.2.2 *Performance Comparison and Impact of Interleaving* — *316*

	8.3	Detailed Assessment of MD PRMA and MD FRMA Performances	317
		8.3.1 Impact of Acknowledgement Delays on MD PRMA Performance	*317*
		8.3.2 MD FRMA vs MD PRMA	*320*
		8.3.3 Performance of MD FRMA in TDD Mode	*321*
		8.3.4 Impact of Voice Model Parameters on MD PRMA Performance	*322*
	8.4	Combining Backlog-based and Load-based Access Control	323
		8.4.1 Accounting for Multiple Access Interference	*323*
		8.4.2 Performance of Combined Load- and Backlog-based Access Control	*325*
	8.5	Summary	326
9	**MD PRMA with Prioritised Bayesian Broadcast**		**329**
	9.1	Prioritisation at the Random Access Stage	329
	9.2	Prioritised Bayesian Broadcast	331
		9.2.1 Bayesian Scheme with Two Priority Classes and Proportional Priority Distribution	*331*
		9.2.2 Bayesian Scheme with Two Priority Classes and Non-proportional Priority Distribution	*332*
		9.2.3 Bayesian Scheme with Four Priority Classes and Semi-proportional Priority Distribution	*332*
		9.2.4 Bayesian Scheme with Four Priority Classes and Non-proportional Priority Distribution	*333*
		9.2.5 Priority-class-specific Backlog Estimation	*334*
		9.2.6 Algorithms for Frame-based Protocols	*335*
	9.3	System Definition and Simulation Approach	336
		9.3.1 System Definition	*336*
		9.3.2 Simulation Approach	*337*
		9.3.3 Traffic Scenarios Considered	*338*
	9.4	Simulation Results for Mixed Voice and Web Traffic	339
		9.4.1 Voice and a Single Class of Web Traffic	*339*
		9.4.2 Voice and Two Classes of Web Traffic	*340*
	9.5	Simulation Results for Mixed Voice and Email Traffic	341
		9.5.1 Performance with Unlimited Allocation Cycle Length	*341*
		9.5.2 Impact of Limiting Allocation Cycle Lengths	*342*
	9.6	Simulation Results for Mixed Voice, Web and Email Traffic	344
		9.6.1 Equal Share of Data Traffic per Priority Class	*344*
		9.6.2 Unequal Share of Data Traffic per Priority class	*344*
	9.7	Summary	347
10	**Packet Access in UTRA FDD and UTRA TDD**		**349**
	10.1	UTRAN and Radio Interface Protocol Architecture	349
		10.1.1 UTRAN Architecture	*349*
		10.1.2 Radio Interface Protocol Architecture	*350*
		10.1.3 3GPP Document Structure for UTRAN	*352*
		10.1.4 Physical Layer Basics	*352*
		10.1.5 MAC Layer Basics	*356*
		10.1.6 RLC Layer Basics	*357*

	10.2	UTRA FDD Channels and Procedures	358
		10.2.1 Mapping between Logical Channels and Transport Channels	358
		10.2.2 Physical Channels in UTRA FDD	358
		10.2.3 Mapping of Transport Channels and Indicators to Physical Channels	362
		10.2.4 Power Control	363
		10.2.5 Soft Handover	364
		10.2.6 Slotted or Compressed Mode	365
	10.3	Packet Access in UTRA FDD Release 99	365
		10.3.1 RACH Procedure and Packet Data on the RACH	366
		10.3.2 The Common Packet Channel	370
		10.3.3 Packet Data on Dedicated Channels	374
		10.3.4 Packet Data on the Downlink Shared Channel	377
		10.3.5 Time-Division Multiplexing vs Code-Division Multiplexing	378
	10.4	Packet Access in UTRA TDD	380
		10.4.1 Mapping between Logical and Transport Channels	380
		10.4.2 Frame Structure and Physical Channels in UTRA TDD	381
		10.4.3 Random Access Matters in UTRA TDD	382
		10.4.4 Packet Data on Dedicated Channels	384
		10.4.5 Packet Data on Shared Channels	384
	10.5	High-Speed Packet Access	386
		10.5.1 Adaptive Modulation and Coding, Hybrid ARQ	386
		10.5.2 Fast Cell Selection	388
		10.5.3 MIMO Processing	388
		10.5.4 Stand-alone DSCH	388
		10.5.5 And What About Increased Data-rates on the Uplink?	389
11	**Towards 'All IP' and Some Concluding Remarks**		**391**
	11.1	Towards 'All IP': UMTS and GPRS/GERAN Release 5	391
	11.2	Challenges of Voice over IP over Radio	394
		11.2.1 Payload Optimisation	395
		11.2.2 VoIP Header Overhead	396
		11.2.3 How to Reduce the Header Overhead	396
	11.3	Real-time IP Bearers in GERAN	399
		11.3.1 Adoption of UMTS Protocol Stacks for GERAN	399
		11.3.2 Shared or Dedicated Channels?	399
		11.3.3 Proposals for Shared Channels	401
		11.3.4 Likely GERAN Solutions	402
	11.4	Summarising Comments on Multiplexing Efficiency and Access Control	403
		11.4.1 TDMA Air Interfaces	404
		11.4.2 Hybrid CDMA/TDMA Interfaces	405
		11.4.3 CDMA Air Interfaces	407
Bibliography			**409**
Appendix			**427**

PREFACE

Hamid Ahgvami, the director of the Centre for Telecommunications Research at King's College London, who supervised numerous research projects on third generation (3G) mobile communication systems at his centre, was intrigued by both code-division multiple access (CDMA) and packet reservation multiple access (PRMA). In the early 1990s, these were two prime multiple access candidates for 3G systems, the latter essentially enhancing an air interface using time-division multiple access (TDMA) as a basic multiple access scheme. When Alex Brand arrived at King's in 1994 as an exchange student to carry out a project in conclusion of his studies at the Swiss Federal Institute of Technology (ETH) in Zurich, his brief was simple: try to combine CDMA and PRMA.

What started as a five-month research project resulted in several publications, a Ph.D. thesis, and quite a few follow-on publications by other researchers on the subject of combined CDMA/PRMA protocols, both at King's and elsewhere. In the following, we refer to these protocols as so-called *multidimensional PRMA* (MD PRMA) protocols, an umbrella term, which accommodates also 'non-CDMA environments'. Nevertheless, we are mostly looking at 'CDMA environments'.

The Ph.D. thesis, while naturally focussing on specific research contributions related to PRMA-based protocols, embedded these results in a quite thorough discussion of multiple access protocols for mobile cellular communications in general, which are the main topic of this book. Accordingly, its starting point was the Ph.D. thesis. It was then substantially expanded to cover multiple access in GSM/GPRS and in UMTS, as well as latest trends in the industry towards the merging of wireless communications and IP-based data communications, including their impact on multiple access strategies for mobile communications. Indeed, in tune with the increasing importance of IP technologies, the main focus of this book is on the support of packet-voice and packet-data traffic on the air interface. Topics of particular interest in this context include matters related to resource utilisation and multiplexing efficiency and probabilistic access control used for access arbitration at the medium access control (MAC) layer.

From an OSI layering perspective, the generic term 'multiple access' spans often both layer 1, the physical layer, and the lower sub-layer of layer 2, the MAC (sub-)layer. We associate '*basic* multiple access schemes' with the physical layer, and 'multiple access *protocols*' with the MAC (sub-)layer. Unlike the few books dealing exclusively with multiple access protocols, a key concern for the present book is the wider framework (of mobile communications) in which they have to operate. Apart from issues associated with the physical layer and with layers above the MAC sub-layer, this includes also general design principles and constraints of mobile communication systems, which have an impact on these protocols.

In its present shape, this book hopes to provide a good balance between specific research results, some disseminated already by the authors in scientific journals and conference proceedings, others published here for the first time, and useful considerations on mobile communications from 2G to evolved 3G systems. The latter includes a discussion of the evolutionary path of the dominant 2G standard, GSM, first to a 2.5G system (mainly through the addition of GPRS), then to the first release of a 'full' 3G system in the shape of UMTS, and finally to subsequent releases adopting more and more IP-based technologies. Possible 4G scenarios are also discussed. This book is therefore a valuable source of information for anybody interested in the latest trends in mobile communications, which is accessible in Chapters 1 to 4, 10 and 11 without having to delve into lots of maths. Chapters 5 to 9 are more geared towards researchers and designers of multiple access protocols and other aspects of air interfaces for mobile communication systems. Accordingly, some of these chapters feature a few mathematical formulas, mostly in the area of probability theory.

Chapter 1 introduces first the main concepts related to mobile cellular communication systems. It then discusses the importance of multiple access protocols and the impact of the emergence of the Internet on cellular communications. Finally, it summarises the specific research contributions of the authors documented in detail in further chapters. They are mainly related to access control in the context of the multiple access protocols investigated.

Expanding on this introductory chapter, Chapter 2 provides more insight into current and future cellular communication systems from 1G to 4G. In particular, it discusses initial requirements on which the design of 3G systems was based, how 2G systems can be evolved to meet 3G requirements, and what drives the further evolution of 3G towards 4G systems. The latter includes a possible convergence between cellular communications, the Internet and the broadcast world. The role which wireless local area networks (WLAN) are expected to play in such scenarios is also reviewed.

In Chapter 3, MAC strategies for cellular communication systems, which help meet the requirements for 3G and beyond, are examined in the context of the general problem of multiple access in cellular communications. A considerable effort is invested in juxtaposing PRMA-based strategies with possible alternatives, assessing the respective advantages and disadvantages qualitatively and/or quantitatively.

Chapter 4 traces the evolution of the GSM air interface standards from the first system release through to release 1999 of the standards, that is from a system designed primarily for voice to one which offers sophisticated support for packet data through an enhanced version of GPRS. The air interface spans roughly the first three OSI layers. Naturally, our main interest lies at the MAC layer, but its description is embedded into an in-depth discussion of physical and logical channels defined for GSM. In fact, for 'plain GSM', the MAC layer is a relatively minor matter anyway, of certain limited relevance for issues such as radio resource utilisation, another topic of interest on which we also present some research results. From a MAC perspective, GPRS is much more interesting and thus featured more prominently than 'plain GSM'. The GPRS MAC layer, in particular the random access protocol, is explained in considerable detail. The description of the latter is complemented by research results we fed into the GPRS standardisation process. The release 1999 additions which are discussed include incremental redundancy and the so-called EDGE COMPACT mode.

To investigate the performance of multiple access protocols, it is necessary to model somehow the physical layer, on the services of which the MAC layer depends. In addition, one has to be aware of what services higher layers expect from the MAC layer. For our research on PRMA-based protocols, as far as higher layers are concerned, the main focus is on the traffic generated, which will be handed down to the MAC layer and which it has to transfer making best possible use of the available physical link. Appropriate models for physical layer performance assessment and traffic generation are discussed in Chapter 5.

Chapter 6 will define MD PRMA in detail, and introduce the two fundamental approaches considered to probabilistic access control at the MAC, namely load-based and backlog-based access control. As mentioned previously, access control is one of the research topics to which we devote particular attention.

In Chapter 7, results of investigations on the benefit of load-based access control in MD PRMA are reported. To this end, for voice-only traffic, the performance of MD PRMA in the presence of intracell and intercell interference is compared with that of a random access protocol and with various benchmarks. The impact of power control errors is also discussed.

In Chapter 8, backlog-based access control for MD PRMA is treated in detail and an assessment of its advantages compared to fixed permission probabilities is provided. This includes a comparison of the multiplexing efficiencies achieved in TDMA-only, hybrid CDMA/TDMA, and CDMA-only environments. The impact of acknowledgement delays and a protocol mode for time-division duplex are discussed. The combination of backlog-based and load-based access control is studied. Again, only packet-voice traffic is considered.

Chapter 9 provides a discussion of approaches to prioritisation at the random access. It then presents simulation results for the chosen prioritised pseudo-Bayesian algorithm, tested in a mixed traffic environment consisting of voice, Web browsing, and email traffic.

In Chapter 10, after having introduced basic concepts of the UMTS air interface and the radio access network architecture, UTRA FDD channels and procedures are reviewed. The main effort is invested in exploring how packet-data traffic can be supported on UTRA FDD according to release 1999 of the standards. Packet access in UTRA TDD is also reviewed. Further, we discuss to what extent some of the research results on access control documented in the previous chapters can be applied to UTRA FDD and UTRA TDD. Finally, the nature of possible enhancements beyond release 1999 providing high-speed packet access is explained.

Chapter 11 concludes the main body of this book. It introduces architectural enhancements to the UMTS packet-switched core network for the support of real-time IP-based traffic, the new IP multimedia subsystem, and enhancements to the GPRS/EDGE radio access network which allow the latter to be connected to the UMTS core network. These endeavours can be viewed as an important step towards 'all-IP'. Challenges relating to the support of real-time IP services over cellular air interfaces are discussed and possible solutions are outlined on how to overcome problems such as the spectral inefficiency associated with standard voice over IP over radio. Enhancements to the GPRS/EDGE air interface, enabling it to support real-time packet-data services, are reviewed. The last section provides summarising comments on multiplexing efficiency and access control, two key topics dealt with extensively throughout this book.

Each chapter is preceded by a short outline of the topics to be treated. Chapters are divided into a number of sections (e.g. Section 4.2), which may in turn be split into several subsections (such as Subsection 4.2.3). With two exceptions, chapters close with a summary and some intermediate conclusions. Readers may find the list of abbreviations useful, and also another list containing the symbols used throughout this book. They can be found after the acknowledgements following this preface. Care was taken to choose the symbols in a manner so that ambiguities are avoided. At times, this required the choice of symbols other than those used in previous publications, possibly untypical ones, such as X for the spreading factor instead of the widely used N, because N is also commonly used to denote the number of time slots per TDMA frame. As far as acronyms are concerned, we made an effort to write them out in full whenever they occurred first in each chapter, but exceptions include regularly recurring acronyms and cases where they are used in passing first, and explicitly introduced soon after. Finally, following a list of references, an appendix provides some useful information on GSM and UMTS standard documents.

ACKNOWLEDGEMENTS

The authors would like to thank the many people at King's College London and elsewhere who either acted in a supporting role, for instance by maintaining the necessary computer infrastructure, or contributed directly to the research efforts documented in this book. The latter include Celia Fresco Diez, David Sanchez, Francois Honore and Jean-Christoph Sindt, all M.Sc. or exchange students involved in research projects relating to topics treated in this book. H. C. Perle and Bruno Rechberger, at the time with the Swiss Federal Institute, Zurich, provided useful input to the early stages of our research efforts, as mentioned in the text. Special thanks go to John Pearson, who reviewed all of our earlier joint publications and Alex Brand's Ph.D. thesis; in part these documents are incorporated in this book. We would also like to express our gratitude to Dr Mark Searle and Prof. Lajos Hanzo for the useful input provided, and to our industrial partners in the LINK ACS research project, NEC, Plextek and Vodafone, for the many stimulating discussions. It was thanks to the involvement in this project that Alex Brand could attend SMG2 GPRS standardisation meetings from 1995 to 1997 as an academic. A special mention goes to John Wiley & Sons, Ltd, particularly for the considerable flexibility shown regarding the submission deadline for this book.

Alex Brand would like to thank his parents for all the support provided, not the least for having wired their house throughout, providing an excellent computing infrastructure which, while visiting them in Switzerland, accelerated the completion of the book, and the parents in Italy for the continuous moral support provided. He would also like to mention his colleagues at BT Wireless, for instance Fred Harrison, Steve Hearnden, Kevin Holley and Steve Mecrow, with whom he had many useful discussions on topics related to this book. Discussions held with Richard Townend proved particularly valuable. Apart from providing some key input to Chapter 10, they have helped by shifting away from the sometimes misleading notion of 'packet-switching vs circuit-switching over the air', and instead focus on dedicated vs common or shared channels.

One person has to be singled out, Monica Dell'Anna, Alex Brand's wife. Among her many roles, she was an invaluable technical consultant, mainly on physical layer issues, a guinea pig as a reader of the text, judging it both in terms of content and presentation, and she acted as an illustrator. She also helped with some of the more tedious jobs, such as text formatting, and besides all this, she ran the household with little support from her husband. Simply put, this book would not have been possible without her. Words cannot express enough gratitude.

ABBREVIATIONS

16QAM	16-Quadrature Amplitude Modulation
2BB-LQC	Two-Burst-Based Link Quality Control (EGPRS)
3G	Third Generation Cellular Communication Systems (also: 1G, 2G, 4G)
3GPP	Third Generation Partnership Project
64QAM	64-Quadrature Amplitude Modulation
8PSK	8-Phase Shift Keying

A

A-Slot	Acknowledgement Slot
AB	Access Burst
ACCH	Access Control CHannel (proposed UTRA channel)
	Associated Control CHannel (GSM)
ACI	Adjacent Channel Interference
ACK	(Positive) Acknowledgement
ACTS	Advanced Communications Technologies and Services
AGCH	Access Grant CHannel (GSM)
AI	Access Indicator (UTRA FDD indicator)
AICH	Access Indicator CHannel (UTRA FDD physical channel)
AIUR	Air-Interface User-Rate
AMC	Adaptive Modulation and Coding
AMPS	Advanced Mobile Phone System
AMR	Adaptive Multi-Rate Voice Codec
AN	Access Network
AP-AICH	CPCH Access Preamble Acquisition Indicator CHannel (UTRA FDD physical channel)
API	Access Preamble acquisition Indicator (UTRA FDD indicator)
APN	Access Point Name (GPRS)
ARIB	Association of Radio Industries and Businesses
ARQ	Automatic Repeat reQuest
ASC	Access Service Class (UMTS)
ASCI	Advanced Speech Call Item(s)
ATM	Asynchronous Transfer Mode
ATDMA	Advanced TDMA
AUC	Authentication Centre
AWGN	Additive White Gaussian Noise

B

BB	Bayesian Broadcast
BCCH	Broadcast Control CHannel (GSM, UTRA logical channel)
BCH	Bose–Chaudhuri–Hocquenghem (Codes), *or* Broadcast CHannel (UTRA transport channel)
BCS	Block Check Sequence (GPRS)
BER	Bit Error Rate
BLER	Block Error Rate (for GPRS)
BMC	Broadcast/Multicast Control (UMTS)
BRMA	Burst Reservation Multiple Access
BPSK	Binary Phase Shift Keying
BRAN	Broadband Radio Access Network
BS	Base Station
BSC	Base Station Controller
BSIC	Base Station Identity Code
BSN	Block Sequence Number
BSS	Base Station System
BSSGP	BSS GPRS Protocol
BTS	Base Transceiver Station

C

C-Plane	Control Plane
C-PRMA	Centralised PRMA
C-Slot	Contention Slot
CA	Cell Allocation (of radio frequency channels) in GSM Channel Assignment in UMTS (for CPCH operation)
CAF	Channel Access Function
CBCH	Cell Broadcast Channel (GSM)
CBR	Constant Bit-Rate
CCCH	Common Control CHannel (GSM, UTRA logical channel)
CCPCH	Common Control Physical CHannel (UTRA physical channel)
CCTrCH	Coded Composite Transport CHannel (UTRA)
CD	Collision Detection
CD/CA-ICH	CPCH Collision Detection/Channel Assignment Indicator CHannel (UTRA FDD physical channel)
CDI	Collision Detection Indicator (UTRA FDD indicator)
CDI/CAI	Collision Detection Indicator / Channel Assignment Indicator (UTRA FDD indicator)
CDM	Code-Division Multiplexing
CDMA	Code-Division Multiple Access
CDPA	Capture-Division Packetised Access
CEPT	Conférence Européenne des Administrations des Postes et des Télécommunications (European Conference of Postal and Telecommunications Administrations)
CFCCH	COMPACT Frequency Correction CHannel
CIR	Carrier-to-Interference Ratio

CLSP	Channel Load Sensing Protocol
CLT	Central Limit Theorem
CM	Connection Management (GSM and UMTS)
CN	Core Network
Codit	Code Division Testbed
CON	Contention State
CPAGCH	COMPACT Packet Access Grant CHannel (GPRS)
CPBCH	COMPACT Packet Broadcast Control CHannel (GPRS)
CPCCH	COMPACT Common Control CHannel (GPRS)
CPCH	Common Packet CHannel (UTRA FDD transport channel)
CPICH	Common Pilot CHannel (UTRA FDD physical signal)
CPRACH	COMPACT Packet Random Access CHannel (GPRS)
CPPCH	COMPACT Packet Paging CHannel (GPRS)
CRNC	Controlling Radio Network Controller
CS	Circuit-Switched
CS-1	Coding Scheme 1 (also: CS-2, CS-3, and CS-4, all GPRS)
CSB	Circuit-Switched Benchmark
CSCF	Call State Control Function
CSCH	COMPACT Synchronisation CHannel (GPRS)
CSICH	CPCH Status Indicator CHannel (UTRA FDD physical channel)
CSMA	Carrier Sense Multiple Access
CTCH	Common Traffic CHannel (UTRA logical channel)
CTDMA	Code-Time-Division Multiple Access
CTS	GSM-based Cordless Telephony System

D

D-AMPS	Digital Advanced Mobile Phone System (IS-136 TDMA)
DAB	Digital Audio Broadcasting
DCA	Dynamic Channel Assignment or Allocation
DCCH	Dedicated Control CHannel (UTRA logical channel)
DCH	Dedicated CHannel (UTRA transport channel)
DCS	Digital Cellular System
DFT	Deferred First Transmission, *also* Discrete Fourier Transform
DLC	Data Link Control
DLL	Data Link Layer
DPAC	Dynamic Packet Admission Control (proposed for UTRA FDD)
DPCCH	Dedicated Physical Control CHannel (UTRA FDD physical channel)
DPCH	Dedicated Physical CHannel (UTRA physical channel)
DPDCH	Dedicated Physical Data CHannel (UTRA FDD physical channel)
DPSCH	Dedicated Physical SubCHannel (GERAN)
DRA	Dynamic Resource Assignment
DRAC	Dynamic Resource Allocation Control (for UTRA FDD)
DRNC	Drift Radio Network Controller
DRX	Discontinuous Reception (GSM)
DS	Direct Sequence
DSCH	Downlink Shared CHannel (UTRA transport channel)

DTCH	Dedicated Traffic CHannel (UTRA logical channel)
DTM	Dual Transfer Mode (i.e. GSM and GPRS)
DTX	Discontinuous Transmission
DVB	Digital Video Broadcasting

E

E	Erlang
E-FACCH	Enhanced Fast Associated Control Channel (GSM)
E-IACCH	Enhanced In-band Associated Control Channel (GSM)
E-TCH	Enhanced TCH (8PSK modulation on GSM physical channels)
ECSD	Enhanced Circuit-Switched Data
EDGE	Enhanced Data Rates for Global (initially: GSM) Evolution
EED	Equal Error Detection
EEP	Equal Error Protection
EGPRS	Enhanced GPRS
EIR	Equipment Identity Register
EPA	Equilibrium Point Analysis
EPRMA	Extended PRMA
ETSI	European Telecommunications Standards Institute

F

FACCH	Fast Associated Control CHannel (GSM)
FACH	Forward Access CHannel (UTRA transport channel)
FB	Frequency Correction Burst (GSM)
FCA	Fixed Channel Assignment
FCC	Federal Communications Commission (of the US)
FCCH	Frequency Correction CHannel (GSM)
FCFS	First-Come First-Serve
FCS	Frame Check Sequence
	Fast Cell Selection (for UMTS HSDPA)
FDD	Frequency-Division Duplexing
FDMA	Frequency-Division Multiple Access
FEC	Forward Error Correction (Coding)
FER	Frame Erasure Rate
FET	First Exit Time (a stability measure)
FH	Frequency Hopping
FN	Frame Number (GSM)
FPLMTS	Future Public Land Mobile Telecommunications System
FP-Slot	Fast Paging Slot
FRAMES	Future Radio Wideband Multiple Access Systems
FRMA	Frame Reservation Multiple Access
FTP	File Transfer Protocol

G

GERAN	GSM/EDGE Radio Access Network
GGSN	Gateway GPRS Support Node

GMM/SM	GPRS Mobility Management and Session Management
GMSC	Gateway MSC
GMSK	Gaussian Minimum Shift Keying
GoS	Grade of Service
GSM	Groupe Spécial Mobile (Special Mobile Group), *or* Global System for Mobile Communications
GPRS	General Packet Radio Service
GTP	GPRS Tunnelling Protocol

H

HARQ	Hybrid ARQ
HCAF	Heuristic Channel Access Function
HCS	Hierarchical Cellular Structures *or* (in EGPRS) Header Check Sequence
HLR	Home Location Register
HSCSD	High Speed Circuit-Switched Data
HSDPA	High Speed Downlink Packet Access
HSS	Home Subscriber Server
HSUPA	High Speed Uplink Packet Access

I

I	In-phase
I-Slot	Information Slot
IEEE	Institute of Electrical and Electronics Engineers
IETF	Internet Engineering Task Force
IGA	Improved Gaussian Approximation
i.i.d.	independent and identically distributed
I/L	Interleaving
IMS	IP Multimedia Subsystem
IMSI	International Mobile Subscriber Identity
IMT-2000	International Mobile Telecommunications 2000
IP	Internet Protocol
IPv4	Internet Protocol version 4 (analogous, IPv6)
IPRMA	Integrated PRMA
IR	Incremental Redundancy
IS-95	Interim Standard 95 (CDMA)
IS-136	Interim Standard 136 (TDMA)
IS-661	Interim Standard 661 (a hybrid CDMA/TDMA system)
ISMA	Inhibit Sense Multiple Access, *or* Idle Sense Multiple Access
ISDN	Integrated Services Digital Network
ITU	International Telecommunications Union

J

JD	Joint Detection

K

KBAC	Known-Backlog-based Access Control

L

LA	Link Adaptation
LHS	Left Hand Side (of an equation)
LINK ACS	A collaborative research project on Advanced Channel Structures for mobile communications funded by the UK government, which was part of phase II of the LINK personal communications programme
LLC	Logical Link Control (GPRS)
LQC	Link Quality Control (EGPRS)

M

MAC	Medium Access Control
MAHO	Mobile Assisted Handover
MA	Mobile Allocation (of radio frequency channels) in GSM
MAI	Multiple Access Interference
MAIO	Mobile Allocation Index Offset (for frequency hopping in GSM)
MBS	Mobile Broadband System
MCS-1	Modulation and Coding Scheme 1 (for EGPRS, MCS-1 to MCS-9)
MD PRMA	Multidimensional PRMA
MD FRMA	Multidimensional FRMA
ME	Mobile Equipment
MGCF	Media Gateway Control Function
MGW	Media GateWay
MIMO	Multiple-Input Multiple-Output
MM	Mobility Management (GSM and UMTS)
MO	Mobile Originated
MPDCH	Master Packet Data CHannel (GPRS)
MPEG	Moving Pictures Expert Group
MRF	Media Resource Function
MS	Mobile Station
MSC	Mobile-services Switching Centre
MT	Mobile Terminated
MVB	Minimum Variance Benchmark
MVBwd	Minimum Variance Benchmark with dropping

N

N-RACH	Normal RACH
NACK	Negative Acknowledgement
NB	Normal Burst
NC-PRMA	Non-Collision PRMA
NMT	Nordic Mobile Telephony System
NRT	Non-Real-Time

nTDD	narrowband TDD (TD/SCDMA UTRA mode)
NWL	Network Layer

O

ODMA	Opportunity Driven Multiple Access
OFDM	Orthogonal Frequency-Division Multiplexing
OSI	Open Systems Interconnection
OVSF	Orthogonal Variable Spreading Factor (UMTS)

P

P-CCPCH	Primary CCPCH (UTRA physical channel)
P-CPICH	Primary CPICH (UTRA FDD physical signal)
PA-Slot	Paging Acknowledgement Slot
PACCH	Packet Associated Control CHannel (GPRS)
PAGCH	Packet Access Grant CHannel (GPRS)
PBCCH	Packet Broadcast Control CHannel (GPRS)
PCCCH	Packet Common Control CHannel (GPRS)
PCCH	Paging Control CHannel (UTRA logical channel)
PCH	Paging CHannel (GSM, UTRA transport channel)
PCM	Pulse Code Modulation
PCMCIA	Personal Computer Memory Card International Association
PCN	Personal Communications Networks
PCPCH	Physical Common Packet CHannel (UTRA FDD physical channel)
PCS	Personal Communications System
PCU	Packet Control Unit (GPRS)
PDC	Personal Digital Cellular
PDC-P	Personal Digital Cellular Packet (Packet Overlay to PDC)
PDCP	Packet Data Convergence Protocol (UMTS)
PDCH	Packet Data CHannel (GPRS)
PDMA	Polarisation-Division Multiple Access
PDN	Packet Data Network
PDP	Packet Data Protocol
PDSCH	Physical Downlink Shared CHannel (UTRA physical channel)
PDTCH	Packet Data Traffic CHannel (GPRS)
PDU	Protocol Data Unit
PHS	Personal Handyphone System
PHY	Physical Layer
PI	Paging Indicator (UTRA FDD indicator)
PICH	Paging Indicator CHannel (UTRA FDD physical channel)
PLMN	Public Land Mobile Network
PNCH	Packet Notification CHannel (GPRS)
PPCH	Packet Paging CHannel (GPRS)
PRACH	Packet Random Access CHannel (GPRS)
	Physical Random Access CHannel (UTRA physical channel)
PRMA	Packet Reservation Multiple Access
PS	Packet-Switched

PS	Puncturing Scheme (EGPRS)
PSI	Packet System Information (GPRS)
PSK	Phase Shift Keying
PSTN	Public Switched Telephone Network
PTCCH	Packet Timing advance Control CHannel (GPRS)
PUSCH	Physical Uplink Shared CHannel (UTRA TDD physical channel)

Q

Q	Quadrature Phase
QAM	Quadrature Amplitude Modulation
QoS	Quality of Service
QPSK	Quaternary or Quadrature Phase Shift Keying

R

R97	Release 1997 of Specifications (similarly R96, R98, R99)
RA	Routing Area
RAB	Radio Access Bearer
RACE	Research and technology development in Advanced Communications technologies in Europe
RACH	Random Access CHannel (GSM, UTRA transport channel)
RAMA	Resource Auction Multiple Access
RAN	Radio Access Network
RAP	Random Access Protocol
RCMA	Reservation-Code Multiple Access
RES	Reservation State
RF	Radio Frequency
RHS	Right Hand Side (of an equation)
RLC	Radio Link Control
RLP	Radio Link Protocol
RNC	Radio Network Controller
RNS	Radio Network Subsystem
RNTI	Radio Network Temporary Identity
RR	Radio Resource Management (GSM)
RRBP	Relative Reserved Block Period (GPRS)
RRC	Radio Resource Control (UMTS)
RT	Real-Time
RTP	Real-Time Protocol (running on top of UDP/IP)
RTT	Radio Transmission Technology
RV	Random Variable

S

S/P	Supplementary/Polling Bit (GPRS)
S-CCPCH	Secondary CCPCH (UTRA physical channel)
S-CPICH	Secondary CPICH (UTRA FDD physical signal)
S-RACH	Short RACH (i.e. using short access bursts and optionally mini-slots)
SACCH	Slow Associated Control CHannel (GSM)

SB	Synchronisation Burst (GSM)
SCH	Synchronisation CHannel (GSM, UTRA physical signal)
SDCCH	Stand-alone Dedicated Control CHannel (GSM)
SDMA	Space-Division Multiple Access
SECAF	Semi-Empirical Channel Access Function
SF	Spreading Factor
SFH	Slow Frequency Hopping
SGA	Standard Gaussian Approximation
SGSN	Serving GPRS Support Node
SHCCH	SHared channel Control CHannel (UTRA TDD logical channel)
SI	Status Indicator (UTRA FDD Indicator)
SID	Silence Descriptor (GSM)
SIGA	Simplified Improved Gaussian Approximation
SIL	Silence State
SIM	Subscriber Identity Module (GSM)
SINR	Signal-to-Noise-plus-Interference Ratio
SIP	Session Initiation Protocol
SIR	Signal-to-Interference Ratio
SMG	Special Mobile Group
SMS	Short Message Service
SM-SC	Short Message Service Centre
SNDPC	Sub Network Dependent Convergence Protocol (GPRS)
SNR	Signal-to-Noise Ratio
SPDCH	Slave Packet Data CHannel (GPRS)
SPSCH	Shared Physical SubCHannel (GERAN)
SRNC	Serving Radio Network Controller
SSMA	Spread Spectrum Multiple Access
SSPRMA	Spread Spectrum PRMA

T

TA	Timing Advance
TACS	Total Access Communications System
TAI	Timing Advance Index (GPRS)
TBF	Temporary Block Flow (GPRS)
TCH	Traffic Channel (GSM)
TCP	Transport Control Protocol (for IP)
TD/CDMA	Hybrid Time-Division Code-Division Multiple Access
TD/SCDMA	Hybrid Time-Division Synchronous Code-Division Multiple Access
TDD	Time-Division Duplexing
TDM	Time-Division Multiplexing
TDMA	Time-Division Multiple Access
TFC	Transport Format Combination (UTRA)
TFCI	Transport Format Combination Indicator (UTRA)
TFI	Transport Format Indicator (UTRA)
TLLI	Temporary Logical Link Identity (GPRS)
TMSI	Temporary Mobile Subscriber Identity

TN	Time-slot Number (GSM)
TPC	Transmit Power Control
TR	Technical Report
TRX	Transceiver or Transmit Receive Unit
TS	Technical Specification
TTI	Transmission Time Interval (UTRA)

U

U-Plane	User Plane
UDD	Unconstrained Delay Data
UDP	User Datagram Protocol (for IP)
UE	User Equipment
UED	Unequal Error Detection
UEP	Unequal Error Protection
UMTS	Universal Mobile Telecommunications System
UTRA	UMTS Terrestrial Radio Access or Universal Terrestrial Radio Access
UTRAN	UMTS (or Universal) Terrestrial Radio Access Network
US	United States
USCH	Uplink Shared CHannel (UTRA TDD transport channel)
USF	Uplink State Flag
USIM	Universal Subscriber Identity Module (UMTS)

V

VAD	Voice Activity Detection
VBR	Variable Bit-Rate
VLR	Visitor Location Register
VoIP	Voice over IP
VRRA	Variable Rate Reservation Access

W

WAP	Wireless Application Protocol
WCDMA	Wide-band CDMA
WLAN	Wireless Local Area Network
wTDD	wideband TDD (original UTRA TDD mode)
WWW	World Wide Web

Z

ZVB	Zero-Variance Benchmark

SYMBOLS

$A[t]$	Slots available for contention in time-slot t
$\overline{A}$	Average number of available slots over a suitable time-window
B	Number of message bits in a block (for block codes)
C	Number of contending terminals
$C[t]$	Number of collision slots in time-slot t
$C_f[t]$	Number of unsuccessfully contending users in time-slot t
C_k	Capture probability of strongest packet when k users transmit simultaneously
$C_s[t]$	Number of successfully contending users in time-slot t
$C_x[t]$	Number of collision slots from time-slot $t - x$ to time-slot $t - 1$
$\overline{D}$	Average access delay for S-ALOHA
D_{acc}	Data access delay
D_{CDM}	Transfer delay for code-division multiplexing
D_d	Packet interarrival time
D_{gap}	Mean talk gap duration
D_{max}	Delay threshold for packet dropping in PRMA and MD PRMA
D_n	Drift in state n
D_{pc}	Reading time between packet calls
D_{slot}	Time-slot duration
D_{spurt}	Mean talk spurt duration
D_{TDM}	Transfer delay for time-division multiplexing
D_{tf}	TDMA frame duration
$D_{transfer}$	Transfer delay
D_{vf}	Voice frame duration
E	Sub-slots per slot, i.e. code-slots per time-slot; also evidence
$F_X(x)$	Cumulative distribution function (cdf) of the random variable X
G	Normalised offered traffic
G_0	Optimum traffic level for S-ALOHA (in terms of maximum throughput S)
G_E	Offered traffic in Erlang
H	Header bits; also hypothesis
$I_{intercell}$	Normalised intercell interference level (random variable)
$\overline{I}_{intercell}$	Average normalised intercell interference level (mean of the respective RV)

$K, K[t]$	Users simultaneously accessing the channel (e.g. in a particular time-slot)
K'	Calls assigned to each time-slot (in case of circuit-switching)
$\overline{K}$	Mean number of users per time-slot (long-term average)
$\overline{K}_f$	Mean number of users per time-slot in a particular TDMA frame
$K'_{\max}$	Maximum value of K' at a certain admissible packet error rate $(P_{pe})_{\max}$
$K_{pe\max}$	Maximum value of K at a certain admissible packet error rate $(P_{pe})_{\max}$
L	Length of data unit or block over which block coding is applied
M	Number of voice conversations
$M_{0.01}$	Maximum number of conversations at $P_{\text{drop}} = 0.01$ or $P_{\text{loss}} = 0.01$
M_x	Maximum number of conversations at $P_{\text{drop}} = x$ or $P_{\text{loss}} = x$
N	Time-slots per TDMA frame, also number of terminals in a simple S-ALOHA system
N_c	Number of nearest base stations among which serving station is selected
N_d	Number of datagrams per packet call
N_f	Reuse factor or cluster size (the inverse is the reuse efficiency)
N_{pc}	Number of packet calls per session
N_t	Number of backlogged terminals at the start of time-slot t
N_{tc}	Number of traffic channels
P	Power level; also a probability or ratio to be evaluated, normally with appropriate index
P_0	Reference power level
P_b	Call blocking probability
P_c	Probability that a datagram or messages assumes the maximum size of c
P_{drop}	Packet or frame dropping ratio
P_e	Probability of bit error or BER due to MAI
P_i	Received power level of user i
P_k	Transmitted power level of user k
$P_K(k)$	Probability distribution function of a discrete random variable K
P_{loss}	Ratio of packets lost with MD PRMA due to MAI *and* packet dropping
$(P_{\text{loss}})_{\max}$	Maximum admissible packet-loss ratio
P_{pe}	Ratio of packets lost with MD PRMA due to MAI
$P_{pe}[K]$	Packet error probability due to MAI in a given slot carrying K packets
$(P_{pe})_{\max}$	Maximum admissible packet error rate
Pr	Probability (of an event in waved brackets)
P_{slot}	Probability of selecting a particular time-slot in the TDMA frame
P_{succ}	Probability of successful packet transmission (in S-ALOHA)
$Q(x)$	Complement of the error function
Q_e	Probability of bit success, $1 - P_e$
$Q_{pe}[K]$	Probability of packet success, $1 - P_{pe}[K]$
R	Number of terminals with a reservation in this TDMA frame, also number of cells and random number for PRACH procedure in GPRS
$R[t]$	Reserved slots in time-slot t or reservation-mode users accessing time-slot t

$R'[t]$	Reserved slots in time-slot t as seen at the end of this slot, i.e. $R[t] + C_s[t]$
$\hat{R}[t]$	Estimation of reserved slots in time-slot t or reservation-mode users accessing time-slot t
R_{av}	Average user bit-rate
R_c	CDMA channel rate
R_{cell}	Total (aggregate) user bit-rate sustained in one cell
R_{ec}	PRMA channel rate after error-coding
R_{max}	Maximum bit-rate for DRAC operation
R_p	PRMA channel rate before error-coding
R_s	Source rate (voice terminal)
S	Throughput (normalised)
S_0	Maximum throughput level for S-ALOHA
$S[t]$	Number of success slots in a time-slot
S_d	Size of datagrams
SNR	Signal-to-noise ratio
$\overline{SNR}$	Average signal-to-noise ratio
T	Number of simultaneously active voice users (RV, same as V)
T_b	Bit duration
T_c	Chip duration
T_{retry}	Time period between unsuccessful access attempts with DRAC
$T_{validity}$	Duration of the 'reservation phase' with DRAC
U	(Code-time-)slots per TDMA frame or resource units, $U = N \cdot E$
V	Number of active voice users
$\overline{V}$	Average number of active voice users
$\overline{V}_C$	Average number of active voice users in a pure CDMA system
$\overline{V}_{CT}$	Average number of active voice users in a hybrid CDMA / TDMA system
W	Random variable for cumulative power level of $K-1$ power-controlled users
X	Processing gain or spreading factor
$Y[t]$	Number of contending users in time-slot t
Z	Random variable for received power level of a power-controlled user
$a_k(t)$	Spreading, signature, or direct-sequence of user k
ac	Allocation cycle length
b	Backoff-rate for exponential backoff
$b_k(t)$	Data sequence of user k
c	Parameter for truncation of Pareto distribution
c_{offset}	Offset between K_{pemax} and $\overline{K}$ at a given $(P_{loss})_{max}$ in multiples of σ_K
d	Hamming distance
d_{il}	Interleaving depth
d_{min}	Minimum Hamming distance
e	Error correcting capability of an (L, B, e) block code; also 2.718 282
f_{CAF}	Channel access function

SYMBOLS

$f_X(x)$	Probability density function (pdf) of the continuous random variable X
i	Priority class
$i_{\text{intercell}}$	Snapshot of the normalised intercell interference level
i_u	Update interval for signalling of p with Bayesian broadcast
k	Prioritisation parameter; also realisations of the random variable K
l	Parameter determining the shape of one type of channel access functions
m	Intermediate parameter for prioritisation
n	System state (backlog)
$n(t)$	Signal of additive white Gaussian noise
n_f	TDMA frame number
n_i	Backlog of priority class i
n_s	Time-slot number (from 1 to N)
p	(Generic) transmission or access permission probability in general: a parameter specifying a probability value (that is, an input value, while P is an output probability value or ratio)
$p(1)$	Initial permission probability (in transmission backoff schemes)
p_0	Probability of packet generation in a given slot (for S-ALOHA)
p_i	Transmission probability for priority class i
$p_{\max}$	Load-based upper limit for access permission probability p
p_v	Voice permission probability
$\overline{p}_v[R]$	Average p_v values classified according to R
p_{vi}	Initial p_v for access function
$p_{v\max}$	Max. voice permission probability
p_x	Generic transmission permission probability
q	Complement of p
r	Distance between mobile and base station; or Random number (from 0 to 1)
$r(t)$	Received signal
r_0	Cell radius of hexagonal cell
r_c	Code-rate for FEC coding
s''	Number of successful contentions in a frame except the first one of each MS
$s[t]$	Success slot
$s_k(t)$	Transmitted signal of user k
t	Time (often discrete time with unit time-slot duration D_{slot})
v	Mean of estimated backlog distribution; also realisation of V
v'	Intermediate mean of v within a TDMA frame (for MD FRMA)
v_i	Mean of estimated backlog distribution of priority class i
w	Estimated number of waiting terminals; also realisation of W
x	Number of forbidden slots after contention due to acknowledgement delay
z	Prioritisation parameter ($z = z_1 + z_2$); also realisation of Z
z_1, z_2	Prioritisation parameters
$\aleph$	Natural numbers

SYMBOLS

Φ_K	Probability distribution of the number of users per time-slot
α	Throughput proportion for prioritised Bayesian broadcast, or propagation attenuation, or normalised sensing (inhibit) delay in CSMA (ISMA)
α_{ca}	First slope in channel access function
α_v	Voice activity factor
β	Parameter of Pareto distribution
β_{ca}	Second slope in channel access function
γ	Probability that a talk spurt ends in a slot
γ_{cr}	Capture ratio
γ_i	Prioritisation parameter
γ_{pl}	Path loss coefficient
δ	Rate of transition from silence to talk state
ε	Rate of transition from talk to silence state
$\varepsilon(t)$	Power control error
ζ	Attenuation in dB due to shadowing (log-normally distributed)
η	Sustained conversations with PRMA per equivalent TDMA channel
η_{map}	Multiple access protocol efficiency
η_{mux}	Multiplexing efficiency relative to perfect statistical multiplexing
η_r	Gross resource utilisation on GSM traffic channels
θ_k	Carrier phase at transmitter
λ	Parameter of exponential and of Pareto distribution
λ_{ar}	Arrival rate (per time-slot)
$\hat{\lambda}_{ar}$	Estimation of the arrival rate
λ_c	Call arrival rate
λ_v	Voice arrival rate
λ_d	Data arrival rate
μ	Generic mean of a distribution
μ_{coll}	Mean number of terminals involved in a collision
μ_{Dd}	Mean packet interarrival time
μ_{Dpc}	Mean reading time duration between calls
μ_{Dsess}	Mean session interarrival time
μ_{Nd}	Mean number of packets per packet call
μ_{Npc}	Mean number of packet calls per session
μ_s	Inverse of the mean service time on traffic channel
μ_{Sd}	Mean of the Pareto distributed message or datagram size
σ	Probability that a talk gap ends in a certain time-slot; also generic standard deviation
σ_d	Probability with which a random data packet is generated in each time-slot
σ_K	Standard deviation of the distribution of RV K
σ_{pc}	Standard deviation of the log-normally distributed power control error $\varepsilon(t)$
σ_s	Standard deviation of log-normal shadowing
σ_{Sd}	Standard deviation of the message or datagram size

τ	Propagation delay
τ_k	Propagation delay of the signal of user k
τ_{p-a}	Delay between uplink and downlink access slots on UTRA FDD PRACH
φ_k	Carrier phase at receiver
ω_c	Carrier frequency
ψ_k	Phase-shift due to fading

1

INTRODUCTION

This book focuses on issues related to multiple access for cellular mobile communications, with a specific interest in access arbitration through multiple access protocols situated at the lower sub-layer of the second OSI layer, namely the medium access control (MAC) layer.

In this chapter, first an introduction to cellular mobile communication systems is provided. This introduction will be further expanded upon in Chapter 2, particularly with respect to the features which distinguish the different generations of mobile communication systems, from analogue first generation (1G) systems to possible fourth generation (4G) scenarios. Next, it is discussed what impact the emergence of the Internet may have on cellular communication systems. The importance of multiple access protocols is also examined, particularly in the context of packet-based systems, and packet reservation multiple access (PRMA) is considered as a case study. Finally, together with some background information, an overview of our own research efforts related to PRMA-based protocols is provided. These efforts are mainly concerned with how to combine PRMA with code-division multiple access (CDMA), when such a combination is beneficial, and more generally with different approaches to access control at the MAC layer and their respective benefits. They are documented in detail in later chapters.

1.1 An Introduction to Cellular Communication Systems

1.1.1 The Cellular Concept

The first land mobile communication systems were based on wide area transmission [1]. Each base station had to provide coverage for large autonomous geographical zones. Calls of customers leaving a zone had to be dropped and re-established in a new zone [2]. Such systems suffered from low-capacity and high-transmit power requirements for mobile transceivers, shortcomings that would not have allowed us to witness the tremendous growth in mobile communications in the past few years, with penetrations now exceeding 70% in many countries. Only the introduction of *cellular* mobile communication systems in the late 1970s made this development possible, by enabling frequencies, used in one cell, to be reused under certain conditions in other cells to increase capacity. Nowadays, *mobile* communication systems are almost by implication *cellular* communication systems as well. We use either of these two terms interchangeably, sometimes also the full term, namely *cellular mobile communication systems*.

Cellular mobile communication systems are designed to provide moving users (from pedestrians to travellers in high-speed trains) with a means of communication. In contrast to (basic) cordless telephones, *cellular telephones* (also referred to as mobile phones, mobile stations, mobile terminals or sometimes simply handsets) are not attached to a particular base station, but may make use of any one of the base stations provided by the company that operates the corresponding network. Each of these base stations covers a particular area of the landscape, called a *cell*. The ensemble of base stations should cover the landscape in such a way that the user can travel around and carry on a phone call without interruption, possibly making use of more than one base station, as shown in Figure 1.1. The procedure of changing a base station at cell boundaries is called *handover* or *handoff*. We prefer the first term, since it implies (unlike the second term) that an effort is made to sustain a call across cell boundaries. Obviously, these systems can also serve stationary users, and do so increasingly, as fixed telephones are more and more substituted by wireless phones.

Communication from the mobile station (MS) to the base station (BS) takes place on the *uplink* channel or *reverse link* and from BS to MS on the *downlink* channel or *forward link* (Figure 1.1). To enable communication, some resources need to be allocated to the base station (these may be frequency bands, time-slots, sets of codes, or any combination of the three), which in turn may assign a portion of them to individual calls to support communication on both uplink and downlink channels. The amount of resources allocated to users will depend on the current resource availability and the particular requirements of each requesting user. As the base station must be able to assign individual portions of its resources to support multiple communications, *basic multiple access techniques* (such as frequency-, time-, or code-division multiple access, with FDMA, TDMA, and CDMA as their respective acronyms) are required together with *multiple access protocols*, which govern access to these resources. The basic multiple access schemes are briefly described further below in this section, and the importance of the multiple access protocols is examined in Section 1.3.

1.1.2 Propagation Phenomena in Cellular Communications

The design of cellular communication systems is particularly challenging because of the adverse propagation conditions experienced on the radio channel. Without discussing the complex underlying physical mechanisms, for which the reader may consult a mobile communications handbook such as that in Reference [3] or a book dedicated to radio

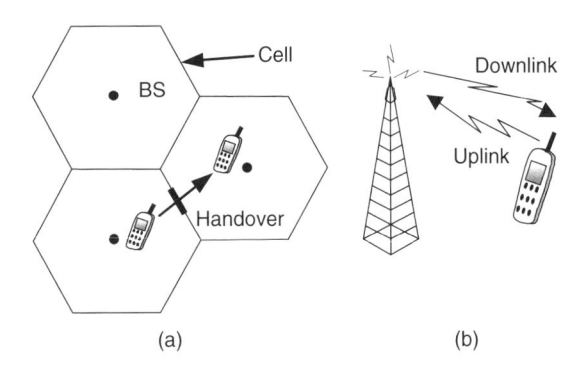

Figure 1.1 (a) Basic principle of cellular communications. (b) Uplink and downlink channel

propagation such as that in Reference [4], three main propagation effects are usually distinguished. These are the pathloss, slow fading or shadowing, and fast fading or multipath fading. The *pathloss* describes the average signal attenuation as a function of the distance between transmitter and receiver, which includes the free-space attenuation as one component, but also other factors come into play in cellular communications, resulting in an environment-dependent pathloss behaviour. *Shadowing* or *slow fading* describes slow signal fluctuations, which are typically caused by large structures, such as big buildings, obstructing the propagation paths. *Fast* or *multipath fading* is caused by the fact that signals propagate from transmitter to receiver through multiple paths, which can add at the receiver constructively or destructively depending on the relative signal phases. The received signal is said to be in a deep fade when the paths add destructively in a manner that the received signal level is close to zero. Fades occur roughly once every half wavelength [3]. Given that we are dealing with wavelengths of 30 cm and less in cellular communication systems, it is clear that multipath fading can result in relatively fast signal fluctuations; exactly how fast depends on the speed of the mobile station and on the dynamics of the surrounding environment.

When designing cellular communication systems and particularly when planning the deployment of such systems (e.g. choosing suitable base station locations), one will have to account for these propagation phenomena appropriately. One way to do this is to use deterministic propagation tools such as ray tracing tools, which will calculate experienced signal levels for every specific location of the planned system coverage area, taking into account every structure which could affect signal levels. Another way is to resort to statistical models, which have to be established by analysing propagation measurements performed in suitably chosen environments, e.g. classified as dense urban, typical urban, suburban and rural propagation environments, to name just a few. For the purposes of some of our investigations, we will deal with distance-independent pathloss coefficients and a so-called lognormal shadowing model, as outlined in detail in Chapter 5.

1.1.3 Basic Multiple Access Schemes

For reasons discussed in detail in Chapter 3, we make a distinction between *basic multiple access schemes*, such as FDMA, TDMA, and CDMA, associated with the physical layer (PHY) on the air interface of a mobile communications system, and *multiple access protocols*, situated at the medium access control (MAC) layer above the PHY. Roughly speaking, the basic schemes provide the capability of dividing the total resources available to a base station into individual portions, which can be assigned to different users, and the protocols govern access to these resource portions, e.g. provide access arbitration.

Analogue first generation cellular communication systems made use of FDMA as a basic multiple access scheme. In digital 2G systems, TDMA is predominant, but a CDMA-based system exists as well. CDMA is the most commonly used form of multiple access for third generation systems, in some cases complemented by a hybrid CDMA/TDMA scheme.

1.1.3.1 Frequency-Division Multiple Access

In FDMA, each communication is carried over one or two (depending on the duplex scheme, see below) narrowband frequency channels. The channel bandwidth and the modulation scheme determine the gross bit-rate that can be sustained. Because of non-ideal

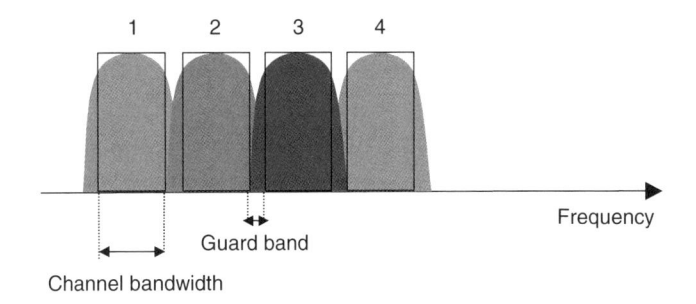

Figure 1.2 FDMA channels and guard bands

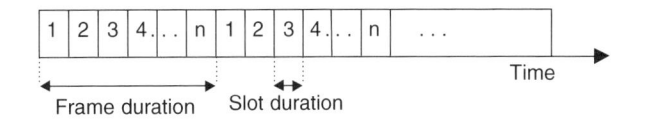

Figure 1.3 TDMA frames, time-slots, and bursts

filters, guard bands must be introduced between these channels to avoid so-called adjacent channel interference. This is illustrated in Figure 1.2.

1.1.3.2 Time-Division Multiple Access

In TDMA, rather than assigning each user a channel with its own frequency, users share a channel of a wider bandwidth, which we shall call a *(frequency) carrier*, in the time domain. This is achieved by introducing a framing structure with each TDMA frame subdivided into N *time-slots*, if N user channels are to be supported. User i is then allowed to access the carrier only during time-slot i, by transmitting a so-called *burst* which fits into this time-slot, as shown in Figure 1.3. In order to sustain a continuous gross source bit-rate of R_s bit/s, the transmission speed during the burst transmission must be at least NR_s bit/s.

Provided that enough spectrum is available, multiple carriers may be assigned to each cell. Therefore, such TDMA systems feature typically also an FDMA element, and are in reality hybrid TDMA/FDMA systems.

1.1.3.3 Code-Division Multiple Access

In CDMA, narrowband signals are transformed through spectrum spreading into signals with a wider bandwidth, the carrier bandwidth. Like in TDMA, multiple users share the carrier bandwidth, but like in FDMA, they transmit continuously during the call or session. The multiple access capability derives from the use of different *spreading codes* for individual users. Because of the spreading of the spectrum, CDMA systems are also referred to as spread spectrum multiple access (SSMA) systems.

Two basic CDMA techniques suitable for mobile communications are distinguished, namely frequency hopping (FH) and direct-sequence (DS) CDMA techniques. 'Proper' FH/CDMA systems have not been specified for mobile communications so far and are not discussed any further, but so-called slow frequency hopping (SFH) can also be applied in TDMA systems. The second generation Global System for Mobile Communications

(GSM) for instance features an SFH option, the benefits of which are discussed extensively in Chapter 4.

For one 2G system called cdmaOne and most 3G systems, e.g. the Universal Mobile Telecommunications System (UMTS), DS/CDMA was chosen as a basic multiple access scheme. In DS/CDMA, a bit-stream is multiplied by a direct sequence or spreading code composed of individual *chips*. They have a much shorter duration than the bits of the user bit-stream, and this is why the original signal's spectrum is spread. The bandwidth expansion factor, often simply referred to as *spreading factor* and in this book denoted by the symbol X, is equal to the duration of a bit, T_b, divided by the duration of a chip, T_c, i.e. $X = T_b/T_c$.

The same codes used at the transmitting side to spread the signals are used at the receiving side to de-spread them again. If codes assigned to different signals or user channels are mutually orthogonal, then these signals can be perfectly separated at the receiving side. In practise, due to multipath propagation, fully orthogonal separation at the receiving side may not be achieved even when the codes are orthogonal at the transmitting side. Provided that appropriate measures are taken, this is not really a problem, but it has an interesting consequence, which is highly relevant for some of the topics discussed in this book. Non-orthogonality creates mutual interference between all users, so-called multiple access interference (MAI). The resource assigned to an individual user in a CDMA system is therefore not so much a code, but rather a certain power level. This is illustrated in Figure 1.4, which shows sharing of resources in the time-domain, the frequency-domain, and in terms of power levels, for FDMA, TDMA and CDMA respectively.

1.1.3.4 Frequency-Division Duplex and Time-Division Duplex

To sustain a bi-directional communication between a mobile terminal and a base station, transmission resources must be provided both in the uplink and downlink directions. This can either happen through frequency-division duplex (FDD), whereby uplink and downlink channels are assigned on separate frequencies, or through time-division duplex

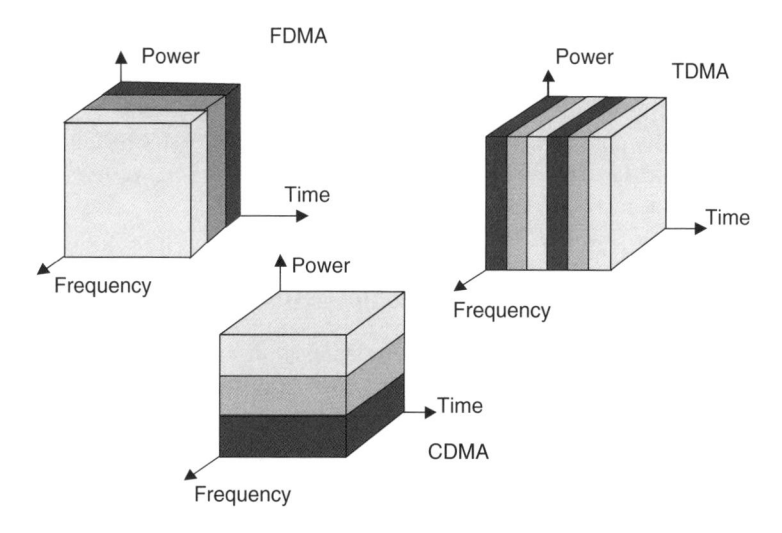

Figure 1.4 Sharing of time-, frequency- and power resources between three users in FDMA, TDMA and CDMA respectively

(TDD), where uplink and downlink transmissions occur on the same frequency, but alternate in time. Both methods can be applied in conjunction with any of the above-described multiple access schemes. 1G and 2G systems apply FDD. In UMTS, a 3G system, both FDD and TDD modes are supported, not least because symmetric uplink and downlink spectrum is normally required for FDD-only systems, but 3G spectrum consists of both so-called *paired bands* (i.e. symmetric spectrum) and *unpaired bands*.

A description of the key features of 1G, 2G and 3G systems is provided in Chapter 2, which also considers possible 4G scenarios. Advantages and disadvantages of the different basic multiple access schemes are examined in more detail in Section 3.2. Approaches to the modelling of the physical layer performance for some of our investigations are discussed in Chapter 5. Chapter 4 on multiple access in GSM and GPRS deals also to quite a considerable extent with physical layer issues.

1.1.4 Cell Clusters, Reuse Factor and Reuse Efficiency

As pointed out earlier, resources used in one cell may be reused in other cells, but this must be done in such a way that ongoing communications experience sufficient quality. Assume for now that we are dealing with frequency resources, and that the main factor affecting the quality is the so-called *co-channel interference*, that is, interference generated by communications in other cells transmitting on the same frequency as a desired communication link in a test cell. The required communication quality, together with other factors, e.g. related to propagation conditions, such as the pathloss coefficient, will determine the minimum distance that must be respected between two co-channel cells, the so-called *reuse distance*. This leads to the concept of *cell clusters* (or cellular reuse patterns), namely a set of neighbouring cells within the reuse distance, any two of which are not allowed to use the same frequency. The frequencies are instead reused in a cell occupying the same relative position in a neighbouring cluster, as illustrated in Figure 1.5. In other words, every cell in a cluster obtains a share of the total bandwidth available to an operator, and the same bandwidth is reused in other clusters. The number of cells within each cluster is called the *(frequency) reuse factor* or *cluster size* N_f [5], which is seven in the example depicted. The *reuse efficiency* is the inverse of the reuse factor, hence $1/N_f$.

With such a cellular approach, it is in theory possible to increase capacity without limit through cell splitting (i.e. deploying multiple small cells in an area previously served by a single big one), but there are certain practical constraints.

1.1.5 Types of Interference and Noise Affecting Communications

The permitted co-channel interference, which depends on various physical layer aspects such as the modulation scheme employed, is a key parameter determining the minimum frequency reuse factor. Communications taking place on adjacent channels can also create notable mutual interference because of non-ideal filters both at the transmit side (resulting in some power being also radiated outside the allocated channel) and at the receive side (due to receive filters not fully rejecting out-of-band signals). This is referred to as adjacent channel interference (ACI). It is strongest between directly adjacent or neighbouring channels, and decreases as channels further apart are being considered owing to the filter

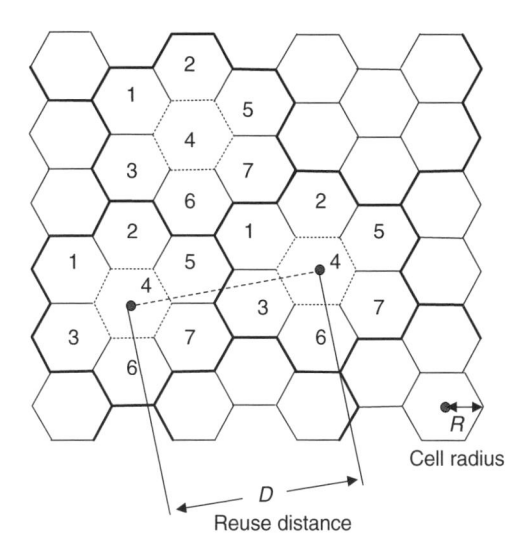

1.1 AN INTRODUCTION TO CELLULAR COMMUNICATION SYSTEMS

Figure 1.5 Cell clusters assuming hexagonal cells with a frequency reuse factor of seven

attenuation. In general, ACI is much less of a problem than co-channel interference. All the same, when several frequency channels are assigned to a cell, if they are neighbouring channels, guard bands should separate them to avoid excessive ACI, otherwise non-neighbouring channels should be assigned. Unfortunately, due to limited 3G spectrum and the fact that wideband channels are used, there are 3G scenarios for such systems where neither sufficient guard band is available nor non-neighbouring channels can be chosen, hence ACI becomes an issue, as discussed in Section 2.3.

Compared to TDMA and FDMA systems, CDMA systems exhibit certain peculiarities. Firstly, the reuse factor can be set to one in CDMA systems (and in fact often is). This is also referred to as *universal frequency reuse*. In this case, mutual interference is generated between all cells, hence rather than referring to this as co-channel interference, the term *intercell interference* is used. Secondly, while user channels within a cell are separated from each other in an orthogonal manner both in TDMA systems (perfect separation between time-slots can be achieved through guard periods) and in FDMA systems (assuming sufficient guard bands to avoid ACI), this is not necessarily the case in CDMA systems. Since spreading codes do not always provide orthogonal separation, interference within a cell, so-called *intracell interference*, can become an issue as well. Therefore, intracell and intercell interference can both be non-negligible components of the total multiple access interference experienced by a communication link in a CDMA system.

On top of interference generated by other users in the system, additional noise sources may affect the quality of a communication, for instance thermal noise. In the following, the term 'interference' refers to noise generated by other cellular users, and 'noise' to thermal noise as well as noise generated by sources outside the considered system. The communication quality in terms of bit error rate (BER) or frame erasure rate (FER) can therefore be expressed as a function of the signal-to-noise ratio (SNR), or the signal-to-interference ratio (SIR), depending on which type of signal disturbance is dominant. Typically, at the beginning of a system build out, when there are few cells, the system is *coverage-limited*, and the SNR is mostly relevant. As cells are added to fill in coverage holes and reduce cell radii, and the user traffic increases, the system becomes *capacity-limited* and the SIR becomes more critical. In situations where neither interference nor noise can be ignored, the signal-to-interference-plus-noise ratio (SINR) may be considered as a channel measure. However, if the nature of the interference is significantly different from that of the noise and affects the signal in a different manner, then it may not be possible to lump the two together and describe the performance as a function of the SINR. Instead, one would have to use, for example, SNR curves parameterised to interference levels or SIR curves parameterised to noise levels.

The signal quality can also be expressed as a function of the ratio of energy per bit E_b either to the noise power per Hertz, N_0, or the interference power per Hertz, I_0. Finally, instead of using signal and interference levels at the 'base-band', the so-called carrier-to-interference ratio (CIR) at the radio frequency level is often used. According to Reference [6],

$$\text{CIR} = \frac{E_b \cdot R_b}{I_0 \cdot B_c}, \tag{1.1}$$

with R_b the bit-rate in bits per second, and B_c the radio channel bandwidth in Hertz.

1.2 The Emergence of the Internet and its Impact on Cellular Communications

In the 'wired world', we are witnessing how traffic of all types is increasingly being carried on packet-switched networks using the connectionless internet protocol (IP) — or rather, the IP protocol suite, which features various other protocols on top of IP, e.g. transport protocols such as the transport control protocol (TCP) and the user datagram protocol (UDP). This development is mainly due to the tremendous success the Internet has enjoyed in recent years (incidentally, not unlike cellular communications). Initially constrained to non-real-time applications such as Telnet, file transfer, email and Web browsing, this move towards IP now embraces audio and video streaming with more stringent delay constraints, and even 'proper' real-time traffic such as Voice over IP (VoIP). Strictly speaking, it is typically voice over RTP (the real-time protocol, an application-level protocol), UDP and IP. It is now widely anticipated that the same will eventually happen in the wireless world as well, which has some serious technical implications on cellular communication systems. In the following, we deal with the general implications; the impact specifically on multiple access protocols is discussed in the next section.

Already in the late eighties (e.g. in Reference [7]), Goodman, whom we will refer to at various other occasions in this text, suggested that both the fixed architecture and the air interface of 3G systems should be based entirely on packet-switching for all types of services. Not only would the available resources be exploited more efficiently, but also certain functions could be decentralised and distributed over many processors, which would improve the scalability of such systems. He also proposed a packet-based multiple access protocol called packet reservation multiple access (PRMA) suitable for the wireless links between mobile and base stations. Although his vision of an all-packet system was probably not driven by the Internet at that time (he suggested that '3G systems, in harmony with *broadband integrated services digital networks*, would use shared resources to convey many information types'), it is in some respects in tune with Internet architecture principles.

Goodman's vision has not really caught on during initial 3G standardisation efforts. True, unlike 2G systems, first releases of 3G systems have incorporated packet-data support right from the start, however, without proper support of real-time packet data services such as packet voice. For instance, the first UMTS release might well provide improved support for packet data over the air interface compared to the GSM General Packet Radio Service (GPRS). However, the packet-based infrastructure in the fixed network, which has evolved from the GPRS infrastructure, was not designed for voice. Instead, it was intended that voice would always be delivered over the circuit-switched infrastructure.

But why this reluctance towards packet-voice and an all-packet system? There was a significant amount of scepticism in the industry regarding the feasibility of an all-packet system which could deliver the high-quality standards required for voice communications. Also, decentralisation, explicitly advocated by Goodman, to some extent inherent in the move to a packet-only system, and certainly consistent with the Internet architecture, means loss of operator control. This is something that operators do not like too much, as they tend to control the types of services delivered through their networks. The Internet, by contrast, is built according to the 'end-to-end principle', where the infrastructure in-between the end nodes is not concerned with the services to be delivered.

In the case of the cellular communications industry, apart from commercial considerations, there are sound technical reasons for this desire to control matters. Take the issue of handovers as an example. Goodman [7] proposed to decentralise them completely, effectively placing them into full control of the mobile terminals, in order to cope efficiently with the large volume of handover-related signalling traffic as a result of smaller and smaller cell sizes due to ever increasing traffic density. Cellular operators, however, like to control which terminal is served by which cell and thus prefer network-controlled handovers. This is firstly because normally only the network has a complete view of the communication quality on *both* up- and downlink (the latter through measurement reports sent by terminals), which may be different. Secondly, the quality of individual communications has to be traded off against system capacity and the quality of other communications, requiring careful admission control and load balancing by the network. In general, centralised algorithms exploiting the global view of a matter perform better than decentralised ones with only local information available. Obviously, they are also more complex, and therefore sometimes inappropriate, but when it comes to efficient use of scarce and precious air-interface resources, it is often worth the effort.

In spite of the initial scepticism by the cellular industry towards packet-based systems, the power of the Internet is proving too strong, and things are moving on. Subsequent releases of 3G systems will be capable of supporting voice over the packet-switched infrastructure as well. This does not necessarily imply a complete decentralisation of the architecture of cellular systems, at least not in the beginning. Operators driving these developments are predominantly interested in the new services they hope to deliver over their networks, as discussed in somewhat more detail in Section 2.4 and again in Chapter 11, and most of them are not (yet?) prepared to give up control. In terms of our main topics of interest for this book, such developments will primarily impact multiple access protocols, less the basic multiple access schemes. However, as decentralisation goes further (assuming that it will eventually), autonomous 'plug-and-play' base stations become desirable, something which could affect the choice of preferred basic multiple access schemes as well. For instance, in CDMA systems, to improve the transmission quality, terminals are often connected to the network via multiple base stations. This implies some co-ordination between these base stations. Furthermore, entities are needed that can process the signals of multiple base stations. Dealing with such matters through packet-based systems is by no means impossible, but it is somewhat of an obstacle to full decentralisation.

1.3 The Importance of Multiple Access Protocols in Cellular Communications

In 2G systems, which were designed to carry voice and some low-bit-rate data services by setting up 'circuits' or dedicated channels for the duration of a call, access arbitration is only required at the time of setting up a call to request such dedicated channels. With the advent of advanced 2G systems, e.g. GSM complemented by GPRS, and first releases of 'true' 3G systems such as UMTS, non-real-time data carried on common or shared channels becomes increasingly important, which calls for more sophisticated multiple access protocols. As just discussed, these systems will further evolve to support real-time IP traffic. What does this mean in terms of choosing appropriate multiple access protocols?

We are continuing to use Reference [7] as a case study, since packet reservation multiple access, the multiple access protocol proposed by Goodman for the air-interface uplink channel, was a subject of extensive research efforts by the authors, as documented in this book (see Section 1.4). Consider a traffic source which alternates between 'off' or 'silent' (no packets are generated) and 'on' or 'active' (packets are generated at a rate matching the channel transmission rate, e.g. one packet per TDMA frame fitting into one time-slot). A typical example would be a voice source subject to *voice activity detection*. By reserving resources on the air interface only during 'on' phases, when packets need to be transmitted, rather than hanging on to them for the entire duration of a call (as would be the case in a 'circuit-switched model'), PRMA attempts to make efficient use of air-interface resources. Compared to a conventional TDMA air interface, where N time-slots can sustain N calls, with PRMA M calls can be multiplexed onto N now *shared* time-slots, with $M > N$; how many exactly depends obviously on the so-called *activity factor*, i.e. the fraction of time the traffic source is in 'on' state. In other words, PRMA allows for a certain degree of *statistical multiplexing* over the air. One could therefore conclude that in conjunction with a packet-switched infrastructure, a 'packet-switched air interface' such as PRMA would also make sense. However, while the split between packet-switching and circuit-switching is fairly evident in the fixed network infrastructure, when dealing with the air interface, the situation is a little bit more complicated.

Essentially, on the air interface, we can distinguish between dedicated channels on one hand and common or shared channels on the other. Typically, *dedicated channels*, which are set apart for the sole use of one communication link between a mobile terminal and a base station, are associated with the 'circuit-switch model'. Conversely, *shared channels* (shared between a limited number of users) or *common channels* (common to the whole cell population), for which appropriate multiple access protocols are crucial, are often associated with the 'packet-switch model'. This is indeed often appropriate, particularly for packet-based services which exhibit very *bursty* traffic characteristics, i.e. traffic sources which alternate between short activity periods (e.g. at high bit-rates)

1.3 THE IMPORTANCE OF MULTIPLE ACCESS PROTOCOLS

and long inactivity periods. However, depending on the type of packet-traffic, the basic multiple access scheme in use, and the frequency planning applied, a statistical multiplexing gain may also be obtained when using dedicated channels over the air interface. In fact, this may even be the better choice in certain circumstances. Roughly speaking, apart from traffic characteristics, this depends on whether the system is blocking-limited or interference-limited.

A *blocking-limited* system is one in which a fixed number of channels are available per cell. The transmission quality is largely independent of the resource utilisation, that is the fraction of available channels which are assigned to ongoing calls. New calls or sessions are admitted if a channel (e.g. a time-slot) is available, and blocked if this is not the case. In such a scenario, using a protocol such as PRMA, which carries all traffic on shared or common channels, provides indeed a performance advantage for all types of packet traffic owing to statistical multiplexing.

In an *interference-limited* system, by contrast, the transmission quality depends on the system load: the more calls are admitted to the system, the worse it gets. Obviously, the aim is to admit calls only if the required quality levels can be met, but exceeding the appropriate load level somewhat results only in a gradual degradation of quality, and may occasionally be tolerated. This is referred to as *soft-capacity* feature and the load level at which the required quality level can just be met is sometimes called *soft-blocking* level. (Incidentally, by applying PRMA on a blocking-limited system, we can get soft capacity as well.)

As we identified earlier, the key resources assigned to users in CDMA systems are power levels, which makes these systems naturally interference limited. Also a TDMA system can be operated in an interference-limited fashion, if it features the option of slow frequency hopping. In such interference-limited systems, statistical multiplexing occurs naturally through interference multiplexing, even when dedicated channels are used. Voice activity detection, for instance, leads to an interference reduction during voice silent phases also on dedicated channels, meaning that the total power budget is shared flexibly and dynamically between users. Whether dedicated or shared channels are the preferred option then depends on the statistical behaviour of a communication source, the available code resources (in the case of CDMA), the overhead that is required to maintain a dedicated link while the source is silent, and the efficiency of the multiple access protocol in assigning and releasing, e.g. shared channels. If the used multiple access protocol is well designed, the balance may be tipped towards shared channels, particularly for bursty sources. However, for packet-voice in a CDMA system, dedicated channels may often be the best choice, which means that from a statistical multiplexing perspective, the performances of 'packet-switched voice' and 'circuit-switched voice' are similar (assuming that the same type of voice activity detection is applied in both cases). This puts 'voice over IP' at a disadvantage compared to optimised circuit-switched voice, because of the additional overheads associated with IP protocol headers.

We are dealing extensively with such topics throughout this book. In Chapter 3, we provide a review of basic multiple access schemes and multiple access protocols. In Chapter 4, where the GSM air interface and its GPRS additions are described, key topics include resource utilisation and blocking-limited versus interference-limited system operation. In Chapter 10, the different options available on the UMTS air interface for the support of packet traffic are described. In Chapter 11, the issue of interference-limited operation versus blocking-limited operation specifically for supporting VoIP in enhanced

GPRS systems is re-examined and the overheads associated with VoIP are discussed. Additionally, our own research efforts relating to PRMA-based protocols, which are summarised in Section 1.4, touch upon the issue of when shared channels are beneficial, and when dedicated channels are advantageous.

1.4 A PRMA-based Protocol for Hybrid CDMA/TDMA

This section provides the motivation for and background information to the authors' PRMA-related research efforts, mostly dealing with enhancements to PRMA to make the protocol suitable for air interfaces which feature also a CDMA component. They are documented in detail in Chapter 6 to 9 (with Chapter 5 providing the necessary channel and traffic models). It also identifies the specific original contributions made by the authors. Some of these research efforts are also relevant for GPRS and for UMTS, as discussed in Chapter 4, 10 and 11. Due to the nature of this section, certain parts of the text are not self-contained and serve mostly as a pointer to further chapters in this book. For instance, to appreciate fully how the combined CDMA/PRMA protocol we propose works, it is required to know how the base PRMA protocol works. This is only discussed briefly in this section, while a thorough description of PRMA is provided in the context of a fairly comprehensive treatment of multiple access protocols for mobile communication systems in Chapter 3.

1.4.1 Why Combine CDMA and PRMA?

PRMA is a multiple access protocol suggested by Goodman *et al.* [8] in the late 1980s for the uplink channel of 3G microcellular communication systems. All traffic including voice communications is carried in the shape of packets, and resources are allocated on the basis of individual packet spurts (as determined by voice activity detectors in the case of voice) rather than calls or circuits. This allows *statistical multiplexing* to be exploited on a TDMA air interface, that is, the number of users supported by the system is not limited by the peak bit-rate of the individual users, instead, it depends on their statistical behaviour. The relevant upper bound is *perfect statistical multiplexing*, where this number is only limited by the average user bit-rate, R_{av}, i.e. if a cell offers a capacity of R_{cell}, then R_{cell}/R_{av} users can be sustained simultaneously.

We have already pointed out earlier how an interference-limited CDMA system features an inherent statistical multiplexing capability even when dedicated channels are used (for voice traffic, this requires voice activity detection or variable-bit-rate voice codecs). In fact, it is often claimed that CDMA offers near-perfect statistical multiplexing [9]. The reader may therefore ask, what is the point of combining CDMA with PRMA?

The answer is the following: with on–off sources, the statistical multiplexing gain depends essentially on the standard deviation, normalised to the mean, of the number of simultaneously active users. What 'simultaneously active' means in this context depends on the system considered. If it features a TDMA element, the relevant time-step is a time-slot, in a 'pure' CDMA system it would rather be a frame. Put more generally, the gain depends on the standard deviation of the normalised instantaneous MAI level. The lower this normalised standard deviation, the higher the gain. At least with Poisson traffic, this is equivalent to saying that the higher the number of sources multiplexed onto a common resource, the higher the multiplexing gain. We discuss this in detail in Chapter 7. Statistical multiplexing is also possible with sources of variable-bit-rate (VBR) nature, and similar considerations apply in this case, but rather than with the number of simultaneously active users, one would have to deal with the aggregate bit-rate.

In pure CDMA, since users transmit continuously during activity phases, the common resource is an entire carrier. In hybrid CDMA/TDMA, on the other hand, the relevant resource is a single time-slot, as long as the time-slots are operated independently from each other, which is the case when users *already admitted* to the system can access arbitrary time-slots without restrictions. Since the average number of sources per time-slot is comparatively low at typical carrier bandwidths, the normalised variance is extremely large in the case of unconstrained channel access, resulting in a low inherent statistical multiplexing gain. By combining CDMA with PRMA, it is possible to control the access of users to the channel on a packet-spurt level in a manner that decreases the variance of the MAI. Essentially, this is achieved because access control provides load-balancing between time-slots. In other words, owing to access control, the whole carrier is now shared as a common resource like in pure CDMA. We will demonstrate in Chapter 7 that, depending on the circumstances considered, the statistical multiplexing gain in hybrid CDMA/TDMA systems can be increased considerably through controlled channel access, which clearly illustrates the benefits of this combination.

The question that arises now is whether access control provides the same benefits in a pure (wideband) CDMA system. As long as the users being served in a cell demand only services requiring relatively moderate bit-rates (i.e. $R_{av} \ll R_{cell}$) and are reasonably 'well behaved', the number of users that can be multiplexed onto a single carrier will be high, providing a substantial amount of inherent statistical multiplexing gain. The capacity cannot be improved much further by controlling the *instantaneous* interference level through access control. Obviously, the *average* interference level must be controlled, which is achieved by admission control. However, for the support of high-bit-rate packet-data users, putting a scheme such as PRMA on top of CDMA, or more generally, using the types of access control techniques proposed here in conjunction with a combined CDMA/PRMA scheme, may still be attractive. In the so-called UTRA FDD mode (where UTRA stands for UMTS Terrestrial Radio Access), for instance, they could be applied both on the Common Packet Channel (CPCH) and on dedicated channels, as discussed in more detail in Chapter 10.

For an air interface based on hybrid CDMA/TDMA, combining PRMA with CDMA is straightforward. The channel structure is as in PRMA; that is, the time axis must be divided into slots, which are grouped into frames. Terminals spread their packets before accessing the channel, such that each time-slot may carry multiple packets, owing to the CDMA feature. In the protocol we propose, time-slots may simultaneously carry traffic packets from users having a reservation on that time-slot and also be used for access attempts of contending users not holding, but wishing to obtain, a reservation. In order to protect the reservation mode users from excessive MAI and to stabilise the protocol, the access permission probability p of contending users is dynamically controlled. The core contribution of our research is the investigation of different load-based and backlog-based access control algorithms in this context, as discussed in the remainder of this text in detail.

1.4.2 Hybrid CDMA/TDMA Multiple Access Schemes

In 1994, when we started our research on PRMA-based MAC strategies for the uplink channel of hybrid CDMA/TDMA-based air interfaces, operational or planned systems relied either on TDMA (e.g. GSM [10]) or on CDMA (cdmaOne) for multiple access. The main research efforts towards 3G were also either directed to TDMA-only systems, such as the European RACE Advanced TDMA (ATDMA) collaborative research project [11], or to CDMA-only systems (e.g. the RACE Code-division test-bed or Codit project [12]). However, hybrid CDMA/TDMA schemes had already been proposed in the late 1980s for GSM, and were eventually being reconsidered for 3G.

Hybrid CDMA/TDMA-based systems were investigated extensively at the University of Kaiserslautern in Germany (e.g. Reference [13]). These investigations focussed mainly on issues related to the physical layer, and the introduction of a TDMA component appeared initially to be driven by the interest of that research group in joint detection (JD) schemes for CDMA receivers. Implementation of JD in pure CDMA systems is rather difficult due to excessive computational complexity of such algorithms in the presence of a large number of simultaneous users. By introducing a TDMA component, which allows for orthogonal separation between users assigned to different time-slots, the number of users to be multiplexed by means of CDMA can be reduced, which significantly reduces the computational complexity of JD.

Another proposal combining certain features of CDMA with TDMA was put forward for personal communications systems in the US. In this proposal, standardised as IS-661 [14], but never deployed commercially outside New York, multiple access within a cell is provided by means of TDMA only, but the TDMA channel is spread in order to decrease the frequency reuse factor, and different codes are assigned to different cells. In some ways, a GSM system can provide the same feature through slow frequency hopping. In fact, a system not unlike GSM in this respect, which was proposed in Reference [15], is referred to as slow frequency-hop TDMA/CDMA system. How reuse factors can be reduced in GSM through slow frequency hopping is explained in Section 4.6.

A third hybrid approach referred to as code-time-division multiple access (CTDMA) was proposed by Massey in 1989 and investigated at the Swiss Federal Institute of Technology (ETH) in Zurich [16]. In CTDMA, again only one code per cell is used, and the time-division element is provided through staggering of users in intervals of a few chips rather than providing 'proper' time-slots. Other publications dealing either exclusively with hybrid CDMA/TDMA systems or comparing their performance with other systems include those in References [17–20].

In the following, when we refer to hybrid CDMA/TDMA, we mean systems using both CDMA and TDMA 'with proper time-slots' within a cell for multiple access purposes, hence excluding systems such as IS-661 and CTDMA.

As discussed in more detail in Chapter 2, a hybrid CDMA/TDMA scheme with origin in Reference [13] was indeed adopted for 3G (for UMTS, to be precise), namely TD/CDMA, albeit only in its TDD mode, referred to now as UTRA TDD mode. Compared to Reference [13], the scheme has experienced substantial revisions, mainly to ensure coexistence with the UTRA FDD mode based on *wideband* CDMA (WCDMA).

1.4.3 Literature on Multiple Access Protocols for Packet CDMA

Spread over the last two decades, a significant number of publications have appeared on the topic of multiple access protocols for packet transmission in spread spectrum systems, which use CDMA as a basic multiple access scheme. For instance, an article on a spread spectrum version of slotted ALOHA appeared as early as 1981 [21]. Similarly, references cited in Reference [22] on unslotted spread spectrum ALOHA date back to 1982. A CDMA random access protocol specifically intended for cellular communications is 'packet CDMA' [9]. Load sensing protocols were mentioned in Reference [23] dating from 1984, and later considered in Reference [24] and [25]. More recently, hybrid CDMA/ISMA was proposed by Prasad *et al.* (see for instance Reference [26, Ch. 10], with references dating back to 1993). In this protocol, ISMA stands for inhibit sense multiple access, but it is also used in the literature for idle sense multiple access.

By contrast, the more specific topic of *reservation*-based multiple access protocols for CDMA systems with some kind of implicit or explicit time scheduling, in particular for hybrid CDMA/TDMA systems, seems to have received very little attention before the mid 1990s. There is an early publication on reservation schemes for spread spectrum systems dating from 1983 [27]. However, in the proposed protocol, spread spectrum is not used to provide a multiple access feature like in CDMA, but only to capture a single packet in the case where more than one user accesses the channel simultaneously. Other than that, judging from the literature research we carried out, our publications on joint CDMA/PRMA (i.e. References [28–30], which are based on Reference [31]) appear to be part of the first wave of publications on such protocols. In one aspect, joint CDMA/PRMA has some resemblance with a channel load sensing protocol proposed in Reference [32], as elaborated upon further in Section 6.4.

Soon after we first published the idea of combining CDMA with PRMA, and most likely independently from our research efforts, Dong and Li proposed a similar protocol called spread spectrum PRMA (SSPRMA) [33] and a variation thereof in Reference [34]. Furthermore, Tan and Zhang proposed a protocol similar to ours in 1996, but without the TDMA feature [35]. In the meantime, the topic of multiple access protocols for systems featuring a CDMA component, which are suitable for the integration of various types of packet traffic, has become a popular research area. Later work specifically building on joint CDMA/PRMA includes that in References [36–45].

1.4.4 Access Control in Combined CDMA/PRMA Protocols

In the following, we provide an overview of the specific aspects of the PRMA-based protocol we investigated, indicate where we first published the respective ideas (where previously published), and in which chapter they will be explained in more detail. We also point at related efforts and publications by other researchers. As we referred to the proposed protocol in most of our publications as 'joint CDMA/PRMA', we will continue to use this name in the following, when referring to such publications. However, for convenience and for reasons outlined in Chapter 6, we will only use the term multidimensional PRMA (MD PRMA) in subsequent chapters.

1.4.4.1 Joint CDMA/PRMA

In conventional PRMA as defined in Reference [8] and described in detail in Chapter 3, resources are allocated on the basis of packet spurts rather than circuits. Users holding a reservation transmit their packets in a TDMA-fashion on their reserved time-slot (such reserved slots are referred to as *I-slots* here, with 'I' for 'Information'), and release the slot upon completion of the packet-spurt transfer. Reservations are obtained through a contention procedure that follows the slotted ALOHA approach. Any currently unreserved time-slot is available for Contention (a so-called *C-slot*). Contending users obtain permission to access such a C-slot with access probability p, as determined by a Bernoulli experiment they must carry out before accessing the slot with p as parameter.

If PRMA is operated in a hybrid CDMA/TDMA environment, individual time-slots may carry several packets. In the protocol we proposed in 1995 [31], which we initially referred to as 'joint CDMA/PRMA' (e.g. Reference [30]), a time-slot may simultaneously carry several 'reserved packets' *and* at the same time be available for access attempts of contending users. This protocol is defined in detail in Chapter 6. In such a scenario, MAI becomes an issue of concern. As in PRMA, for real-time traffic, users not obtaining a reservation within a certain delay limit (due to lack of resources or unsuccessful contentions) will have to drop packets. On top of that, owing to the CDMA component, packets may be erased due to MAI. The overall performance measure of the protocol for real-time traffic will be the packet-loss ratio P_{loss} as a function of the traffic load, where P_{loss} is the sum of the probability P_{pe} of packets being erased due to MAI and the packet dropping probability P_{drop}. For non-real-time traffic, packets need not be dropped, erased packets may be retransmitted, and adequate performance measures are access delay and transmission delay. The protocol performance is significantly affected by the approach to controlling the access permission probability p of contending users.

1.4.4.2 Load-based Access Control with Channel Access Functions

In References [28–31], we were concerned with the control of the MAI through load-based dynamic access control using the concept of 'channel access functions' (CAFs). Simply put, these relate the number of users in reservation mode R using a certain time-slot n_s to the probability p with which contending users are allowed to access this slot in the following frame. For instance, for the case illustrated in Figure 1.6 $p(n_f + 1, n_s = 3) = f(R[n_f, n_s = 3])$. In reality, it is slightly more complicated, as explained in Section 6.4.

When the average number of packets per spurt (i.e. per reservation period) is $\gg 1$, choosing appropriate CAFs results in a reduction of slot-to-slot load fluctuations (in other words, the variance of the MAI is reduced) and, as a consequence, P_{loss} is reduced as well. This is achieved by restricting access for contending users to time-slots with high load (by choosing low p values) and setting p (close) to one for time-slots with low load, which, compared to unconstrained channel access, balances the load between time-slots. In these investigations, no hard limit was assumed for the number of packets that can be carried in a single time-slot, and distinct codes were not distinguished. Instead, the packet error probability P_{pe} (or the corresponding success probability $Q_{pe} = 1 - P_{pe}$) as a function of the number of users accessing a time-slot K (or in other words, as a function of the MAI) is determined, assuming random coding, and using so-called Gaussian approximations for the MAI. This approach could also be viewed as having an infinite number of non-orthogonal codes, with the packet error probability limiting implicitly the number of codes that can be used simultaneously. The packet error probability affects both the success

1.4 A PRMA-BASED PROTOCOL FOR HYBRID CDMA/TDMA

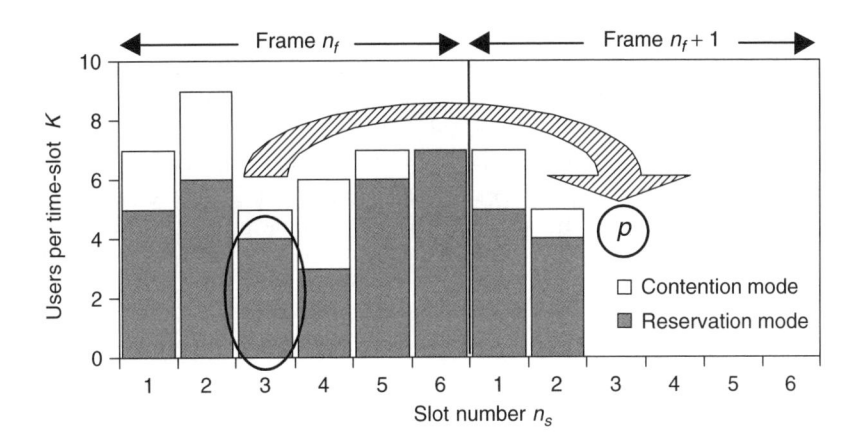

Figure 1.6 Control of access permission probability p for slot 3 of frame $n_f + 1$ based on the load of reservation mode users in slot 3 of frame n_f

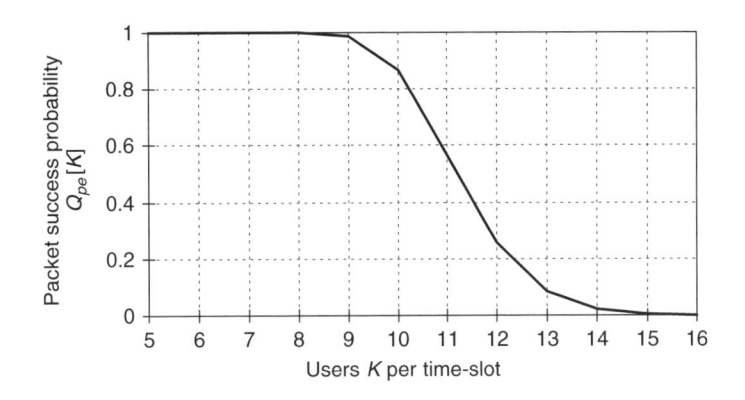

Figure 1.7 $Q_{pe}[K]$ for $X = 7$ and a (511,229,38) BCH code

probability of contending users (and thus P_{drop}) and the quality of service of users holding a reservation, the latter by erasing some of their packets. As an example, $Q_{pe}[K]$ for a (511, 229, 38) BCH code and a spreading factor $X = 7$ is shown in Figure 1.7. The BCH code is used for forward error correction coding. For more explanations and details on the exact conditions considered, refer to Chapter 5. In such a scenario, clearly, the notion of a slot having either status I or C needs to be abandoned.

The results presented in Reference [30] and [31] were based purely on heuristic channel access functions. For References [28] and [29] these functions were optimised to reduce P_{loss} based on statistics gathered. The packet-loss performance of joint CDMA/PRMA was compared with that of a scheme without access control (where $p \equiv 1$), which will be referred to as *random access protocol*, and with that of a few theoretical benchmarks. These results are discussed in detail in Chapter 7.

It was found that this type of load-based access control used with joint CDMA/PRMA works extremely well for voice-only traffic and its performance comes very close to that of the benchmarks being used. It can therefore be improved only insignificantly with more elaborate adaptive methods. This is also confirmed by looking at the results for voice-only traffic provided by Mori and Ogura in References [36] and [37]. However, in the case of mixed voice/data traffic, when data is sent only in contention mode (that is, reservations are never granted), the performance achieved with a single channel access function is not satisfactory. In this case, the performance can be improved by treating voice and data in different ways. Mori and Ogura, for instance, proposed in Reference [36] to base the data permission probability on the number of voice users in reservation mode R plus the estimated number of contending voice users. Another approach, suggested by Wang *et al.* in Reference [40], essentially consists in admitting data users only to very lightly loaded time-slots.

Apart from the fact that joint CDMA/PRMA improves the packet-loss performance compared to random access CDMA, our investigations also revealed that, compared to conventional PRMA, multiplexing efficiency is increased. This is due to increased trunking efficiency, as the population of users multiplexed onto a shared resource[1] is much larger than that in a conventional PRMA scheme with the same number of time-slots per frame. This is consistent with the observations in Reference [46], where it was found that the multiplexing efficiency of a similar protocol called PRMA++ is increased when the number of time-slots per frame increases. It was therefore suggested in Reference [46], that several frequency carriers could be pooled together for multiplexing purposes, should this protocol be operated on an air interface with a low number of time-slots per frame. In fact, an extended PRMA scheme with four carriers pooled together was already proposed earlier in Reference [47].

1.4.4.3 MD PRMA with Backlog-based Access Control

If the graceful degradation or soft-capacity feature of CDMA illustrated in Figure 1.7 is ignored, and one simply assumes that, owing to the CDMA feature, a fixed number of code-slots per time-slot are provided, as was assumed in Reference [33] for SSPRMA, then such a protocol becomes essentially equivalent to a conventional PRMA protocol operating on several carriers. The only fundamental difference is that, instead of several frequency-slots per time-slot, there are several code-slots per time-slot.

We adopted such an approach for a more detailed investigation on the multiplexing efficiency achieved through PRMA-based protocols with different combinations of code- or frequency-slots per time-slot and time-slots per frame. We chose multidimensional PRMA (MD PRMA) as an umbrella name for these protocols in References [48,49]. In such a scenario with a rectangular grid of *mutually orthogonal* resource units, we can again introduce the notion of C-slots and I-slots from conventional PRMA, now however on the basis of individual *code*-time-slots rather than time-slots. Contention-mode packets may now collide if they are transmitted with the same code on the same time-slot, as shown in Figure 1.8. On the other hand, when perfect orthogonality between codes is assumed, no matter how many contending terminals use available codes, reservation mode users transmitting with other codes in the same time-slot are not adversely impacted. In this scenario, load-based access control does not make much sense. Instead, to limit the collision probability, stabilise the protocol and optimise the performance, investigations related to backlog-based access control in homogeneous and heterogeneous

---

[1] Due to access control in joint CDMA/PRMA, the shared resource is the *ensemble* of time-slots in a frame rather than a single time-slot, as pointed out in Subsection 1.4.1.

1.4 A PRMA-BASED PROTOCOL FOR HYBRID CDMA/TDMA

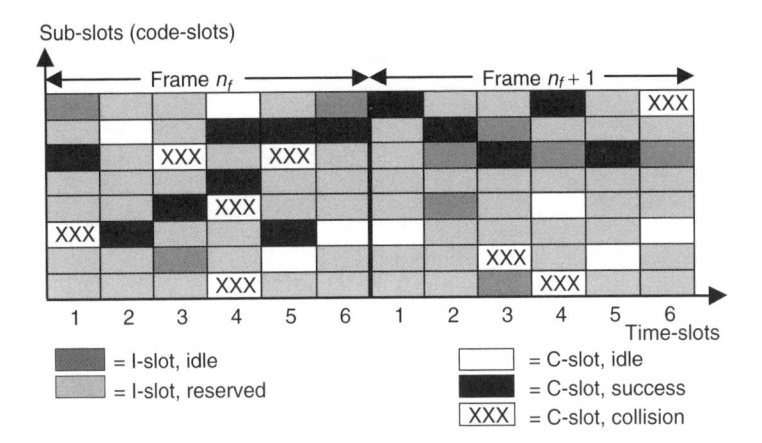

Figure 1.8 MD PRMA on orthogonal resource units. C-slots are available for contention, while terminals holding a reservation transmit on I-slots

traffic scenarios were carried out. With backlog we mean the number of users wanting to transmit, but not holding a reservation (i.e. the contending users).

Among a class of similar algorithms used to estimate the backlog in slotted ALOHA, which were compared in Reference [50], we chose *pseudo-Bayesian broadcast control* proposed by Rivest [51] and adapted this algorithm to MD PRMA [49]. A detailed derivation of the algorithm is provided in Chapter 6. In Reference [52], we investigated the impact of acknowledgement delays, with and without Bayesian broadcast, and the effect of interleaving on the voice dropping performance of MD PRMA. Also, building on frame reservation multiple access (FRMA [53]), we investigated the performance of a TDD mode referred to as multidimensional FRMA (MD FRMA) with Bayesian broadcast. These results are presented in Chapter 8.

The choice of the pseudo-Bayesian algorithm for backlog estimation was motivated by earlier investigations we carried out on the random access for the General Packet Radio Service (GPRS [54]) based on this algorithm. Some of the results from these investigations on random access stabilisation and prioritisation can be found in Chapter 4, others were published in Reference [55]. In GPRS, discrimination between the access delay performance experienced by different services is possible through computation of different permission probabilities for contending terminals according to the delay-class they belong to, e.g. by using a prioritised version of pseudo-Bayesian broadcast. In MD PRMA, on top of access delay discrimination for non-real-time services, the prioritised algorithm allows trading of voice dropping performance against data access delay according to the chosen value of a prioritisation parameter, as discussed in detail in Chapter 9.

The algorithm is tested in a mixed traffic environment, where on top of the usual on–off voice traffic, Web browsing and email traffic is also generated. For Web traffic generation, the traffic model from the ETSI UMTS selection criteria [56] is used with the parameter values specified therein, while for email traffic, the same model structure is adopted, but with adapted parameter values derived from data of large email log files. Statistics on these email log files were obtained from industrial partners participating in a LINK collaborative research project on Advanced Channel Structures for mobile communication systems (LINK ACS [57]). Most results presented in Chapter 9 for MD PRMA with mixed traffic were first published in Reference [52].

Also in Reference [52], we have looked at a scenario where distinct code-slots are discerned, but the codes are not perfectly orthogonal any more. Instead, the packet error probability due to the MAI caused by the total number of packets transmitted in a time-slot is accounted for. Thus, in a sense, we combined the random coding scenario illustrated in Figure 1.7 with the orthogonal code-time-slot scenario considered above. First, for voice traffic, we investigated to what extent P_{drop} suffered from MAI when using the pseudo-Bayesian algorithm, since its adaptation for MD PRMA was based on the assumption of orthogonal code-slots. However, under the circumstances considered, P_{drop} alone is not the figure of merit, instead, P_{loss} is relevant, which includes also the packets erased due to MAI. The more interesting question is therefore, whether it is possible to reduce P_{loss}, possibly at the expense of increased P_{drop}. We found that this is indeed possible by combining pseudo-Bayesian broadcast with load-based access control. The reader is referred to Chapter 8 for detailed results.

1.4.4.4 Protocol Versions for Pure CDMA Systems

Given the adoption of wideband CDMA for the UTRA FDD mode, it would be interesting to know to what extent backlog- and load-based access control are useful for such an air interface. Since random access attempts and user traffic are code-multiplexed rather than time-multiplexed [58], i.e. have to share the same power budget, the situation is similar to that in the hybrid CDMA/TDMA system considered, where random access attempts may be made on time-slots carrying user data. In fact, Cao has adapted pseudo-Bayesian broadcast for wideband CDMA [59]. Furthermore, a load-based access control scheme inspired by Reference [30] has been proposed for the UTRA FDD mode in Reference [60], which was initially referred to as dynamic packet admission control, but is now called *dynamic resource allocation control*. It can be used on dedicated channels. Centralised probabilistic access control can also be applied to the random access channel and the optional common packet channel in UTRA. UTRA access control is discussed in detail in Chapter 10.

Tan and Zhang have proposed a protocol called reservation-code multiple access (RCMA) in Reference [35] suitable for pure (wideband) CDMA, which is essentially PRMA with code-slots instead of time-slots, or MD PRMA with only one time-slot per frame, but a large number of code-slots. They claim that their protocol is much more efficient than PRMA already with a 'median number of codes'. But they completely ignore the fact that supporting a large number of users with CDMA requires a large spreading factor and consequently a large bandwidth, hence every code to be provided comes at a cost.

It is shown in Chapter 8 that PRMA, MD PRMA and RCMA essentially provide the same multiplexing efficiency with appropriate access control, provided that the number of time-slots in PRMA, code-time-slots in MD PRMA and code-slots in RCMA is the same. All slots in a frame are assumed to be orthogonal, hence this comparison does not take soft-capacity into account, nor does it attempt to quantify the bandwidth required to properly support the number of time-, code-time-, or code-slots considered. This may appear to be a very abstract scenario, but at least it is a fair comparison. Other than reporting the fundamental outcome in the conclusion of Reference [52], results of these investigations have so far only been published in Reference [61]. For completeness, we point at Reference [17], where CDMA, TDMA and hybrid systems are compared from a packet queuing perspective. This is probably a bit more abstract in terms of MAC layer aspects than our comparison of PRMA-based protocols, and in particular does not deal with access control, but unlike our comparison, it accounts for physical layer effects in a fairly thorough manner.

1.4.4.5 Impact of Power Control Errors

The impact of power control errors is an important matter in systems with a CDMA feature and needs also to be investigated in the case of joint CDMA/PRMA, as already noted in Reference [30]. In Reference [38], Hoefel and de Almeida reproduced our results from References [29] and [30] and investigated the impact of power control errors on joint CDMA/PRMA, using 'semi-empirical' channel access functions as we did in Reference [29]. They compared these results with those reported in the literature for 'circuit-switched CDMA' (e.g. Reference [26]) and found that the loss in capacity due to power control errors, although slightly higher with joint CDMA/PRMA, was of the same order as that with 'circuit-switched CDMA'.

Our findings are similar, but on top of the specific case looked at in Reference [38], we made an attempt to assess in a more generic manner the potential benefit of load-based access control compared to the random access protocol as a function of both power control errors and the spreading factor. This is accomplished by resorting to a fairly straightforward analysis of the random access protocol (to save simulation time), through simple benchmarks for controlled access already used in Reference [30], and through enhanced benchmarks derived for Reference [61]. The spreading factor affects the size of the population multiplexed onto a common resource and thus the multiplexing efficiency. Again, these results were so far only published in Reference [61].

For completeness, it is reported that Hoefel and de Almeida carried out further investigations on joint CDMA/PRMA to assess in more detail some physical layer aspects. In Reference [62], the protocol performance with mixed voice/data traffic on a Nakagami-m frequency selective fading channel is assessed. In Reference [63], the impact of slow and fast Rayleigh fading is investigated, and further results on the impact of power control errors are provided. Some of these results were obtained from a system-level simulator they implemented, which simulates mobiles in 19 different cells (a centre test cell and two tiers of interfering cells). In our investigations, when accounting for intercell interference, we resorted to other means to assess intercell interference level, as discussed in detail in Chapter 5. Other work building on joint CDMA/PRMA includes Reference [43], where dynamic access control in a somewhat modified scheme is performed by 'access probability controllers' based either on fuzzy logic or neural network interference estimators. Finally, in Reference [45], a so-called 'non-collision' version of the protocol is investigated, where terminals request resources frame by frame on a separate time-slot without risking collisions. This request mechanism is also exploited to combat the negative impact of power control errors on the protocol capacity. We will reconsider this proposal in Chapter 7, where we examine power control issues.

1.4.5 Summary

In summary, one of the key topics of our research efforts documented in this book, which are mostly related to PRMA-based multiple access protocols, concerns the control of the access of contending users to:

- provide stabilisation (see Chapter 7 and 8);
- protect reservation-mode users from multiple access interference (again Chapter 7 and 8);
- and provide prioritisation at the random access (see Chapter 9).

The first and the third aspects are relevant wherever protocols with a random access element are applied, and thus also for example in GPRS. Accordingly, in Chapter 4, where a detailed description of GSM/GPRS multiple access principles is embedded in the wider framework of air interface matters, such issues are dealt with as well. The second aspect is mostly relevant for hybrid CDMA/TDMA air interfaces, when time-slots may simultaneously be accessed by contending users and carry traffic of users holding a reservation. However, it is also relevant for UTRA FDD, where users holding a reservation (e.g. on dedicated channels) are not separated in time from contending users. Therefore, in Chapter 10, where the multiple access concepts relevant for UTRA are introduced, such matters are also touched upon.

2
CELLULAR MOBILE COMMUNICATION SYSTEMS: FROM 1G TO 4G

The basic principles of cellular communications were explained in the introductory chapter, and terms such as cluster size and reuse efficiency introduced. In the following, some more considerations on the advantages and limitations of the cellular concept will be made before reviewing first generation (1G) and second generation (2G) cellular communication systems. Moving on to 3G, we will first discuss the initial requirements according to which 3G systems were designed, and then to what extent they are likely to be satisfied by first releases of 'true' 3G systems on the one hand, and evolved 2G systems on the other. While special attention is paid to the GSM evolution route and to UMTS, the initially European, but now global proponent for 3G cellular communications, evolutionary routes from cdmaOne to cdma2000 are also reviewed.

The emergence of the Internet witnessed in the 1990s is expected to have a fundamental impact on the telecommunications industry, including cellular communications. We will examine how this affects 3G, what new requirements arise and how these can be met through subsequent releases of UMTS. Finally, looking further into the future, possible manifestations of 4G systems will also be discussed, including the role wireless LAN technologies may play in this context.

2.1 Advantages and Limitations of the Cellular Concept

It was outlined in the previous chapter how the cellular concept allows use of the same set of frequencies in multiple cells, and how it is in theory possible to arbitrarily increase capacity to match growing demand for wireless communication services through cell splitting. In practice, however, there are certain limitations. With smaller and smaller cells as a result of cell splitting, it becomes increasingly difficult to place base stations at the locations that offer the necessary radio coverage [1]. Furthermore, as the cell radius decreases, the handover rate increases. This will place a costly burden onto the network in terms of increased signalling load and, given the non-zero probability of handover failures, the call-dropping probability may increase, particularly for fast moving mobiles. Finally, it would be wasteful to deploy a large number of base stations covering small cells in areas with low traffic density.

All these considerations call for a network topology where different cell types are deployed concurrently, to provide capacity, where required, and at the same time ensure universal coverage. According to the topography and other conditions, base stations may cover a *macrocell* (also called *umbrella cell*), a *microcell* (typically to be found in city centres, on highways or even indoors), or a *picocell* (usually deployed indoors), with diameters of several kilometres, up to one kilometre and possibly as little as a few tens of metres respectively. These cells of different types, each cell-type constituting a separate layer of a multi-layered system, will serve overlapping coverage areas. In this way, a fast moving mobile terminal, for instance, can be served by a macro- or umbrella cell, in order to limit the number of handovers per call, while slower moving mobiles in the same coverage area are normally supported by microcells.

For traffic management purposes, these multiple layers can be organised in a *hierarchical cellular structure* (HCS, see e.g. Reference [64]), in which overflow traffic from microcells is handed 'up' to the hierarchically higher macrocellular layer. A space segment may be added to the terrestrial segment, where satellite spot beams overlay clusters of terrestrial macrocells [65], as illustrated in Figure 2.1. GSM, the most important 2G cellular system in operation today, is one example of existing cellular systems that can support multiple cell layers.

From these considerations, a first set of requirements for the efficient operation of cellular communication systems can be inferred:

- the system should use available resources (i.e. frequency spectrum) as efficiently as possible, to limit infrastructure deployment and prevent some of the problems listed in relation with small cell radii;

- handover procedures should be fast, require a minimum amount of signalling (preferably to be exchanged on dedicated signalling resources to avoid an impairment of ongoing communications), and be robust to avoid dropped calls; and

- the system should be able to support multiple cell layers, for instance in the shape of hierarchical cellular structures, to provide high-speed mobility and hot-spot capacity.

This list will be expanded in the following, first with respect to the initial 3G requirements identified in the early 1990s, according to which first releases of 3G systems were designed, then with respect to additional requirements identified later on, mainly as a result of the Internet revolution. Before we do that, the older 1G and 2G systems will be reviewed briefly.

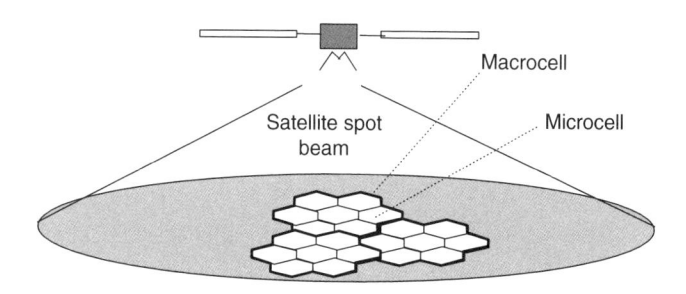

Figure 2.1 Microcells, macrocells, and a satellite spot-beam

2.2 1G and 2G Cellular Communication Systems

2.2.1 Analogue First Generation Cellular Systems

In the late 1970s and early 1980s, various first generation cellular mobile communication systems were introduced, characterised by analogue (frequency modulation) voice transmission and limited flexibility. The first such system, the *Advanced Mobile Phone System* (AMPS), was introduced in the US in the late 1970s[1]. Other 1G systems include the *Nordic Mobile Telephone System* (NMT), and the *Total Access Communications System* (TACS). The former was introduced in 1981 in Sweden, then soon afterwards in other Scandinavian countries, followed by the Netherlands, Switzerland, and a large number of Central and Eastern European countries, the latter was deployed from 1985 in Ireland, Italy, Spain and the UK.

While these systems offer reasonably good voice quality, they provide limited spectral efficiency. They also suffer from the fact that network control messages — for handover or power control, for example — are carried over the voice channel in such a way that they interrupt speech transmission and produce audible clicks, which limits the network control capacity [7]. This is one reason why the cell size cannot be reduced indefinitely to increase capacity.

Such constraints did not prevent these systems from enjoying considerable success with the public, so that subscriber numbers were still growing in the mid 1990s. In Italy, for instance, they peaked at four million users in March 1998, corresponding to a penetration of roughly 7%, and (while less impressive in absolute figures) penetration exceeded 10% in most Scandinavian countries. However, they have been increasingly thrust aside by 2G systems in most parts of the world. In the meantime, after closing down 1G systems, spectrum refarming from 1G to 2G has taken place in several countries.

2.2.2 Digital Second Generation Systems

Capacity increase was one of the main motivations for introducing 2G systems in the early 1990s. Compared to the first generation, 2G offers [69]:

- increased capacity due to application of low-bit-rate speech codecs and lower frequency reuse factors (the cluster size can be as low as three compared to seven in analogue systems for example, see also Subsection 2.3.2 on evolved 2G systems);
- security (*encryption* to provide privacy, and *authentication* to prevent unauthorised access and use of the system);
- integration of voice and data owing to the digital technology; and
- dedicated channels for the exchange of network control information between mobile terminals and the network infrastructure during a call, in order to overcome the limitations in network control of 1G systems (note though, that handover-related signalling still steals into the traffic channel in GSM).

---

[1] According to Lee, the FCC released frequencies for cellular communication systems in the 800 MHz band in 1974, and AMPS served 40 000 mobile customers in 1976 [66]. Other sources indicate later launch dates though, for instance 1979 for a pre-operation AMPS system in Chicago [67]. See also Reference [68], which provides a detailed history of cellular communications.

At the time of writing, the major force driving the growth in cellular communications are still such 2G systems, which were first introduced in Europe, Japan, and the US, and are now in operation worldwide.

Initially, these systems operated only at 900 MHz (800 MHz in the US and Japan), with up-banded versions at 1800 MHz (1900 MHz in the US, 1500 MHz in Japan) coming soon after. These up-banded systems are aimed primarily at people moving around in cities at pedestrian speeds with hand-held telephones. They are referred to as Personal Communications Networks (PCN) in Britain and Personal Communications Systems (PCS) in the US, to distinguish them from the 'classical' cellular systems operating below 1 GHz.

One of these 2G systems needs to be singled out, GSM, a TDMA-based system with optional slow frequency hopping. The acronym stood initially for 'Groupe Spécial Mobile', but fortunately lends itself conveniently to 'Global System for Mobile Communications'. With a subscriber number close to 500 millions in early 2001, a footprint covering virtually every angle of the world, and a share of the digital cellular market close to 70% in February 2001 (according to figures published by the GSM Association [70]), it truly deserves this name. Introduced in early 1992, already by the end of 1993, GSM networks had been launched in more than 10 European countries, and outside Europe for instance in Hong Kong and Australia. From 1994, GSM gradually conquered the markets in the remaining European countries and large parts of the rest of the world.

The pan-European standardisation effort leading to GSM was initiated by the *Conférence Européenne des Administrations des Postes et des Télécommunications* (CEPT) in 1982 with the formation of the Groupe Spécial Mobile (GSM) [3]. In 1988, the European Telecommunications Standards Institute (ETSI) was founded, which was from then on responsible for the evolution of the GSM standards [10]. ETSI is still formally responsible for the GSM standards, but the technical work relating to system evolution is now carried out within the framework of the Third Generation Partnership Project (3GPP), a body created in 1998 to standardise UMTS. The GSM evolution work was added in the year 2000. ETSI is a 3GPP member along with other standardisation bodies and interest groups from around the world. In the context of cellular systems, ETSI's role is now to transform technical specifications created in 3GPP into regional standards for Europe, whether this is for evolved GSM or for UMTS.

While GSM is currently the uncontested standard for 2G digital cellular communications in Europe, there is no clear dominance of a single digital standard in most of the rest of the world. In the US, for instance, there are essentially two types of 2G *cellular* systems operating at 800 MHz, which are incompatible with each other. The first is a TDMA system called North American Digital Cellular or Digital AMPS (D-AMPS), and sometimes simply referred to as TDMA, according to the basic multiple access scheme it is based on. This can create confusion, since there are other TDMA-based cellular systems, notably GSM. The second system, which was launched later, is cdmaOne, the first operational CDMA system [71]. The relevant air-interface specifications are the so-called interim standards IS-136 (for D-AMPS) and IS-95 (for cdmaOne).

The fragmentation observed in the US was accentuated when the Federal Communications Commission (FCC) sold frequencies in the 1900 MHz band for PCS without mandating the technology to be used. On top of up-banded D-AMPS and cdmaOne systems, this allowed a 1900 MHz version of GSM to enter the US market. Similar developments could also be observed in the rest of the Americas, although Brazil, rather than selling spectrum at 1900 MHz, decided in the year 2000 to auction 1800 MHz spectrum for PCS. This decision favours GSM because, at least at the time of writing, 1800-MHz versions of D-AMPS and cdmaOne equipment were not available.

The first and most popular Japanese 2G digital standard is Personal Digital Cellular (PDC). It was later complemented by the Personal Handyphone System (PHS), somehow a hermaphrodite between mobile and cordless system, which caters only for low mobility, but is popular for certain applications owing to relatively high data-rates (32 kbit/s, later enhanced to 64 kbit/s). Both standards are TDMA-based and have not seen wide deployment outside Japan. PDC has received some attention outside Japan, though, owing to the tremendous success enjoyed by the '*i-mode*' service since its launch in February 1999. This is a service similar to WAP (wireless application protocol) services, but (at least at the time of writing) rather more popular. It enables access to some form of Internet through mobile handsets, and runs on top of the packet-overlay added to PDC, referred to as PDC-P. In the late 1990s, a cdmaOne system was launched in Japan. CdmaOne systems enjoy considerable success also elsewhere in the Far East, most notably in South Korea, where an IS-95 derivative dominates the 2G market.

2.3 First 3G Systems

2.3.1 Requirements for 3G

Already before the launch of 2G systems, the research community started to think about requirements for a new, third generation of mobile communication systems and about possible technological solutions to meet them (see for instance References [1,7,12,72,73]). Before 3GPP was established, ETSI was one of the major players regarding the standardisation of 3G systems. It called its 3G representative Universal Mobile Telecommunications System (UMTS) and established a number of requirements, according to which such a system should be designed. Those of interest here are the ones relating to the air interface or radio access, the so-called UMTS Terrestrial Radio Access (UTRA), which are listed in Reference [74]. This list appears to capture most of the 3G requirements stated in the literature of the early 1990s, when 3G emerged as a mainstream research topic. It is therefore summarised in the following. The UTRA requirements pertain to four different categories, namely to bearer capabilities, operational requirements, efficient spectrum usage, and finally complexity and cost.

2.3.1.1 Bearer Capabilities

(1) UMTS has to deliver services with bit rates up to 2 Mbit/s indoors, at least 384 kbit/s in suburban outdoor and at least 144 kbit/s in rural outdoor environments.

(2) UMTS should be flexible in terms of service provision, in particular:

— negotiation of bearer service attributes should be possible;

— parallel bearer services to enable service mix should be possible;

— circuit-switched and packet oriented bearers should be provided;

— variable-bit-rate real-time capabilities should be provided; and

— scheduling (and pre-emption) of bearers according to priority should be possible.

(3) UMTS has to provide services with a wide range of bit-rates in a variety of different environments with bit error rates (BER) as low as 10^{-7} for certain services. To deliver optimum performance in all of these, a very flexible air interface is required.

(4) UMTS should provide seamless handover between cells of one operator, possibly even between cells of different operators. Seamless mean in this context, that the handover must not be noticeable to the user.

2.3.1.2 Operational Requirements

(5) Compatibility with services from the following existing core networks must be provided: ATM bearer services, GSM services, Internet Protocol (IP) based services, and ISDN services.

(6) Automatic radio resource planning should be provided, if such planning is required.

2.3.1.3 Efficient Spectrum Usage

(7) The air interface should make efficient use of the radio spectrum for typical mixtures of different bearer services.

(8) Given the asymmetric UMTS frequency allocation and the likely overall traffic asymmetry (due to Web browsing, for instance), the air interface should support operation in unpaired frequency bands or, according to Reference [74], 'variable division of radio resource between uplink and downlink resources from a common pool [must be provided]'.

(9) UMTS should allow multiple operators to use the band allocated to UMTS without coordination (this includes public, private, and residential operators).

(10) UMTS should allow flexible use of various cell types and relations between cells (e.g. indoor cells, hierarchical cells) within a geographical area without undue waste of radio resources.

2.3.1.4 Complexity and Cost

(11) Handportable and PCMCIA-card-sized UMTS terminals should be viable in terms of size, weight, operating time, range, effective radiated power and cost.

(12) The development and equipment cost should be kept at a reasonable level, taking into account the cost of cell sites, the associated network connections, signalling load and traffic overhead.

Goodman stated in the early 1990s a vision for 3G as to 'create a single network infrastructure, that will make it possible for all people to transfer economically any kind of information between any desired locations', a statement which appears to cover the essence of the above listed requirements. He added that 'a unified wireless access [will replace] the diverse and incompatible 2G networks with a single means of wireless access to advanced information services' [1]. The latter is something which may well fail to materialise, as will be pointed out later on.

Two questions, which will be discussed in the following, arise here:

(1) To what extent can evolved 2G systems meet these requirements?
(2) Do currently specified 3G systems meet these requirements?

2.3.2 Evolution of 2G Systems towards 3G

With the large subscriber base and the considerable investment in 2G infrastructure in mind, it cannot come as a surprise that operators of such systems are keen to protect this investment. Assuming that there is a market need for 3G systems meeting the above requirements list (one would assume that operators having paid billions of dollars for 3G licenses must have made this assumption), there appear to be two ways to achieve this:

- Standardise 3G in a manner so that at least part of the 2G network infrastructure can be reused. In the case of GSM and UMTS, this has materialised to some extent. Certain GSM core network nodes can potentially be reused for UMTS. Also, the UMTS handover requirements state that handover to 2G systems, e.g. GSM, should be possible. Furthermore, it was at least attempted to choose design parameters for UTRA which ease implementation of dual-mode GSM/UMTS handsets; dual-mode operation is expected to be a standard feature of most UMTS handsets. Correspondingly, it is possible to deploy UMTS gradually in a GSM system, where in a first phase only selected sites are equipped with UMTS base stations, while universal coverage is provided by GSM.

- Evolve capabilities of 2G systems to meet 3G requirements, for instance through enhancements to the air interface.

Given the importance of GSM and the large number of advanced features which have been or are still being standardised for this system, it will be discussed briefly to what extent such an evolved GSM system may fulfil 3G requirements from an air-interface perspective. The main air-interface related enhancements to GSM, which are already standardised (as discussed in detail in Chapter 4), are:

- higher data-rates for circuit-switched services through aggregation of several time-slots per TDMA frame with High Speed Circuit-Switched Data (HSCSD) [75];

- efficient support of non-real-time packet-data traffic with the General Packet Radio Service (GPRS), which entails enhancements to both the air interface [54] and the network [76];

- higher data-rates on individual GSM physical channels through use of higher order modulation schemes within the existing carrier bandwidth of 200 kHz, referred to as Enhanced Data Rates for Global Evolution (EDGE). 'Plain GSM', HSCSD, and GPRS can then exploit these higher data-rates (see Reference [77]).

Ignoring network constraints, with HSCSD, GSM could in theory offer user data-rates of at most 8×14.4 kbit/s $= 115.2$ kbit/s for circuit-switched data traffic. With GPRS, a data-rate of up to 8×21.4 kbit/s $= 171.2$ kbit/s can be provided for packet traffic for the case where no forward error correction coding is used [54]. Neither do these data-rates meet the UMTS requirements, nor can they be achieved in an economical manner, since aggregation of eight time-slots in a TDMA frame would result in rather power-thirsty handsets and raise other issues such as how to dissipate the additional heat being generated. A detailed discussion on realistic data-rates is provided in Chapter 4. By applying EDGE to GPRS, without error coding, the data-rates per slot can be increased to 59.2 kbit/s, hence with eight time-slots up to 473.6 kbit/s could be achieved, which exceeds the UMTS requirements for all environments but indoors. Note though that apart from requiring time-slot aggregation, due to lack of error protection, such throughput levels can only be achieved at very high carrier-to-interference ratios, which has obviously repercussions on cell planning and capacity. With mobile terminals capable of aggregating eight time-slots, the suburban requirements of 384 kbit/s could be met while allowing for moderate error protection, which would reduce the required CIR a bit.

GSM was initially designed with specific tele-services in mind, each of them being individually standardised. GPRS and the general bearer service introduced mainly, but not only for HSCSD [78] provide increased flexibility, as do other standardisation work items related to services[2]. However, the constraints of the GSM air interface will not allow the second and third requirement listed above to be satisfied fully. For instance, GPRS was not designed for real-time packet-data services and in its early versions, it is only suitable for real-time services with severe restrictions (if at all). Proper real-time capabilities are being added to GPRS, but provision of high and variable bit-rate real-time service will remain restricted.

Automatic resource planning is normally not provided, although dynamic channel assignment (DCA) and dynamic resource assignment (DRA) could be used to ease the planning process. Alternatively, a combination of slow frequency hopping with fractional loading may allow the deployment of a one site/three sector (1/3) reuse-pattern for carriers not carrying broadcast channels, eliminating the need for frequency planning for these carriers [79,80]. A system scenario for GSM with two hierarchical layers is described in Reference [81], where 1/3 reuse is applied to hopping channels of macrocells, 1/3 or even 1/1 reuse to those of microcells, and carriers with broadcast channels for microcells are planned adaptively. The GPRS COMPACT mode [82], a stand-alone data-only solution (i.e. without support of GSM circuit-switched services) relies also on a 1/3 reuse. Such matters are discussed in detail in Chapter 4.

An evolved GSM system will fail to meet 3G requirements on two more counts. Firstly, asymmetric frequency allocation is not possible (it is possible to provide asymmetric services, but the total resources managed by base stations are always symmetric). Secondly, for public operation, GSM certainly does not allow multiple operators to use the *total* allocated band for GSM without co-ordination (i.e. without reserving for each operator a dedicated part of the available spectrum). However, if the respective requirement (number (9) in the list) is interpreted in this manner, it will also not be met by UMTS. Residential (or private) use without co-ordination is a different matter; a GSM-based cordless telephony system is a part of the GSM evolution story. As far as efficient use of the radio spectrum is concerned, refer to the discussion of multiple access schemes

---

[2] Tele-services are fully specified end-to-end services providing the complete capability, including terminal equipment functions, for communication between users. Bearer services, by contrast, provide only the capability for the transmission of signals between user–network interfaces; they provide bearers used by tele-services.

in Chapter 3 for comments on spectral efficiency of TDMA used in GSM, and CDMA, the main alternative.

In conclusion, 2G systems continue to evolve and the boundary between advanced 2G and 'true' 3G systems is increasingly being blurred. GSM, for instance, with its wealth of implemented and imminent features, is now rightly referred to as an advanced 2G system (or alternatively, a generation 2.5 system [72]), and may, in certain manifestations, become an integral part of 3G systems. However, as stand-alone systems, 2.5G systems will struggle to meet all 3G requirements. In particular (at least in the case of an evolved GSM system), they will only to a limited extent be able to provide efficient support of high and variable bit-rate multimedia services.

2.3.3 Worldwide 3G Standardisation Efforts

These high and variable bit-rate multimedia services are, at least from today's perspective, exactly those services that may create market demand for new systems. This is one reason why the mobile communication communities in Europe, the Far East, and the US specified 'true' 3G systems. There are other reasons, as well. For instance, handset and infrastructure manufacturers must naturally be interested in deployment of new systems.

Japan was pushing particularly hard for 3G and leads on 3G deployment, mainly because of overcrowded 2G systems, but likely also motivated by a wish to break out of a technological isolation in which its industry found itself with PDC and PHS.

The International Telecommunications Union (ITU), which refers to 3G systems as either Future Public Land Mobile Telecommunications System (FPLMTS) or, more handy, International Mobile Telecommunications 2000 (IMT-2000), initially had the intention of controlling the 3G standardisation process in a manner such that a single system would emerge. This would have allowed, for the first time, worldwide roaming with a single handset, as envisaged by Goodman in Reference [1]. The idea was that the different regions of the world would submit system proposals capable of meeting a given set of requirements. The proposal best meeting these requirements would then be selected or, if this was not possible, an attempt would be made to merge different proposals into a single one in a consensus building phase.

While all bodies standardising such systems actually submitted their proposals to the ITU in the middle of 1998 [83], it became clear that the ITU would not be in a position to enforce a unified system. Instead, it would essentially have to approve all viable proposals meeting the core ITU 3G requirements, which the regional bodies intend to implement. To complicate matters, while ETSI in Europe and the Association of Radio Industries and Businesses (ARIB) in Japan put in place their own procedures to select one such proposal, there were no concerted efforts in the US towards 3G standardisation. As with 2G systems, it was believed that the marketplace should choose a system, resulting in several 3G proposals from the US. For these reasons, ITU then advocated 'a "*family of systems*" concept, defined as a federation of systems providing IMT-2000 service capabilities to users of all family members in a global roaming offering' [83]. The latter entailed efforts during the consensus building phase at least to enable worldwide roaming, for instance by facilitating the implementation of multi-mode terminals, given that a complete harmonisation seemed unachievable.

Eventually, two main camps (with various sub-streams) formed. The first one is united in the original Third Generation Partnership Project (3GPP) mentioned previously, dealing with the standardisation of UMTS and the evolution of GSM, the second one in a similar structure named 3GPP2, dealing with cdma2000, an evolution of cdmaOne. The systems being developed by these two organisations are based on different core network standards. Moderately successful air-interface harmonisation efforts have taken place in the framework of an 'operator harmonisation group', but although similar air-interface technologies are considered for UMTS and cdma2000, the systems remain essentially incompatible. Other harmonisation efforts in the same framework resulted in the introduction of 'hooks' and extensions in the relevant standards, allowing the 3GPP air interface to be deployed on a 3GPP2 network infrastructure and vice versa, as agreed in the second quarter of 1999. What relevance this option will have in practise remains to be seen.

2.3.4 The Third Generation Partnership Project (3GPP)

In Europe, several radio interface proposals were seriously considered within ETSI for UMTS. Five concept groups were set up in ETSI SMG2 during 1997, classifying the different proposals according to the basic multiple access schemes employed. These were wideband CDMA (WCDMA), wideband TDMA, hybrid TDMA/CDMA (referred to as TD/CDMA), orthogonal frequency-division multiplexing (OFDM) with a TDMA element for multiple access, and opportunity driven multiple access (ODMA). Strictly speaking, ODMA is not a basic multiple access scheme, but rather a technique that transforms mobile terminals into relay stations. Signals from a terminal far away from a base station can be relayed by other terminals nearer to it (potentially over multiple hops) to improve coverage and lower transmission power, thereby reducing interference and thus increasing spectral efficiency.

The strongest contenders were WCDMA and TD/CDMA, while ODMA was eventually suggested as an option on top of whatever basic multiple access scheme would be chosen. After a heated debate, it became evident that a unanimous decision for only one proposal was not possible. While a majority preferred WCDMA, it was appreciated that TD/CDMA would lend itself better to time-division duplexing suitable for operation in unpaired frequency bands. As a compromise choice, rather than picking both frequency-division duplex (FDD) and time-division duplex (TDD) mode of one of the two proposals, it was decided to choose the WCDMA FDD mode together with the TD/CDMA TDD mode. This is reflected in the ETSI candidate submission [84] to ITU. In the following, these two modes are referred to as UTRA FDD and UTRA TDD respectively.

Japan, in its own selection process, was also considering various proposals and finally opted for a WCDMA based system [85] very similar to the ETSI WCDMA mode. Since Japanese companies had also contributed to the WCDMA concept in ETSI and were eager to avoid finding themselves in similar technological isolation as with PDC, it was only natural that they decided to join forces with ETSI. This led eventually to the creation of 3GPP in 1998, which took over detailed standardisation of the UTRA FDD and TDD modes from ETSI (with UTRA now standing for *universal* terrestrial radio access rather than UMTS terrestrial radio access). Apart from ETSI and Japanese bodies, 3GPP was also joined by some of the US and the South Korean WCDMA proponents. Further information on this topic can be found in Reference [86].

The core network to be used for the 3GPP system is an evolved GSM core network. As a result, work on specifications dealing with protocols and network components common to GSM and UMTS was transferred from ETSI to 3GPP in 1998. 3GPP can therefore be viewed as providing a 3G evolution route mainly for GSM operators, with Japanese PDC operators also adopting this route to 3G.

The most important D-AMPS operators decided also to join forces with the GSM camp, albeit in a somewhat different manner. They decided to enhance their 2G digital voice capability with a GPRS-based high-speed packet-data capability. To meet the ITU data-rate requirements for 3G (which are more relaxed than the respective UMTS requirements), GPRS would have to be enhanced through EDGE, leading to enhanced GPRS (EGPRS). This is why EDGE now stands for enhanced data-rates for *global* rather than GSM evolution, as it did earlier. Initially, it was planned that EGPRS would be closely coupled to D-AMPS, for instance by enabling D-AMPS signalling messages to be directed to dual-mode D-AMPS/EGPRS terminals via EGPRS. Recently, however, one key D-AMPS operator has reconsidered its evolution route to 3G and decided to deploy full GSM/GPRS first, with a view to introduce UMTS later on. In such a scenario, EGPRS may still be deployed as an enhancement to the existing GPRS infrastructure, but D-AMPS would be relegated to a more or less 'stand-alone' 2G technology.

GPRS and first release EGPRS radio access networks need to be connected to a 'plain GSM' core network. Since further releases can be connected to the evolved GSM core network used for UMTS, most of the remaining standardisation work related to GSM evolution was transferred in the year 2000 from ETSI to 3GPP as well.

A further addition to 3GPP took place in late 1999, namely that of a Chinese partner with its own air-interface technology submitted earlier to ITU as an IMT-2000 candidate, namely a UTRA TDD derivative. This is not included in the first release of the UMTS standards (namely release 1999), but added as another UTRA mode later on.

2.3.5 The Universal Mobile Telecommunications System (UMTS)

A main architectural design principle of the universal mobile telecommunications system (UMTS) is the split of the fixed UMTS infrastructure into core network (CN) and access network (AN) domains. An additional design principle is the logical split of the global architecture into a so-called 'access stratum', containing equipment and functionality specific to the access technique (e.g. radio-related functionality), and 'non-access strata', as shown in Figure 2.2. The access stratum includes protocols between the mobile terminal and the access network, and between the access network and the serving core network. While the former support the transfer of detailed radio-related information, the latter are independent of the specific radio structure of the access network. This is important, it means that the CN should not be affected by the choice of radio transmission technologies in the access network, such that new types of access networks can be defined as and when required and attached to the existing core network.

The only suitable access network type defined in release 1999 specifications is the UMTS or Universal Terrestrial Radio Access Network (UTRAN), consisting of a set of radio network subsystems. These in turn are composed of a Radio Network Controller (RNC) and a number of base stations, the latter (for lack of agreement) somewhat oddly termed *Node B* in UMTS. The radio technologies featured by UTRAN release 1999 are the UTRA TDD and the UTRA FDD mode. The CN consists of a 'circuit-switched domain' or CS-domain, which is composed of Mobile services Switching Centres (MSC) very similar to those already used in GSM, and a 'packet-switched' or PS-domain, which

2 CELLULAR MOBILE COMMUNICATION SYSTEMS: FROM 1G TO 4G

```
┌─────────────────────────────────────────────────┐
│              Non-access strata                  │
└─────────────────────────────────────────────────┘
       ┌─────────────────────────────────┐
       │         Access stratum          │
       └─────────────────────────────────┘
┌──────────────┐    ┌──────────┐    ┌──────────────┐
│Mobile station│    │  UTRAN   │    │ Core network │
└──────────────┘    └──────────┘    └──────────────┘
         $U_u$ (radio) interface    $I_u$ interface
```

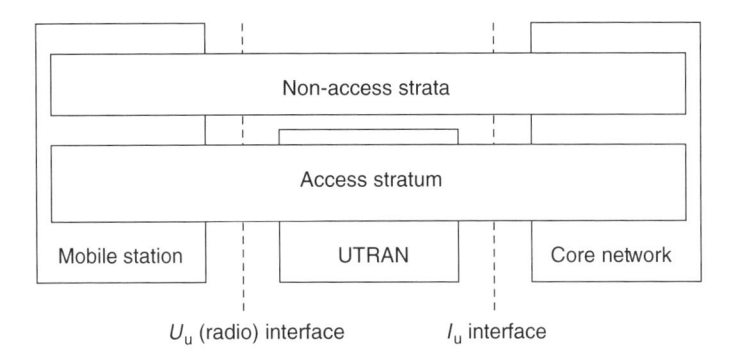

Figure 2.2 Basic logical UMTS architecture

is an evolution of the GPRS core network. Accordingly, there are two variants of the I_u interface between AN and CN shown in Figure 2.2, namely I_u-CS and I_u-PS.

The fundamental UMTS service principle is to standardise service capabilities rather than the services themselves, which helps achieving flexibility in service provision. With an appropriate set of service capabilities, users, service providers, and network operators should be in a position to define services themselves according to their specific needs [87].

WCDMA was investigated as one of the two final multiple access modes in the ACTS FRAMES project (alongside TD/CDMA). WCDMA built also on concepts evaluated during the RACE Codit project [12]. Furthermore, during the concept development phase taking place in ETSI SMG2, features of the Japanese WCDMA proposal [85] were also adopted. This led to the final concept adopted as a basis for the UTRA FDD mode [88], which was further enhanced during detailed standardisation work in 3GPP, for instance with features of some of the American WCDMA submissions to ITU.

UTRA FDD makes use of direct-sequence wideband CDMA operating at a basic carrier spacing of 5 MHz (which can be reduced down to 4.4 MHz, if required). In order to provide high data-rate services in an efficient manner (e.g. with some degree of trunking or multiplexing efficiency) and to benefit as much as possible from frequency diversity, the carrier bandwidth should be as wide as possible, preferably even wider than 5 MHz. On the other hand, the spectrum situation did not allow for more, in fact, even 5 MHz carriers will limit deployment flexibility considerably, as will be discussed below.

Initially, a chip-rate of 4.096 Mchip/s was envisaged, and 16 time-slots were supposed to fit into a 10 ms frame. Harmonisation efforts with cdma2000 lead to a reduced chip-rate of 3.84 Mchip/s and the elimination of one time-slot per frame, such that now 15 time-slots of 0.666 ms length form a frame of 10 ms. On dedicated channels, a link between mobile station and base station is continuously maintained at some minimal bit-rate. The purpose of the time-slots is therefore not to provide a TDMA feature, but to structure the transmitted data, for instance to specify the periodicity of power control commands and other overhead related to the physical layer. In UTRA TDD, by contrast, these are 'proper' time-slots (in the TDMA sense). A good and readily available overview of WCDMA is provided in Reference [89].

The TD/CDMA proposal [90] adopted as a basis for the UTRA TDD mode is to a large extent based on a scheme proposed in Reference [13], and was further investigated as mode 1 of the two final multiple access modes considered in the European ACTS

FRAMES research project [91] (mode 2 being WCDMA). It employs a combination of CDMA and TDMA as basic multiple access methods. The total number of time-slots per frame can be shared flexibly (alas, with certain deployment constraints) between uplink and downlink, thereby enabling asymmetric use of the total spectrum resources. Initially, the suggestion was to apply the GSM time-slot/frame structure and to adopt an integer multiple of the 200 kHz used in GSM for carrier spacing. This would (at least in theory) have allowed TD/CDMA to be deployed on a per-time-slot basis in an existing GSM system, provided that six or eight carriers are allocated to a single GSM cell. Such an approach could be viewed as an evolutionary approach similar to those described in Subsection 2.3.2, but just a little bit more radical. However, having been adopted in conjunction with WCDMA as the other UTRA mode, the main worry was then harmonisation between these two modes. As a result, the TD/CDMA scheme adopted some fundamental design parameters such as chip-rate, carrier spacing and slot/frame structure from the WCDMA scheme [84]. First, the harmonised parameter set was based on a chip-rate of 4.096 Mchip/s, but following harmonisation efforts with cdma2000, also TD/CDMA ended up with a chip-rate of 3.84 Mchip/s and 15 time-slots per 10 ms frame.

2.3.6 The Spectrum Situation for UMTS

Before being able to discuss whether we can expect UMTS to meet the requirements listed earlier, we need to examine the UMTS spectrum situation, as this has a quite considerable impact on this discussion. The initial spectrum identified for 3G systems is situated around 2 GHz. It consists of 2×60 MHz of paired spectrum set aside for 3G in Europe, Japan, South Korea and other parts of the world, but not in the US, where a part of this spectrum was auctioned by the federal government for PCS. This paired band with two equally sized portions for the uplink and the downlink is for instance suitable for UTRA FDD. In addition to that, in Europe, 20 MHz of unpaired band was set aside for public operation of 3G systems, and another up to 15 MHz of unpaired band for 'license-exempt' or 'self co-ordinated' use, e.g. for residential telephony, the latter not of interest in the following. Unpaired band is not suitable for UTRA FDD, but can be used by the UTRA TDD mode.

In most countries having licensed this 3G spectrum, the available bandwidth is carved up between four to six licensees (i.e. operators), so that with few exceptions, the best an individual operator could get was 2×15 MHz of paired band plus 5 MHz of unpaired band. In the UK, for instance, there are five licensees, resulting in differently composed spectrum packages, as shown in Figure 2.3. Three of these packages consist of 2×10 MHz plus 5 MHz, one of 2×15 MHz and the biggest one, set aside for a new entrant (i.e. an operator not owning a 2G network in the UK), of 2×15 MHz plus 5 MHz of spectrum. Similarly, in Germany, where six licensees share the spectrum, each is assigned only 2×10 MHz of paired band, and four of them an additional 5 MHz of unpaired band. This is rather miserable, particularly when taking into account that some GSM 1800 operators hold licenses for 2×30 MHz of paired band. In the year 2000, more spectrum was identified for 3G, which will eventually be made available to operators, although it is up to individual countries to either improve the spectrum situation of existing 3G license holders or assign this spectrum to new licensees.

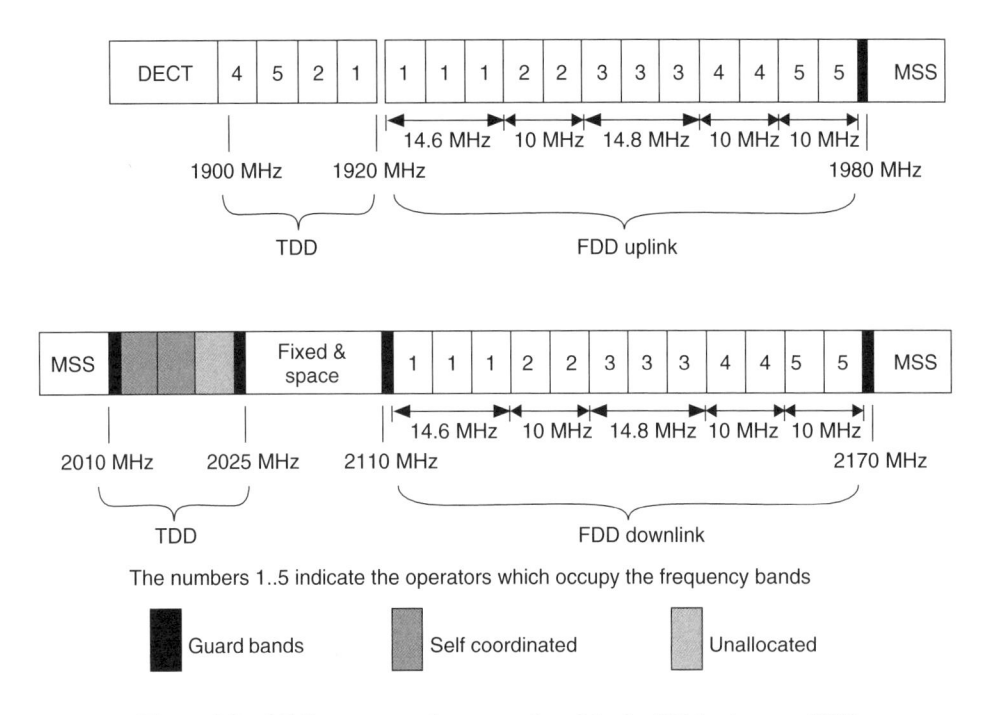

Figure 2.3 3G Spectrum packages auctioned in the UK in the year 2000

In the following, we will consider 2 × 10 MHz plus 5 MHz of spectrum as an example allocation for an individual operator, as quite a few operators will have to deal with such a meagre allocation at least for the next few years to come. A possible deployment scenario could look as follows: 2 × 5 MHz are used for the macrocell layer, and another 2 × 5 MHz for the microcell layer, in both cases using UTRA FDD at a frequency reuse of one. The additional 5 MHz of unpaired spectrum could be used for UTRA TDD (again at a frequency reuse of one), either to provide additional capacity at the microcell layer, or to deploy picocells. Using UTRA TDD at the microcell layer raises a few issues in terms of adjacent channel interference (ACI), though, since the unpaired spectrum is right next to the paired uplink spectrum. For instance, it may be rather difficult to deploy TDD carriers at the same cell sites as FDD carriers, since the TDD transmitter could cause detrimental interference to the FDD receiver. This and other interference scenarios are discussed in more detail in Reference [86].

2.3.7 UTRA Modes vs UTRA Requirements

Without going into too much detail, let us examine to what extent we can expect a UMTS system featuring the two UTRA modes (as per release 1999) to meet the requirements listed in Subsection 2.3.1. In accordance with this list, requirements pertaining to bearer capabilities, operational requirements, spectrum usage, and finally complexity issues are discussed in this sequence.

Both UTRA modes were obviously designed to provide the required bit-rates up to 2 Mbit/s, and include error coding schemes and automatic repeat request (ARQ) strategies which allow the required low BER to be achieved. However, with the little spectrum available to individual operators, providing a single user with 2 Mbit/s has quite considerable implications, such as complete lack of trunking efficiency and very tricky interference management. For a start, offering a 2 Mbit/s service involves dedicating an entire carrier to the requesting user. At the macrocell layer, this would be clearly unacceptable: one cannot cease coverage of an entire macrocell only to serve a single user, but then, Reference [74] required 2 Mbit/s indoors only, so this is not really relevant. Even at the microcell layer, 2 Mbit/s can only be offered to a user who is very close to the base station (and hence preferably stationary), since otherwise detrimental interference would be inflicted onto neighbouring cells because of the universal frequency reuse. In fact, the total bit-rate which can be supported by a single-carrier WCDMA cell, averaged over the user distribution in space, may often fall below 1 Mbit/s (this depends on various factors, such as the service mix in the considered cell). This does not prevent operators from providing 2 Mbit/s indoors (as required by Reference [74]) through picocells. However, given the little spectrum available (particularly relative to the targeted bit-rates) and, as a result, the deployment constraints and the limited trunking or multiplexing efficiency, it is going to be rather expensive.

In essence, should there be a significant market for 2 Mbit/s services, then additional 3G spectrum will have to be assigned to existing 3G license holders and further improvements to UMTS will have to be made to be able to provide them at a reasonable cost. Such improvements to UTRA FDD and UTRA TDD are already in the pipeline (see further below and Chapter 10 for details).

The UMTS design principles discussed earlier allow for integration of different radio interface solutions. This is on the one hand in view of heritage systems, mainly GSM, and on the other with respect to the integration of the planned satellite component and the Mobile Broadband System (MBS) intended for very high bit-rates. An alternative solution would therefore be to deliver 2 Mbit/s services through a new air interface suitable for operation at non-3G frequencies well above 2 GHz, where more bandwidth is available, while still making use of the UMTS CN infrastructure. In terms of supported data-rates, the new air interface could fit in-between the current UTRA modes and MBS. We will reconsider such ideas in Sections 2.4 and 2.5.

Whether the chosen UTRA modes are well adapted to the different bearer types to be provided will be discussed in more detail in later chapters. For now, note that irrespective of whether the air interface supports packet oriented bearers adequately, the UMTS packet-switched CN domain as per release 1999 does not feature mechanisms to provide the necessary Quality of Service (QoS) for real-time packet-data services.

To complete the discussion on requirements relating to bearer capabilities, seamless handovers between cells of one operator supporting the same UTRA mode are possible. In general, there is a trade-off to be made between QoS on the one hand and signalling requirements, which affect system complexity and cost, on the other. For instance, a particularly seamless type of handover between UTRA FDD cells, namely soft handover, places a considerable burden on the infrastructure. Handovers between the two UTRA modes and to GSM are possible, to what extent they will be seamless remains to be seen.

Regarding the operational requirements, specifically radio resource planning, while universal frequency reuse obviates the need for frequency planning, the same cannot be said about cell planning. The opposite is rather true: choosing cell locations and determining reference power levels and other radio-related parameters to manage interference is rather tricky in CDMA-based systems, particularly if a wide range of services is to be supported.

With respect to efficient spectrum usage, several concerns exist, most of them having rather little to do with the design of UMTS, and much more with 3G spectrum limitations. The issue of trunking and multiplexing efficiency, already alluded to above, will be a recurring topic throughout this book. When dealing with high bit-rates, we need as 'thick' pipes (i.e. as wideband carriers) as possible. Unfortunately, particularly at the limited spectrum available, this is in direct conflict with the need for efficient support of hierarchical cell layers. Since layer separation in HCS is preferably achieved through frequency-division, a better frequency granularity would have been very helpful. With the deployment scenarios considered in Subsection 2.3.6 for 2×10 MHz plus 5 MHz of spectrum, either only two layers with a rigid 10 MHz to 15 MHz share of spectrum resources per layer can be provided, or three layers with an equally rigid 10 MHz-10 MHz-5 MHz resource split. This is not exactly satisfactory.

Another concern has to do with efficient support of asymmetric traffic. Assuming that the UTRA TDD mode will only be deployed in the unpaired frequency band, the degree to which the split of radio resources between uplink and downlink can be asymmetric is limited. Again based on the same spectrum allocation, given that at least one time-slot (out of 15) needs to be assigned to the uplink and two to the downlink in the UTRA TDD mode, the maximum possible asymmetry ratio is $10 + 5 \times (14/15)$ to $10 + 5/15$, or 59% to 41%. Consequently, resources would go wasted, should the total traffic asymmetry exceed this ratio. One might be tempted to deploy UTRA TDD also in the paired band (where there are no regulatory constraints), but this does not solve all problems either. Ideally, we would like to share resources flexibly between uplink and downlink on a cell-by-cell basis, such that the resource split can be adapted precisely to the experienced traffic asymmetry ratio in each cell. However, co-channel and adjacent channel interference may require time-slot synchronisation and co-ordinated choice of uplink and downlink time-slots across cells of a single operator, and in the worst case even between operators using adjacent 5 MHz slots for their TDD deployment. This would eliminate any possibility for swift adjustment of the split between uplink and downlink resources according to the experienced traffic situation. Obviously, ACI issues relating to coexistence of UTRA FDD and UTRA TDD discussed in the previous subsection apply also in the case where TDD is deployed in the paired band.

Regarding size, weight, operating time and cost of terminals, where UTRA will be deployed gradually (which is the most likely scenario in a GSM environment), UMTS handsets will have to be dual-mode GSM/UMTS capable, adding complexity and cost to handsets. Having to implement two fairly distinct UTRA modes will further add to complexity and cost of both handsets and infrastructure.

At this point in time, most operators are busy deploying UTRA FDD infrastructure. Systems have been launched in the Isle of Man and in Japan, with other countries expected to follow later in 2002 or in 2003, depending on sufficient availability of handsets. If and when UTRA TDD infrastructure will be deployed is currently unclear.

2.3.8 3GPP2 and cdma2000

The cdmaOne operators have their own evolution route to 3G in the shape of cdma2000. Copying more or less the 3GPP structure, American, South Korean and other standardisation bodies got together to form 3GPP2 as a framework, in which the various cdma2000 releases providing an evolution to 3G are defined.

In a first step, cdmaOne operators have the possibility to upgrade the air interface from IS-95 to cdma2000 1×RTT (for Radio Transmission Technology). This is a narrowband CDMA air interface using the same chip-rate and carrier spacing as IS-95 (1.2288 Mchip/s and 1.25 MHz respectively), but provides improved spectrum efficiency (owing for instance to faster power control) and increased data-rates up to 144 kbit/s. With data-rates somewhere in-between GPRS and EGPRS, such an upgraded system must clearly be positioned as a 2.5 G system. Since cdma2000 1×RTT infrastructure can serve (most) cdmaOne handsets and cdma2000 1×RTT handsets work also with cdmaOne infrastructure, a gradual upgrade in existing spectrum is possible. This is particularly convenient for US operators, as currently no spectrum is set aside specifically for 3G in the US. Obviously, to benefit from the improvements, both handsets and infrastructure need to be cdma2000 1×RTT compliant. Most cdmaOne operators are expected to deploy cdma2000 1×RTT infrastructure eventually. For instance, operators in the US, Japan, and South Korea are planning to do so.

The wideband counterpart to cdma2000 1×RTT is cdma2000 3×RTT, an air-interface mode employing multi-carrier transmission on three 1.25 MHz carriers (at 1.2288 Mchip/s per carrier) on the downlink, and wideband direct-sequence CDMA (with up to 3.6864 Mchip/s) on the uplink. Accounting for the necessary guard bands, deploying a cdma2000 3×RTT (multi-)carrier is expected to require the same 5 MHz slots as a UTRA carrier. In theory, it is possible to transmit on up to 12 1.25 MHz carriers on the downlink [86], and with according multiples of 1.2288 Mchip/s on the uplink. One of the most important constraining factors will be the spectrum situation, as usual. Particularly for US operators, it will often be difficult to find the necessary 5 MHz slots (the same is true for US operators wishing to deploy UTRA), let alone the 15 MHz-plus-guard-band slots required for '12×RTT'.

Initially, a direct-sequence or direct-spread wideband CDMA option similar to UTRA FDD was also considered for cdma2000. As a result of the harmonisation efforts with UTRA FDD, this option was eventually dropped, leaving UTRA FDD as the only direct-spread wideband CDMA standard for cellular communications [86]. The main advantage of the multi-carrier option is that it can be introduced in existing cdmaOne spectrum (provided that there are multiple adjacent 1.25 MHz carriers), while still allowing cdmaOne (and cdma2000 1×RTT) terminals to be served. This is because individual carriers can be operated independently (and thus demodulated independently by the terminals). Apart from increased data-rates made possible by multi-carrier transmission (similar to those of UTRA FDD, i.e. 384 kbit/s outdoors and 2 Mbit/s indoors), a cdma2000 3×RTT terminal demodulating all three carriers together will benefit from additional frequency diversity compared to transmission and reception on a single 1.25 MHz carrier.

The spectrum situation in some of the cdma2000 target countries together with the fact that cdma2000 1×RTT offers potential for further evolution (as discussed below) make the future of cdma2000 3×RTT look somewhat uncertain. It is predominantly of interest for countries such as South Korea and Japan with 2G CDMA systems (i.e. cdmaOne), where additional spectrum specifically for 3G is available. In Korea, however, some operators indicated a preference for deploying UMTS in the new spectrum while upgrading cdmaOne to cdma2000 1×RTT in the existing spectrum.

2.4 Further Evolution of 3G

Significant enhancements to the GSM specifications were only seriously considered once the basic system was deployed in numerous countries and proved functional. By contrast, work on subsequent releases of 3G specifications was already well underway before first release systems were even launched. This includes both improvements to existing features and introduction of new features. From a system perspective, the most important new feature is probably the support of IP-based multimedia services, which has also repercussions on the air interface. Other features discussed in the following relate to the air interface, namely evolutions of cdma2000 1×RTT and existing UTRA modes providing increased data-rates on the one hand, and additional UTRA modes on the other.

2.4.1 Support of IP Multimedia Services through EGPRS and UMTS

While packet-data support in GSM was added at a relatively late state through GPRS to a system designed initially for circuit-switched services, packet-data is supported in UMTS right from release 1999, the first system release. As far as the core network is concerned, this is achieved through evolved GPRS core network nodes. On the air interface, UMTS (or rather the two UTRA modes) offers several options for the support of packet data, which are discussed in Chapter 10. All the same, it is probably fair to say that ETSI and 3GPP members involved in the early phases of UMTS standardisation approached the subject matter very much with a 'circuit-switched mindset'. For instance, as mentioned earlier, the evolved GPRS core network used in release 1999 does not provide any means for QoS control, so that delay-critical real-time services such as voice have to be delivered via the circuit-switched core network domain. This conflicts with Goodman's vision of 3G cellular packet communications promoted in Reference [7]. He advocated packet transmission over the air interface and a packet-switched fixed network infrastructure with decentralised and distributed call control and call management functions, for reasons already discussed in the introductory chapter.

In the meantime, the standardisation community has reconsidered matters, although for somewhat different reasons than Goodman's, namely the emergence of the Internet, which runs on a packet-switched infrastructure using packet-based Internet protocols. In early 1999, a few operators and infrastructure manufacturers got together to form 3G.IP [92], an 'industrial lobby group' intended to influence 3GPP (for UMTS) and ETSI (for EDGE/GPRS) towards adoption of what was then termed an 'all IP' network architecture for *release 2000* of the relevant specifications. This further evolved GPRS architecture, based on packet technologies and IP telephony, would function as a common core network to access networks based on both EDGE and WCDMA radio access technologies.

The main driving force was the desire to provide all possible types of IP-based real-time and non-real-time services simultaneously and efficiently through cellular access. Any IP-based service proving popular on the Internet and accessible to wireline users should be made accessible to wireless users as well. In some sense, this would lead to a convergence between these two worlds, as the same set of services could be delivered irrespective of the access technology (e.g. wireless, 'conventional wireline', and cable television installations enhanced for bi-directional delivery of IP-based packet traffic), providing personal mobility and interoperability between mobile networks and fixed networks. The creation of new IP-based services specifically for wireless users should also be facilitated. Additionally, some operators wanted to get into a position, in which they could cease investment in the circuit-switched core network infrastructure as quickly as possible and focus only on the packet-switched infrastructure instead. This would not only facilitate network management, but also allow operators to 'ride the Internet cost curve', which implies that the performance of infrastructure equipment doubles every so many months, while at the same time the costs per unit halve. Unfortunately, at least as far as 'release 2000' is concerned, the 'Internet cost curve' is likely only to apply to a few core network entities, but not to the radio access network infrastructure, which is the costliest part of the cellular network infrastructure. In summary, therefore, the main requirements driving this type of evolutionary effort on top of those listed in Subsection 2.3.1 are:

- efficient support of IP-based real-time and non-real-time services via both WCDMA and EDGE radio access technologies;
- personal mobility and interoperability between mobile and fixed networks;
- further increased service flexibility by removing dependencies between applications and underlying networks; and
- service creation at 'Internet speed'.

3GPP eventually adopted the top-level architecture devised by 3G.IP during regular meetings held from May 1999 onwards. However, due to delays with release 1999 and the required effort for detailed standardisation of this new architecture, its delivery was split into two releases, namely releases 4 and 5. Most features would only be delivered in release 5 initially scheduled for completion by the end of 2001, but now further delayed. Release 1999 equates to release 3 in this context. In general, release numbering in GSM and UMTS is a confusing matter, onto which the appendix attempts to shed some light.

Regarding the core network, this release 5 architecture uses a few new functional entities such as call servers performing IP-based call control using protocols defined by the Internet Engineering Task Force (IETF, the body standardising IP protocols). Additionally, GPRS core network nodes need to be further evolved to be capable of providing the necessary QoS. From an air-interface perspective, the release 1999 UTRA modes require a few enhancements, particularly to make sure that real-time services can be supported efficiently (e.g. header compression over the radio link). The main effort, however, is going into EDGE radio access, since GSM was initially not designed at all for packet data, and the first GPRS release was not intended to support real-time packet-data services. Work on the required enhancements, first discussed in 3G.IP, started in ETSI in 1999, but as the release 5 core network specified in 3GPP is common to WCDMA and EDGE radio access, EDGE radio work was also transferred to 3GPP in the year 2000. The enhancements are discussed in Chapter 11.

2.4.2 Improvements to cdma2000 1×RTT, UTRA FDD and TDD

Two evolutionary steps are currently envisaged for cdma2000 1×RTT. The first one, initially referred to as HDR for high data-rates, is now referred to as *1×EV data-only*

(EV standing for evolution). Marketing people will hopefully come up with a better name eventually. This evolutionary step enables data-rates of up to 2.4 Mbit/s to be provided on the usual 1.25 MHz carriers. As the name suggests, 1×EV data-only carriers do not support real-time services such as voice, hence the total frequency resources have to be split into conventional cdma2000 1×RTT carriers and 1×EV data-only carriers. This complicates system deployment and does not make efficient use of spectrum resources, since a fixed amount of resources needs to be set apart for data traffic regardless of temporal and spatial fluctuations of the traffic mix experienced. In a second step, *1×EV voice and data* will provide even higher data-rates (5.2 Mbit/s are currently aimed for, but there are still competing proposals with differing bit-rate capabilities), while also allowing voice users to be served on the same 1.25 MHz carriers.

Most improvements to the UTRA FDD and TDD modes are related to High-Speed Downlink Packet-Data Access (HSDPA). Marketing experts seem to be called for again to identify a less cumbersome acronym, but at least the name is self-explanatory. Data-rates of 10 Mbit/s are envisaged, perhaps even 20 Mbit/s. HSDPA is to be provided on existing UTRA FDD and TDD carriers (that is, alongside voice traffic, for example). Additionally, a stand-alone HSDPA solution, which could be deployed in the unpaired spectrum, may be considered as well, although 3GPP had not yet committed to this at the time of writing. In such a solution, if it materialises, uplink requests for data to be delivered on stand-alone downlink carriers would have to be placed on bi-directional UTRA FDD or TDD carriers. Eventually, HSUPA, an uplink equivalent to HSDPA, may be specified as well.

All data-rates mentioned in this subsection are theoretical *peak* rather than *average* data-rates. They account only for lower layer overhead, if any at all. Some of the techniques employed to boost data-rates in cdma2000 and UMTS and to improve spectrum efficiency are similar to those used by EDGE, such as adaptive modulation schemes. This implies, for instance, favourable propagation conditions to reach the highest possible data-rates, meaning that users located close to cell centres will benefit most from these enhancements.

Figure 2.4 summarises the possible evolution scenarios, from an air-interface perspective, for GSM/UMTS on the one hand, and cdmaOne/cdma2000 on the other and provides a *rough* idea regarding chronology. As shown, several variants of the cdmaOne air interface exist, IS-95A being the base specification, which supports circuit-switched data up to 14.4 kbit/s, and IS-95B offering packet-data support up to 64 kbit/s. Two UTRA TDD modes are shown, UTRA wTDD is the 'original' UTRA TDD mode specified in release 1999, UTRA nTDD an additional UTRA TDD mode mentioned in the next subsection.

2.4.3 Additional UTRA Modes

Before harmonisation with UTRA FDD, the original UTRA TDD mode featured a chip-rate of 2.167 Mchip/s, with a suggested carrier spacing of 1.6 MHz. For release 4 of the specifications, a new UTRA TDD mode is introduced, namely TD/SCDMA, a scheme championed by the Chinese, which is based on hybrid time-division *synchronous* code-division multiple access. This scheme is currently envisaged to operate at a chip-rate of 1.28 Mchip/s with the carrier spacing intended originally for UTRA TDD, namely 1.6 MHz. Because of this relatively narrow carrier spacing, it is sometimes referred to as

2.4 FURTHER EVOLUTION OF 3G

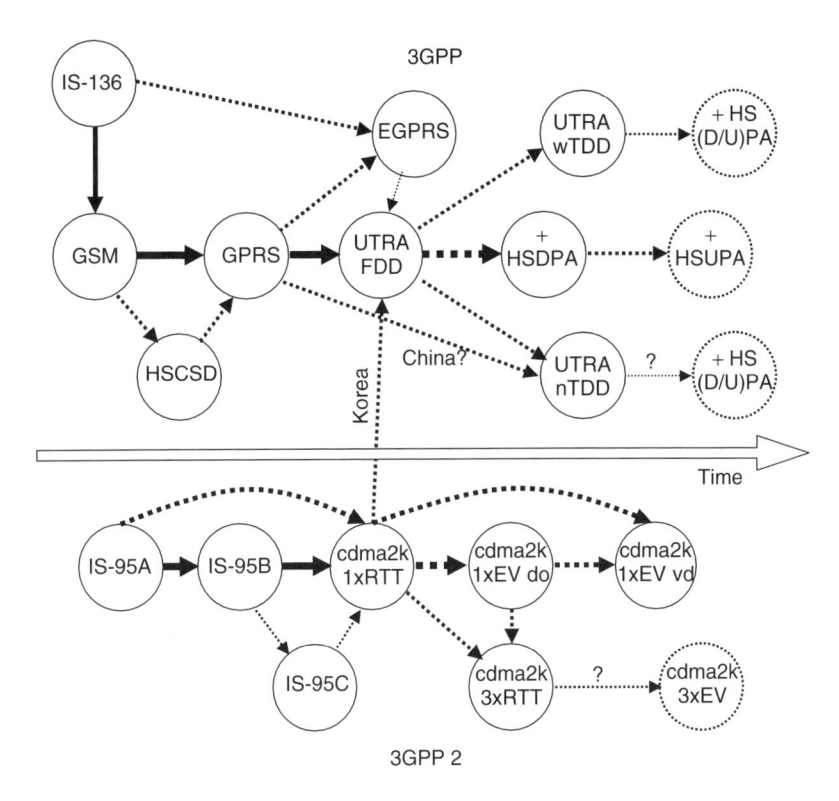

Figure 2.4 Evolution scenarios for GSM/UMTS and for cdmaOne/cdma2000 from an air-interface perspective

narrowband TDD or *nTDD*, and the 5 MHz TDD mode accordingly as *wideband TDD* or *wTDD*. UTRA nTDD features a frame-slot structure optimised for the use of smart antennas, which is different from that of the other two UTRA modes.

The narrower carrier spacing provides better frequency granularity, allowing for three carriers to be deployed in a 5 MHz slot (e.g. at a 1/3 reuse). Unfortunately, at least in the context of the 3G spectrum allocation discussed in Subsection 2.3.6, it does not necessarily facilitate system deployment, because the TDD/FDD interference scenarios remain similar. Additionally, due to the different frame structure, if operator 1 were to adopt wTDD in the unpaired 5 MHz spectrum slot and operator 2 were to opt for nTDD in the adjacent 5 MHz slot, they could not even synchronise frames to combat ACI.

We mentioned earlier that MBS could potentially be connected to the UMTS core network. It is not clear whether such a thing, if it were going to happen, would be referred to as another UTRA mode. A more likely development is that of the addition of a 'UMTS convergence layer' to the HIPERLAN type 2 system, an ETSI wireless local area network (WLAN) standard featuring 20 MHz carriers and operating at 5 GHz [93]. This could then be attached to the UMTS core network and qualify as another UTRA mode. Obviously, one would have to make sure that the core network could cope with the high data-rates on offer, as HIPERLAN 2 is designed for gross data-rates up to 54 Mbit/s.

2.5 And 4G?

2.5.1 From 1G to 3G

The fundamental shift from analogue 1G to digital 2G technology required a completely new infrastructure and new mobile terminals. Since 2G GSM technology was the common denominator for the whole of Europe, which enabled for the first time roaming throughout the continent and later throughout the world, there was no market demand for analogue/digital dual mode terminals and no need for a smoothened transition from 1G to 2G. In the US, by contrast, analogue AMPS, not digital 2G technology, was the common denominator enabling nationwide roaming, hence dual-mode terminals proved popular.

The transition from 2G to 3G was also supposed to consist of a major paradigm shift, namely from mostly voice with some low-bit-rate data (at 9.6 kbit/s, at best 14.4 kbit/s) to a multitude of personalised multimedia services at bit-rates up to 2 Mbit/s, with data being dominant over voice. With hindsight, this step has been less revolutionary than initially envisaged, due to the tremendous success of some 2G systems and continued efforts to evolve their capabilities towards better data support and increased bit-rates, blurring the boundaries between 2G (or rather 2.5G) and 3G.

2.5.2 Possible 4G Scenarios

2G came roughly 10 years after 1G, and 3G roughly 10 years after 2G. 3G started to be a mainstream research topic shortly before first 2G systems were launched in the early 1990s. Similarly, at the turn of the millennium, 4G is increasingly being talked and written about in conferences (e.g. [94]), scientific publications [95] and for instance in certain ITU meetings. The time-gap between generations tends to shorten, initially envisaged for 2010, one Japanese operator announced an anticipation of the first 4G system launch to 2006. At the time of writing, there is no agreed view yet on what exactly will constitute a 4G system, what features it will offer that distinguishes it significantly from a 3G system and justifies finally the 4G label. Different scenarios are being talked about, some of which could well materialise before 2006, but while associated with 4G by some, they look much more like evolved 3G or 3.5G to others.

Some people associate 4G with the merger of cellular communications and the Internet, referring to 'all IP' systems as 4G systems. Remember that 3G.IP called its architecture initially an 'all IP' architecture. Although delayed with respect to the original schedule, coming fairly soon after the first release of UMTS and clearly an evolution of UMTS, this '3 G.IP architecture' is firmly in 3G or 3.5G territory. It is important to note that 'all IP' means different things to different people. The '3 G.IP architecture' is designed to offer IP-based services efficiently. However, it does not replace existing cellular mechanisms, for instance for mobility management, by equivalent IP-based protocols, such as *Mobile IP* [96] and mechanisms being talked about in the IETF 'seamless mobility' (or 'seamoby') working group [97]. Instead, it adds IP-based technologies and protocols, where appropriate and beneficial rather than for the sake of an IETF badge, gradually to the existing UMTS system.

Soon after 3G.IP, a similar body with a somewhat longer-term perspective was established, namely the Mobile Wireless Internet Forum (MWIF [98]). Rather than focussing on the evolution of 3GPP networks only, this forum promoted a 'full merger' of cellular communications and the Internet, thereby also bridging protocol differences between 3GPP and 3GPP2 systems. Again, depending on personal taste, this could either be viewed as a 3.5G or a 4G effort.

An alternative view of 4G is just bigger, better and mainly faster than 3G. While 3G is characterised by data-rates up to 2 Mbit/s, 4G could offer for instance 155 Mbit/s, as targeted by MBS, through new air interfaces. Whether these are then attached to the existing UMTS core network (evolved to cope with these data-rates) or a new core network is another question.

We have mentioned HIPERLAN 2 as a representative of a WLAN system, which could be attached to the UMTS core network. To achieve this, a UMTS convergence layer is needed, which translates HIPERLAN protocols into UMTS protocols and vice versa (i.e. provides the necessary interworking). An alternative view is that such interworking is not required, but disparate systems could be glued together through IP protocols such as Mobile IP, enabling sessions initiated on a WLAN to be continued through 3G coverage when leaving the coverage area of WLAN access points (the equivalent of base stations). Session continuity alone does not provide the QoS required to offer real-time services across different access technologies. Instead, seamless so-called 'vertical' handovers between, for example, a WLAN access point and a UMTS cell are needed (as opposed to 'horizontal' handovers between cells featuring the same radio technology, e.g. two UMTS cells). For seamless vertical handovers to be possible, either an integration through the UMTS convergence layer is required, or new IP-based mechanisms currently being looked at in various IETF working groups would have to be adopted both by UMTS and by WLAN systems based on HIPERLAN 2. This second approach could not only work between UMTS and WLANs, it could include any access technology, e.g. short-range radio technologies such as Bluetooth. How HIPERLAN 2 can inter-operate with UMTS to provide IP-based services is a research topic in the European collaborative research project BRAIN and its successor MIND [99,100]. Further information can also be found in a draft ETSI report on interworking between HIPERLAN 2 and 3G cellular systems [101].

An extreme view is that once these IP-based mechanisms are in place, and provided that most buildings have flat-rate broadband connections enabling WLAN access points to be connected to the Internet, close to contiguous coverage could be provided in urban areas through cheap WLAN equipment deployed by private users. In this scenario, 4G would effectively be ruled by WLANs. However, WLANs cater only for low degrees of mobility (e.g. limited speed, which is fine for a lot of applications requiring high data-rates). More importantly, QoS cannot always be guaranteed, because they operate in unlicensed spectrum, often shared with other sources of interference (e.g. microwave ovens) on top of competing WLANs. Finally, arrangements would have to be found that encourage owners of private systems such as WLANs to provide mutual coverage, as discussed for instance in Reference [94].

A more popular view is that 4G will be characterised by the integration of a multitude of existing and new wireless access technologies in a manner that, at any given time, a user (or rather her terminal) may select the best suited of all access technologies that are available at her current location. These could include short-range technologies such as Bluetooth and WLANs as well as various types of cellular access technologies and even access through satellites. The choice between them could be based on parameters affecting QoS such as data-rates and delay performance, cost, and other factors, as determined by a personal user profile. Again, the integration of the various systems would be provided by IP-based technologies. The question that arises is whether such a scenario implies the same type of radical change in service capabilities and base technology, which characterised the transition from 1G to 2G, and to a lesser extent from 2G to 3G. In fact, we are witnessing similar developments already with first 3G systems. UMTS release 1999 networks, for instance, are often being built out only gradually, hence coverage is shared between GSM and UMTS (this implies obviously dual-mode GSM/UMTS terminals). In areas with both GSM and UMTS coverage, if GSM (or GPRS) meets the requirements imposed by the requested service, either of the two can be chosen. Admittedly, having a choice between GSM and UMTS only, which may even be constrained by the operator due to operational concerns such as traffic balancing, is a far cry from this particular 4G vision of user choice between many technologies. It does not end there though. We could come significantly closer to this vision through a scenario introduced earlier, namely the addition of HIPERLAN 2 coverage through HIPERLAN 2 access points which are connected to a UMTS network.

One increasingly popular vision of 4G stipulates the convergence or interoperability of cellular technologies and digital broadcast technologies such as Digital Audio Broadcast (DAB) and Digital Video Broadcast (DVB) as a main 4G ingredient. This would indeed imply a fundamental paradigm shift justifying the step from 3G to 4G. As a trivial example, requests for video on demand could be placed on the uplink of a cellular system, and the service then delivered to users via DVB. In such a scenario, being able to use DAB or DVB alongside other access technologies would help dealing with the expected downlink-biased traffic asymmetry. However, issues such as how many users would have to request a specific service simultaneously for it to be delivered with preference by broadcast technologies would have to be investigated. More complex scenarios could be envisaged as well, and interoperability could occur at various levels, not only at that of the wireless delivery mechanism. This type of convergence would go hand in hand with the integration of other wireless technologies as discussed above. According to Pereira, '4G will encompass all systems, from heterogeneous, hierarchical, competing but complementary, broadband networks (public and private, operator driven or *ad-hoc*) to personal area and *ad-hoc* networks, and will inter-operate with 2G and 3G systems, as well as with digital (broadband) broadcasting systems' [94]. He then goes on to state that 'reconfigurable radio is seen as the engine for such integration, and IP as the integrating mechanism, the *lingua franca*'.

2.5.3 Wireless Local Area Network (WLAN) Standards

Various evolutionary scenarios to 4G have been described, and WLANs play an important role in most of them, hence it is appropriate to list briefly current and emerging WLAN standards.

Wireless LAN technologies, for instance based on frequency hopping or direct-sequence spread spectrum schemes, have been around for a few years, but their uptake was initially hindered by interoperability problems between products from different manufacturers. The turning point came with the availability of products that could inter-operate with each other readily, irrespective of which manufacturer they were from, owing to compliance with a common standard. The standard in question is one of the 802.11 family of standards from the Institute for Electrical and Electronics Engineers (IEEE), namely IEEE 802.11b. To be precise, IEEE 802.11b specifies only the physical layer, which makes use of direct-sequence spread spectrum, and provides data-rates of up to 11 Mbit/s. The common MAC layer used by all physical layers within the 802.11 family of standards is defined in the base IEEE 802.11 base specification. It is a distributed MAC based on carrier sensing multiple access with collision avoidance (CSMA/CA) and an optional centralised component for 'time bounded' services — see Chapter 3 for more details on CSMA.

Peer-to-peer communication between clients with IEEE 802.11b equipment (e.g. laptop computers) is possible (in the so-called 'independent mode') as well as communication between multiple clients and a central 802.11b access point ('infrastructure mode'). In either scenario, if multiple users share a link (i.e. a channel with a bandwidth of 22 MHz), the maximum (gross) data-rate of 11 Mbit/s is shared between all users of the link. This maximum data-rate can only be achieved under good propagation conditions, which limits the range (i.e. the maximum propagation distance) to as little as 25 m in a 'closed office' environment. Transmission modes with reduced data-rates down to 1 Mbit/s exist as well, through which the range can be increased. IEEE 802.11g is an evolution of the 802.11b physical layer, also based on a direct-sequence spread spectrum scheme, which should provide further increased data-rates.

IEEE 802.11b is becoming increasingly popular, but it is designed for operation in the frequency band from 2.4 to 2.5 GHz (as is IEEE 802.11g), a limited piece of spectrum which, being shared with other users such as Bluetooth and microwave ovens, happens to become increasingly crowded. An alternative band available for unlicensed operation is that between 5 GHz and 6 GHz, where more spectrum is available (potentially more than 600 MHz in certain countries) and less non-WLAN interference sources are expected. A physical layer suitable for operation in that band is IEEE 802.11a. It uses OFDM as a modulation scheme, operates at a carrier spacing of 20 MHz, and provides data-rates from 6 to 54 Mbit/s (again involving a trade-off between data-rate and range). IEEE 802.11a competes with other WLAN technologies designed for the same frequency band, such as HIPERLAN 2 standardised by ETSI, which was mentioned earlier. To avoid the type of market fragmentation experienced with earlier IEEE 802.11 standards, harmonisation efforts between IEEE 802.11a and HIPERLAN 2 have taken place, resulting in a virtually identical physical layer. However, the MAC layer solutions adopted are different, reflecting the different mindsets with which the standardisation of these two systems was approached. IEEE 802.11a makes use of the same MAC layer as 802.11b, which is well suited for data communications in computer networks. By contrast, HIPERLAN 2, standardised by ETSI and thus an effort supported predominantly by players in the telecommunications industry wishing to provide ATM-like features such as good quality of service support, makes use of a multiple access scheme combining TDMA/TDD with reservation ALOHA. Apart from supporting different QoS levels according to the requirements of various service categories, HIPERLAN 2 offers features such as dynamic frequency selection and transmit power control, which are a regulatory requirement for the 5 GHz band outside the US. The base 802.11a standard does not provide these features, but they are gradually being added through supplementary IEEE 802.11 specifications. Efforts to achieve a more comprehensive harmonisation between 802.11a and HIPERLAN 2 are underway.

2.6 Summary

The commonly accepted requirements for 3G cellular mobile communication systems, such as flexible and efficient provision of a large variety of services ranging from low to high bit-rates, anywhere at any time, cannot be met by 1G and 'traditional' 2G systems. Evolved 2G systems, such as GSM enhanced with GPRS, HSCSD and EDGE, will better fulfil these requirements, effectively blurring the boundaries between 2G and 3G systems. However, the efficient support of high and variable bit-rate multimedia services will remain challenging also with evolved 2G systems.

3G was initially supposed to become a single worldwide standard, but this vision has not fully materialised. We are now essentially dealing with two 'standards families', namely one produced by 3GPP providing an evolutionary route for GSM systems to UMTS, and another produced by 3GPP2, leading from cdmaOne to cdma2000. 3G systems can in theory sustain high bit-rates over the air interface owing to relatively wideband carriers (although not in all manifestations of cdma2000). In practise, however, with the currently available 3G spectrum, the targeted 2 Mbit/s bit-rates can essentially only be provided to stationary users close to microcell base stations or served by picocells.

As with the transition from 2G to 3G, initially intended to represent a step change from voice services with a little bit of low-bit-rate data to high-bit-rate multimedia services, but in reality occurring more gradually, 3G systems are expected to evolve gradually to meet the 4G vision. To be precise, at this point in time, a single common 4G vision does not yet exist, but a common thread seems to be that of convergence, be it between cellular communications and the Internet, between cellular communications and the broadcast world, or all three worlds together.

3
MULTIPLE ACCESS IN CELLULAR COMMUNICATION SYSTEMS

To examine the problem of multiple access in cellular communications, first the relevant OSI layers need to be identified, which is not necessarily straightforward. A split into *basic multiple access schemes* such as CDMA, TDMA, and FDMA, associated with the *physical layer*, and *multiple access protocols*, situated at the *medium access control layer*, is adopted here.

After a discussion of basic multiple access schemes, approaches chosen for medium access control in 2G cellular communication systems are briefly reviewed. The main effort will be invested in the identification of medium access control strategies suitable for systems that serve a substantial amount of packet-data users, starting with 2.5G systems such as GPRS, but mainly focussing on 3G and beyond.

It was pointed out in the introductory chapter that, in the specific case of CDMA systems, certain types of packet traffic might be best served on dedicated channels. We will briefly reconsider this issue here, but defer a more detailed discussion on this topic to later chapters. Here, the main focus is on multiple access protocols for common or shared channels. A case is made for reservation ALOHA-based protocols. As a representative of this family of protocols, PRMA is considered in more detail, and possible enhancements to PRMA are discussed, leading to the identification of design options available in the wider reservation ALOHA framework. Appropriate design choices are made and an outline is provided of the extent to which they will be investigated in subsequent chapters.

3.1 Multiple Access and the OSI Layers

A company wishing to operate a licensed cellular communications system will normally have to obtain from a national regulator (through a beauty contest or an auction, for instance) a certain amount of frequency spectrum in which it can operate its system. This spectrum constitutes the global communications resource for that system.

Consider a conventional cellular communications system, where communication over the air interface takes place between base stations and mobile handsets[1]. Each base

---
[1] In UMTS, there is the option for suitably enhanced mobile handsets to act as a relay for calls of other handsets, in which case communication over the air also takes place between handsets (this is referred to as Opportunity Driven Multiple Access (ODMA) [90]).

station will usually manage a part of this global resource (possibly dynamically together with other base stations) and assign individual resource units to multiple ongoing calls according to the availability of resources and current requirements of these calls. To be able to do so, means must be provided to split the resources assigned to a base station (for instance a part of the total spectrum assigned to an operator) into such small resource units and rules must be established which govern the access of users to them.

From these considerations and with reference to the terms used in previous chapters, it appears that the problem of multiple access can readily be split into the sub-problems of:

(a) providing a *basic multiple access scheme* such as frequency-division multiple access (FDMA), time-division multiple access (TDMA) or code-division multiple access (CDMA); and

(b) choosing a suitable set of rules, a so-called *multiple access protocol* on top of that.

The basic scheme would be associated with the first and lowest OSI layer, the physical layer, while the multiple access protocol is commonly situated at the lower sub-layer of the second layer, the so called medium access control (MAC) layer (Figure 3.1).

However, this split is not necessarily evident and, in fact, often not done in literature. Rom and Sidi, for instance, situate the protocols at the MAC (sub-)layer [102], but their protocol classification includes TDMA and FDMA (see Figure 3.2). In Reference [26], a similar classification is made, which includes also CDMA as a 'protocol', but Prasad does not care much about layering and uses 'multiple access techniques' and 'multiple access protocols' interchangeably. In Reference [103], the terms 'MAC layer' and 'multiple access protocols' do not even exist, however, the problem of network access is identified by Schwartz. It is pointed out that in the case where a common medium is used for access by users, provision for fair access must be made, either through polling by a centralised controller (controlled access) or through random access (also referred to as contention).

Bertsekas and Gallager refer to media where the received signal depends on the transmitted signal of two or more nodes (as is the case on a radio channel) as *multi-access* media and indicate that in such case a MAC sub-layer is required [104], as opposed to point-to-point links, where the signal received at one node depends only on the signal transmitted by a single other node. They do not explicitly introduce the term 'multiple

Layer 3	Network layer (NWL)
Layer 2	Data link control sub-layer (DLC)
	Medium access control sub-layer (MAC)
Layer 1	Physical layer (PHY)

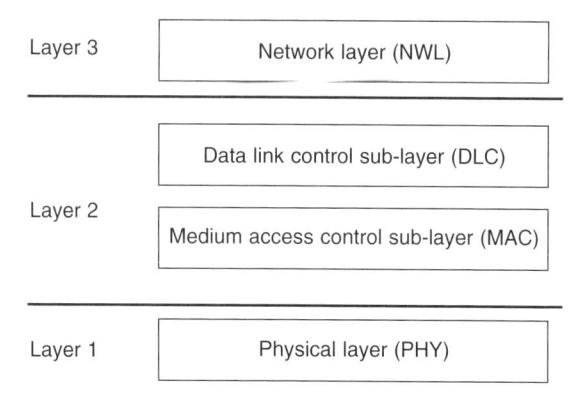

Figure 3.1 OSI layers relevant for the air interface

3.1 MULTIPLE ACCESS AND THE OSI LAYERS 51

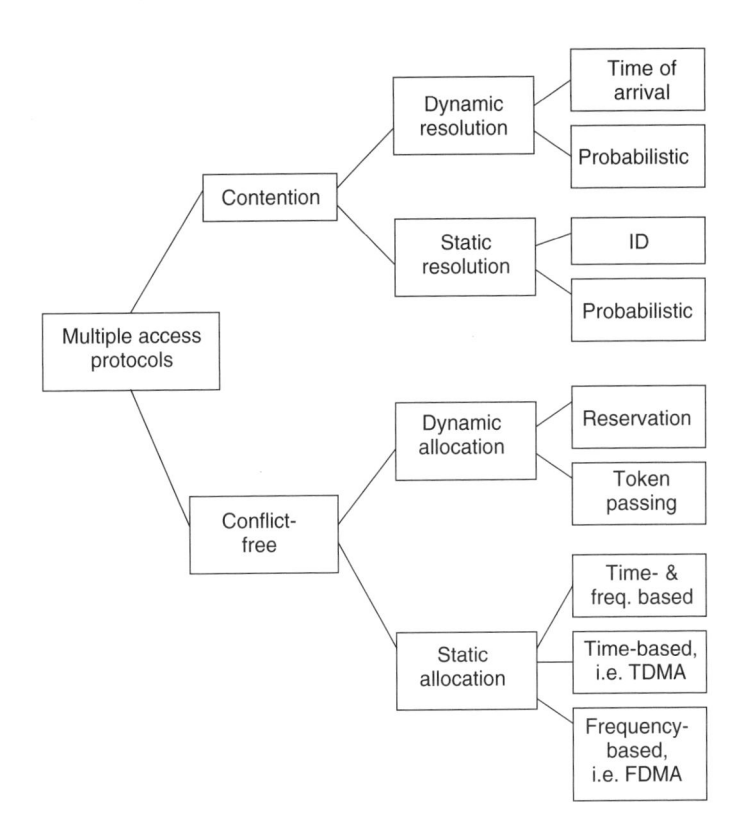

Figure 3.2 Multiple access protocol classification according to Rom and Sidi

access protocol'. Interestingly, for our purposes, time-division and frequency-division multiplexing are treated in Reference [104] as part of the physical layer of point-to-point links and it is pointed out that on a broadcast channel such as a satellite channel, such multiplexing can be used to provide a collection of virtual point-to-point links.

Similarly, Lee identifies five currently known 'multiple access schemes on physical channels', on top of FDMA, TDMA and CDMA mentioned previously, adding polarisation-division multiple access (PDMA) and space-division multiple access (SDMA) [66]. These can be associated with the first or physical layer in the OSI reference model. In his terminology, multiple access protocols appear to be 'multiple access schemes on virtual channels', and these are treated separately.

There are arguments *against splitting* the multiple access problem into a basic scheme determined by the choice of a physical layer and a protocol on top of that. Firstly, if a rigid division was possible, and basic multiple access schemes and multiple access protocols could each be classified separately, it would essentially be possible to select each of them independently. However, this is clearly not the case, as there are interdependencies, and the boundaries get easily blurred. For instance, in the case of pure ALOHA, the physical layer is a broadband broadcast channel, which *per se* does not provide any particular means for multiple access. The multiple access capability is entirely provided by the protocol. On the other hand, CDMA can rightly be considered as a hybrid between conflict-free basic multiple access schemes (dedicated codes) and contention protocols (common interference budget, resulting in potential 'collisions'), as does Prasad. Given the above, it would be convenient to consider TDMA and CDMA as much as protocols as for instance slotted ALOHA.

With the exception of Prasad, one could argue that those authors previously listed who did not split the multiple access problem were mainly concerned with computer networks, in which the effort to be invested in the physical layer is rather limited (and for which, incidentally, OSI layering was devised). In this case, the physical layer is often simply a virtual bit-pipe, that is, a virtual link for transmitting a sequence of bits. It translates incoming bit-streams into signals appropriate for the transmission medium through use of a modem [104]. It does not normally include means to provide a certain reliability. These means need to be provided by the data link control (DLC) layer, which is responsible for provision of a virtual link for reliable packet transmission, and which is the higher of the two sub-layers of the second OSI layer, as indicated in Figure 3.1.

In a mobile communications system, however, the transmission medium is the error prone radio channel, a medium subject to shadowing and fast fading. Designing a MAC layer on top of such an unreliable physical layer would prove rather difficult. Therefore, considerably more effort needs to be invested in the physical layer. In GSM for instance, the physical layer entails means for detection and correction of physical medium transmission errors [105]. This means that the burden of error control, usually attributed to the DLC layer, is now shared between the physical layer, which provides *forward* error control, and the DLC layer, which provides *backward* error control. Forward error control implies the addition of redundancy at the transmit-side through forward error correction (FEC) coding in a manner that the receive-side can (at least to a certain extent) correct errors introduced on the radio channel. Backward error control means that when the receiver detects errors that it cannot correct, it requests the transmit side to retransmit the erroneous data. This is also referred to as Automatic Repeat reQuest (ARQ). What is particularly important here is that the GSM physical layer specifies inherently a TDMA scheme through specification of bursts which need to be transmitted within time-slot boundaries. These bursts include for instance training sequences necessary for equalising channel distortions. Interestingly, in the GSM specification 05.05 [105], which is entitled 'physical layer on the radio path', Chapter 5 is on 'multiple access and time-slot structure'.

Correspondingly, and as highlighted in the previous chapter, the major struggle regarding the definition of the air-interface technology for UMTS was to agree on a physical layer which provides means for multiple access. Everything else (such as MAC layer issues) was, at least initially, considered to be of secondary importance. Obviously, the choice of a certain set of physical layer technologies imposes constraints on the design of MAC strategies.

In the light of these considerations, the approach adopted here assumes that the physical layer has to provide means for the support of multiple users, that is the possibility to split a global resource into small resource units, which can be assigned to individual users. This is termed a *basic multiple access scheme*. On top of that, a *multiple access protocol* situated at the MAC sub-layer is required which specifies a set of rules on how these resources can be accessed by and assigned to different users. These rules may be complemented by rules relating to admission control. Furthermore, the rules governing resource allocation are not always associated with the MAC, they may be associated, fully or partially, with a separate resource allocation algorithm.

3.2 BASIC MULTIPLE ACCESS SCHEMES 53

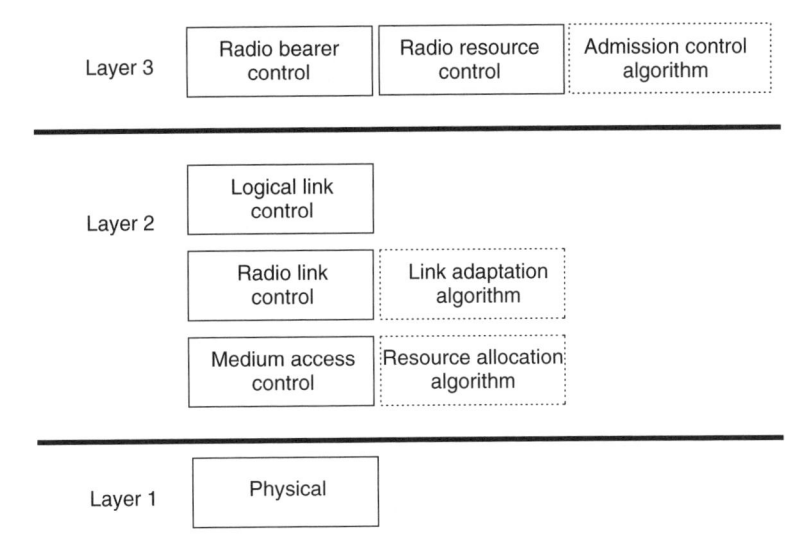

Figure 3.3 Layered structure of the UTRA TD/CDMA radio interface

Figure 3.3 shows, somewhat simplified, the layered structure used for the specification of the UTRA TD/CDMA proposal in Reference [90]. The layers relevant for the air interface are layers 1, 2, and those parts of layer 3 that are radio-related. Solid boxes represent protocols, while dotted boxes represent algorithms in Figure 3.3. The resource allocation algorithm and the admission control algorithm are associated with layer 2 and layer 3 respectively. Note that resources are in general allocated by layer 2 if requested or authorised by the radio resource control (RRC) entity situated at layer 3; one could therefore argue that the resource allocation algorithm should be part of the RRC. Note further that the DLC is split in this proposal into radio link control (RLC) and logical link control (LLC) in the same manner as in GPRS. In the end, the LLC was found to be redundant for UMTS and did not make it into the relevant specifications (see Section 10.1).

3.2 Basic Multiple Access Schemes

Lee identified five basic multiple access schemes, namely FDMA, TDMA, CDMA, PDMA, and SDMA, as already listed above. PDMA is not suitable for multiple access in cellular communication systems due to cross-polarisation effects arising as a result of numerous reflections experienced on the typical signal path in the propagation channel of such systems. Instead, orthogonal polarisations can be exploited to provide polarisation diversity (see for example Reference [106]). A significant amount of research effort has been invested in SDMA (a collection of articles can be found in Reference [107]) and there are endeavours to enable the deployment of this scheme in cellular communication systems. SDMA may impose particular requirements on a medium access scheme, and there are indeed proposals for multiple access protocols which take SDMA explicitly into account [108]. However, one could argue that since SDMA will normally be used on top of other multiple access schemes such as CDMA, TDMA and/or FDMA to increase capacity, it is not a 'full' multiple access scheme in its own right. SDMA will not be considered in the following, and we will restrict our attention to FDMA, TDMA, and CDMA, which were already introduced in Section 1.1.

FDMA is the oldest multiple access scheme for wireless communications and was used exclusively for multiple access in first generation mobile communication systems down to individual resource units or physical channels. Although plain FDMA is not an interesting choice any more for the provision of individual resource units for cost and efficiency reasons (limited frequency diversity, required guard bands), second and third generation systems include an FDMA element. In the relatively narrowband TDMA-based 2G systems with a small number of slots per frame (D-AMPS: 30 kHz carrier, three users per carrier; GSM: 200 kHz carrier, eight full-rate users per carrier) FDMA still fulfils a role in providing multiple access, although not down to individual channels. In 3G systems with wideband carriers, on the other hand, it is predominantly used to assign parts of the total bandwidth available for such systems to individual operators, and to separate the different hierarchical layers of a system belonging to a single operator.

TDMA was an obvious choice in the 1980s for digital mobile communications, since it is very suitable for digital systems; it is cheaper than FDMA (no filters are required to separate individual physical channels), and provides somewhat more frequency diversity. It also lends itself very well to operation with slow frequency hopping (SFH), as demonstrated in GSM. This provides additional frequency and interference diversity, which is discussed in detail in Subsection 4.2.3. Furthermore, a TDMA/SFH system can be operated as an interference-limited system (see Subsection 4.6.5), such that it exhibits a soft-capacity feature normally associated with CDMA [79,81].

Spread spectrum techniques were initially used in military applications due to their anti-jamming capability [6], the possibility to transmit at very low energy density to reduce the probability of interception, and the possibility of ranging, tracking, and time-delay measurements [110]. Spread spectrum *multiple access*, or rather CDMA[2], did not appear to be suitable for mobile communication systems because of the so-called *near–far effect*. Recall from Section 1.1 that the shared resource in a CDMA system is the signal power. For the system to work properly, signals from different users must be received at the base station at roughly equal power levels. If no special precautions are taken, then a terminal close to a base station may generate lethal interference to the signals from terminals far away. However, it was eventually possible to overcome this near–far problem through fast power control mechanisms, which regulate the transmit power of individual terminals in a manner that received power levels are balanced at the base station.

CDMA has a number of advantages compared to TDMA, such as inherent frequency and interference diversity (which are less inherent to TDMA, but can be provided as well when adding SFH, as discussed above). Furthermore, it exploits multipath diversity through use of RAKE receivers in a somewhat more elegant way than TDMA through equalisers. The key question is, however, whether CDMA can provide increased capacity or, rather, increased spectral efficiency in terms of bits per second per Hertz per cell. In the following, when we refer to capacity, we mean effectively spectral efficiency.

The capacity in a CDMA system is interference limited and, therefore, any reduction in interference converts directly and (more or less) linearly into increased capacity [111].

---

[2] In Reference [109], spread spectrum multiple access (SSMA) is referred to as a broadband version of CDMA, hence not every CDMA system is necessarily a spread spectrum system. Conversely, spectrum spreading does not necessarily imply that a multiple access capability is provided.

This is the main reason for claims made in References [6] and [111] that CDMA (specifically the 2G system cdmaOne) offers a four- to six-fold increase in capacity compared to competing digital cellular systems based on TDMA. However, in these references, the CDMA capacity evaluation is based on equally loaded cells (a favourable condition, CDMA systems are known to suffer particularly badly from unequal cell loading, see for example Reference [112]). Furthermore, power control errors, which reduce the capacity, are only to a limited extent accounted for. Finally, in Reference [6], the capacity gain due to voice activity detection is assumed to amount to the inverse of the voice activity factor, namely three-fold. In other words, only average interference levels are accounted for, which results in a too generous capacity assessment, as there is a non-negligible probability that an above average number of users are talking at once [111]. On the other hand, the TDMA capacity assessment in these references is based on very plain blocking-limited systems with a reuse factor of four in Reference [6], and even worse, seven in Reference [111].

As outlined above and discussed in detail in Subsection 4.6.5, an advanced TDMA system such as GSM with a SFH feature allows for interference-limited operation, in which case voice activity detection translates also more or less directly into capacity gains. In Reference [113] it is claimed that interference-limited GSM (with a one site/three sector or 1/3 reuse pattern, see Subsection 2.3.2) offers better coverage efficiency and capacity than CDMA-based PCS, while CDMA outperforms blocking-limited GSM (with a 3/9 reuse pattern).

In Reference [114], it is found that in CDMA-based PCS with a rather narrow carrier bandwidth of 1.25 MHz and therefore limited frequency diversity, capacity for slow mobiles is limited by the downlink (since only FEC coding and interleaving counteract multipath fading, while on the uplink, antenna diversity can also be applied). For fast mobiles, on the other hand, capacity is limited by the uplink (as power control is too slow to track the fast power fluctuations perfectly). Due to this imbalance, the system capacity with only one class of mobiles is lower than that of GSM even with a 3/9 reuse-pattern, where this imbalance is not experienced with SFH owing to the better frequency diversity. Only with a mixture of fast and slow mobiles can the capacity of CDMA-based PCS match or slightly exceed that of blocking-limited GSM. Note also that the support of hierarchical cellular structures is easier with (narrowband) TDMA systems than with wider band CDMA systems [114,115], due to better frequency granularity (see also Section 2.3 on this topic).

Clearly, we did not provide the ultimate answer to whether 2G CDMA systems are spectrally more efficient than 2G TDMA systems. It is true that interference-limited systems should in general provide higher capacity than blocking-limited systems, due to (wasted) excess CIR experienced in the latter on certain channels, as discussed in Section 4.6. However, apart from the fact that interference-limited operation is not limited to CDMA systems, if non-real-time data users are to be served, this deficiency of blocking-limited systems can be compensated through link adaptation and incremental redundancy. Refer to Sections 4.9 and 4.12 regarding the application of these techniques in GPRS and EGPRS respectively. In essence, therefore, for 2G systems, matters are not as clear-cut as some people might think they are.

One way to meet the high and variable bit-rate requirements for 'true' 3G systems, which may require the allocation of considerable bandwidths to individual users, is to adopt wideband versions of the existing TDMA or CDMA schemes, which have carrier bandwidths of a few MHz. Wideband TDMA schemes, however, exhibit several disadvantages. Since the TDMA frame duration should not exceed a few milliseconds due to delay constraints of real-time services, when the carrier bandwidth is large, bursts for low-bit-rate services have to be so short that the relative overhead for training sequences and guard periods becomes excessive [109]. Furthermore, according to Reference [86], achieving the necessary cell ranges would have been difficult with a wideband TDMA system, requiring a narrowband option as a companion solution. Therefore, unlike for 2G systems, wideband CDMA schemes have emerged as the preferred solution for 3G systems, as already discussed in detail in the previous chapter.

A plain FDMA scheme would not be suitable to provide low and high bit-rates simultaneously, since either the bandwidth would have to be kept variable, resulting in complex filter design, or high bit-rates would have to be provided by aggregating numerous frequency slots, requiring multiple transmit-receive units. However, there is one way to allow for a cheap (in terms of implementation complexity and therefore costs) and efficient aggregation of numerous narrowband carriers to provide the resources required for high-bit-rate services: orthogonal frequency-division multiplexing (OFDM).

In OFDM, transmission occurs on a large number of narrowband sub-carriers, but instead of multiple transmit-receive units required for conventional FDMA, owing to the application of inverse discrete Fourier transform operations at the transmitter and discrete Fourier transform operations at the receiver, the use of a single such unit will do [116]. Interestingly, these sub-carriers can overlap partially without losing mutual orthogonality, thereby ensuring high spectral efficiency.

OFDM alone is essentially only a modulation scheme, it does not provide means for multiple access. It must therefore be combined with a suitable multiple-access scheme, such as TDMA (as proposed for UTRA), or CDMA. Owing to TDMA, flexible support for low and medium bit-rate services is provided, while keeping the number of sub-carriers fixed (the filter complexity is therefore comparable to GSM). Only for very high bit-rate services, for which more expensive handsets can be justified, would the number of sub-carriers assigned to a user need to be increased. OFDM-based schemes were seriously considered in Europe and Japan for 3G cellular systems, but the time did not yet appear to be ripe for their use in cellular communications. However, it is very likely that we will encounter OFDM-based systems in the context of 4G, if not in the shape of a new air interface for cellular communication systems (which is possible as well), then in that of WLANs such as HIPERLAN 2 and IEEE 802.11a, which are expected to play an important role in 4G scenarios. Recall also from Section 2.5 that 4G might entail convergence between cellular and digital broadcast technologies. Since OFDM-based schemes were selected for digital audio and video broadcasting, this would add another OFDM-based component to 4G.

As outlined above, any CDMA or TDMA system will normally include an FDMA component, and can therefore be considered as a hybrid CDMA/FDMA or TDMA/FDMA system. Furthermore, as discussed in the first chapter, CDMA can also be combined with TDMA, resulting in a hybrid CDMA/TDMA(/FDMA) scheme. In such a scheme, variable bit-rates can be offered with a constant spreading factor by pooling multiple codes in a single time-slot, multiple time-slots in a TDMA frame or any combination thereof. Alternatively, like in wideband CDMA schemes, variable spreading factors can be used. Advantages of this hybrid scheme are, at least in theory, the following.

- The complexity of joint detection algorithms is reduced due to the reduced number of users multiplexed by means of CDMA.

- The introduction of a TDD mode is made easier, since the scheme, unlike pure CDMA, inherently uses discontinuous links.

- Soft handovers, which add considerable burden to the infrastructure, are not required. Furthermore, to assist the base station in the handover decision procedure, a mobile terminal can monitor neighbouring cells in time-slots during which it neither transmits nor receives without requiring an additional receiver. With pure CDMA, at least two receiver branches would be required for this [109][3].

- Frequency diversity provided by the CDMA component can be further increased by slow (i.e. burst-wise) frequency hopping, a well proven feature in TDMA systems such as GSM. This is beneficial when the coherence bandwidth exceeds the carrier bandwidth, which may happen in micro- and picocells [109].

- Finally, the evolution from GSM to 3G would not only be possible from the GSM network infrastructure, but also from the GSM air interface, using the same TDMA slot/frame structure and integer multiples of the GSM carrier bandwidth.

In the UTRA TDD mode, which is indeed based on hybrid CDMA/TDMA, due to harmonisation with UTRA FDD, the GSM slot/frame structure was eventually abandoned. For the same reason (i.e. since the same 5 MHz carrier spacing is used), given the current 3G spectrum situation outlined in Section 2.3, slow frequency hopping is not really possible. Furthermore, as discussed in Subsection 5.1.3, multi-user detection schemes are quite fundamental, if not a necessity in hybrid CDMA/TDMA systems, which increases the receiver complexity considerably. While such schemes would be even more complex in pure wideband CDMA systems, they are not really required. They can be introduced at a later stage to squeeze the most out of the spectrum, possibly after having deployed other less complex capacity enhancing techniques.

3.3 Medium Access Control in 2G Cellular Systems

3.3.1 Why Medium Access Control is Required

If we were to consider a system with point-to-point links only, there would be no need for a MAC layer and a multiple access protocol. Although radio channels are by nature broadcast or multi-access channels, it would in theory be possible to provide virtual point-to-point links from the base station to all users and vice versa through time- or frequency-division multiplexing. However, it is not possible in a cellular communications system to provide such point-to-point links to all potential connections, since radio resources are scarce, users move between coverage areas of different cells and normally only a small fraction of users dwelling in a cell will actually want to make a call.

In such systems, a multi-access or shared channel and consequently a multiple access protocol are required at least:

---

[3] UTRA FDD overcomes this problem through a so-called slotted mode described in Section 10.2.

- to allow mobile users to register in the system (e.g. when switching their handset on);

- for mobile users to send occasional location update messages. This enables the network to track users and to limit sending pages (i.e. notifications of incoming calls) in cells of the appropriate location area rather than all cells of the network; and

- to allow users (or rather terminals) to place a request for resources to make a call. This could be either a user initiated or *mobile originated* call, or as a response to a page, i.e. a *mobile terminated* call. Upon reception of such requests, the base station will attempt to reserve the required resources and notify the user of the resources to use and any potential temporal restrictions regarding the use of these resources.

By far the most important service in first and 'plain' second generation systems is circuit-switched voice. In such systems, resources are split in every cell into a small part of common resources such as broadcast and common control channels, which include the multi-access channel on the uplink, and a much larger part of dedicated resources, that is traffic and dedicated control channels. The multi-access channel is essentially only used for the purposes outlined above, while all other activities (in particular transfer of user data during a call) take place on (virtual) point-to-point links.

3.3.2 Medium Access Control in GSM

In GSM, the set of broadcast and common logical channels required, the latter referred to as Common Control CHannel (CCCH), is usually mapped onto one physical channel (one time-slot per TDMA frame). The CCCH consists of the multi-access channel (or RACH for Random Access CHannel) on the uplink, and a number of logical channels on the downlink, including the Access Grant CHannel (AGCH), and the Paging CHannel (PCH). On the AGCH, assignment messages are sent by the base station in response to channel request messages received on the RACH.

The PCH is usually the bottleneck in the system, as for every mobile terminated call a page needs to be sent in every cell of the location area in which the intended recipient currently dwells. The resources allocated for the PCH and the other common downlink channels will also determine the resources available for the RACH, since an equal amount of resources needs to be allocated to the uplink and downlink of these common channels. Therefore, abundant resources are normally available on the RACH. Consequently, efficient use of the RACH is not of prime concern and a simple implementation of one of the first random access techniques introduced in literature, the slotted ALOHA or S-ALOHA algorithm proposed in 1972 [117], was an appropriate choice for the multiple access protocol in GSM.

With respect to the terminology introduced earlier, one can state that the TDMA-based physical layer in GSM provides a physical channel or time-slot to the RACH (in other words, to the MAC layer), on which S-ALOHA is used as the multiple access protocol. Actually, since the RACH is used to place channel request messages to set up a circuit (either on a dedicated control channel to exchange some signalling messages, or on a traffic channel for a voice or data call), the multiple access protocol used in GSM could be considered as a variant of reservation ALOHA or R-ALOHA, a protocol family which will be discussed in more detail below.

For further details on physical channels, logical channels, and the random access procedure in GSM, refer to Chapter 4.

3.4 MAC Strategies for 2.5G Systems and Beyond

3.4.1 On the Importance of Multiple Access Protocols

In systems that support predominantly circuit-switched voice, not much effort needs to be invested in the design of suitable multiple access protocols. However, where packet-data plays a significant role, multiple access protocols are required to allow mobiles to place requests for resources to transmit *individual* packets. Thus, on top of an initial request to set up a call or session, numerous other requests will follow during the lifetime of such a call. Consequently, the traffic load on the multi-access or shared channel increases and the resource allocation entity will have considerable work to do to provide the requested individual reservations.

In the recent past, we could witness the tremendous success enjoyed by i-mode, a service launched in Japan in February 1999, which runs over PDC-P, the packet overlay to the Japanese 2G PDC system. At the time of writing, quite a few GSM operators have launched GPRS services, and most of those who have not are in the process of doing so. Unfortunately, we are not yet in a position to confirm the success of GPRS, mainly due to lack of GPRS handsets in significant quantities. However, the industry is clearly expecting that the demand for data services over cellular communication systems will finally take off outside Japan as well and, since this is almost exclusively in the shape of packets, that GPRS will play an important role at least in the first phase of this data explosion.

One could argue that packet-data traffic is most efficiently supported by carrying it *only* on common or shared channels (rather than on circuit-like dedicated channels). In reality, this is not necessarily the case for all types of packet traffic, as briefly pointed out in the next subsection in the context of CDMA systems, and examined in more detail in later chapters. All the same, it may apply to a significant share of the data traffic, and it is therefore worthwhile to invest more thought into efficient MAC strategies suitable for common and shared channels. In the following, possible alternatives are discussed.

The interested reader will already have observed that the notion of 'requests' and 'reservations' constrains the focus here to reservation-based multiple access protocols. For completeness, it should be mentioned though that protocols have been proposed for mobile communication systems, which do not rely at all on reservations. For instance, in Reference [118], a scheme for packet-voice transmission in cellular communications entirely based on S-ALOHA is discussed, which is claimed to provide high capacity since it can operate at a frequency reuse factor of one (provided that the average normalised traffic load per cell is low). This scheme exploits the capture effect discussed in more detail below. In such a scheme, due to a significant risk of packet erasure (both due to collisions within cells and temporarily high loads in neighbouring cells), a fast ARQ scheme would be required for real-time services such as packet-voice. In general, however, cellular communication systems are designed in a manner that ARQ is not required for real-time services, because it would be very difficult to achieve the required delay performance and to avoid jitter (i.e. delay variations). For non-real-time services, by contrast, ARQ is much less of an issue. GPRS for instance applies a selective ARQ scheme, as discussed in

Section 4.10, and the so-called COMPACT mode described in Section 4.12 is very much based on ARQ as an enabling technique for tighter frequency reuse (although not down to a reuse factor of one).

3.4.2 Medium Access Control in CDMA

The discussion of different MAC strategies provided in the remainder of this chapter, while intended to be general, will also consider their applicability in a CDMA context, where appropriate. However, given the importance of CDMA in 3G systems, and due to the peculiarities of this multiple access scheme, certain aspects pertaining specifically to medium access control in CDMA will first be discussed separately.

It was mentioned earlier that CDMA could be considered as a hybrid between conflict-free basic multiple access schemes and contention protocols. It is conflict-free, since every user is assigned dedicated codes, which allow the base station to distinguish between users. On the other hand, given the fact that all users are multiplexed onto a shared wideband channel, and that spreading codes cannot provide orthogonal separation between users on the uplink mobile communication channels, even users within one cell will create mutual interference (i.e. multiple access interference). As a result, the performance will degrade with increasing number of users, and packets or frames of individual users may be erased. This can be viewed as a collision, something typical for contention-based multiple access protocols. Furthermore, it was pointed out in the introductory chapter that CDMA provides inherent statistical multiplexing by averaging the interference of a large number of users, and therefore exhibits a feature which, again, we would normally attribute to the multiple access protocol rather than to the basic multiple access scheme. One might draw the conclusion, therefore, that the MAC layer (or more specifically, access control) is less important in a CDMA-based cellular communications system than for instance in a TDMA-based system.

The number of codes per cell available for user separation may be limited, in which case codes cannot be provided on a per-user-basis, but only on a per-call-basis. Even when plenty of codes are available, the base station cannot decode user signals without having a rough idea on what codes it is expected to use to do so. Therefore, an access mechanism is required for mobiles to request codes. The logical channel used for this purpose is typically some type of random access channel, on which a requesting user may pick one of a limited number of codes known to the base station (the allowable codes could for instance be signalled by the BS). However, once users have accessed the system and obtained dedicated codes, CDMA 'automatically provides' multiplexing of the different users, and one could argue that, for packet-data traffic too, a user should be allocated a dedicated channel (i.e. keep the allocated code and have free access to the channel during a session, obviously provided that enough codes are available). Thus, the focus shifts from the MAC layer to the admission control level, where algorithms are required to calculate the total admissible interference level given the different service requirements and the statistical behaviour of users already admitted as well as new users. The reader may refer to Reference [119] and references cited therein for further information. Power control is also an important matter in this context. In a multi-service environment, where individual services have different requirements, service-specific reference power levels should be chosen to maximise the capacity (e.g. Reference [120]). Since the chosen power control strategies affect interference levels, it may be advantageous to consider admission control and power control jointly.

Where does this leave access arbitration, for example through channel access control? There are several reasons why an approach entirely relying on admission control and service-specific power control may not be adequate.

First, in order to carry out closed-loop power control, a dedicated control channel must be set up together with the dedicated traffic channel, which is a rather slow process. This does not matter for circuit-switched services, where such a set-up is only required at the beginning of a call. For packet-data services, on the other hand, it is rather inconvenient to repeat this procedure for the transfer of every individual packet, particularly if the packet is short.

To limit the access delay, two fundamental alternatives are provided for uplink packet-data services in WCDMA [84]. Either, the dedicated control channel is maintained for the entire duration of a call or a session, and only the traffic channel is released during silence periods. This constitutes an unnecessary overhead load affecting the system capacity, particularly if no data is transmitted during a large fraction of the session duration. Alternatively, in the case of very short packets, rather than waiting for the set-up of dedicated channels following a random access message, the packets are more or less directly appended to this message, although without the possibility of closed-loop power control, which will again affect the capacity of the system. As an intermediate approach, a third possibility may be available, the so-called Common Packet CHannel (CPCH, an optional feature in UTRA FDD). In this case, user data is only sent following a random access message after an additional collision resolution interval. The CPCH is paired with a dedicated physical control channel on the downlink, which can be used for fast power control. However, if the message transmission starts immediately after the collision resolution interval, power control will not have converged, which can again affect the system capacity, particularly if user data transmission occurs at high data-rates. To ensure convergence before starting the message transmission, the network may order the terminal to send first a power control preamble. This adds some delay and introduces a certain overhead. If the CPCH is only used for packets with a certain minimum size, this overhead may well be acceptable.

Summarising the above, whether user data is transmitted on common channels such as RACH and CPCH or on dedicated channels, the access delay can only be reduced at the expense of capacity. Also, since common channels and dedicated channels are code-multiplexed, they are subject to a common interference budget. Given that closed-loop power control is not performed for short packets, admission control alone may not be sufficient, if short packets make up a significant proportion of the total traffic. Instead, admission control should be complemented by common channel access control, to limit the performance degradation due to common channel traffic.

Second, even if closed-loop power control is provided, the inherent statistical multiplexing capability of CDMA may be affected significantly if the service mix to be supported contains a few high-bit-rate users. Therefore, access control may not only be required for the common channels, but also for packet-data users for which a dedicated channel was set up, to limit interference fluctuations and increase the multiplexing gain. This is indeed possible in UTRA FDD.

Third, while instability problems at the random access are normally less significant in a CDMA context than in a TDMA context (see next section for details), it is still advantageous to ensure stability in all circumstances. Cao proposed in Reference [59] backlog-based access control for WCDMA in a manner similar to that proposed by us in References [49] and [52] for MD PRMA on TD/CDMA, to ensure stability in a wide range of circumstances.

Packet data support on UTRA FDD will be discussed in detail in Chapter 10, including the issue of channel access control both for common and dedicated channels. Other interesting contributions to access control for CDMA systems include References [121] and [122]. Both consider mixed packet-switched data traffic, for which spread S-ALOHA is used as the multiple access protocol, and packetised, but 'circuit-switched' voice (that is, voice carried on dedicated channels). Access control is only applied to data traffic. The approach proposed in Reference [122] is load-based, and resembles in certain aspects the scheme we proposed in Reference [30].

3.4.3 Conflict-free or Contention-based Access?

For the time being, consider just how mobiles should be provided access to the common or shared channel(s) to place requests. Bertsekas and Gallager refer to two extreme approaches for this problem. The first one is the 'free-for-all' approach, in which nodes normally send new packets immediately, hoping for no interference from other packets, but with the risk of collision of packets sent at the same time. The second is the 'perfectly scheduled' approach, where no collisions can occur. Classifying multiple access protocols as either following the 'free-for-all' approach or the 'perfectly scheduled' approach is not possible, since there are approaches that are in-between these two extremes. An alternative is to split multiple access protocols into random access protocols and polling protocols, as did Schwartz, or, roughly equivalent, contention-based and conflict-free protocols, as suggested by Rom and Sidi in Reference [102] and illustrated earlier in Figure 3.2[4]. Essentially, conflict-free protocols avoid collisions, but require some scheduling, while contention-based protocols do not require scheduling at the expense of collisions, which may occur. However, contention-based protocols do not need to follow the 'free-for-all' approach, as access can be controlled in various ways, e.g. through probabilistic measures to reduce the collision probability, which is discussed in detail throughout this text[5].

Conflict-free protocols have the advantage of using the available resources efficiently during high-load periods, but exhibit poor delay performance at low load. There are a few other issues to be taken into consideration for cellular communication systems when deciding between the two. The population of subscribers dwelling in a cell coverage area is subject to considerable fluctuation due to user movement, and is often only known on the basis of a location area spanning several cells, and not on a per-cell-basis. Also, only a small fraction of these dwellers may actually want to access the system. Therefore, any form of scheduling makes no sense at least for providing access to the system for registration, location update messages, or call establishment request messages (we call these activities *initial access* for further reference purposes).

---

[4] Rom and Sidi exclude centralised protocols such as polling (a conflict-free protocol) in their classification. 'Token passing' shown in Figure 3.2 is simply the decentralised version of polling. For protocol classification, 'polling' usually includes 'token passing'.
[5] Note that Schwartz uses 'controlling access' for token passing or polling protocols, while in this book, 'controlling access' is usually intended to mean probabilistic access control in contention-based protocols.

It may be possible to use scheduled approaches to place subsequent request messages, for instance in ongoing packet-data sessions, again, however, with the inconvenience that a user with an ongoing packet-data session may change cell. Another disadvantage of conflict-free schemes is that idle users do consume a portion of the channel resources, hence may be inefficient if a large number of users has to be served [102]. This makes them most appropriate for systems with a moderate and constant user population, conditions typically not satisfied in cellular communication systems. It can therefore be no surprise that all of the numerous 'multiple access schemes on virtual channels' listed in Reference [66] use contention for gaining initial access. The only scheme in which subsequent transmissions are scheduled is a hybrid reservation/polling scheme termed capture-division packetized access (CDPA, described for instance in Reference [123]), which can be viewed as a refinement of the S-ALOHA-based concept proposed in Reference [118] and briefly discussed in Subsection 3.4.1. In this scheme, rather than granting reservations for a certain period of time, each uplink transmission unit is scheduled individually by means of scheduling commands sent on the downlink. Such an approach provides complete flexibility in the choice of the scheduling algorithm, with centralised PRMA being one option, which we will discuss briefly in Section 3.6. However, this flexibility comes at the expense of complexity and control overhead. Control messages need to be protected by strong error correction coding to provide the required protocol robustness. This is the case in cellular systems in general, but particularly an issue for polling protocols, and with CDPA even more so due to universal frequency reuse.

In light of the above and in order to provide a universal access scheme applicable to initial and subsequent request messages, which can be implemented easily, scheduled approaches will not be considered in the following. We will concentrate instead on protocols that use contention to gain both initial and subsequent access to the system. No rule without exception, however. While the GPRS MAC uses reservation ALOHA as a multiple access protocol, it also features a scheduling element during reservation periods.

3.5 Review of Contention-based Multiple Access Protocols

The 'multiple access schemes on virtual channels' are listed in a rather arbitrary manner in Reference [66], but can essentially be associated with two fundamental approaches to channel access:

- access based on some form of the ALOHA protocol, mostly slotted, with or without reservations, with random or deterministic approaches to collision resolution; and

- access based on some form of channel sensing, including listening to a busy tone, idle or inhibit signal.

In the former class of schemes, the focus is on how to resolve collisions, once they occur, through appropriate retransmissions. They are referred to as *random access schemes*, because access attempts are essentially random (the 'degree of randomness' depends on the precise scheme being considered). In the latter, the effort is on avoiding collisions as much as possible by evaluating all available information before accessing the channel.

Collisions may occur also in the latter class of schemes occasionally, hence some effort needs to be invested in resolving them as well.

Note that the split into these two classes is orthogonal to the one in dynamic and static contention resolution shown in Figure 3.2. Both resolution types can in theory be applied to either of these two classes of schemes. Dynamic probabilistic resolution, for instance, can be achieved by controlling 'permission probabilities' dynamically according to the current system state.

3.5.1 Random Access Protocols: ALOHA and S-ALOHA

The so-called ALOHA protocol appears to be the first random access protocol described in the literature. According to Bertsekas and Gallager, it was proposed by Abramson in Reference [124]. In this scheme, when a packet arrives in the sending queue of a mobile terminal, it transmits the packet immediately on a resource shared between all mobile users admitted to the system. If no other MS accesses this shared resource at the same time, the base station can receive the packet successfully, otherwise, all packets sent simultaneously will collide and need to be retransmitted. To avoid repeated collisions, each MS involved in a collision will back off for a random time interval before attempting to retransmit its packet. This protocol is particularly simple to implement, but suffers from low throughput. Consider a *perfect collision channel*, on which packets are always erased when they collide (even if they overlap only for the tiniest instant of time), but are always received correctly if no collision occurs. In this case, for unit-length packets, the throughput S per unit time as a function of the offered traffic G amounts to $S = G \cdot e^{-2G}$, with a maximum value of $S = S_0 = 1/2e \approx 0.18$ at $G = G_0 = 0.5$. For derivations of this result, refer for example to References [102] or [104].

An improved version of 'pure' ALOHA is slotted ALOHA or S-ALOHA [117]. Here, the time axis is divided into slots of equal length, into which packets must fit, implying that packet transmission must be synchronised to the slot boundaries. Three types of slots are distinguished, namely *idle slots* (in which no terminals try to access the channel), *success slots* (exactly one terminal accesses the channel) and *collision slots* (two or more terminals access the channel). If a collision occurs, the terminals involved in the collision will again back off for a random period of time. A possible implementation in this case is that terminals retransmit packets in slots following the collision slot according to the outcome of Bernoulli experiments with a fixed retransmission probability value p as parameter. Unless otherwise mentioned, this is the approach we will adopt throughout the remainder of this text.

While packets collide with pure ALOHA even if they overlap only partially, with S-ALOHA packets either overlap completely or not at all. In other words, the so-called vulnerable period (in which no other terminal should transmit to avoid collision) is reduced from double the length of a packet to exactly the length of a packet, which is illustrated in Figure 3.4. This in turn doubles the maximum throughput from $1/2e$ to $1/e \approx 0.37$ on a perfect collision channel. A TDMA-based air interface lends itself naturally to slotted ALOHA, provided that guard periods are introduced to cater for the propagation delay. As mentioned previously, the access algorithm on the GSM RACH is based on a relatively plain implementation of S-ALOHA.

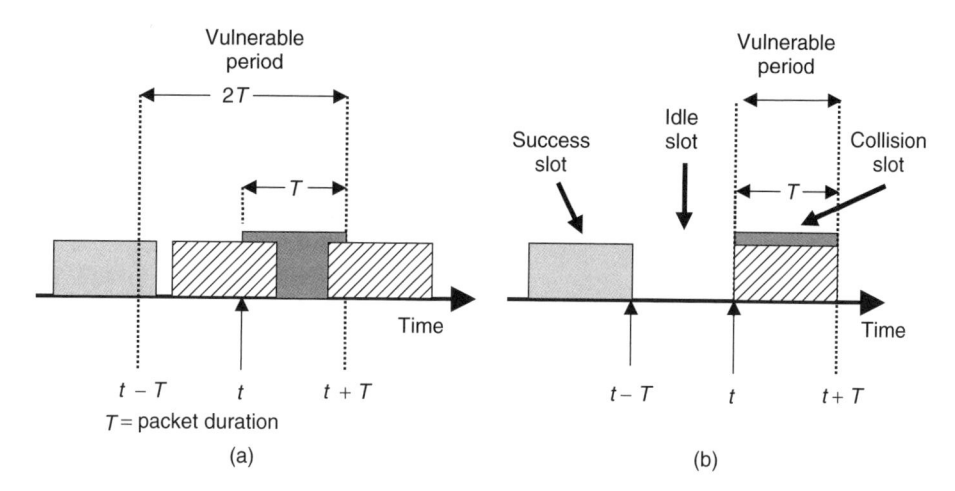

3.5 REVIEW OF CONTENTION-BASED MULTIPLE ACCESS PROTOCOLS

Figure 3.4 (a) 'Vulnerable period' in the pure ALOHA protocol during which no other packet transmission must initiate for the packet starting at time t to be received successfully. (b) 'Vulnerable period' and slot types with S-ALOHA

3.5.1.1 Throughput and Stability of Slotted ALOHA

References [102] and [104] both provide a detailed treatment of pure ALOHA and S-ALOHA protocols, including analytical studies on the throughput and the delay performance under various conditions. Here, we content ourselves with a very rough S-ALOHA throughput analysis, which follows in most aspects [104].

To establish the throughput behaviour, we have to analyse (or rather model) the distribution of the total traffic G offered to the multi-access channel, which is expressed in terms of packets per slot. G, also referred to as 'attempt rate' in Reference [104], is composed of newly arriving (or generated) packets and retransmitted packets. Assume first that, irrespective of the number of terminals we are dealing with, the *total* number of packets generated at the different terminals behaves according to a Poisson process with rate λ, such that the probability of k packets being generated per slot amounts to

$$P_k = \frac{\lambda^k e^{-\lambda}}{k!}. \tag{3.1}$$

Obviously, since this excludes retransmitted packets, $G > \lambda$. Assume now that the total traffic is again Poisson, namely at rate G. The probability of successful transmission, P_{succ}, is simply the probability that exactly one packet is offered to the channel in each slot, which is obtained by replacing λ with G in the above formula and setting $k = 1$. This success probability happens to be the normalised throughput we are looking for as well (or the departure rate according to Reference [104]), hence

$$S = Ge^{-G}. \tag{3.2}$$

The maximum throughput is $S_0 = 1/e$ at $G = G_0 = 1$.

This is an extremely rough analysis, which ignores completely the dynamics of the system. To gain more insight, assume that we are dealing with a finite number of terminals N, each terminal being either in *origination mode*, during which packets may be generated, or in *backlogged mode*, which a terminal enters when the first transmission attempt with a new packet was unsuccessful and retransmissions are required. For simplicity, it is assumed that terminals ignore new packet arrivals while they are busy trying to transmit a packet. The *system state n* is the number of terminals in backlogged mode.

Assume further that each terminal in origination mode generates packets according to Poisson arrivals at rate λ/N, hence the probability p_0 that a terminal generates a packet in a slot is $p_0 = 1 - e^{-\lambda/N}$ [104]. Strictly speaking, this is the probability that it generates at least one packet, but recall that it ignores any subsequent arrivals until it transmits the packet successfully. What complicates the analysis now is that both the total arrival rate λ_{ar} and the rate of retransmitted packets obviously depend on the system state n, so that G is now

$$G = (N - n) \cdot p_0 + np, \tag{3.3}$$

with p the probability with which backlogged terminals retransmit a packet in any given slot. To eliminate this state dependence, assume for the sake of argument that $p = p_0$. Again, the throughput is the same as the success probability per slot, which is the probability that exactly one packet is transmitted per slot. This probability can be calculated through the binomial formula, it is

$$S(p_0) = \binom{N}{1} p_0 (1 - p_0)^{N-1} = Np_0 (1 - p_0)^{N-1}. \tag{3.4}$$

If N is sufficiently large, at $G = Np_0$, the throughput behaviour according to Equations (3.2) and (3.4) is virtually the same, as shown for $N = 40$ in Figure 3.5.

Realistically, it is neither easily feasible nor reasonable to set $p = p_0$, the latter since p_0 is typically low, while p should be as high as possible to reduce retransmission delays. This is where the problems start. Carleial and Hellman reported a so-called bi-stable behaviour in Reference [125] for this type of system with a fixed transmission probability p different from p_0. It means that there are two stable operating points (at which the total

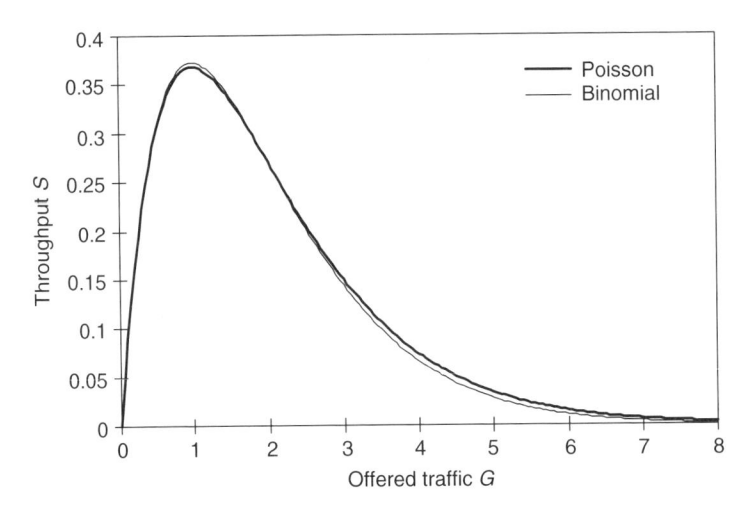

Figure 3.5 S-ALOHA throughput according to the Poisson formula and the binomial formula for $N = 40$

3.5 REVIEW OF CONTENTION-BASED MULTIPLE ACCESS PROTOCOLS

arrival rate equals the departure rate or throughput). At the desired one, most terminals are in origination mode and the system provides reasonable throughput, at the undesired one, most terminals are in backlogged mode, the throughput is low and, accordingly, the delay is high. In-between the two stable equilibrium points, there is also a third equilibrium point which is unstable, as shown in Figure 3.6.

To illustrate this, note first that even if $p_0 \neq p$, as long as both p_0 and p are small, according to Reference [104], the probability of successful packet transmission can still be approximated by

$$P_{\text{succ}} = G(n)e^{-G(n)}. \tag{3.5}$$

We do not equate P_{succ} to $S(n)$ here, defining a throughput valid only for a particular system state does not seem to be very useful. Define now the *drift* in state n, D_n, as the expected change in backlog from one slot to the next slot, which is the expected number of arrivals, i.e. $(N - n) \cdot p_0$, less the expected number of departures P_{succ}, that is

$$D_n = (N - n)p_0 - P_{\text{succ}}. \tag{3.6}$$

Figure 3.6 shows the state-dependent arrival rate (the straight line) and the departure rate according to Equation (3.5) for $p > p_0$. System equilibrium points occur where the curve and the straight line intersect. If the drift, which is the difference between the straight line and the curve, is positive (symbolised by arrows pointing towards the right), then the system state tends to increase, while it decreases when the drift is negative. This explains immediately why the middle equilibrium point is unstable and the other two are stable. In the words of Bertsekas and Gallager, the system 'tends to cluster around the two stable points with rare excursions between the two'.

Clearly, we would like to avoid the undesired operating point. However, if p is set to a fixed value, and the system experiences a number of successive collisions, leading to growing n, it may happen that suddenly $np \gg 1$, thus $G \gg 1$, at which point P_{succ} is low. This is exactly how the system can get caught in the undesired stable operating point, from which it will find it difficult to escape.

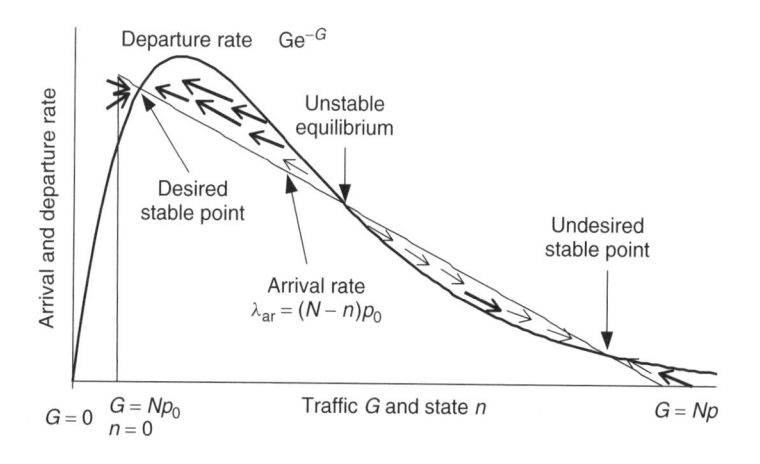

Figure 3.6 Drift and equilibrium points in S-ALOHA

3.5.1.2 Stabilising S-ALOHA

Often, trying to avoid the undesired operating point is referred to as *stabilising* the protocol, although, strictly speaking, the system is also stable in the undesired operating point. However, if an infinite number of terminals is considered, and Poisson arrivals at rate λ are assumed, then the straight line in Figure 3.6 becomes horizontal. As a result, the undesired stable operating point disappears and, instead, when $np \gg 1$, the system state just grows without bound. Based on the definition provided in Reference [104], namely that a multi-access system is stable for a given arrival rate if the expected delay per packet is finite, this particular system would indeed be unstable, thus calling for *stabilisation*. Other contention-based protocols, including pure ALOHA, exhibit similar stability problems.

One approach that alleviates the problem of instability somewhat, but does not guarantee stable operation in all circumstances, is to let retransmitting users reduce the probability p with which they access a slot with every further retransmission attempt (see for example Reference [126]). This is termed transmission backoff, and *exponential* transmission backoff is particularly well known. To stabilise S-ALOHA 'properly', probabilistic control mechanisms were proposed in the literature, which alter p in a dynamic manner (slot by slot) to maximise the probability of successful transmission in each slot. However, as long as the same retransmission probability value applies to all backlogged terminals, the maximum throughput remains constrained to $S_0 = 1/e$. Such schemes, in which the value for p is not controlled individually for each terminal, are referred to in the following as *global*[6] probabilistic control schemes. Obviously, a system in which the arrival rate exceeds the maximum departure rate must necessarily be unstable, hence the best we can hope for is stable operation for $\lambda_{ar} \leq S_0$.

Different approaches to retransmission control are considered in the context of investigations on the GPRS random access discussed in Section 4.11, including a global probabilistic control scheme. The latter will also be examined in more detail in Chapter 6.

3.5.2 Increasing the Throughput with Splitting or Collision Resolution Algorithms

With *global* probabilistic control schemes, new users may also be granted access to the channel at instants in which other users try to recover from an earlier collision. Another approach is to resolve collisions immediately by controlling access of each user *individually* based on its own history of retransmissions and the channel feedback, and not to allow new users to access the system during such a *collision resolution period*. This not only stabilises the system, but can also result in increased throughput. Such schemes are referred to as *splitting algorithms* in Reference [104], since the set of users involved in a collision is split into smaller subsets until individual users are singled out, which then can transmit without the risk of a collision. In Reference [102], on the other hand, they are called *collision resolution protocols*, since it is attempted to resolve each collision individually and ensure that all the users involved in this collision will transmit their packet successfully before allowing new users to access the channel[7].

---

[6] One might be tempted to term them 'centralised' rather than 'global', but they can be implemented in a decentralised fashion.
[7] This may lead to some confusion, since any retransmission in a plain S-ALOHA scheme can be viewed as an attempt to resolve a collision, although without success guarantee. By contrast, for a finite user population, collision resolution protocols guarantee successful transmission in a finite amount of time.

3.5 REVIEW OF CONTENTION-BASED MULTIPLE ACCESS PROTOCOLS 69

In the basic splitting algorithm, a set of users involved in a collision is split into j subsets, the first of which transmits in the next slot, following the collision slot, whilst the others must wait until all the users in the first subset have transmitted successfully. This is indicated either by a success slot, which means that there was only one user in the first subset, or an idle slot, which means that the first subset was empty. Subsequent collisions require a further split into j subsets. These schemes can be visualised in the shape of a tree or a stack, and are referred to as *tree algorithms* or *stack splitting algorithms*. A tree algorithm with $j = 2$ is referred to as a binary tree algorithm. According to Reference [104], when optimising j, a stable throughput of 0.43 can be achieved with tree protocols, and a further improvement to avoid unnecessary collisions suggested by Massey allows an increase to 0.46. The drawback of these schemes is that a mobile station is forced to monitor the channel feedback continuously to keep track of the end of each collision resolution period. In cellular communication systems, where battery life is almost everything, and mobile handsets are sent to 'sleep mode' whenever possible, this is clearly not desirable. One way to overcome this problem is to let new arrivals join the subset of nodes currently allowed to transmit. However, this has to be paid for by limiting the maximum throughput to 0.40. Such an approach is referred to as an *unblocked* stack algorithm, as opposed to a *blocked* stack algorithm where new users have to wait for the beginning of a new collision resolution period.

Further enhancements include the splitting of packets according to their arrival time, allowing packets in the earliest arrival interval to transmit first, such that a first-come first-serve (FCFS) policy is adopted, and choosing the allocation interval (i.e. the arrival interval of packets allowed to access the next slot) in a manner that maximises the chances of success given the available information. With such an FCFS splitting algorithm, a stable throughput of up to 0.4871 can be achieved, with optimum size of sub-intervals even 0.4878. The problem of the FCFS policy is again that mobiles are required to monitor the channel feedback continuously. This can be overcome by an approach which is approximately last-come first-serve, which still allows a throughput of up to 0.4871 to be achieved, however, at the expense of higher delay compared to the FCFS policy.

For further information, the reader is referred to References [102] and [104], where the different protocol variants are discussed in considerable detail and references to a large number of relevant articles are provided. A discussion of collision resolution algorithms specifically for CDMA systems can be found in Reference [127].

3.5.3 Resource Auction Multiple Access

In Reference [128], Amitay and Nanda proposed the use of resource auction multiple access (RAMA, originally proposed for fast handovers and resource allocation) for statistical multiplexing of speech in wireless personal communication systems. In RAMA, resources (traffic channels on normal carriers) are assigned to users through an auction process, taking place on a special RAMA carrier. The auction consists of a symbol-wise announcement of the identities[8] of bidding users (through signalling on the uplink), alternating with acknowledgements by the base station, such that only one user survives at the end of an auction (e.g. the user with the highest identity). This guarantees that in

---

[8] A MAC address could for instance be used as an identity, or any other suitable name or address identifying uniquely the requesting mobile terminal.

each assignment cycle, consisting of auction and subsequent resource assignment, one user will be assigned a resource (subject to availability of the resource, of course). This is a partial justification for the fact that the authors term this scheme a *deterministic* access scheme as opposed to random access schemes. The scheme bears close resemblance to the logarithmic search scheme described in Reference [104, p. 343], which in turn is related by Bertsekas and Gallager to the splitting algorithms just described.

Amitay and Nanda claim higher multiplexing efficiency for RAMA than can be achieved with PRMA, which is a scheme based on random access. However, they neither seem to carry out a comparison based on the same user population size, nor do they account for the overhead in the shape of the special RAMA carrier, which needs to be put aside for the assignment cycles. Whether RAMA can really provide capacity gains compared to schemes such as PRMA depends very much on the exact amount of this overhead, which is determined by various factors. In addition, due to the nature of the auction process, the structure of the RAMA carrier will have to differ from that of normal carriers (in terms of slots, frames, and possibly modulation schemes), which adds undesired complexity to the system.

3.5.4 Impact of Capture on Random Access Protocols

At this point, it is important to mention that the radio channel in a mobile communications system is far from a perfect collision channel. Due for instance to fast fading, or to excessive co-channel interference, the base station may not receive a packet correctly even if it did not collide with another packet. On the other hand, and more interestingly, due to only partially correlated fading processes and the near–far effect, packets may arrive at the base station with significantly different power levels. If several packets collide in a given time-slot, the base station might still be able to capture and correctly receive the strongest one, which is referred to as *capture effect*. A simple way to model the ability of a receiver to capture a packet is by means of a *capture ratio* γ_{cr} (see for example Reference [129]), also referred to as *capture factor* in Reference [130]. With P_i, $i = 1..k$, the power level received at the base station for packet i, the jth of k simultaneously transmitted packets can be captured, if

$$P_j > \gamma_{cr} \sum_{i=1, i \neq j}^{k} P_i, \tag{3.7}$$

where γ_{cr} depends for instance on the error correction coding scheme employed. If the distribution of received packet power levels is known, which in turn depends on the propagation condition and the spatial distribution of the mobiles, then γ_{cr} can be translated into the probability C_k of capturing one packet in the presence of k simultaneously transmitted packets.

In the case of an S-ALOHA scheme, if the total offered traffic (composed of newly arriving packets and retransmitted packets) is Poisson with rate normalised per slot G, then the throughput can easily be calculated as

$$S = Ge^{-G} + \sum_{k=2}^{\infty} C_k \frac{G^k e^{-G}}{k!}, \tag{3.8}$$

3.5 REVIEW OF CONTENTION-BASED MULTIPLE ACCESS PROTOCOLS

where the first term is the throughput without capture according to Equation (3.2) derived earlier. From Equation (3.8), it is immediately evident that the capture phenomenon translates directly into higher throughput without requiring any modifications to the S-ALOHA protocol[9]. In scenarios typical for a mobile communications system, assuming no or limited power control[10], the throughput can assume values up to 0.6, as found in Reference [131]. In Reference [132], where we evaluated the performance of S-ALOHA with various retransmission schemes using the capture model established for the standardisation of the GSM GPRS service [133], we reported similar values. Some of the results obtained for Reference [132], mainly on the delay performance of the considered schemes, are reported in Section 4.11. For now, Figure 3.7 juxtaposes the throughput achieved when capture is accounted for according to Equation (3.8) with that achieved on a perfect collision channel.

In the case of splitting algorithms, the effect of capture on the achievable throughput cannot be identified that easily, since it might have repercussions on the operation of the algorithm. In Reference [134] for instance, variations of the basic binary tree algorithm are discussed for the case in which separate feedback is available for a capture slot and a 'normal' success slot, and also for the case where the base station is not able to recognise whether the successfully received packet is one captured among several transmitted packets or not. To make the analysis tractable, a simplified scenario is considered in Reference [134] with a dominant and a non-dominant group of mobiles. Only packets of the dominant group may be captured, and only if exactly one dominant user accesses a given time-slot, irrespective of the number of non-dominant packets sent in this time-slot. The case with separate feedback allows higher throughput values to be achieved, as expected. In Reference [130], similar investigations on tree-based algorithms are conducted, but

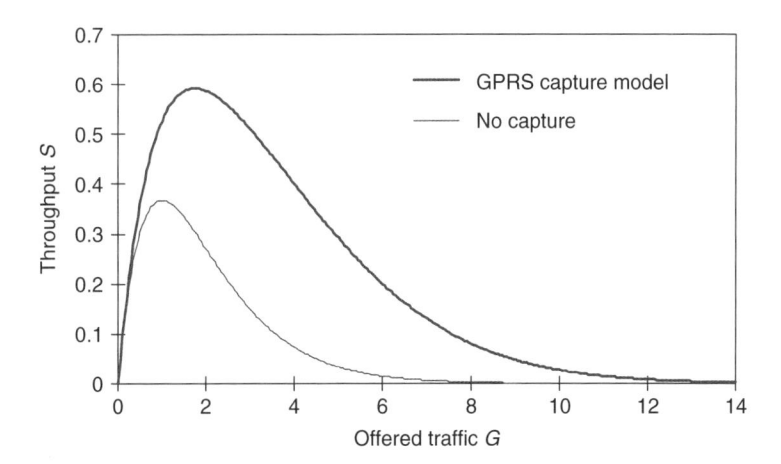

Figure 3.7. Throughput of S-ALOHA with capture according to the GPRS capture model and without capture. Traffic and throughput are indicated in terms of packets per slot

---

[9] Note though, that adaptive control of the transmission probability to stabilise the protocol requires taking capture into account, as discussed in Reference [131]. Furthermore, the rate of the offered traffic at which the throughput is maximised, G_0, is not $G_0 = 1$ anymore, instead, it depends on the C_k values.

[10] For random access, only open-loop power control is used (if at all) which, with the exception of specific TDD scenarios, is quite inaccurate. Tight power control, typically considered desirable in mobile communications, would have an adverse effect on the capture probabilities.

this time with probabilistic capture models, including a model using the capture factor introduced in Equation (3.7). According to Reference [135], certain collision resolution algorithms applied to PRMA are sensitive to capture in that their performance at low traffic is negatively affected. In general, we would expect the capture effect to translate better into increased throughput with plain and stabilised random access protocols than with splitting protocols. However, further investigations would have to be carried out to verify this conjecture and to assess the exact performance differences in realistic scenarios for mobile communication systems.

3.5.5 Random Access with CDMA

Ignoring potential issues related to receiver complexity, both S-ALOHA and pure ALOHA lend themselves easily to operation in a system providing code-division multiple access (References [21] and [25] respectively). In such schemes, several packets can be received simultaneously, how many exactly depends on various parameters, such as the spreading factor and the amount of FEC coding applied. Interestingly, while the version of the protocol without slots still performs worse than the slotted version, the maximum achievable throughput degrades by much less than the 50% observed in the case without spreading. For instance, in the scenario we considered in Reference [136], with 1024-bit packets, perfect power control and no error coding, the degradation was found to be 20% and only 10% for spreading factors of 15 and 31 respectively. This is due to the 'soft-collision' feature of CDMA: a partial packet overlap will not necessarily result in the loss of the respective packets. Also, for instance in Reference [137], it was shown that, with the right amount of FEC coding, the bandwidth-normalised throughput of the CDMA version of S-ALOHA is higher than that of conventional narrowband S-ALOHA. However, this does not account for capture. Incidentally, if capture is accounted for, some sort of soft-collision feature may also be obtained without CDMA.

In a CDMA system, the random access channel could either be time-multiplexed with other channels, in which case the adoption of a slotted version of ALOHA is near at hand. Alternatively, code multiplexing is also possible, which is the solution adopted for UTRA FDD. In this case, both unslotted and slotted ALOHA could be used. A slotted version was chosen for UTRA FDD, where a random access time-slot lasts double the time of the time-slots known from traffic channels.

For references on access control for spread S-ALOHA, refer to Subsection 3.4.2, for details of the UTRA random access, to Chapter 10.

3.5.6 Protocols based on some Form of Channel Sensing

To increase the throughput further (both in systems with and without spread spectrum), protocols were proposed in which transmitters 'sense' the shared channel before they attempt to transmit. In other words, rather than accessing the channel at random, they listen before they talk. This does not require the detection of information sent by other users, it is enough if users are capable of sensing the existence of a carrier, which indicates that the channel is busy. For this reason, these protocols are referred to as carrier sense multiple access (CSMA) protocols.

Fundamentally, with these protocols, a terminal that senses the channel to be idle will transmit its packet, while if the channel is sensed to be busy it will refrain from

3.5 REVIEW OF CONTENTION-BASED MULTIPLE ACCESS PROTOCOLS

transmitting, thereby reducing the chances of collisions. If a collision occurs, the terminal will again sense the channel after a randomised backoff period and, depending on the outcome, transmit its packet. Given the non-zero transmission delay and thus a non-zero *sensing delay*, a user may sense that the channel is idle and start to transmit even though another user has already started transmission, hence collisions may occur.

Several variations of these protocols exist, which differ in the behaviour of the users before sensing and when they find the channel busy. Furthermore, they may be enhanced by collision detection (CSMA/CD protocols) and collision avoidance (CSMA/CA). With collision detection, a terminal interrupts its transmission as soon as it detects that it is involved in a collision, thereby reducing the time during which the channel goes wasted because of collisions. With collision avoidance, the channel is first probed by sending a mini-packet, and only if this is transmitted successfully, is a proper data packet sent, making CSMA/CA in essence a reservation protocol.

Consider packets of equal duration T and assume that the sensing delay depends only on the transmission delay. Clearly, the lower the *normalised sensing delay* α, which is the ratio of transmission delay τ to packet duration (that is, $\alpha = \tau/T$), the lower the risk of collisions and the higher the throughput. For instance, in the case of non-persistent CSMA without collision detection, the maximum throughput is $(1 + \sqrt{2\alpha})^{-1}$ for the slotted version of the protocol and $(1 + 2\sqrt{\alpha})^{-1}$ in the case of the version without slots [104]. It is important to recognise that the concept of slots is not the same here as with slotted ALOHA. The slot size is taken as the maximum transmission delay τ experienced in the considered system, it is therefore typically much shorter than the packet duration. As a matter of fact, neither slotted CSMA nor CSMA without slots could easily be accommodated on a TDMA-based air interface such as GSM. However, to get an idea of the potential of CSMA with some representative values for α, take the GSM slot-length of 577 µs as an indication for the possible packet duration, and consider the maximum radius of a cell in GSM, which is normally 35 km [105]. With a propagation speed of $3 \cdot 10^8$ m/s and taking into account that the maximum propagation distance between mobiles is twice the cell radius, the worst case α amounts to 0.4, resulting in a maximum throughput of 0.53 and 0.44 respectively, for slotted and unslotted versions of the protocol. With a cell radius of 3 km, this increases to 0.79 and 0.73 respectively, and with 1 km to 0.85 and 0.82. Depending on α, these values can be increased further with collision detection.

While the problem is less severe than with ALOHA, CSMA is inherently unstable and therefore needs to be stabilised, for instance with appropriate probabilistic control mechanisms similar to those used for ALOHA. Combining CSMA with splitting algorithms to resolve collisions, on the other hand, appears not to make sense according to Reference [104], since it is not possible to increase the throughput by doing so.

Apart from limited suitability for TDMA-based air interfaces, there are a number of generic problems regarding the use of such schemes in cellular communication systems. A transmitting terminal may be hidden for a sensing terminal due to the shadowing effect or because of range restrictions when mobile stations are located at opposite cell edges. Furthermore, the fast fading and noisy mobile radio channel will make it difficult for a mobile terminal to sense a busy channel quickly, resulting in an increase of the normalised sensing delay α. Finally, collision detection is difficult to implement in radio networks because a transmitting node's own signal would drown out any other signal arriving at the receiver of that node. However, collision avoidance can be implemented, for instance by letting the base station acknowledge the probing mini-packet through sending its own mini-packet, which is then interpreted by the probing terminal as a reservation.

Solutions have been suggested for the hidden terminal problem, such as a 'busy-tone' transmission by the base station on the downlink as soon as it starts to receive a packet from an MS on the uplink. Instead of sensing the shared uplink channels, other mobiles have to sense the busy tone on the downlink. Together with most of the basic variations of the CSMA protocol, this *busy tone* solution was first introduced and analysed in a series of papers by Kleinrock and Tobagi [138–140]. Further variations were proposed in the literature which cater for the hidden terminal problem, such as inhibit or idle sense multiple access (ISMA). In inhibit sense multiple access, the base station sets intermittently inhibit bits rather than transmitting continuously a busy tone when it finds the uplink channel busy. Rather than for sensing delay, α stands for the inhibit delay, which is the total delay between the start of the packet transmission at one terminal and the detection at other terminals of the inhibit bit set by the base station to indicate that the channel is now busy. It does therefore not only consist of the propagation delay, but also of the 'inhibit bit scheduling delay' and possibly processing delays.

It was just pointed out that CSMA was not suitable for a TDMA-based air interface with time-slots of equal size. The same is also true for the solutions suggested to cater for the hidden terminal problem. On top of that, collision detection by listening to a busy tone or inhibit signal while transmitting an own packet would require a mobile handset with two transmit-receive units. In certain systems, however, handsets supporting basic services do not need to be able to transmit and receive simultaneously, allowing the design of very cheap handsets with only one such unit. GSM is one example, as discussed in detail in Section 4.2.

In References [129] and [141], the impact of capture on ISMA is assessed and compared with that on S-ALOHA. According to Reference [129], at a capture ratio γ_{cr} of 6 dB, which results in an S-ALOHA throughput of 0.6 under the conditions considered, the throughput of slotted ISMA is larger than 0.9 for a relative inhibit delay α of 0.01, but smaller than 0.6 for $\alpha = 0.2$, at which value the throughput of conventional slotted CSMA without capture is still significantly larger than that of S-ALOHA without capture. At $\gamma_{cr} = 6$ dB, the situation is even worse for non-slotted ISMA, which is outperformed by S-ALOHA for $\alpha > 0.1$. Clearly, for a fair comparison, one would also have to account for the impact of the transmission delay on S-ALOHA, as guard periods may be required, reducing the usable slot portion.

A variant of a CSMA/CA protocol with optional provisions for the hidden node problem is applied successfully in wireless LAN systems complying with the IEEE 802.11 series of standards. WLANs use short-range radio technology, therefore, transmission delays are not a major issue. Incidentally, IEEE 802.11b, the most popular WLAN standard at the time of writing, features a physical layer based on direct-sequence spread spectrum technology, but without a CDMA element. The multiple access capability is entirely provided by the multiple access protocol at the MAC.

3.5.7 Channel Sensing with CDMA

The application of channel sensing is also possible in CDMA systems. However, since multiple users can be supported simultaneously in CDMA, it is not sufficient to sense the channel busy, e.g. by sensing a carrier. Correspondingly, one refers to *channel*

3.5 REVIEW OF CONTENTION-BASED MULTIPLE ACCESS PROTOCOLS

load sensing rather than *carrier sensing* in such situations. The purpose of channel load sensing is to limit the number of users accessing the channel simultaneously to a certain threshold value, above which multiple access interference causes too much performance degradation. This requires the sensing terminals to establish the number of users on the channel, or at least the current level of MAI. In Reference [24], it is reported that Tobagi and Storey suggested the estimation of the received 'pseudonoise power' in Reference [23].

An interesting contribution to channel load sensing for CDMA-based mobile communication systems, which will be discussed in more detail in Subsection 6.4.3, can be found in Reference [32]. The suggested channel load sensing protocol is based on spread slotted ALOHA, but to enable sensing, the slot size is smaller than the packet size. Judging from the results presented, the sensing delay α has to be below 0.1 to obtain some throughput increase compared to spread slotted ALOHA.

A further sensing protocol for CDMA is hybrid CDMA/ISMA, which is described and analysed in detail in Reference [26, Chapter 10].

3.5.8 A Case for Reservation ALOHA-based Protocols

Alternatives for contention-based channel access were discussed at length above. These alternatives range from pure and slotted ALOHA, through elaborate collision resolution mechanisms, to protocols in which the channel is sensed before a packet is transmitted. It is now time to remind the reader that the interest here is in reservation protocols, where, once the channel is successfully accessed, terminals will obtain a reservation of a certain amount of resources for the transmission of user data.

Since a channel request message will, for most traffic types, be short compared to the user data to be transmitted in the following reservation period, resources required for access purposes will amount only to a small fraction of the total resources required. In fact, due to the fast fluctuations of the radio channel, individual transmission units, which include physical layer overheads such as pilot symbols or training sequences, must be rather short, and channel request messages fit in general onto a single such unit. In GSM, for instance, time-slots with a duration of 577 μs carry either normal bursts with a payload of 116 bits including overhead for FEC, or access bursts with a reduced 'gross payload' of 36 bits due to extended guard periods.

Even a short email message or an Internet datagram will normally have to be transported over several tens of bursts, longer messages over hundreds of bursts. By choosing appropriate reservation intervals, therefore, the overall efficiency of a reservation-based protocol is affected only to a small extent by the throughput that can be achieved on an access slot. Thus, it does not make much sense to complicate the system design to achieve an 'access throughput' of say 0.9 with ISMA, capture, and (possibly unrealistically) short inhibit delays, when values up to 0.6 can be achieved with little effort using slotted ALOHA and exploiting the capture effect. It cannot come as a surprise that S-ALOHA is indeed the preferred approach to channel access in mobile communication systems, being used in GSM (including GPRS), as well as chosen for both modes of UTRA. The most obvious choice of a multiple access protocol for cellular systems supporting packet-data traffic is indeed a reservation ALOHA or R-ALOHA protocol.

In the following, the terms *reservation ALOHA* and *R-ALOHA* are used in a broad sense, that is, rather than an individual protocol, they denote:

A family of protocols, in which mobiles place access requests on certain resources according to the S-ALOHA policy and information is subsequently transferred on reserved or dedicated resources.

The dedicated resources to be used are typically specified in an assignment or reservation message sent by the base station. Several reservation ALOHA protocols are described in Reference [26], including those proposed by Roberts in 1973, by Crowther *et al.* also in 1973, and by Binder *et al.* in 1975. With the above definition, the approach used in 'plain' GSM can also be considered as R-ALOHA, where reservations consist of entire 'circuits'. A large number of design choices can be made within the framework of R-ALOHA, which should, as far as there are no constraints imposed by the chosen physical layer solution, be optimised with regards to the services to be provided. Before examining in more detail requirements that multiple access protocols have to meet in a packet-traffic environment and identifying the available design options, *packet reservation multiple access*, an example of an R-ALOHA protocol, will be introduced and discussed. The reasons for this are manifold:

(1) several original concepts discussed and investigated in Chapters 6 to 9 are either investigated in the context of PRMA or consist of PRMA enhancements;

(2) for the further evolution of the GPRS MAC, which is based on an R-ALOHA protocol, PRMA concepts could potentially be adopted;

(3) it is easier to discuss design options based on an example, which can be used for illustration purposes.

With PRMA in mind, but applicable to any R-ALOHA protocol operating on a time-slotted channel, the resources required to place access requests will be referred to as *C-slots* for 'contention slots', while resources allocated for information transfer will be referred to as *I-slots*.

3.6 Packet Reservation Multiple Access: An R-ALOHA Protocol Supporting Real-time Traffic

3.6.1 PRMA for Microcellular Communication Systems

PRMA was proposed by Goodman *et al.* in Reference [8] as a technique for organising the flow of information from dispersed terminals to a central base station in microcellular communication systems. Thus it controls the *traffic on the uplink channel* in a given cell. The downlink traffic, on the other hand, consists of a continuous stream of packetised information. The protocol allows for statistical multiplexing of packetised voice and other packet-based traffic by assigning resources on the basis of packet spurts rather than circuits.

Goodman and his co-researchers published a number of papers on PRMA. In the context of cellular and cordless communications, PRMA is briefly described in Reference [1]. In Reference [7], PRMA is proposed, together with a cellular packet switch network, in order to meet some of the requirements of 3G wireless networks. In Reference [142], the efficiency of PRMA for voice-only traffic is investigated in detail by presenting various simulation results obtained for different sets of design parameters. In Reference [143], an equilibrium point analysis for PRMA with voice-only traffic is provided, whereas in Reference [144], mixed voice/data traffic is considered in a similar analysis, and some simulation results for this kind of traffic are presented. In Reference [145], the impact of slow and fast fading channels is assessed and it is suggested that packet headers should be error protected. Finally, in Reference [146], an integrated packet reservation multiple access protocol (IPRMA) is introduced, and simulation results for mixed voice and data traffic are presented for both conventional PRMA and IPRMA. It is shown that IPRMA improves the data traffic performance.

Numerous publications by other research teams on PRMA itself and possible extensions to or modifications of PRMA have appeared in the meantime. Some of the modified protocols will be described briefly in later subsections. Here, a selection of publications concerned with PRMA, as defined in Reference [8], is listed. In Reference [147], the performance of PRMA with fixed and dynamic channel assignment schemes is investigated. In Reference [148], PRMA performance is compared with that of CDMA through system simulations. In Reference [149], a Markov analysis for PRMA with voice-only traffic is provided, and in Reference [150] a similar analysis for mixed voice and data traffic. In Reference [151], an implementation of PRMA for a cordless system in office microcells using higher order modulation schemes is discussed. This includes considerations on speech quality impairments already published separately [152].

The PRMA protocol was introduced in Reference [8] as a combination of slotted ALOHA and TDMA. However, in accordance with the terminology introduced earlier, where TDMA is associated with the physical layer, we refer to PRMA simply as a variant of an R-ALOHA protocol (as actually also done in Reference [143]). Judging from the descriptions provided in References [26] and [104], PRMA is closely related to the reservation ALOHA protocol suggested by Crowther *et al.* for satellite systems. In order to cater for the delay sensitivity of the voice service, a packet dropping mechanism is added to the Crowther scheme. Furthermore, unlike in satellite systems, propagation delays are small in a microcellular environment, justifying the assumption of immediate acknowledgements, which is often made in the context of investigations on PRMA.

3.6.2 Description of 'Pure' PRMA

PRMA was initially designed for the transfer of packetised information pertaining to one of two different information categories, namely *periodic* and *random* information. Periodic information sources produce individual packets at regular intervals (i.e. periodically) during an activity phase or *packet spurt*. An example may be a voice terminal employing voice activity detection (VAD), which generates packets regularly during activity phases of the speaker, but stops packet generation during silence or idle phases. 'Periodic terminals' send packets first in *contention mode*, attempting to obtain a reservation. Once successful, they continue sending packets in *reservation mode*. We can therefore distinguish three states for a periodic terminal, IDLE, CONTENTION, and RESERVATION, as illustrated in Figure 3.8. Random information sources do not require resources in a periodic manner and transmit packets only in contention mode.

For the uplink channel of PRMA, N time-slots of fixed length, each able to carry a single packet, are grouped into frames in a manner such that the frame rate matches the packet arrival rate of the basic service (e.g. voice). The frames may be referred to

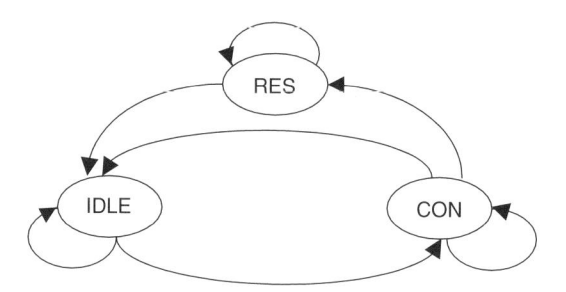

Figure 3.8 State diagram for PRMA, mobile terminal side

as 'TDMA frames' in order to avoid confusion with the term 'voice frames'. The slots are either available for contention (C-Slots) or reserved for the information transfer of a particular periodic terminal (I-Slots), as indicated by the base station.

At the start of a packet spurt, the terminal switches from idle to contention mode and tries to obtain permission to send a contention packet on the next available C-Slot by carrying out a Bernoulli experiment with some *permission probability* p_x. The outcome of such a Bernoulli experiment is one with probability p_x and zero with probability $1 - p_x$. In the case of a positive outcome (i.e. one), it transmits the first packet of the spurt. If this packet is received correctly by the BS, it will send a positive acknowledgement, which implies a reservation of the same slot (now an I-Slot) in subsequent frames. In the case of a negative outcome of the random experiment, a collision on the channel with another contending terminal, or unsuccessful transmission due to bad channel conditions, the contention procedure is repeated in subsequent slots, until the MS is successful, in which case it switches to reservation mode. Holding a reservation means uncontested access to the channel for the remainder of the current packet spurt on the assigned I-slot. If the MS leaves its assigned I-slot empty, this is interpreted by the BS as the end of the respective packet spurt, and it will revert the status of the corresponding slot to a C-slot.

This mechanism is illustrated in Figure 3.9, which shows four consecutive frames. For contending terminals, time 'progresses' slot by slot, and the diagram should be read row by row, while once a terminal gains a reservation, the time relevant for PRMA progresses vertically, since its reserved time-slots are all in the same column.

With delay sensitive, but loss insensitive services, such as voice communications, contention packets not successfully transmitted within a delay threshold value D_{max} are *dropped*, and the contention process is continued with the next packet in the spurt. The dropping mechanism will result in a so-called front-end clipping, that is, the first few packets of a packet spurt may be suppressed. This will obviously cause a deterioration of the perceived voice quality. Subject to the required quality, some maximum admissible *packet dropping ratio* P_{drop} will have to be specified. Goodman *et al*. [8] reported that the voice quality was satisfactory up to a dropping level of 1%, and this value is commonly adopted in literature concerned with PRMA performance assessment.

Note that the transition from CON to IDLE depicted in Figure 3.8 occurs only if a terminal has to drop an entire packet spurt, which should normally not happen. In the case of services that are loss sensitive, but delay insensitive, henceforth referred to as non-real-time (NRT) data services, packets are (within implementation constraints) never dropped and therefore this transition is not possible.

3.6 PACKET RESERVATION MULTIPLE ACCESS

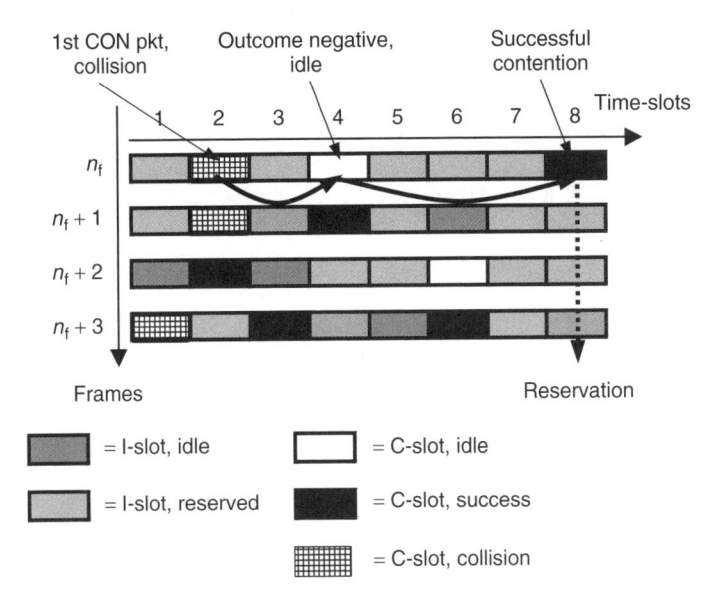

Figure 3.9 Illustration of PRMA operation

For PRMA to work in the manner described above, the base station must send an acknowledgement after each C-slot in a manner that it is available to the mobile stations before the next uplink slot starts. A positive acknowledgement (ACK) implies a reservation of this particular slot in subsequent frames for the successfully contending MS, while a negative acknowledgement (NACK) is sent following an idle C-slot or a collision. This approach, where a simple acknowledgement specifies the reserved resource, is also referred to as *implicit resource assignment*.

Due to 'packet-switched' transmission, every packet is preceded by a packet header. The header size required depends very much on the system and protocols considered and on what overhead is accounted for, i.e. only MAC headers or also higher layer headers. In publications associated with Goodman, the headers are assumed to contain the packet destination address (although this may not always be required), and the header length is normally 64 bits. Therefore, when the efficiency of PRMA is compared with that of a 'circuit-switched' TDMA system, the header overhead is explicitly accounted for. For instance, in References [142] and [146], with P_{drop} at most 1%, $M = 37$ and $M = 36$ conversations respectively can be supported on $N = 20$ time-slots, which would result in $\eta = 1.85$ and $\eta = 1.8$ conversations per time-slot or TDMA channel. However, accounting for the relative header overhead, which is larger in the second case due to a lower voice coding rate being considered, these values reduce to $\eta = 1.64$ and $\eta = 1.29$.

3.6.3 Shortcomings of PRMA

The suggested approach for the signalling of acknowledgements, namely immediately and in binary form (ACK or NACK), is very resource efficient in theory, but exhibits a number of drawbacks which will prevent it from being implemented in practice.

(1) From a *physical point of view*, immediate acknowledgement is assumed on the basis that the propagation delay is negligible in microcellular mobile communication systems. However, in a microcellular scenario, the combined processing delay required at the base station (to detect and decode information sent on the uplink and start signalling on the downlink) and at the mobile station (to detect and decode the ACK or NACK and prepare for sending another contention packet in the case of a NACK) will — at least from the current perspective — be dominant. Therefore, if immediate acknowledgement should really be required, for the transmission of a single bit on the downlink, there will have to be a gap on the uplink (i.e. between the end of a time-slot and the beginning of the following time-slot, see Figure 3.10). This would be rather inefficient. Furthermore, while 3G systems may well rely primarily on small cells, that is micro- and even picocells, larger cells will also have to be deployed, for instance umbrella-cells within a hierarchical cellular structure, as discussed in the previous chapter. In these cells, the propagation delays are non-negligible, hence even if the processing delays shrink towards zero owing to ever-faster CPUs, immediate acknowledgement would again require guard periods, and thus be inefficient. Using PRMA only in small cells and another protocol for large cells to overcome this problem is not convenient from a complexity perspective.

(2) From an *implementation point of view*, note that PRMA requires essentially a TDMA-based physical layer for the uplink channel, where information is transmitted in packets or bursts fitting into individual time-slots. Goodman suggested that acknowledgement bits should be multiplexed onto the continuous stream of downlink information, implying a downlink channel which is not necessarily based on TDMA. However, using different physical layer concepts for up- and downlink increases system complexity and thus cost. If, on the other hand, the same TDMA physical layer specified for the uplink is also used for the downlink, all information needs to be transmitted using an appropriate burst format. This would mean that acknowledgements have to be carried on bursts, possibly piggybacked onto user information. Depending on the implementation, for a mobile station to decode acknowledgements, it might have to process the whole burst, making immediate acknowledgements in most scenarios relevant for cellular communications conceptually impossible. We will thus normally have to live with delayed acknowledgements, that is, acknowledgements will only be available in time for the next but x slots, where x depends on the processing delay, the propagation delay, and the structure of the downlink channel. This may, depending on the approach adopted for access control, have a

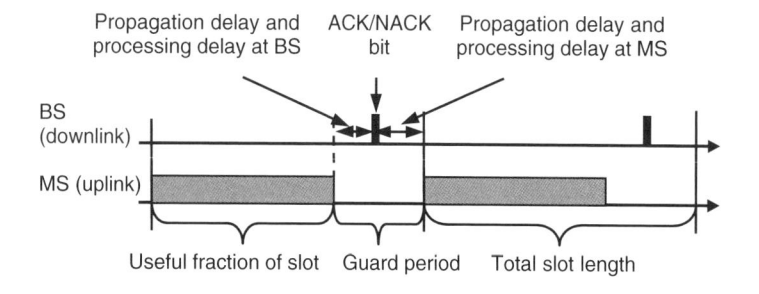

Figure 3.10 Impact of propagation and processing delays on acknowledgements

significant impact on the protocol performance, as discussed in detail in Chapter 8 for an enhanced PRMA protocol. There are exceptions, however, namely certain TDD scenarios outlined in Section 6.3, where immediate acknowledgements are possible.

(3) From a *content point of view*, binary acknowledgements are insufficient in the presence of the capture effect. This is because a positive acknowledgement not containing any further information is not sufficient to identify the successful terminal in the event of a packet being correctly captured in the presence of other contending packets. As a result, multiple mobile stations may try to transmit on the respective reserved resource. Requesting the base station to mirror in the acknowledgement a random number included by the mobile terminal in the contention packet, as in GSM, can provide a partial solution to this problem. This will necessarily consume several bits. The full GSM solution is discussed in Section 4.4, the one for GPRS, which is similar, in Section 4.10. Ideally, the mobile station is assigned a short, but in the considered context unambiguous identity, which can be mirrored back by the BS in the acknowledgement.

(4) On a somewhat different note, it should be reported that the implicit assignment strategy is only suitable if the traffic to be supported is of an on/off type requiring a constant, *a priori* known bit-rate during on-phases or activity periods. Means that are somewhat more elaborate are required for the support of variable-bit-rate services, as discussed in Subsection 3.6.4 and in Section 3.7.

Another shortcoming of PRMA as defined in Reference [8] is that of identifying the end of a packet spurt simply by leaving a slot idle. This is not an advisable approach on a channel subject to deep fades. The last packet in a spurt should be identified explicitly to avoid losing reservations prematurely, as for instance proposed in Reference [153].

Finally, some researchers voiced reservations regarding the fact that C-slots and I-slots share the same pool of physical slots. At high load, there may be intervals during which no C-slots are available, making it impossible to meet the Quality of Service (QoS) requirements of high priority services. Where this is a justified concern, one could for instance provide dedicated C-slots, as discussed in the following subsection.

3.6.4 Proposed Modifications and Extensions to PRMA

In the recent past, several extensions and modifications to PRMA have been proposed in the literature, to enhance the protocol performance or to adapt it to different environments or different services. In PRMA, contention packets carry as much user information as normal information packets, thus precious resources are wasted when collisions occur. One of the first fundamental modifications to PRMA was therefore to let contending users transmit only signalling information (e.g. the information normally contained in the packet header) to obtain a reservation. This allows C-slots to be split into several mini-slots to increase capacity for contention-mode users (see Figure 3.11). Mini-slotted protocols based on PRMA were for instance considered in References [47,135,154–157]. Because more than one user may successfully contend on a C-slot split into several mini-slots, implicit channel assignment with acknowledgements on a per-full-slot basis is not normally possible in this case. One solution is for the base station to send *explicit* channel

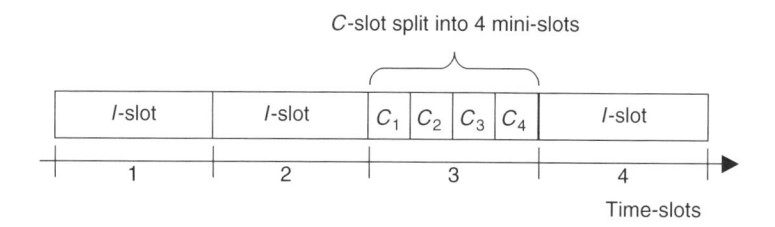

Figure 3.11 Frame/slot structure in mini-slotted protocols

assignment messages. In the extended PRMA (EPRMA) protocol proposed in Reference [47], not only mini-slots are considered, but also four carriers are pooled together to increase trunking efficiency for multiplexing.

In Reference [135], PRMA is classified as an *in-slot* protocol, where every slot may either be a C-slot or an I-slot. In contrast, in *out-slot* protocols, I-slots and C-slots are mapped onto distinct, separate time-slots. The PRMA derivative proposed in Reference [154] could be considered as a hybrid, since a minimum number of C-slots is always provided, while idle I-slots can also be used temporarily as C-slots. PRMA++ [46], another PRMA derivative that was developed during the RACE ATDMA project, is a pure out-slot protocol. Certain uplink time-slots always carry C-slots (which are referred to as R-slots in Reference [46]), while the remaining time-slots (except for one per frame reserved for paging acknowledgements) are always I-slots. The C-slots are paired with dedicated Acknowledgement slots (A-slots) on the downlink, in a manner that leaves enough time for propagation and processing delay, as shown in Figure 3.12. Notice also that the Paging Acknowledgement slot (PA) has a counterpart on the downlink, namely the Fast Paging slot (FP). Compared to pure PRMA, this arrangement of resources has the advantage of:

- guaranteeing a minimum amount of resources for random access;
- allowing the base station centralised control over resource allocation;
- ease of implementation owing to a fixed timing relationship between the random access slots on the uplink and the acknowledgement slots on the downlink;
- support of operation in macrocellular environments, where immediate acknowledgements are not possible.

This has to be paid for by reduced multiplexing efficiency. According to Reference [46], 28 conversations can be multiplexed onto 17 time-slots and 139 onto 71, while we report in Reference [49] 30 and 145 conversations respectively for PRMA with the same design

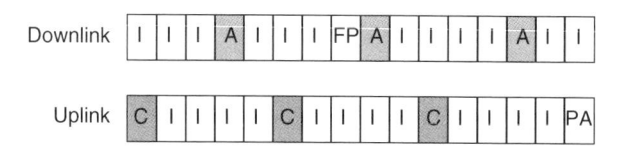

Figure 3.12 Frame structure in PRMA++

3.6 PACKET RESERVATION MULTIPLE ACCESS

parameters, the same maximum P_{drop} of 1%, but a shorter delay threshold D_{max} at which packets are dropped[11]. In Reference [135] on the other hand, where several frame-based in-slot and out-slot protocols are compared, it is found that out-slot protocols outperform in-slot protocols. However, this is mainly due to the fact that C-slots in the out-slot protocols considered are always split into several mini-slots, while those of the in-slot protocols considered are never split. As will be discussed later, most of the advantages PRMA++ provides compared to pure PRMA can also be provided with in-slot protocols.

Proposals were also made to better integrate different types of traffic with different service requirements. As an example, the integrated PRMA protocol [146] mentioned earlier supports random data traffic better than PRMA. A further protocol version building on IPRMA, which supports a mixture of voice, data, and video traffic, is examined in Reference [158]. In aggressive PRMA discussed for instance in Reference [159], voice users can contend on slots reserved by data users to provoke a collision which causes the data users to relinquish the respective reservation. This decentralised pre-emption scheme allows the voice dropping probability to be reduced at the expense of increased data delay, it thus increases the voice multiplexing gain in a mixed traffic scenario.

The centralised PRMA or C-PRMA protocol described in Reference [157], which is a hybrid reservation/polling protocol proposed in the context of CDPA (see Subsection 3.4.3), enables efficient service integration owing to explicit polling of every single packet. This requires additional overheads due to the command messages that precede every downlink slot and is, at least in the way proposed in Reference [157], constrained to small cells. Owing to dynamic scheduling, C-PRMA provides also for retransmission of erroneous packets even in the case of real-time traffic, since retransmissions can be scheduled immediately. This immediate retransmission capability is provided on both uplink and downlink. Another approach to service integration was proposed and investigated in Reference [160], namely a resource allocation strategy for voice/data/video integration, which is based on PRMA++, but provides conflict-free access for resource request messages of data and video services.

It should by now have become evident that implicit resource assignment is not suitable for the integration of services with different bit-rate requirements. Instead, explicit assignment messages will have to be sent, to specify precisely all the resources to be allocated. These messages must contain all the parameters identifying the channels to be allocated and must also indicate for how long they are allocated. Alternatively, as proposed for C-PRMA, every single packet can be scheduled and polled according to the specific service requirements.

Several extensions to PRMA are concerned with the provision of a TDD transmission mode, such as the so-called frame reservation multiple access (FRMA) protocol proposed in Reference [53] and an adaptive PRMA TDD protocol proposed in Reference [161]. FRMA is discussed in more detail in Chapter 6, as it provides the basis for the TDD mode of the MD PRMA protocol proposed in Chapter 6 and investigated in Chapter 8.

Finally, while access control in PRMA is based on fixed permission probability values, different types of adaptive control for pure or modified PRMA schemes have been suggested for instance in References [154,162–164]. Furthermore, collision resolution or splitting algorithms were studied for C-PRMA in Reference [157] and for the various frame-based in-slot and out-slot protocols compared in Reference [135]. For the PRMA

---

[11] In Reference [46], 18 and 72 time-slots per frame were considered for PRMA++. Ignoring the slot carrying paging acknowledgements on the uplink leaves 17 and 71 respectively.

TDD protocol proposed in Reference [161], service-specific permission probability values are computed adaptively based on an estimate of the current number of contending users. In Reference [52], we proposed to do fundamentally the same with MD PRMA, but both the algorithm used for backlog estimation and the approach to prioritisation discussed in Chapters 6 and 9 respectively, are different. Similar to pre-emption used in aggressive PRMA, prioritisation allows the voice dropping probability to be reduced at the expense of increased data delay.

3.6.5 PRMA for Hybrid CDMA/TDMA

In Reference [31] and a series of articles listed in the introductory chapter, we suggested using a PRMA-based protocol, here referred to as multidimensional PRMA, on a hybrid CDMA/TDMA air interface. Soon after, Dong and Li had similar ideas [33,34]. On a CDMA/TDMA air interface, one has to account for the fact that several packets can be transmitted successfully in a single time-slot. The available resources can for instance be modelled as time-slots subdivided into several code-slots, resulting in *code-time-slots* as the basic resource units (Figure 3.13). For the time being, assume that we are dealing with an in-slot protocol where each code-time-slot can be a C-slot or an I-slot.

In this case, the main and straightforward extension to PRMA required is that if a contending user obtains permission to access a time-slot in which several code-slots are available for contention, one of these code-slots has to be selected by the mobile station at random. Furthermore, in the acknowledgement, the relevant code-slot has to be identified. In Reference [33], Dong and Li restricted their investigations more or less to this basic scenario (for instance, they considered only fixed values for the permission probability p), and provided simulation and analytical results for the protocol. Our contributions, on the other hand, extend beyond this in that different approaches to dynamic control of p are considered, namely service-specific access control based on backlog estimation similar to the proposals in Reference [161] and load-based access control.

One interesting aspect in a CDMA scenario, which was not addressed by Dong and Li (nor, for that matter, by Tan and Zhang in Reference [35] for their code-slots-only protocol), is the impact of MAI. Rather than considering only the packet dropping probability P_{drop} as a protocol performance measure, the packet erasure rate due to MAI, P_{pe},

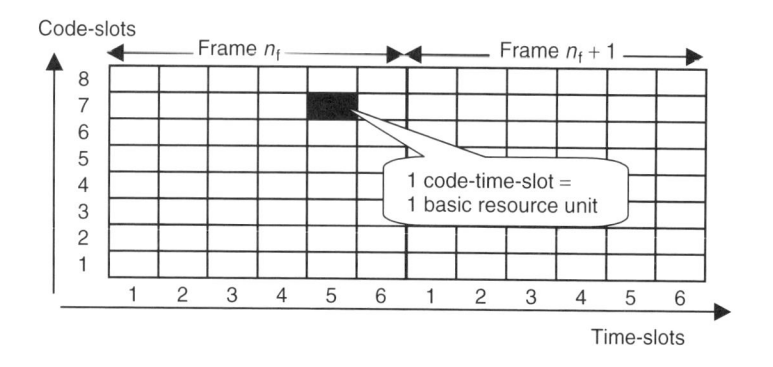

Figure 3.13 Time-slots subdivided into code-slots

needs to be accounted for as well. The overall performance measure for real-time traffic is now the packet-loss ratio P_{loss}, which is $P_{drop} + P_{pe}$, as a function of the traffic load.

With in-slot protocols, the issue is how to protect reservation-mode users from excessive MAI stemming from contending users, since these two user categories are allowed to transmit at the same time (e.g. in the same time-slot). For this purpose, we proposed and explored the concept of load-based access control in References [28–31], to which additional contributions were provided by Mori and Ogura [36,37], Wang et al. [40], Lee et al. [42], and Chang et al. [43].

Another way of protecting reservation-mode users from MAI is simply to separate I-slots and C-slots in time, that is, certain time-slots carry only C-slots and others only I-slots. This looks very similar to PRMA++, so we could consider this type of out-slot version of a hybrid CDMA/PRMA protocol to be its '++' version. Such a '++' version of the protocol was investigated in detail in Reference [165]. As in the case of PRMA versus PRMA++, everything else being equal (e.g. not considering mini-slots in both cases), we found that the out-slot version of the protocol reduces the multiplexing efficiency in terms of the number of users sustained at a P_{loss} of 1%. In fact, it is somewhat worse in the hybrid CDMA/TDMA case due to the fact that entire time-slots with multiple code-slots have to be set apart for contention purposes, reducing the granularity with which resources can be split between C-slots and I-slots. On the plus side, due to the fact that the number of users accessing time-slots with I-slots can be controlled tightly, the MAI can be controlled as well (how well depends on the degree of channel fluctuations), allowing P_{loss} to be reduced at low load.

There can be reasons dictated by the physical layer that may tip the balance towards out-slot protocols. For instance, in an intermediate version of the UTRA TDD proposal [90], the possibility of splitting a time-slot used exclusively for contention into two half-slots was considered, thus effectively allowing for mini-slots only if an out-slot or a hybrid protocol is adopted, but not with a pure in-slot protocol. Additionally, there could be problems associated with joint detection applied in UTRA TDD, if contending users access the same time-slots as reservation mode users (see discussion in Section 5.3), which may point again towards out-slot protocols.

A hybrid between an out-slot and an in-slot protocol was examined in Reference [42]. In this protocol, some time-slots may only be accessed by users holding reservations, other time-slots are accessible to both contending users and users holding a reservation, and one time-slot is used exclusively for contention. This last time-slot is split into several mini-slots. A pure out-slot version of a hybrid CDMA/PRMA protocol termed 'Joint CDMA/NC-PRMA' is proposed in Reference [45], where the time-slot used for contention is split into several very short mini-slots, each mini-slot carrying several signatures or tones. This protocol variant is a so-called *non-collision* protocol (hence NC-PRMA). Every admitted voice or video user has its assigned signature and mini-slot, so that users contending for resources do not collide on the request mini-slot. However, neither is a solution provided for non-real-time packet-data users, for which no dedicated signatures are available, nor for initial access, which would have to occur on separate resources. Also, this 'non-collision' feature comes at the expense of significantly increased system complexity, the mini-slots are based on a completely different physical layer solution than the normal slots. Apart from transmitter and receiver complexity implications, this also means that cells have to be synchronised, to ensure temporal separation of request slots from traffic slots across cells.

Building on the in-slot protocol they investigated in Reference [33], Dong and Li proposed another mini-slot protocol in Reference [34], which is a hybrid between in-slot and out-slot protocols. Certain code-time-slots are reserved for the transmission of short request packets containing no user information. The lower data-rate required for these packets compared to regular packets means that the spreading factor can be increased, which in turn reduces MAI, allowing these C-slots to be split into several 'mini-code-slots' (as opposed to 'mini-time-slots').

Expanding further on the topic of temporal separation between various classes of code-time-slots on hybrid CDMA/TDMA interfaces, Akyildiz *et al.* [166] proposed a protocol for hybrid CDMA/TDMA which does 'BER scheduling'. This means that users are assigned to time-slots in a manner that services with similar requirements in terms of bit-error-rates or packet erasure rates share the same time-slots. Recall Figure 1.7 shown in Section 1.4. If there are different services with different error-rate requirements, the number of users allowed to access a single time-slot without violating the service requirements will be determined by the service needing the lowest error rate. Say a time-slot can serve only four users if one of them is a video user, but nine users if all of them are voice users. To maximise the capacity, video users should be grouped together on separate time-slots and so should voice users. The basic concept was already proposed in Reference [18] by Kautz and Leon-Garcia, there termed 'service partitioning'.

What BER scheduling boils down to is to ensure that every signal is received at an SINR which just satisfies the required error performance. Instead of BER scheduling, this could also be achieved through differential power control or through variable-rate FEC coding. Akyildiz *et al.* [166] compared their BER scheduling protocol with a system for multimedia traffic proposed in Reference [167] employing differential power control and found that, under the conditions considered, BER scheduling performed better.

One problem with BER scheduling in a cellular environment is the matter of fluctuating intercell interference, which should ideally also be taken into account for the scheduling process. In fact, BER scheduling itself will amplify interference fluctuations experienced by other cells, pointing towards the need for cell synchronisation and co-ordinated scheduling across cells, which has considerable complexity implications.

3.7 MAC Requirements vs R-ALOHA Design Options

3.7.1 3G Requirements Relevant for the MAC Layer

A long list of initial 3G requirements, namely those specifically established for UTRA, were already discussed in Section 2.3. These pertain to one of four groups, requirements relating to complexity and cost, operational requirements, requirements on efficient spectrum usage and those on bearer capabilities. In the following, we will attempt to identify which of these requirements are relevant for the MAC layer, and what should be taken care of when selecting a multiple access protocol to help meeting these requirements. More precisely, as we have already seen why R-ALOHA protocols are particularly suitable for cellular communications, given these requirements, we should identify relevant R-ALOHA design options and choose the appropriate ones.

One might object that it does not make sense to identify MAC layer requirements based on the initial set of UTRA requirements relevant for the first release of UMTS, now that 3G systems are evolving to support IP multimedia services, and 4G systems are increasingly

3.7 MAC REQUIREMENTS VS R-ALOHA DESIGN OPTIONS

being talked about. However, from a MAC perspective, the main additional requirement in this respect, on top of the relevant UTRA requirements listed in Section 2.3, would appear to be that of efficient support of IP-based real-time and non-real-time services. This is to some extent already captured by the second UTRA requirement on bearer capabilities. In any case, we will certainly keep this requirement in mind in the following discussion, and the support of different types of packet data is anyway our main area of concern.

It turns out that almost all of these UTRA requirements have an impact on the choice of the MAC layer. As far as requirements on complexity and cost are concerned, R-ALOHA protocols are fairly simple to implement, as proven by their use in existing cellular communication systems and by the choice of an R-ALOHA scheme for GPRS. They should certainly not constitute insurmountable obstacles for the provision of small and cheap handsets, and easy infrastructure deployment. Also, by respecting certain restrictions when designing the protocol, battery saving measures can readily be incorporated[12]. Regarding operational requirements, note that dynamic channel or resource assignment, which facilitates automatic resource planning, was for instance discussed for PRMA in Reference [147].

When it comes to efficient spectrum usage, the relevant requirements appear to be:

- allowing for efficient operation when supporting mixed services;
- allowing for efficient operation both in paired and unpaired frequency bands; and
- being able to operate in various different cell types.

In order not to waste resources in one link direction, it should be possible to split the total resources available for uplink and downlink channels in an asymmetric manner according to the asymmetry of the total traffic load. This is rather difficult in systems using frequency-division duplex, as these systems do not easily lend themselves to operation in unpaired frequency bands[13]. To use resources efficiently under asymmetric traffic load (e.g. due to a large fraction of downlink traffic generated by Web browsing), and to support operation in unpaired frequency bands, the MAC layer should support time-division duplexing (TDD).

Regarding the efficient use of resources with mixed services in general, a distinction between blocking-limited and interference-limited systems is needed. For blocking-limited systems, a MAC strategy suitable for common and shared channels, where resources are only assigned to users in the intervals in which they actually have packets to send, is in general more efficient for packet-data traffic than the use of dedicated channels. The achievable resource utilisation with R-ALOHA protocols depends on the amount of resources on which the protocol operates (this determines the trunking or multiplexing efficiency), on how the split into resources available for access and those used for information transfer is performed, on the efficiency of the access mechanism, and that of the resource allocation algorithm. In interference-limited systems, matters become a little bit

---

[12] This will also depend on the organisation of the traffic flow on the downlink. From a battery saving point of view, to maximise the sleep mode ratio, it would be ideal if the mobile terminals knew exactly when to expect relevant signalling messages and downlink traffic.
[13] An option being discussed is referred to as 'variable duplex spacing'. An uplink carrier would be paired with more than one downlink carrier. Due to the large carrier bandwidth in WCDMA and spectrum constraints, the granularity of potential asymmetry ratios would be quite limited, though.

trickier. Rather than in 'resource utilisation', which is difficult to define, we are interested in spectral efficiency. Certain types of packet-data traffic, e.g. voice, might be supported more efficiently on dedicated channels, as discussed in Chapters 10 and 11. As in the case of GSM, one could call a protocol using a random access scheme to set up a dedicated channel still an R-ALOHA protocol. Burstier types of traffic, on the other hand, are likely supported more efficiently using a 'proper' R-ALOHA protocol on common or shared channels, so similar considerations as in the case of blocking-limited systems apply.

Finally, with respect to the required support of different cell types, it would be advantageous to have a single protocol being able to cater for all cell types ranging from pico- to macrocell. In the case of R-ALOHA protocols, this will mainly have an impact on the assumptions that can be made regarding acknowledgements.

There remain the requirements related to bearer capabilities, which are somewhat rephrased compared to the previous chapter and ordered according to their subsequent treatment. 3G systems must:

- support both circuit-switched and packet-switched services;
- provide seamless handovers;
- be able to satisfy a wide range of bit-rate and QoS requirements;
- support variable-bit-rate real-time services; and
- be able to schedule (and pre-empt) bearers according to priority.

A MAC solution that supports packet-switched services efficiently should normally provide all the means needed to support circuit-switched services as well. From a MAC point of view, the main requirement for circuit-switched services, assuming that they are served on dedicated channels, is the provision of a suitable access mechanism, which is provided by R-ALOHA protocols. On a time-slotted system, it may be open to debate whether dedicated resources for *initial* random access are necessary, to start a session in the packet-data case, or to set up a circuit-switched call, on top of the C-slots available to place requests for transmission of individual packets or packet spurts. This may depend on physical layer capabilities and constraints. For instance, it is possible to synchronise contention packets tightly to slot-boundaries during packet-voice calls, such that they can be treated in the same manner as normal packets. On the other hand, request packets sent to set up a session or call are rarely synchronised, and may have to be treated differently on separate resources.

Regarding handovers, Goodman suggested in Reference [7] that the combination of an R-ALOHA protocol providing a fast access mechanism (such as PRMA) with a distributed packet-based architecture, where decisions are localised and thus signalling exchanges and handover delays minimised, would be beneficial. However, there are also drawbacks associated with such decentralised handovers, as discussed in Section 1.2. One could also look at this issue from a different angle. We are trying to identify suitable multiple access protocols for common and shared channels. We could argue that seamless handovers apply mostly to dedicated channels, hence this requirement is of little relevance here. In fact, for instance on UTRA FDD, the so-called soft handover type, which implies simultaneous connections to multiple cells and is therefore particularly seamless, is only supported on dedicated channels, which is one more reason for using those for voice-type traffic in such a WCDMA-based system. Abstracting from CDMA, the need for seamless handovers depends on the services to be supported on the shared channels. If voice is to be supported through a PRMA-like protocol, then seamless handover must be possible on shared channels as well. Probably the most important requirement to be satisfied by the multiple access protocol in this respect, irrespective of the precise nature of the handover algorithm to be used, is that of a fast signalling capability. R-ALOHA protocols can provide fast access to resources required for signalling, particularly when using access prioritisation. Whether signalling then interrupts an ongoing user traffic stream (as in GSM) depends less on the protocol as on the capabilities of radio transmitters and receivers, e.g. whether they support signalling streams in parallel to user data streams.

The remaining three bullet points all deal with QoS or service requirements in general. These deserve treatment in a separate subsection.

3.7.2 Quality of Service Requirements and the MAC Layer

QoS requirements are normally specified in terms of reliability (e.g. bit-error-rate), data-rates and/or delay performance. Broadly speaking, services can be classified in terms of loss sensitivity and delay sensitivity. Voice for instance is very sensitive to delay, but only moderately sensitive to loss of some data. Video *communication* (or *interactive* video, as opposed to downloading or streaming of video films) is sensitive to both delay and loss. Delivery of certain types of data services, which were referred to earlier as non-real-time (NRT) services, is insensitive or only moderately sensitive to delay, but loss sensitive (delivery of email messages for instance is, at least with respect to the time scales relevant for the MAC, almost completely insensitive to delay).

3.7.2.1 Considerations on Delay Sensitivity

As the MAC layer is responsible for access arbitration, it will determine the *access delay* experienced which, together with the resource allocation delay[14] and the delay for the transmission of a certain amount of data on the allocated resources, determines the total delay performance of the different services. In the case of NRT packet-data services, QoS requirements may be defined in terms of the total admissible delay for the transmission of a packet with a defined length (an example is GPRS, see Section 4.7). In this case, the amount of resources that need to be allocated depends on the access and allocation delays already experienced. The faster the access and allocation procedures, the lower the transmission data-rate required to meet a certain delay limit.

Such a trade-off does not exist for real-time services. In the case of a packet-voice source which generates on–off traffic patterns, for instance, the transmission data-rate, and thus the amount of resources required to be allocated, is determined by the nominal bit-rate of the voice codec and the protocol overheads. The delay sensitivity relates to the delay that can be tolerated until these resources are allocated and ready for use. *Fast access mechanisms* must be provided for such services, furthermore, it might be necessary to *drop data units* that exceed a certain delay limit.

In the case of real-time variable-bit-rate (VBR) packet-video or multimedia streams, similar considerations apply, that is, access mechanisms must be fast and packets may be

---

[14] The resource allocation delay is the delay between reception of a resource request by the network and the instant at which the allocated resources are ready for the requesting terminal's use.

dropped (although interactive video is normally considered to be more sensitive to packet loss than voice communications). The resources required, however, vary with time, and the delay sensitivity relates mainly to the admissible delay from the instant at which the required amount of resources increases to the instant at which the additionally required resources must be provided. With video traffic, off or silent phases typical for packet-voice traffic subject to voice activity detection are less likely to occur. This means that during the entire call, a terminal will almost always have at least some resources assigned, which could potentially be used to signal requests for further resources. Therefore, the focus shifts from enabling efficient access to resources to algorithms that manage efficiently the pool of available resources and divide these between the competing users according to the individual service requirements. Such algorithms can be fairly complex and are a research topic of their own (see for example Reference [160]).

3.7.2.2 Reliability

The major burden for providing a certain reliability is carried by the physical layer, through FEC coding, and by the RLC layer, through automatic repeat request (ARQ) mechanisms, although not necessarily for all services. ARQ mechanisms may additionally also be provided by layers above the RLC. The MAC layer can also have an impact on reliability, for instance if there is a dropping mechanism.

3.7.2.3 The Issue of (Channel) Access Control

It was already pointed out earlier that, at least from a throughput point of view, it does not make sense to invest much effort in elaborate splitting algorithms for reservation-based protocols. Also, while the delay performance of FCFS algorithms is somewhat better at high load than that of S-ALOHA stabilised through access control, it should be noted that, for delay sensitive services, the load on the access resources or C-slots must be kept low in either case, and the main factor affecting the access delay will be the appropriate provision of C-slots. We will therefore focus on *access control* in the sense of control of the permission probability, that is, the probability with which a contending terminal may access a C-slot. From the above, it should be clear that by access control, we mean controlling access to the channel for individual reservation requests (e.g. through a permission probability, as discussed in the context of S-ALOHA and of PRMA), in order to ensure fair and efficient usage of resources. This is different from access or admission control based on security policies, for example.

With respect to the quality of service, one could ask whether access control should be used only to stabilise S-ALOHA and to keep the access delay as low as possible, regardless of the service considered, or whether access control should be service dependent. The latter would allow prioritisation of certain services already during the random access phase.

At first glance, since information on priority included in higher layer protocol headers is normally not interpreted by the MAC, the common approach appears to be not to provide service-dependent access control. It is then up to the radio resource control entity, for instance, to provide enough resources to a user that needs to transfer a packet quickly (that is, to schedule bearers according to priority). However, consider for instance a train traffic control system, which is a possible GPRS application. In such a system, short packets need to be transmitted quickly to ensure safe operation of trains. Given these stringent delay requirements, access delay becomes crucial and is normally the dominant factor of the total transfer delay due to the shortness of the packets. The RRC cannot make good for access delay values larger than the total admissible delay.

If service-specific access control is not possible, guaranteeing a low access delay for train control messages means guaranteeing low access delays for all services supported by the system in question. This in turn requires provision of abundant access resources (i.e. C-slots in the context relevant here), which means inefficient resource utilisation. If, however, service-specific access control is applied, and the system carries a mix of high-priority users and users that request delay insensitive services, it is possible to trade low access delay for high-priority users against higher access delay for low-priority users. Depending on the exact traffic mix, delay constraints of high-priority users can even be met at moderate-to-high total C-slot traffic load.

It is therefore suggested that *priority-class specific access control* should be provided, as is the case in GPRS. This allows prioritisation not only to be performed through admission control and resource allocation control, but also during the random access phase. There are several means of letting the MAC obtain priority information. In GPRS, for instance, users subscribe to certain QoS profiles, which may include the chosen service delay-class. Priority may be negotiated at session set-up, and thus be selected on a per-session basis. In protocols like PRMA, different traffic types (e.g. random and periodic traffic) are defined, and information on the traffic type, which is relevant for the access priority, must be included in the MAC protocol headers anyway. Finally, if priority information is contained in higher layer overheads (for instance in the type-of-service field and the traffic-class field of packet headers as in IP version 4 [168] and version 6 [169] respectively), this information can in theory be made known to lower layers through layer primitives (the means for interaction between adjacent layers). If IP header compression is performed over the radio link, then the IP headers must be interpreted in any case before transmission.

For further information on random access prioritisation through probabilistic access control, refer to Section 4.11, Chapter 9 and Section 10.3. For completeness, we report here that prioritisation at the random access is possible without controlling the permission probability in the service-specific manner suggested here and, for instance, in Reference [161]. An alternative approach is to provide separate access resources to the different priority classes, that is different classes of C-slots, as proposed in Reference [160]. A similar trade-off as with access control can then be made through appropriate scheduling of such C-slots. For high priority users, both the delay until occurrence of the next C-slot should be kept low and the normalised traffic on the respective class of C-slots should be low to ensure high success probability. By contrast, both the normalised load on low-priority C-slots can be higher and their timely occurrence is less critical. However, access control is required anyway to stabilise S-ALOHA, and service-specific access control can easily be combined with a stabilisation algorithm, as will be shown later. This appears therefore to be the more elegant approach.

3.7.2.4 The Issue of Resource Allocation

Whether the MAC layer is fully responsible for resource allocation depends on the system considered. In PRMA with implicit resource assignment, for instance, it is, and the resource allocation delay is essentially zero (once the resource is accessed successfully, it will automatically be assigned to the accessing user). However, in cellular communication systems such as GSM and UMTS, a *radio resource control* entity situated at the third OSI layer is responsible for resource allocation. Normally, the MAC layer will have to request resources from the RRC before assigning them to the requesting user. In this case, the MAC can only control the access delay, but neither resource allocation delay nor transfer delay. Also, sophisticated resource allocation algorithms required to support mixed traffic types may be handled by a separate entity, as shown in Figure 3.3, to ease the burden of the MAC. All these considerations point to the requirement for the MAC to have centralised control over those resources that can be allocated for information transfer, which, as just outlined, is not the case with the original PRMA protocol.

Some of the requirements discussed above are conflicting and a suitable trade-off will have to be found. For instance, it is difficult to ensure high resource utilisation and to support services that are both delay and loss sensitive at the same time. To provide the desired QoS, such services require essentially the reservation of resources according to their peak bit-rate (or at least close to it) for the duration of the respective call.

3.7.3 A few R-ALOHA Design Options

In the following we identify, without claiming completeness, design options available in the framework of R-ALOHA protocols. Some of these options were initially identified with time-slotted air interfaces in mind, but it turns out that they are equally relevant in a 'pure CDMA environment'. With one exception, hybrid reservation/polling protocols such as C-PRMA are not considered, since they complicate the system design, impose constraints on the cell size, and introduce overheads in the shape of poll commands. These have to be heavily error protected to ensure correct and efficient protocol operation, such that any gains which may be obtained through dynamic scheduling would be offset at least partially. The exception is GPRS which, although applying an R-ALOHA protocol, provides also means for scheduling in the shape of the Uplink State Flag (USF).

(1) *Split of channel resources* between those used for contention (e.g. C-slots) and dedicated or reserved resources (e.g. I-slots):

— In a *static split*, dedicated resources are put aside for contention, the remaining resources are used for information transfer. In a TDMA environment, these protocols may also be referred to as *out-slot* protocols. An example is PRMA++. If a hybrid CDMA/TDMA air interface is considered, a static split could mean that one or a few code-slots per time-slot are set aside for contention, in which case there is no temporal separation between C-slots and I-slots. Alternatively, a temporal separation could be achieved by splitting the total resources at the level of entire time-slots.

— In a *dynamic split*, also referred to as *in-slot* protocol on air interfaces featuring a TDMA element, any resource currently not used for information transfer is available for contention. Pure PRMA for instance is an in-slot protocol.

— In a *hybrid approach*, a minimum level of resources is put aside for C-slots (these might include special slots for initial access), while the remaining resources carry normally I-slots, but can also carry C-slots. TDMA examples include the GPRS MAC solution (which may also operate in a purely dynamic

3.7 MAC REQUIREMENTS VS R-ALOHA DESIGN OPTIONS

fashion), and the PRMA derivative proposed in Reference [154]. Pure CDMA air interfaces are typically also hybrid in this respect, dedicated codes are used for random access, but contending users are not separated in time from users holding reservations, meaning that the interference budget is shared between the two.

(2) *Type of resource assignment:*

— *Implicit resource assignment*: if an MS contends successfully on a given resource, it can use the same resource (for instance a code-time slot, but also simply a code in CDMA) in the future in reservation mode. The base station needs only to send a simple acknowledgement to notify the MS of the successful reception of its resource request, which assigns implicitly this resource unit to the MS. However, for reasons discussed in Subsection 3.6.3, this acknowledgement must contain an unambiguous MS identity rather than a single ACK bit as in the case of 'plain' PRMA.

— *Explicit resource assignment*: upon reception of a resource request message, the base station decides whether, when, and how many resources to allocate. It has to specify these resources explicitly in an assignment message.

(3) *Duration of the reservation phase*:

— *Packet or packet spurt*: once an MS has contended successfully for resources, it may keep the allocated resources until its packet spurt is successfully transferred or, in more general terms, its transmission queue is empty. The base station may have the possibility to pre-empt certain users, if it needs to re-allocate resources to higher priority users. This pre-emption mechanism could for instance be in the shape of the *uplink state flag* used in GPRS. Such a USF allows the base station to multiplex several mobiles dynamically onto one reserved resource [54].

— *Allocation cycle*: the reservation period for resources is limited to the duration of an allocation cycle. Its length may be service-specific. After the expiration of an allocation cycle, the MS has to request further resources, either by contending again for resources or by piggybacking resource extension requests onto ongoing information transfer packets prior to expiration of the current cycle (see next design option). Use of allocation cycles was proposed for NRT data in TD/CDMA [90].

(4) *Signalling* of modified resource requirements or extension requests during the reservation phase:

In the case of VBR real-time data traffic, resource requirements vary during a call and the amount of allocated resources may need to be upgraded or downgraded. This needs to be signalled to the base station. The same applies for resource extension requests discussed above. When no dedicated control channel is available permanently for this purpose, the alternatives are:

— *Piggybacking*: to avoid waste of resources due to collisions in contention mode, these signalling messages are piggy-backed onto information packets, thereby using reserved resources for signalling.

— *No piggybacking*: resource modification or extension requests are sent in contention mode in the same way as initial resource requests.

(5) Implementation of *access control*:

✓ *Centralised control*: permission probability values for different services are calculated and signalled by the base station (e.g. GPRS MAC solution).

— *Decentralised control*: relevant permission probability values are calculated by each MS individually, based on the feedback, e.g. in terms of idle slot, success slot, or collision slot (which can either be determined by the MS or signalled by the base station).

(6) Content of *contention packets*:

✓ *Signalling and user data*: contention packets use the same format as normal packets and contain the same amount of user data as normal packets. Pure PRMA is an example. On time-slotted air interfaces, in large cells, synchronisation or *timing advance* during silence-periods may have to be maintained for this to be possible (using for instance mechanisms adopted in GPRS, see Sections 4.9 and 5.3).

— *Signalling only*: the contention packet contains only signalling information required for the base station to allocate resources according to the service requirements. On time-slotted air interfaces, provided timing advance is available, C-slots may be split into several mini-slots, since such contention packets should be relatively small. If timing advance is not available, for large cells, long guard periods are required. As a consequence, C-slots cannot be split into mini-slots, even if only a limited number of signalling bits is conveyed in the contention packet.

3.7.4 Suitable R-ALOHA Design Choices

3.7.4.1 Split of Channel Resources and Type of Resource Assignment

Because of the higher multiplexing efficiency, a dynamic split of channel resources is in general preferred. However, depending on factors such as the nature of the physical layer to be used, there may be reasons for adopting a static split. In most cases, this would be to provide temporal separation between contention resources and reserved resources, for instance if this is required to introduce mini-slots, or for other reasons discussed in Section 3.6.5 for the specific case of hybrid CDMA/TDMA interfaces.

The resource assignment should normally be explicit to enable centralised control over resource allocation and to allocate the appropriate amount of resources in the case of high-speed data services. In Reference [154], for a protocol exhibiting a hybrid resource split, which was designed for a TDMA-based physical layer, it was proposed to optimise the distribution of C-slots within a TDMA frame to minimise the access delay. If resource assignment is explicit to maintain centralised control, this is also possible when a dynamic split is applied[15]. Optionally, two variations to resource split and assignment may be considered in a TDMA environment:

---

[15] One might argue, however, that while the quantity of C-slots obviously has an impact on access delay, with the short TDMA frames considered, their exact distribution within a frame is irrelevant.

- adoption of a hybrid resource split to provide a guaranteed minimum amount of C-slots per TDMA frame. These C-slots can be mapped onto specific resources (for instance some RACH-like channel for session set-up or set-up of circuit-switched calls). Alternatively, at the expense of blocking some low-priority resource requests, when C-slots are about to run out, they are simply not converted to I-slots anymore;

- use of implicit resource assignment for packet-voice users to simplify downlink signalling, provided that the base station does not have any particular reason to allocate a different resource unit to such a voice user than the C-slot in which it contended successfully.

3.7.4.2 Duration of the Reservation Phase

As far as the duration of the reservation phase is concerned, note first that the base station should normally refrain from pre-empting real-time users to guarantee that their QoS requirements are satisfied. This is the reason why no such pre-emption feature exists for voice in pure PRMA. The network may decide not to admit a voice user, but once it is admitted, the network should make every effort to provide sufficient QoS (this is also why implicit resource assignment can be appropriate). For the same reason, allocation cycles do not make sense for real-time services. On the other hand, either the allocation cycle concept or a pre-emption mechanism should be adopted for NRT services, such that resources assigned to low priority NRT services can potentially be reallocated to higher priority (e.g. real-time) users.

The choice between these two approaches will be influenced by several factors. On the one hand, it appears that the pre-emption approach is simpler. It requires less signalling overhead and base station CPU resources, since in terms of resource allocation, actions by the base station are only required when resources are about to run out. Pre-emption can even be decentralised (that is, initiated by the mobile terminal), as is the case in the aggressive PRMA scheme mentioned earlier. On the other hand, the allocation cycle approach is more robust in that an MS not having been granted an extension request will certainly cease to use resources upon expiration of the allocation cycle, allowing the base station to reallocate these resources immediately to another MS. This may not be the case after pre-emption, since due to channel impairments, the pre-empted MS might not receive the pre-emption message and so continues to use the resource. To avoid conflicts in such cases, timers can be introduced that must expire before the base station is allowed to re-allocate the respective resource.

3.7.4.3 A Case for Centralised Access Control

To minimise signalling overhead on the downlink, obviously fully decentralised access control would be desirable. However, in a mobile communications environment, feedback evaluated at the mobile station is extremely unreliable, and would complicate handset implementation. To improve feedback reliability, in the case of backlog-based access control, the base station could signal explicitly idle, success and collision slots, while access control could still be carried out in a decentralised manner. However, this would require mobile stations to monitor the feedback continuously to maintain a reliable backlog estimation, which would conflict with the requirement of saving battery power during inactivity phases. In addition, at least in the 'pre-software-radio-era', this would also require the estimation and prioritisation algorithms to be specified fully in the standards.

It is therefore advantageous to let the base station signal permission probability values rather than feedback.

Such centralised access control, while consuming some signalling resources on the downlink, provides the operator with full flexibility regarding choice of access control scheme (load-based, backlog-based, or a combination thereof) and prioritisation algorithms. Furthermore, it is also advantageous in terms of battery life. In Section 4.11, the required amount of signalling resources is examined in more detail, and it is found that it is quite moderate.

3.7.4.4 On the Content of Contention Packets

On time-slotted air interfaces, if mini-slots are not provided, one could argue that contention packets might well contain user data. However, note that user data is normally protected against fast fading by FEC coding and interleaving over several packets (or bursts), e.g. over the duration of a voice frame, while for contention packets, interleaving is only applied within the packets. Therefore, extra FEC protection is required, which reduces the available payload for user data. If for instance parts of a voice frame were to be transmitted in the access burst, the first voice frame would have to receive special treatment, complicating system design to an extent not justified by the little extra capacity which may be gained. It is therefore often more convenient to restrict the content of contention packets to the necessary signalling data. Ideally, this should include an MS identity which is unambiguous in the relevant context. In GSM and GPRS, this is not the case, requiring extra precautions to be taken when allocating resources, as explained in Sections 4.4 and 4.10.

3.8 Summary and Scope of Further Investigations

To be able to compare and assess different approaches to multiple access in cellular communications, first some relevant terminology was discussed and a case was made for splitting the multiple access problem into provision of a *basic multiple access scheme* such as CDMA or TDMA, and a *multiple access protocol*. The basic scheme is normally associated with the first OSI layer or physical layer, and the multiple access protocol is situated at the MAC layer, which is the lower sub-layer of layer 2.

After a few considerations on basic multiple access schemes, several multiple access protocols known from literature were discussed, with emphasis on their suitability for mobile communication systems. A case was made for multiple access protocols based on reservation ALOHA. PRMA was introduced, its restrictions examined, and modifications and extensions proposed in the PRMA literature reviewed.

Some of the design options available in the framework of R-ALOHA were identified, and it was suggested:

- that C-slots and I-slots share the total resources dynamically, possibly with a guaranteed minimum number of C-slots;
- to provide the possibility of assigning resources explicitly (an implicit mechanism may be provided optionally for certain services such as packet-voice);
- to use service-tailored algorithms to determine the amount of reserved resources and the duration of reservations (if this functionality is not catered for by a separate

3.8 SUMMARY AND SCOPE OF FURTHER INVESTIGATIONS

resource allocation entity), including pre-emption mechanisms or allocation cycles for NRT data;

- to use piggybacking for signalling, where useful and possible;
- to use centralised access control; and
- to use a dedicated burst format for access bursts which carries only request messages, thus to provide mini-slots for contention, if the physical layer allows it.

In the way they are phrased, the first and the last item in this list apply only when the air interface considered features a TDMA element (matters are somewhat different in a 'pure CDMA environment', the concept of mini-slots, for instance, becomes somewhat fuzzy). In Chapters 6 to 9, a MAC strategy based on an evolved PRMA protocol, assuming a TDMA element, is proposed and investigated, which caters conceptually for all these requirements. However, results are only provided for a subset of the listed design options. The main focus is on access control, and it is assumed that resource allocation for other than basic services is handled by a separate entity. It is therefore sufficient to consider on/off type services requiring only one resource unit or I-slot during activity phases, and investigations are restricted to implicit resource assignment. The traffic scenarios considered include only services that are either delay insensitive (i.e. NRT data) or (moderately) loss insensitive (i.e. packet-voice), but not services sensitive to both loss and delay. Also, admission control is not investigated, the traffic load generated is assumed to stem from users already admitted.

The resource split is fully dynamic. There is no need to reserve a minimum number of C-slots either for initial access purposes or for high-priority users, since, with the traffic scenarios considered, total resource utilisation approaches rarely 90%, normally leaving enough C-slots available[16]. The allocation cycle concept will be investigated, but without piggybacking of resource extension requests. A dedicated access burst format is considered for some of the investigations, for others it is assumed that the access bursts carry also user information, as is the case in pure PRMA.

Most of the investigations are carried out for a hybrid CDMA/TDMA multiple access scheme, but pure TDMA and pure CDMA schemes are looked at as well. In particular, issues relating to access control are discussed for all these schemes. Mini-slots are not considered. The TD/CDMA physical layer assumed for most of the results presented in Chapters 8 and 9 would not allow for mini-slots in the case of a dynamic resource split, and the final TD/CDMA specifications for UTRA do not allow for mini-slots at all.

It is important to point out that these investigations were performed before and during the UMTS standardisation process, i.e. at a time when the multiple access scheme to be used for UTRA, let alone the multiple access protocol at the MAC layer, were not known. It is now possible to assess whether, and if so to what extent the proposed concepts were adopted for UTRA. This discussion, as far as not already dealt with in Section 3.4, is left to Chapters 10 and 11. But first, multiple access in GSM and in particular in GPRS is discussed in detail in the next chapter.

---

[16] Temporary lack of C-slots is possible. If a service class has stringent requirements on worst case delay rather than average delay, the above statement may not apply and a minimum amount of C-slots would need to be reserved.

4

MULTIPLE ACCESS IN GSM AND (E)GPRS

This chapter discusses features of the GSM air interface from phase 1 recommendations through to the specifications released in 1999. This entails GSM voice, circuit-switched data, and High Speed Circuit-Switched Data (HSCSD) services. A key topic is the matter of radio resource utilisation, hence on top of summarising the relevant system features, we present also some research results dealing with resource utilisation under heterogeneous GSM and HSCSD traffic load.

The main focus, however, is on the General Packet Radio Service (GPRS), since from the perspective of multiple access protocols, this is the most interesting aspect of an evolved GSM system. The MAC layer, and in particular the random access protocol, are explained in considerable detail. Again, this includes the presentation of some research results, which were fed into the GPRS standardisation process and influenced the design of the employed random access algorithm. Additions to GPRS contained in the 1999 release of the specifications, known under the heading 'EGPRS', are also discussed. The further evolution of the GSM system beyond release 1999 is a topic of Chapter 11.

4.1 Introduction

4.1.1 The GSM System

Various incompatible analogue first generation cellular systems emerged in Europe during the 1980s. By contrast, a concerted effort was made to arrive at a single standard for 2G digital cellular telephony. This pan-European standardisation effort was initiated by the *Conférence Européenne des Administrations des Postes et des Télécommunications* (CEPT) in 1982 with the formation of the Groupe Spécial Mobile (GSM) [3].

Initially, nine radio technology candidates were submitted to GSM, two proposing hybrid CDMA/TDMA, six TDMA, and one FDMA as basic multiple access schemes. At the beginning of 1987, based on simulation and trial results, GSM selected a narrowband TDMA system with a carrier spacing of 200 kHz, eight time-slots per frame, Gaussian minimum shift keying (GMSK) as modulation scheme and a speech codec operating at 13 kbit/s. The GSM duplex scheme is frequency-division duplex (FDD). Additional features introduced to provide good transmission quality include forward error correction coding (FEC) using half-rate convolutional codes combined with interleaving, and slow frequency hopping (SFH) as an option. Furthermore, slow power control can be applied to reduce co-channel interference.

The first release of GSM recommendations was published in April 1988 [170]. Ignoring the preamble, these recommendations consisted of 12 series of documents. Also in 1988, the European Telecommunications Standardisation Institute (ETSI) was founded, with its Special Mobile Group (SMG) taking responsibility for the evolution of the recommendations. By the end of 1993, operators in more than 10 European countries, Hong Kong and Australia had launched their GSM networks. The phenomenal success GSM has enjoyed since then will certainly be known to the reader. The acronym GSM stands no longer for Groupe Spécial Mobile, but rather for Global System for Mobile Communications, and the system has evolved significantly, with numerous new releases following the initial set of recommendations, as detailed further below.

Although ETSI is still formally responsible for the GSM standards, much of the air-interface-related technical work was transferred in the year 2000 from ETSI to the Third Generation Partnership Project (3GPP). The latter is not a standardisation body on its own, but rather, as the name suggests, a partnership of a collection of various regional standardisation bodies from China, Europe, Japan, South Korea and North America. It was set up to develop the specifications for the third generation Universal Mobile Telecommunications System (UMTS), which are then transferred into regional standards by the respective constituting member organisations. This transfer of work from ETSI SMG to 3GPP has taken place to ensure a synchronised evolution of GSM and UMTS, which is important for two reasons. Firstly, UMTS makes use of an evolved GSM core network, so 3GPP took over the responsibility for several GSM specifications related to the core network already for release 1999 (the first UMTS release). Secondly, later releases of the GSM radio access network are designed to be attached to either the 'original' GSM core network (via the A-interface for the circuit-switched and the G_b-interface for the packet-switched part of the core network, see Figure 4.1) or the evolved UMTS version (via the I_u-interface).

Figure 4.2 shows the fundamental building blocks of the initial GSM system, namely the Mobile-services Switching Centre (MSC), to which numerous Base Station Systems (BSS) are attached. These in turn are composed of a Base Station Controller (BSC) in charge of several Base Transceiver Stations (BTS, often referred to as base stations for simplicity).

Additionally, each MSC is equipped with a Visitor Location Register (VLR), which interacts with a central database, namely the Home Location Register (HLR) with associated Equipment Identity Register (EIR) and AUthentication Centre (AUC). The MSC shown in the figure is a special MSC, namely the Gateway MSC (GMSC), which is connected to the Public Switched Telephone Network (PSTN) and which interrogates the HLR. Other components not shown in the figure include the Short Message Service Centre (SM-SC). See GSM 03.02 [171] for a list of building blocks which also includes those added with later releases.

Due to the introduction of new features, a GSM system (or PLMN for Public Land Mobile Network) may now be composed of many more functional entities. Most notably, the specification of the General Packet Radio Service (GPRS) has led to the introduction of two new important components, namely the Serving GPRS Support Node (SGSN) and the Gateway GPRS Support Node (GGSN). These are the building blocks of the packet-switched core network of GSM, and complement the circuit-switched core network

4.1 INTRODUCTION

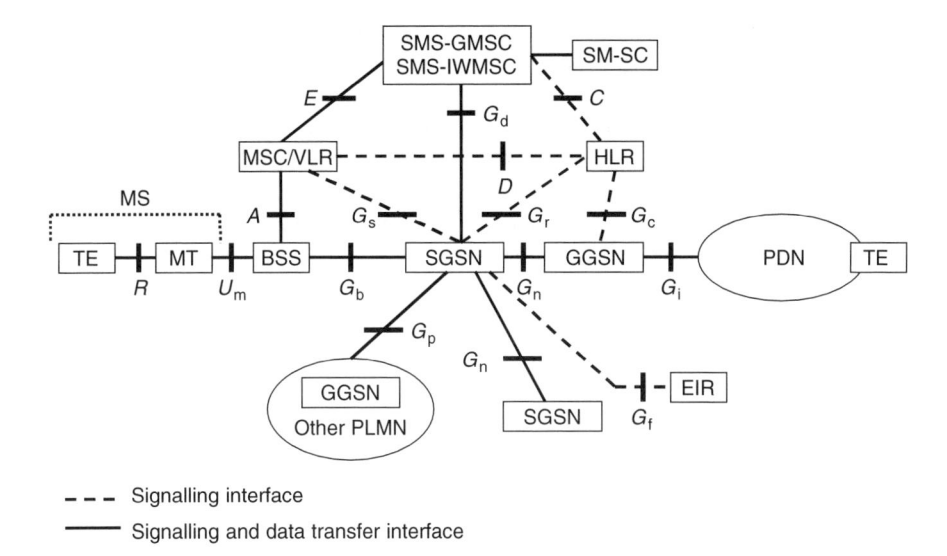

Figure 4.1 BSS connected to circuit-switched and packet-switched core network

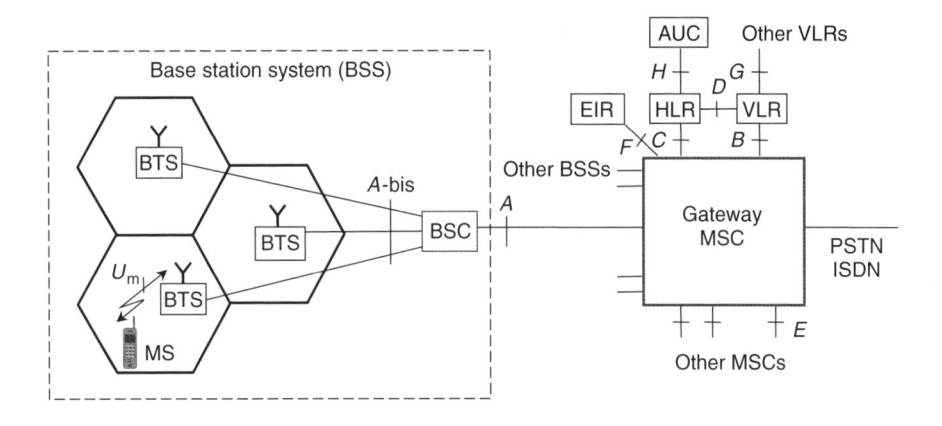

Figure 4.2 Basic building blocks of GSM

composed of MSCs and gateway MSCs. Figure 4.1 illustrates how a BSS is simultaneously connected to these two core networks and shows all pertinent interfaces.

In the following, we will deal predominantly with the air interface, denoted U_m in the two figures shown above.

4.1.2 GSM Phases and Releases

4.1.2.1 Phases 1 and 2

Judging from the available information, a two-phased approach to GSM must have been planned from the outset. To make sure that phase 1 mobiles could be supported in phase 2 networks, care had to be taken that features planned for phase 2 would not result in modifications of the air interface which could have an impact on phase 1 mobiles. Indeed, the logical channels (both traffic and control channels) and their mapping onto physical channels, as defined in the 05 series of the GSM recommendations, remained unaltered. Although the half-rate voice codec was not yet supported in phase 1, the relevant traffic and control channels were already defined. From an air-interface perspective, the only relevant additions in phase 2 appear to have been the extension of the 900 MHz band and the introduction of the 1800 MHz band for GSM operation.

As mentioned earlier, phase 1 recommendations were first published in 1988, but corrections were made to these recommendations in later years. All phase 1 recommendations carry version numbers 3.x.y (0, 1 and 2 were used for draft specifications in early stages of the standardisation process). For instance, the latest phase 1 version of GSM 05.01, which provides an overview of the physical layer of the air interface, is version 3.3.2, dating from December 1991. All these final versions of the recommendations are still available on the ETSI FTP server, which is however only accessible to ETSI members. Fortunately for those ready and eager to read through these not always very reader-friendly documents (they were not really meant to be, after all, they are specifications), there is now an alternative. As a result of the transfer of work from ETSI to 3GPP, all GSM specifications were copied onto the 3GPP FTP server [172]. At least at the time of writing, this server was openly accessible.

Phase 2 *specifications* (note the change in terminology), first published around September 1994, carry version numbers 4.x.y, and their final versions are also available on the two servers.

4.1.2.2 Phase 2+ with Yearly Releases

Phase 1 and 2 systems provided good support for conventional voice and associated supplementary services, circuit-switched data up to 9.6 kbit/s, and the now enormously popular two-way Short Message Service (SMS). With time, a desire grew to extend the GSM system and to allow for new services to be offered, which were not envisaged when GSM was conceived. These include:

(1) circuit-switched data services at higher data-rates;

(2) advanced speech call features (e.g. group calls);

(3) use of GSM in cordless telephony; and

(4) the introduction of a packet-data service.

All these items were initially subsumed under the heading *phase* 2+. The first phase 2+ specifications carry version numbers 5.x.y. Initial 5.0.0 versions of the 05-series were released in early 1996. However, due to the large number of features being considered for phase 2+, and the considerable time required to complete the standardisation of these features, a new concept had to be introduced, namely that of yearly releases. This would enable the phased introduction of these features, with each release introducing a consistent set of new features which could be deployed on their own, i.e. without depending on developments in subsequent releases. Correspondingly, specifications with version number 5.x.y are now referred to as release 1996 (R96) specifications, and every new yearly release up to release 1999 (R99) results in an increment of the first digit of the version number by one, that is, release 1997 (R97) carries version numbers 6.x.y, and so on. The appendix summarises issues related to the terminology, version numbers, and releases of ETSI and 3GPP specifications, in the latter case both for GSM and UMTS.

Phase 2+ features included in release 1996, which affect the air interface, cover items (1) and (2) listed above. The introduction of a traffic channel enabling data-rates up to 14.4 kbit/s and the possibility of traffic channel aggregation, i.e. transmission and reception on multiple time-slots per TDMA frame, provide increased circuit-switched data-rates[1]. The service provided by this time-slot aggregation is referred to as High Speed Circuit-Switched Data (HSCSD). Additional speech call features were standardised under the heading 'Advanced Speech Call Items' (ASCI), enabling:

- multi-level call precedence (i.e. accelerated call set-up for high-priority users) and pre-emption (i.e. seizing of resources in use by a low priority call for a higher priority call, if no idle resources are available at the required time) [173];
- voice group calls (i.e. calls between a predefined group of service subscribers) [174]; and
- voice broadcast calls (i.e. the distribution of speech generated by a service subscriber into a predefined geographical area to all or a group of service subscribers located in this area) [175].

The most notable impact of the introduction of these call features onto the air interface is a new control channel, the notification channel.

Release 1997 had a quite fundamental impact on the air interface, due to the introduction of GPRS. The new features and enhancements are discussed in detail in Sections 4.8 to 4.11, following a brief overview of GPRS in Section 4.7.

From an air-interface perspective, the most relevant item in *release 1998* concerned the introduction of the GSM Cordless Telephony System (CTS), which required a whole host of new logical channels to be introduced. Since this book is dealing with cellular communications rather than cordless telephony, the respective enhancements will not be discussed here. Another item included in release 1998 is the Adaptive Multi-Rate voice codec (AMR).

Release 1999 contains again features that affect the air interface significantly. While all previous enhancements of GSM could be supported on existing physical channels, through the introduction of higher order modulation schemes, release 1999 altered for the first time fundamental aspects of the physical RF layer. The respective work item, EDGE, stood initially for Enhanced Data-rates for *GSM* Evolution, it now stands for Enhanced Data-rates for *Global* Evolution, for reasons outlined in Chapter 2, and re-iterated in Section 4.12. The resulting increase in data-rates can be used in conjunction with circuit-switched data (both single-slot and high speed multi-slot variants), referred to as Enhanced Circuit-Switched Data (ECSD) as well as for GPRS (Enhanced GPRS, EGPRS). Additionally, the new EDGE COMPACT mode of GPRS allows a system to be deployed with as little as 1 MHz of spectrum per link available, which requires changes in the mapping of some control channels onto physical channels. Finally, the requirement for inter-system handover between GSM and UMTS (and also between GSM and cdma2000) required additions to broadcast information and measurement reporting.

---

[1] Strictly speaking, the 14.4 kbit/s service was included in the 05 series only in release 1997. However, specifications of other series included this feature already in release 1996.

The next release after R99 will again result in enhancements to the air interface. At the time of writing, these were not yet finalised. However, the requirements to be satisfied by the air-interface enhancements are known, and likely solutions will be discussed in Chapter 11.

While new features are introduced through new releases, old releases need to be maintained continually to eliminate errors. For instance, as manufacturers started to implement GPRS, they discovered inconsistencies, which needed to be sorted out, requiring numerous change requests to release 1997 specifications throughout 1998, 1999 and even the year 2000. This is not really surprising, considering the substantial additions contained in this release.

4.1.3 Scope of this Chapter

Not so long ago, the reader not fully satisfied with the few early books available on GSM, which were restricted to basic GSM features, had essentially to delve directly into the specifications. The latter obviously provided that she could access them; they were rather expensive for non-ETSI-members. Fortunately, several books dealing with GSM, either exclusively or in the context of mobile communications in general, have been published in recent years. The present book adds to this collection, albeit with a comparatively narrow focus, since it is a book on multiple access in mobile communications rather than a book on GSM.

Ideally, we would like to focus exclusively on multiple access issues in the following, or to put it differently, on the resources provided by the air interface and on how these resources are used. Of particular interest is how logical channels are mapped onto the physical channels, how the MAC layer arbitrates access to those logical channels which require such arbitration, and how well the available resources are utilised. However, to understand general system and air-interface constraints affecting the multiple access protocols, the discussion of the 'GSM MAC layer' is embedded in a wider discussion of air interface issues. As far as GSM phase 2 is concerned, the MAC is a rather minor matter anyway, essentially limited to the S-ALOHA-based multiple access protocol used for arbitration on the random access channel. From a MAC perspective, the GPRS additions are significantly more interesting.

The GSM specifications providing most of the information used for this chapter include the GSM 05 series describing the physical layer of the GSM air interface, selected specifications of the GSM 04 series dealing with the protocols between the mobile station and the BSS, and GSM 03.64 [54]. This last document provides an overall description of the GPRS air interface and is one of the rare GSM specifications containing a compact, but comprehensive overview of the alterations required to the 'standard' GSM specifications (that is, those dealing with air-interface issues) because of the introduction of a new service. In fact, certain clauses are only informative in GSM 03.64, while the normative text is contained in the 05 series, interleaved with phase 2 text and other phase 2+ features affecting the air interface. The reader should be warned that the informative text is sometimes out of synchronisation with the normative text, as continued changes to the 05 series do not always filter through to GSM 03.64 immediately, and if they do, it is sometimes only partially.

Since the 05 series specifications span several hundred pages, and single 04 series specifications can measure a few hundred pages, this chapter will omit a lot of the details not directly related to the MAC layer. The reader needing more details and interested in background information will have to resort to other publications dealing with the topics of interest more thoroughly. For instance, Reference [3] provides a considerable amount of background information on physical layer matters such as modulation schemes, and on speech coding, which are barely dealt with here.

Eventually, those who need to know about certain features of GSM to the level of single bits, be it for professional or research purposes, will have to refer to the specifications themselves. As pointed out earlier, these are now openly accessible — thanks to the power of the Internet! Given their style and the way in which relevant information is spread over numerous documents, it is probably more convenient to gain an overview of the system features elsewhere (as far as the air interface is concerned, why not here?) and only to resort to the standards later in search for all details. However, those readers wishing to familiarise themselves with GSM directly through in-depth reading of the specifications might be well advised to start first with phase 2 versions, to appreciate which of the system components are required for the provision of the basic services. Once these are understood, they can be compared with the latest versions, to identify the additions made to support the new services and features.

It is hoped that thanks to these hints and the referencing of relevant specifications throughout the following text, the reader will gain maximum benefit from this chapter.

4.1.4 Approach to the Description of the GSM Air Interface

The GSM air interface is denoted with the symbol U_m in the GSM specifications, and also referred to as the 'MS–BSS interface on the radio path' in the 04 series of these specifications.

Generally speaking, the relevant OSI layers on the air interface are the lowest 3 layers, and GSM largely conforms to the OSI approach. However, depending on what aspect of the air interface is considered, these layers manifest themselves in different guises, if at all. For instance, for *'plain GSM' signalling*, in accordance with OSI terminology, the lowest two layers are called physical layer (PL) and data link layer (DLL). The third layer carries the generic name 'layer 3', in Reference [176] it is also referred to as radio interface layer 3 (RIL3). Layer 3 functions in GSM include radio resource management (RR), mobility management (MM), and connection management (CM). For GPRS, on the other hand, the second layer is referred to as RLC/MAC, with the medium access control layer (MAC) being the lower sub-layer, and the radio link control layer (RLC) the upper sub-layer. In an additional twist, the RLC/MAC message format conforms to RIL3. For non-transparent circuit-switched data, a radio link protocol (RLP) is required at layer 2. Finally, for circuit-switched voice, there is no specific reference to any layers above the physical layer. This is illustrated in Figure 4.3, which was inspired by Figure 2.1 in GSM 04.04 [177]. The 'other functional units' shown in this figure are those supported by the application, e.g. the voice codec. The RLP for circuit-switched data is also associated with these other units.

The approach to the description of the GSM air interface is bottom up. We first describe in Section 4.2 the physical channels available, in OSI terms thus physical layer issues, and then in Section 4.3 the logical channels and how they are mapped onto these physical channels. According to GSM 04.04, these logical channels are supported on the interfaces between the physical layer and the other layers shown in Figure 4.3. For instance, control

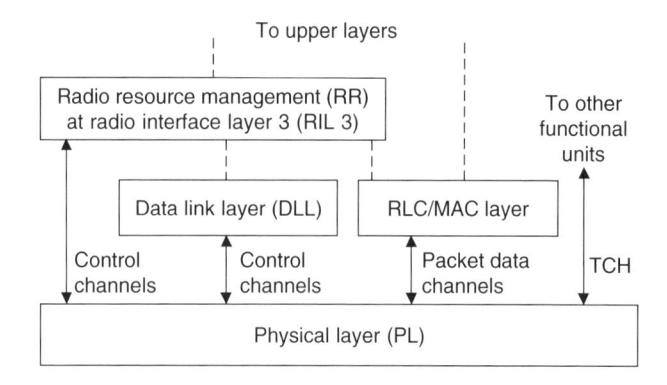

Figure 4.3 Interface between physical layer and higher layers

channels are supported on the interface between the PL and the DLL, while packet data channels (both control and traffic channels) are supported on the interface between PL and RLC/MAC. The RR entity controls directly certain aspects of the physical layer, for instance the channel measurements to be made, which explains the direct interface between these two entities.

The S-ALOHA MAC protocol described in detail in Section 4.4 makes use of one of these logical channels, namely the RACH. Section 4.5 introduces the enhancements required to provide the HSCSD and the ECSD service. The third OSI layer on the radio interface, dealing with radio resource management, mobility management, and call control will not be discussed systematically. However, the purpose of certain procedures associated with these entities, such as the location updating procedure required for MM, will be explained in the context of discussions on the utilisation of GSM air-interface resources provided in Section 4.6. For a detailed description of these procedures, see GSM 04.08 [178] for releases up to R98. For R99, see also its newer ETSI and 3GPP 'spin-off' documents. Section 4.6 provides the necessary pointers.

An introduction to GPRS is provided in Section 4.7. GPRS makes use of the same physical channels as GSM as well as new additional logical channels. These additional channels and their mapping onto the physical channels is described in Section 4.8. Section 4.9 deals with physical layer aspects of GPRS. The fact that the same physical channels as in GSM are used does not mean that other aspects of the physical layer have not been modified. For instance, new coding schemes enabling link adaptation were introduced. Section 4.10 provides a fairly detailed description of the GPRS RLC/MAC layer. Particular attention is given to the GPRS random access algorithm, which is described separately in Section 4.11. This description is accompanied by some research results, which were produced by the authors for the GPRS standardisation process. Finally, Section 4.12 deals with additions to GPRS introduced in R99, most of them related to EGPRS.

4.2 Physical Channels in GSM

The information contained in this section is from the 05 series of the GSM specifications. In particular, GSM 05.01 [105] provides a general description of the 'physical layer on the radio path', and points to related specifications containing more details.

4.2 PHYSICAL CHANNELS IN GSM

Those of relevance here are GSM 05.02 [179], entitled 'multiplexing and multiple access on the radio path' (and thus highly relevant), and GSM 05.04 [180], which deals with modulation.

4.2.1 GSM Carriers, Frequency Bands, and Modulation

4.2.1.1 Carrier Spacing and Frequency Bands

The GSM carrier spacing is 200 kHz. A carrier is also referred to as *radio frequency channel* in GSM. Frequency-division duplexing (FDD) is applied with the duplex spacing dependent on the band, in which GSM operates. At the time GSM was conceived, this used to be the 900 MHz band only, i.e. on the uplink (from mobile to base station) the band from 890 to 915 MHz, and on the downlink from 935 to 960 MHz. For phase 2, a 10 MHz extension band (from 880 to 890 MHz and from 925 to 935 MHz respectively), referred to as E-GSM band, was added. The duplex spacing is 45 MHz.

GSM 900, as the system operating in this band is now also referred to, was initially targeted for mobile communications, i.e. for use in cars. Mainly as a result of a UK initiative promoting so-called personal communications networks (PCN), using truly portable small and low-power handsets suitable for pedestrians, GSM was subsequently enhanced to operate also in the 1800 MHz band. The system operating in this band was initially referred to as DCS 1800, with DCS standing for *digital cellular system*, but it is now mainly known as GSM 1800. It operates from 1710 to 1785 MHz on the uplink, and from 1805 to 1880 MHz on the downlink. The duplex spacing is 95 MHz. Almost all countries in Europe and also most countries in Asia Pacific with GSM coverage have both GSM 900 and GSM 1800 systems in operation. Recently, Brazil, where no GSM 900 coverage exists, opted for the introduction of GSM 1800. Some operators obtained spectrum allocations in both bands, with dual-band mobiles being able to switch seamlessly from one band to the other, blurring the boundaries between mobile and personal communication systems.

As the US prepared for the auctioning of frequencies in the 1900 MHz band for personal communication systems (PCS) during the 1990s, GSM was again enhanced. The respective system, covering 60 MHz in each link direction, is known as PCS 1900 or GSM 1900, and is now deployed in several countries in North and South America, competing with D-AMPS and cdmaOne system operating in the same band. Finally, recent additions include a band for use by railways (R-GSM) with 4 MHz in each direction in the 900 MHz band (just underneath E-GSM), two relatively small bands between 450 and 500 MHz (jointly referred to as GSM 400, to replace first generation analogue systems still operating in these bands), and two times 25 MHz in the 850 MHz band.

GSM, as a result of its being capable of operating in most bands ever made available for cellular communications in the world (that is, excluding recent additions for 3G), combined with the early commercial success of GSM 900 in Europe and then in Asia, now covers almost all populated areas of the globe. The most notable exceptions are Japan and South Korea with no GSM coverage whatsoever, while some white spots on the American continent are expected to be of temporary nature only, particularly now as major D-AMPS operators are considering to go for GSM/GPRS as a stepping stone towards UMTS. At the time of writing, several GSM handset manufacturers offered tri-band mobiles suitable for seamless GSM 900, 1800, and 1900 operation.

4.2.1.2 Modulation Schemes: GMSK and 8PSK

The modulation scheme used in GSM is Gaussian minimum shift keying (GMSK) at a modulation symbol rate of 1625/6 ksymbols/s, that is, approximately 270.833 ksymbols/s, which also corresponds to a bit-rate of 270.833 kbits/s. For a detailed description of GMSK and use of this modulation scheme in GSM, refer to Reference [3].

Release 99 of the specifications brought the introduction of an additional, higher order modulation scheme to provide enhanced data-rates, namely 8-phase shift keying (8PSK). Since each 8PSK symbol contains 3 bits, using the same symbol rate and carrier spacing, the raw bit-rate can be tripled to 812.5 kbits/s. This comes obviously at a cost, namely the requirement for higher signal-to-interference-plus-noise ratios (SINR). The enhanced data-rates can be used both for circuit-switched data and GPRS, which are then referred to as Enhanced Circuit-switched Data (ECSD) and Enhanced GPRS (EGPRS) respectively. By contrast, the acronym E-GSM was not available for this purpose, since it is already occupied, denoting the extended 900 MHz band.

4.2.2 TDMA, the Basic Multiple Access Scheme — Frames, Time-slots and Bursts

4.2.2.1 Time-slots and Frames

The basic multiple access scheme of GSM is TDMA, providing eight basic physical channels per GSM carrier. Therefore, eight time-slots, indexed with Time-slot Numbers (TN) from 0 to 7, are grouped into a TDMA frame. The duration of such a frame is exactly 120/26 ms, which is approximately 4.615 ms. As a consequence, a time-slot lasts 15/26 ms, or roughly 577 µs including guard periods. At the symbol rate of 270.833 ksymbol/s, this corresponds to 156.25 symbol periods. The useful duration of a time-slot, i.e. the duration of bursts transmitted in a time-slot, is shorter. How much shorter depends on the burst format used, which is discussed in more detail below.

For a mobile terminal to be able to receive and transmit bursts on slots with the same time-slot number, without having to be able to transmit and receive simultaneously, the uplink and downlink time-slots are staggered at the base station. More precisely, the start of a TDMA frame on the uplink is delayed by the fixed period of three time-slots from the start of the TDMA frame on the downlink. The performance requirements of basic mobile terminals in terms of adaptive frame alignment, transceiver tuning, and receive/transmit switching are such that a mobile terminal can receive a burst, transmit a burst, and monitor adjacent cells in the same frame. This is illustrated in Figure 4.4.

Mobiles that can entertain bi-directional communication in this manner, without being able to transmit and receive simultaneously, are termed *half-duplex* mobiles.

From the figure, it can be seen that for the base station to receive bursts frame-aligned, the mobile station has to anticipate the burst transmission by a certain period, to account for the transmission delay. This period must be calculated continuously by the base station based on the timing of bursts received from the mobile station, and then signalled to the MS. It is referred to as Timing Advance (TA). The standard TA range in GSM is from 0 to 63 symbols, each symbol corresponding to a two-way transmission distance of approximately 1100 m. The maximum two-way transmission distance is 70 km, hence the maximum cell radius 35 km. Accounting for this maximum value, the net time available

4.2 PHYSICAL CHANNELS IN GSM

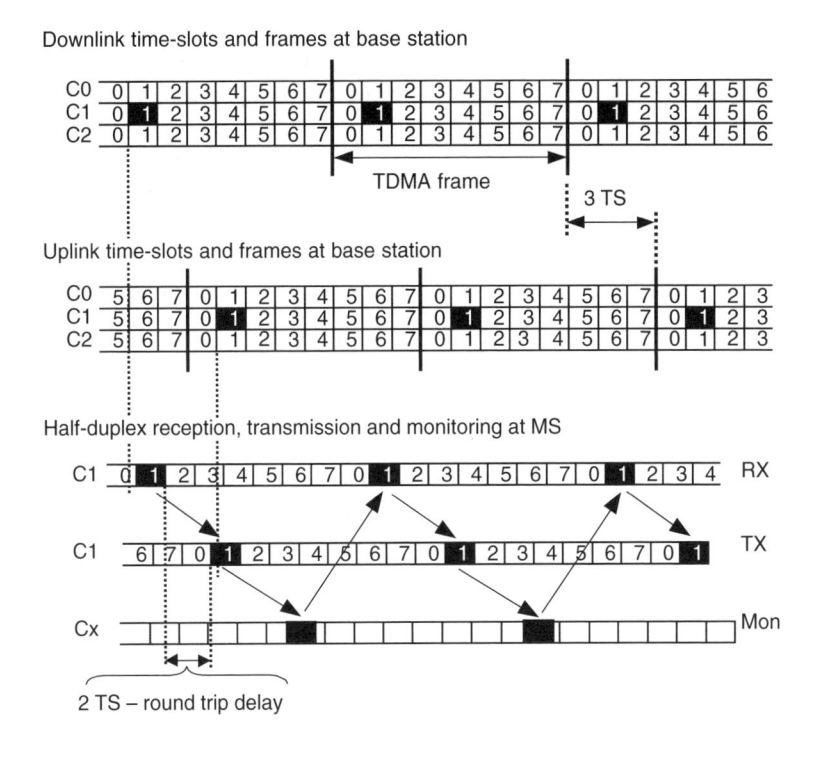

Figure 4.4 Frames, time-slots, and half-duplex transmission (TS = time-slot)

for the MS to switch from receive to transmit and retune the transceiver is approximately one second (two times the slot duration minus the duration of 63 symbols).

In certain cases, it is desirable to have cells with a larger radius than the 35 km default design limit (e.g. in very scarcely populated areas with a flat terrain, or at coast lines). A cell radius of up to 120 km is possible by leaving certain time-slots unused. In GSM 400, the solution is somewhat simpler owing to an extended TA of up to 219 symbols (see Reference [181] for details).

The monitoring period shown in Figure 4.4 is required to perform power measurements of neighbouring cells, which are then reported to the base station and allow the network to decide when a handover is required. This type of handover, which is assisted by the MS through these measurement reports, but fully controlled by the network, is also termed mobile assisted handover (MAHO).

Note that the frames of multiple carriers at a single base station are aligned, as shown in the figure, but that with the exception of the EDGE COMPACT mode discussed in Section 4.12, there is normally no inter-BTS synchronisation, such that frames of different base stations are not aligned.

4.2.2.2 Burst Formats

Bursts are the physical content of a time-slot. There are four different burst formats. Three of them are 'full' burst formats, namely the Normal Burst (NB), the Frequency correction Burst (FB), and the Synchronisation Burst (SB), with a useful duration of 147 symbols each. Together with the 8.25 symbols guard period at the end of the burst, this covers almost completely the time-slot duration of 156.25 symbols. The missing symbol (which is part of the burst tail symbols) is accounted for in Reference [180]. The guard period is required, before all, for power ramping. On the uplink, at the end of the useful part of the burst, the MS using that time-slot will ramp down the power, while the MS using the next time-slot will start ramping up the power. The base station is not required to perform power ramping between adjacent bursts on the downlink, but must be capable of ramping down and up during non-used time-slots. Additionally, the guard period also caters for TA inaccuracies.

Apart from the tail symbols at the beginning and the end of the burst, the normal and the synchronisation burst carry a certain number of *encrypted symbols*, split in two 'half bursts' by a *training sequence*, sometimes also referred to as *midamble*. This training sequence contains known symbols, which allow the equalizer in the receiver to estimate and counteract the distortions experienced by the bursts on the radio propagation channel. The frequency burst, on the other hand, carries only fixed (i.e. known) bits. These three burst formats are shown in Figure 4.5.

The fourth format, namely the Access Burst (AB), is a short one with a duration of only 88 symbols, composed of eight extended tail symbols at the beginning of the burst and three tail symbols at the end, a *synchronisation sequence* of 41 symbols, and 36 encrypted symbols. For similar reasons as for the full burst, the useful part spans only 87 symbols. This burst format, which is shown in Figure 4.6, features an extended guard period of 68.25 symbol periods.

The access burst format is (among other things) used for accessing the system on the random access channel, for instance to set up a call, as discussed in detail in Section 4.4. At the time of call set-up, the TA is not known. The MS will schedule transmissions according to the perceived time-slot boundaries, i.e. assuming that the round-trip (or two way) delay were zero. This will result in the access burst being received with a delay with respect to the time-slot boundaries at the BTS, namely twice the transmission delay from MS to BTS, which is why an extended guard period is required. The 60 additional

1 time slot = 156.25 symbol durations (15/26 or 0.577 ms)
(1 symbol duration = 48/13 or 3.69 µs)

| 3 TS | 58 encrypted symbols | 26-symb. training sequence | 58 encrypted symbols | 3 TS | 8.25 GP |

Normal burst

| 3 TS | 142 fixed symbols | 3 TS | 8.25 GP |

Frequency correction burst

| 3 TS | 39 encrypted symbols | 64-symbol synchr. sequence | 39 encrypted symbols | 3 TS | 8.25 GP |

Synchronisation burst

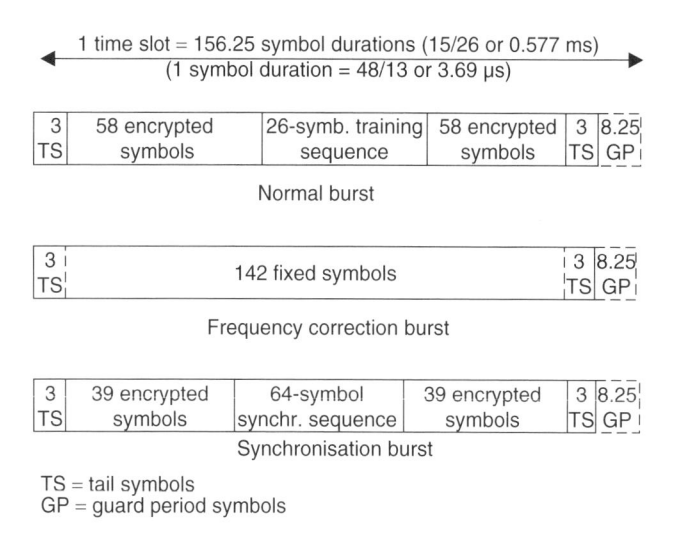

TS = tail symbols
GP = guard period symbols

Figure 4.5 Full burst formats in GSM

4.2 PHYSICAL CHANNELS IN GSM

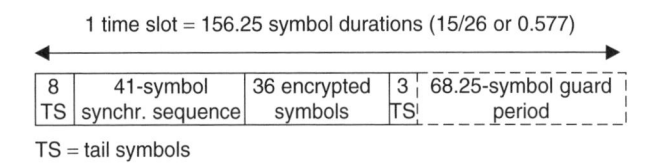

```
           1 time slot = 156.25 symbol durations (15/26 or 0.577)
   ┌────┬───────────────┬────────────┬─────┬────────────────────┐
   │ 8  │  41-symbol    │36 encrypted│  3  │ 68.25-symbol guard │
   │ TS │synchr. sequence│  symbols  │ TS  │      period        │
   └────┴───────────────┴────────────┴─────┴────────────────────┘
   TS = tail symbols
```

Figure 4.6 Short burst format used for the access burst in GSM

symbol periods available for this purpose correspond roughly to the maximum timing advance, i.e. the duration of 63 symbols, which in turn corresponds to the cell radius of 35 km discussed earlier.

The fact that the base station does not know exactly when it is going to receive an access burst within a time-slot boundary is why special synchronisation sequences, rather than the shorter training sequences defined for the normal burst format, are needed. There are several such sequences, which are assigned for use in different cells in a similar manner as frequencies. The same sequence should not be reused in neighbouring cells. This is one of the means for a base station to detect whether a random access burst was addressed to itself rather than to a base station in another cell.

The access burst is not only used on the random access channel, but also in other cases, when the TA is not or not precisely known, e.g. following a handover.

4.2.3 Slow Frequency Hopping and Interleaving

From the very beginning, GSM provided the optional feature of slow, frame-wise frequency hopping, i.e. at a hop-rate of 216.67 hops/s. In the GSM context, optional often means that all handsets must have the capability implemented, while the network operator is free to choose whether to use the feature or not. This is also the case here.

Due to the fact that GSM handsets are typically half-duplex, and therefore need to switch from receive- to transmit- to monitor-frequency within the duration of a TDMA frame anyway, slow frame-wise frequency hopping does not really affect the operation of the handset significantly. It just means that receive- and transmit radio frequency channels change from frame to frame, however, the duplex spacing (as determined by the band in which GSM operates) remains the same. At the BTS, the impact of frequency hopping is a different matter, as discussed in more detail in Subsection 4.6.5.

According to Reference [105], the main advantages of this feature are two-fold. Firstly, diversity on one transmission link is provided, especially to increase the efficiency of coding and interleaving for slowly moving mobile stations. This type of diversity can be termed frequency diversity. Secondly, it allows the quality on all the communications to be averaged through interference diversity. Diversity is achieved when multiple replicas of a given signal are processed, which exhibit a cross-correlation lower than one, as a result of having experienced different channel conditions. For example, in the case of frequency diversity, the low correlation is due to a small correlation bandwidth, as explained in the following.

4.2.3.1 Frequency Diversity

Interleaving is typically applied in cellular communication systems to randomise the propagation channel, such that the probability of a bit being in error is independent from that of previous and subsequent bits being in error. Assuming no interleaving, due to the temporal correlation of the fast fading process, if a bit is in error because of a fading dip, the subsequent bit will also be in error with high probability. This error dependence leads to error bursts, which in turn affect the efficiency of FEC coding negatively. Interleaving (i.e. shuffling around the sequence of bits) after error coding at the transmitter, and de-interleaving before error decoding at the receiver can eliminate this error dependence, i.e. randomise the channel. However, due to delay constraints, the interleaving period is also constrained. For instance, some signalling messages and GPRS data blocks are transmitted over 'radio blocks' of four bursts (one per TDMA frame, thus normally roughly 20 ms in total), and interleaving is limited to shuffling around individual bits within such a block. For fast moving mobiles, which will experience fast channel fluctuations, the time diversity obtained through interleaving and error coding is typically sufficient. For slow moving mobiles, on the other hand, a fading dip may extend over more than 20 ms, hence randomisation is not possible, and a block affected by such a dip would almost certainly be lost.

Frequency hopping exploits the fact that the correlation between the fading processes experienced on two carriers far enough apart in frequency is low, such that the probability of two bursts sent in consecutive TDMA frames — but on different carrier frequencies — being both affected by the same fading dip is low as well. If only one or two bursts in a block are badly affected by fading, then this block may be recovered owing to redundancy provided by FEC coding (which could be viewed as if multiple replicas of user bits were sent over the air interface) and interleaving. For a carrier frequency of 900 MHz and a mobile speed of 3 km/h (e.g. pedestrian speed), two independent single-path fast fading processes with so-called Rayleigh distributed envelope levels generated according to Jakes' model [182, p. 68] are shown in Figure 4.7. This figure illustrates both how the duration of fading dips can exceed 20 ms and how hopping over two frequencies helps to provide the desired randomisation or diversity.

For frequency hopping to be effective, the spacing between radio frequency carriers which are hopped over must be larger than the correlation bandwidth of the propagation channel, such that the fast fading processes are indeed independent. Depending on the propagation environment, e.g. in picocells, the correlation bandwidth can exceed 10 MHz [109]. In cases where several operators have to share a particular GSM band, the individual operator allocation can be smaller than that. Therefore, the often-made assumption of *perfect* frequency hopping resulting in completely uncorrelated burst-to-burst behaviour may be optimistic in certain cases. Mouly and Pautet [176] report a frequency diversity gain of around 6.5 dB. A similar value, namely 5 dB at a frame erasure rate of 1%, is reported in Reference [183], although these results were obtained for proposed EGPRS voice bearers (discussed in Chapter 11) rather than GSM bearers.

4.2.3.2 Interference Diversity and Interference Averaging

The dominant interference contribution is co-channel interference, which comes from users (or base stations, if on the downlink) outside the considered cell. If no frequency hopping is applied, there will be repeated 'collisions' of the same set of co-channel interferers, i.e. those experiencing full or partial time-slot overlap (recall that cells are normally not synchronised in GSM) and transmitting/receiving on the same radio frequency carrier. Depending on relative positions and movement of the desired user and interferers, fluctuations in the propagation conditions, etc., this can lead to constellations where a user is affected by particularly bad interference for an extended period of time.

4.2 PHYSICAL CHANNELS IN GSM

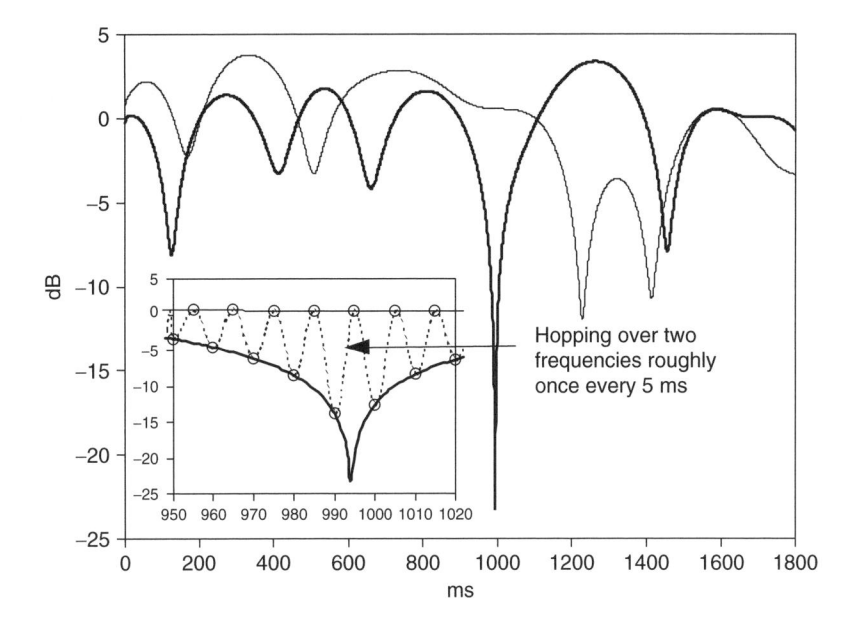

Figure 4.7 Hopping over two frequencies experiencing independent fast (Rayleigh) fading

Provided that different hopping sequences are used in co-channel cells, SFH results in two effects, which alleviate the impact of undesired interference patterns. Both are based on the fact that the interference will randomly vary for each burst, due to avoidance of repeated collisions with the same interferers.

On a microscopic level, i.e. within an interleaving period, similar benefits are obtained as in the case of frequency diversity. If only one or two of eight bursts (over which a voice frame is interleaved) are badly affected because of heavy interference, the frame might still be recovered [80].

On a more macroscopic level, i.e. the system level, under certain conditions, frequency planning can now be carried out taking into account the average co-channel interference rather than the worst-case interference. In particular, the interference reduction due to discontinuous transmission (DTX, i.e. not transmitting during voice inactivity phases) can now be translated into increased capacity as a result of tighter frequency reuse being possible. This effect is also referred to as *interference averaging* rather than interference diversity.

In Reference [80], where these two effects are discussed in more detail, it is demonstrated how hopping over only three different frequencies provides already a significant gain through microscopic interference diversity. Under the considered conditions, which include perfect frequency diversity, a gain of around 1.5 dB could be achieved at a system load of 25%. Increasing the number of frequencies being hopped over to 12, increases the gain relatively moderately to 2.5 dB. Similarly, results presented in Reference [184] would suggest that at a load of 12.5%, the performance difference between eight and 12 hopping frequencies is moderate, while the quality improvement was found to be marginal when more than 12 frequencies were used. By contrast, for frequency planning to be based on average rather than worst-case interference, it is expected that the number of frequencies which are hopped over must be substantial. Only in this case can the size of the population of mutually interfering users be increased to a level, which ensures that the probability of significant excursions from the average interference level is sufficiently low, such that the permitted outage probability level is not exceeded. For more details, refer to Subsection 4.6.5, and in particular, to Section 7.2, where the same problem is considered from a slightly different angle, namely in the context of investigations on multiplexing efficiency in a hybrid CDMA/TDMA system.

4.2.3.3 Hopping Sequences in GSM

As pointed out earlier, SFH has no significant impact on basic handset operation, as the handset needs to switch three times per TDMA frame between frequencies anyway to receive, transmit, and monitor. What is required, essentially, is an algorithm that, given a few parameters, maps the TDMA Frame Number (FN) to a radio frequency channel, or in other words, determines the hopping sequence. An essential requirement for this algorithm is that orthogonality between physical channels within a cell is maintained, i.e. intracell collisions are avoided.

Clearly, in a given cell, the maximum possible number of radio frequency channels over which an MS could hop in theory is the total number of channels allocated to the respective BTS, i.e. the number of channels contained in the so-called Cell Allocation (CA). This number in turn is determined by the total frequency spectrum available to the operator, whether this spectrum is sub-divided to deploy several hierarchical cell layers, and finally the frequency reuse factor within the relevant layer. One of these radio frequency channels is used to carry synchronisation information and the broadcast control channel on TN 0. On this particular time-slot (and potentially on other time-slots on this carrier), frequency hopping is not allowed, as outlined in further detail in Subsection 4.3.6. The Mobile Allocation (MA), that is the set of frequency channels over which the MS hops effectively (both for transmit and receive), is therefore a subset of the CA. The MA may contain up to 64 channels. If SFH is not applied, then the MA contains only one channel.

There are two types of hopping sequences: a cyclic sequence and 63 different pseudo-random sequences. In order to avoid intracell collisions between hopping mobiles, all mobiles using the same time-slot in a given cell must use the same hopping sequence, but with different frequency offsets (the so-called Mobile Allocation Index Offset (MAIO)). When the cyclic sequence is used, the frequency channels of the MA are hopped over one by one in cycles, as the sequence name would suggest. The sequence length corresponds therefore to the number of frequency channels contained in the MA. Cyclic sequences provide frequency diversity, but no 'proper' interference diversity, since all interfering cells use the same hopping sequence [80]. When pseudo-random sequences are used, referred to in the following as *random hopping*, different co-channel cells use different uncorrelated sequences, resulting in the desired interference diversity. According to Reference [81], these sequences have a length of 84 864. Details of the algorithm determining the sequence of frequency channels to be used are provided in GSM 05.02.

In Reference [185], it is stated that cyclic sequences perform better than the pseudo-random sequences for up to 10 hopping frequencies per cell, since they provide better frequency diversity. Furthermore, with sophisticated frequency planning, even cyclic hopping may provide some interference diversity gain. When the frequency reuse is tight, and there are a large number of hopping frequencies, on the other hand, pseudo-random frequency hopping performs better.

At this point, we have finally identified all parameters required to describe a basic physical channel or a *resource unit*. Before summarising them, the different frame structures are discussed briefly.

4.2.4 Frame Structures: Hyperframe, Superframe and Multiframes

The longest recurrent time period on the GSM air interface is called a *hyperframe*, which consists of 2 715 648 TDMA frames. The latter are numbered modulo this hyperframe, thus carry frame numbers ranging from 0 to FN_MAX = 2 715 647. Hyperframes are used to support some cryptographic mechanisms; other than mentioning that they are divided into 2048 *superframes*, they will not be discussed any further in this chapter. A superframe in turn is 1326 TDMA frames or 6.12 seconds long. It carries either 51 multiframes comprising 26 TDMA frames, referred to as a *26-multiframe*, or 26 multiframes comprising 51 TDMA frames, i.e. a *51-multiframe*. Not exactly fitting into the superframe concept, for GPRS and CTS, a new *52-multiframe* was also introduced. The reasoning behind these groupings will become more evident later on when discussing how logical channels are mapped onto physical channels and what information they carry.

4.2.5 Parameters describing the Physical Channel

A physical channel uses a combination of time and frequency division multiplexing and must therefore be defined as a sequence of time-slots and radio frequency channels.

A given physical channel will use the *same* time-slot number in every TDMA frame. A *basic physical channel* is one that makes use of this time-slot in *every* TDMA frame, it is a *full-rate channel*. There are also channels that do not make use of this time-slot in every frame, as indicated by a frame number sequence other than 0, 1, ... FN_MAX, for instance *half-rate channels*.

The radio frequency channel sequence is determined, as discussed above, by the mobile allocation, the MAIO, and a hopping sequence number. With the frame number as input, the radio frequency channel to be used can then be calculated according to an algorithm specified in GSM 05.02.

4.3 Mapping of Logical Channels onto Physical Channels

Logical channels in GSM are grouped into two categories, traffic channels on the one hand, and signalling or control channels on the other.

4.3.1 Traffic Channels

In GSM phases 1 and 2, three types of Traffic CHannels (TCH) are defined, namely full-rate traffic channels (denoted TCH/Fx), half-rate traffic channels (TCH/Hy), and the Cell Broadcast CHannel (CBCH). The full-rate channels occupy one basic physical channel, i.e. transmit one burst per TDMA frame. Two half-rate channels can share a basic physical channel, making alternate use of the TDMA frames.

The symbols 'x' and 'y' are either replaced by 'S' for speech, or by the supported data-rate for circuit-switched data. For instance, the half-rate speech traffic channel is denoted TCH/HS, while the 9.6 kbit/s full-rate data traffic channel is denoted TCH/F9.6.

According to GSM 04.03 [186], these traffic channels are the radio resources on which *user channels* are carried. User channels in turn are intended to carry a wide variety of user information streams. A distinguishing characteristic of these channels is that, in contrast to control channels, they do not carry signalling information for connection management, mobility management, or radio resource management.

Ignoring training sequences, guard bits, etc., and only accounting for the 116 encrypted bits carried on a normal burst, a full-rate channel supports a raw bit-rate of 25.13 kbit/s, a half-rate channel half of that. Deducting further the two stealing bits or flags per burst (see Subsection 3.4.3) and accounting for the fact that only 24 bursts are assigned to a full-rate TCH in a 26-multiframe, as explained later on, further reduces the raw data-rate to 22.8 kbit/s. Depending on the amount of redundancy added for FEC coding, a full-rate data channel can support user data-rates of 9.6 kbit/s, 4.8 kbit/s, or ≤ 2.4 kbit/s, the half-rate data channel only 4.8 kbit/s, or ≤ 2.4 kbit/s.

In release 1996, the TCH/F14.4, a full-rate channel supporting 14.4 kbit/s, was added. Furthermore, for HSCSD, the notion of *multi-slot configuration* was introduced, allowing several (up to eight) TCH/F to be aggregated for a single user. Since asymmetric downlink-biased configurations are supported in HSCSD (offering higher data-rates on the downlink than the uplink), some traffic channels allocated to a single user may be operated in the downlink direction only; these are denoted TCH/FD. However, the HSCSD user allocation must contain at least one conventional, bi-directional traffic channel.

The packet data traffic channel, introduced in release 1997 for GPRS, will be discussed later in more detail. Release 1998 added support for the AMR codec on the full-rate channel (denoted TCH/AFS), supporting eight different data-rates, and on the half-rate channel (denoted TCH/AHS), supporting six different data-rates. Along the way, the TCH/EFS was introduced (for the enhanced full-rate codec). Finally, with release 1999, enhanced traffic channels (E-TCH) carrying user rates of 28.8 kbit/s, 32.0 kbit/s, and 43.2 kbit/s respectively (using the 8PSK modulation scheme defined for EDGE) were added.

The CBCH is used for the short message service cell broadcast. For details of this service, refer to GSM 03.41 [187]. Table 4.1 lists all GSM traffic channels used for voice and circuit-switched data communications, and the CBCH. For the data channels, the data-rate indicated in the third column includes RLP protocol overhead, thus may be higher than the indicated user rate.

4.3.2 Signalling and Control Channels

The signalling or control channels are primarily intended to carry signalling information for connection management, mobility management, and radio resource management. They can also be used to carry other data, for instance short messages [186]. Ignoring CTS control channels, there are three categories of control channels, namely broadcast channels, common control-type channels and dedicated control channels.

4.3.2.1 Broadcast Channels

Broadcast channels convey information that is relevant for all mobiles served by the cell, some even for mobiles served by neighbouring cells.

4.3 MAPPING OF LOGICAL CHANNELS ONTO PHYSICAL CHANNELS

Table 4.1 Traffic channels supported in GSM (excluding packet channels)

Type of channel	Acronym	Data-rate [kbit/s]
Full-rate speech TCH	TCH/FS	13.0
Enhanced full-rate speech TCH	TCH/EFS	12.2
Half-rate speech TCH	TCH/HS	5.6
Adaptive full-rate speech TCH (12.2 kbit/s)	TCH/AFS12.2	12.2
Adaptive full-rate speech TCH (10.2 kbit/s)	TCH/AFS10.2	10.2
Adaptive full-rate speech TCH (7.95 kbit/s)	TCH/AFS7.95	7.95
Adaptive full-rate speech TCH (7.4 kbit/s)	TCH/AFS7.4	7.4
Adaptive full-rate speech TCH (6.7 kbit/s)	TCH/AFS6.7	6.7
Adaptive full-rate speech TCH (5.9 kbit/s)	TCH/AFS5.9	5.9
Adaptive full-rate speech TCH (5.15 kbit/s)	TCH/AFS5.15	5.15
Adaptive full-rate speech TCH (4.75 kbit/s)	TCH/AFS4.75	4.75
Adaptive half-rate speech TCH (7.95 kbit/s)	TCH/AHS7.95	7.95
Adaptive half-rate speech TCH (7.4 kbit/s)	TCH/AHS7.4	7.4
Adaptive half-rate speech TCH (6.7 kbit/s)	TCH/AHS6.7	6.7
Adaptive half-rate speech TCH (5.9 kbit/s)	TCH/AHS 5.9	5.9
Adaptive half-rate speech TCH (5.15 kbit/s)	TCH/AHS5.15	5.15
Adaptive half-rate speech TCH (4.75 kbit/s)	TCH/AHS 4.75	4.75
Full-rate data TCH (14.4 kbit/s)	TCH/F14.4	14.5
Full-rate data TCH (9.6 kbit/s)	TCH/F9.6	12.0
Full-rate data TCH (4.8 kbit/s)	TCH/F4.8	6.0
Full-rate data TCH ($\leq$2.4 kbit/s)	TCH/F2.4	3.6
Half-rate data TCH (4.8 kbit/s)	TCH/H4.8	6.0
Half-rate data TCH ($\leq$2.4 kbit/s)	TCH/H2.4	3.6
Enhanced full-rate data TCH (43.2 kbit/s)	E-TCH/F43.2	43.5
Enhanced full-rate data TCH (32.0 kbit/s)	E-TCH/F32.0	32.0
Enhanced full-rate data TCH (28.8 kbit/s)	E-TCH/F28.8	29.0
Cell Broadcast Channel	CBCH	0.782

The Frequency Correction CHannel (FCCH) carries information that is used by the radio subsystem of mobile stations for frequency correction. This information is carried on the frequency correction burst.

The Synchronisation CHannel (SCH) carries information required by mobile stations for frame synchronisation and identification of a base transceiver station. For this purpose, the synchronisation burst format used on this channel contains a 6-bit Base transceiver Station Identity Code (BSIC) and a reduced format of the TDMA frame number.

The Broadcast Control CHannel (BCCH) is used to broadcast system information messages to the mobile stations in a cell. These messages contain information required for proper operation of the system. For instance, while an MS can find the BCCH by finding first an FCCH and then an SCH, since they are all mapped onto the same basic physical channel, it needs to be told where the other channels are. It needs to know how many CCCHs there are and whether they share a basic physical channel with an SDCCH. Other required broadcast information includes cell allocation, access parameters for the random access procedure, the BCCH frequency list of neighbouring cells, location area identification, and cell selection/reselection parameters. This broadcast information is signalled in the shape of BCCH *system information messages*.

In GSM phase 2, there are only relatively few different system information message types, namely 10 (including sub-types), of which two are not transmitted on the BCCH, but rather on the SACCH. In release 1999, these numbers have roughly doubled. This is due to the addition of GPRS and the introduction of UMTS networks operating alongside GSM networks, the latter requiring the broadcasting of non-GSM information on the GSM BCCH (for instance the BCCH frequencies of neighbouring UMTS cells). A detailed list of system information message types and their content can be found in GSM 04.18 [188] for R99, GSM 04.08 for earlier releases.

4.3.2.2 Common Control Type Channels

As do broadcast channels, on the downlink, common channels convey information which may be relevant for *any* MS served by the cell. However, the information which is transmitted at a given time is directed to a *specific* MS, as identified in the relevant message. Depending on the state in which an MS is, it will therefore have to listen to all occurrences of these common channels (or a subset thereof), in order not to miss out on information directed to it. Similarly, the uplink is for common use of all mobile terminals, but at any given time, there can be only one MS actually using it.

Access Grant CHannel (AGCH), Paging CHannel (PCH), and Notification CHannel (NCH), used on the downlink only, and the Random Access CHannel (RACH), used on the uplink only, are all common control channels. In fact, when combined, they are referred to as the Common Control CHannel (CCCH).

For mobile terminated traffic, mobiles are paged on the PCH, following which they will attempt to access the system on the RACH. For mobile originated activities, access attempts are made on the RACH without prior paging. If the access attempt is successful, and the necessary channel resources are available, the base station will indicate on the access grant channel which type of channel (e.g. SDCCH, TCH) is allocated for further signalling exchange.

The NCH is used to provide voice group call and voice broadcast call notifications to a mobile station.

4.3.2.3 Dedicated Control Channels

Dedicated control channels are assigned exclusively for communication between the network and one specific MS. There are two types of dedicated control channels, namely the Stand-alone Dedicated Control CHannel (SDCCH) and the Associated Control CHannel (ACCH).

An SDCCH is allocated temporarily to a specific MS for signalling exchange with the network. This could be for call set-up signalling (prior to the allocation of a TCH), or for location updating procedures. Short messages addressed to a specific user (rather than broadcast on the CBCH) are also carried on the SDCCH, if this user is not currently engaged in a call.

Associated control channels are of a supporting nature. They are allocated either in conjunction with a TCH, or an SDCCH, hence *associated*. The ACCH can be continuous stream (*Slow* ACCH, SACCH), i.e. mapped onto recurrent TDMA frames, as outlined further below, or operate in burst stealing mode (*Fast* ACCH, FACCH), i.e. temporarily taking away resources from a TCH.

The SACCH is used to transmit information on timing advance and power control (in a 2-octet physical layer header of a SACCH frame, see GSM 04.04). In the main part of the SACCH frame, the network will signal through system information messages the neighbouring BCCH frequencies, on which the MS should perform power measurements (alongside other information, such as whether DTX should be used or not). Correspondingly, in the uplink, these frames are used by the MS to report the measurements made, both on the serving cell and on neighbouring cells. While SMS are normally carried on the SDCCH, they can also be delivered on the SACCH, which allows delivery during an ongoing call, but will reduce the frequency of measurement reporting.

The FACCH is used for handover-related signalling exchanges, for which, given the time constraints, the SACCH would be too slow.

4.3.2.4 Naming and Mapping

Depending on how these logical channels are mapped onto physical channels, they carry various suffixes. For instance, a SACCH associated with a TCH/F is referred to as SACCH/TF, while a SACCH associated with one of eight SDCCHs carried on a separate basic physical channel, i.e. an SDCCH/8, is referred to as SACCH/C8. Further below, we will discuss selected examples for illustration. For a full list of logical channels, the reader is referred to GSM 05.01, and for details on how all these channels are mapped onto physical channels, to GSM 05.02. Table 4.2 summarises the control channels discussed above and their respective acronyms.

Table 4.2 List of GSM control channels (excluding CTS and GPRS)

Broadcast channels	Acronym
Frequency Correction Channel	FCCH
Synchronisation Channel	SCH
Broadcast Control Channel	BCCH

Common Control Channels	Acronym
Access Grant Channel	AGCH
Paging Channel	PCH
Notification Channel	NCH
Random Access Channel	RACH
Common Control Channel (= AGCH + RACH + NCH + PCH)	CCCH

Dedicated Control Channels	Acronym
Full-rate Fast Associated Control Channel	FACCH/F
Half-rate Fast Associated Control Channel	FACCH/H
Slow Associated Control Channel (associated to TCH)	SACCH/T
Slow Associated Control Channel (associated to SDCCH combined with CCCH)	SACCH/C4
Slow Associated Control Channel (associated to SDCCH not combined with CCCH)	SACCH/C8
Slow Dedicated Control Channel (if four of them mapped onto physical channel with CCCH)	SDCCH/4
Slow Dedicated Control Channel (if eight of them mapped onto separate physical channel)	SDCCH/8

Apart from the notification channel introduced in release 1996, all control channels listed above were already contained in GSM phase 1. Further control channels were introduced for EDGE (associated to the E-TCH listed in Table 4.1, see Section 4.5), for GPRS (essentially the packet equivalent of all these channels, as discussed in more detail in Section 4.8), for evolutions to GPRS (refer to Section 4.12), and finally for CTS.

4.3.3 Mapping of TCH and SACCH onto the 26-Multiframe

TCH/F and SACCH are multiplexed as follows onto a 26-multiframe, with frames numbered from 0 to 25: the TCH makes use of the first 12 TDMA frames from 0 to 11, and then again of frames 13 to 24. The SACCH is mapped onto the 13th frame, and the 26th frame is left idle. This is illustrated in Figure 4.8. The idle frame is vital, since it provides the MS with an extended period for the monitoring of neighbouring cells. In this period, it has enough time to receive a frequency correction burst and decode the BSIC contained in a subsequent synchronisation burst from one of the cells it is ordered to monitor.

In the case of half-rate channels, two TCH/H and their respective SACCH are mapped onto the same basic physical channel, as illustrated in Figure 4.9. These two channels are referred to as subchannel 0 and 1 respectively. The SACCH associated with subchannel 0 is mapped onto the 13th frame, that with subchannel 1 onto the 26th frame. In this case, the two mobiles have plenty of extended monitoring time, since active frames alternate with idle frames (exception: frames 12/13).

4.3.4 Coding, Interleaving, and DTX for Voice on the TCH/F

All voice codecs that can be used in GSM deal with voice frames of 20 ms length. Not all bits in a coded voice frame have the same importance. The important bits, which are vital for speech intelligibility and thus must be received correctly, need to be protected well against transmission errors, while other bits can do with less protection to save precious bandwidth. In the following, we discuss coding and interleaving for the GSM full-rate

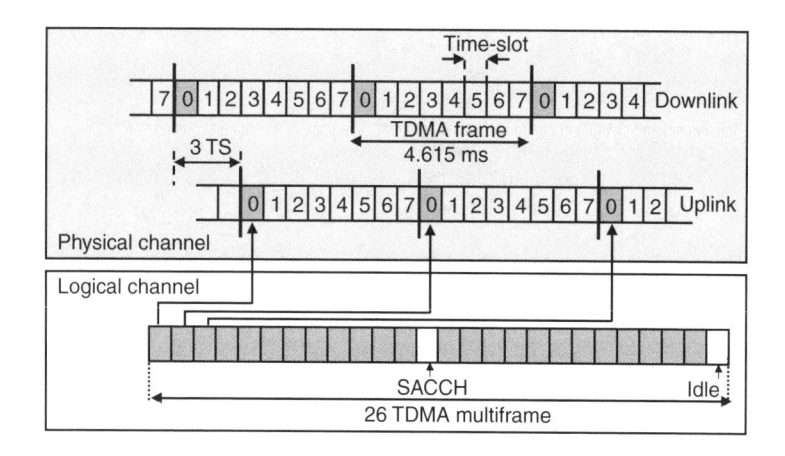

Figure 4.8 Mapping of TCH/F and SACCH on TDMA frames

4.3 MAPPING OF LOGICAL CHANNELS ONTO PHYSICAL CHANNELS 121

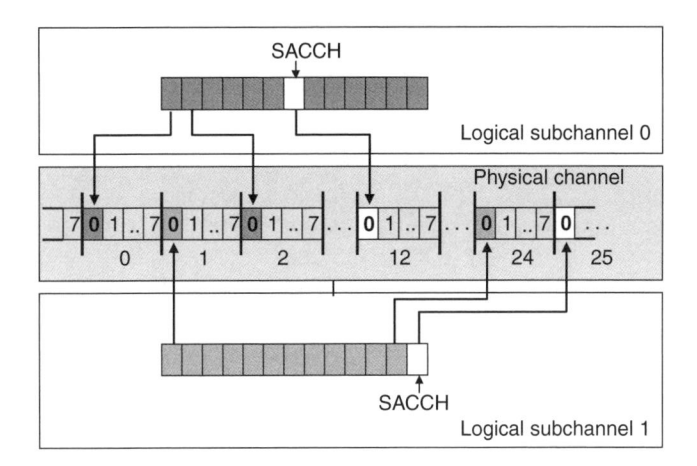

Figure 4.9 Mapping of TCH/H and SACCH on TDMA frames

voice codec, the only codec supported when GSM was launched. A release 1999 system supports also the enhanced full-rate codec, the half-rate codec, and the adaptive multi-rate voice codec, which features eight different modes with different data-rates, allowing the most appropriate mode given current conditions (e.g. the level of interference) to be selected for each frame. The AMR codec can be used on both full-rate and half-rate channels, but not all modes are supported on the half-rate channel, as can be deduced from Table 4.1.

4.3.4.1 Coding and Interleaving for Standard Full-rate Voice

The full-rate voice codec operates at a net bit-rate of 13 kbit/s, it thus generates 260 bits per 20 ms. These are split into two classes of bits, 182 belong to class 1, 78 to class 2, class 1 being more important and thus more error sensitive than class 2. In fact, no error correction coding is applied at all to class 2 bits. Class 1 bits are split into two sub-classes, namely 50 class 1a bits, and 132 class 1b bits. A GSM full-rate voice frame is considered to be in error if one or more of the class 1a bits are in error. To be able to detect whether this is the case, a cyclic redundancy check sequence (CRC) of three bits is calculated over these 50 class 1a bits. All class 1 bits together with these three CRC bits and four tail bits are fed into a convolutional encoder with code-rate $r = 1/2$, thus generating 378 bits per frame. Adding the unprotected class 2 bits results in 456 bits per 20 ms voice frame, corresponding to a gross bit-rate of 22.8 kbit/s. This process is illustrated in Figure 4.10. Coding for the enhanced full-rate codec is essentially the same, but since it operates at a net bit-rate of 12.2 kbit/s, it generates only 244 bits per 20 ms, hence a preliminary coding is required to generate 16 additional bits, before the frame can be subjected to the procedure illustrated in this figure.

The 456 bits per voice frame fit onto eight GSM *half-bursts*, i.e. the encrypted bits either before or after the training sequence of a normal burst. The spare 58th bit available on each half-burst is used to signal whether the transmitted frame is really a voice frame or 'stolen' by the FACCH, which pre-empts the TCH if the need for fast signalling arises, such as in the case of a handover. Accordingly, these 'spare' bits are referred to as *stealing flags*. The interleaving is carried out in a *diagonal* manner, that is, rather than stuffing all bits into four bursts, they are spread over eight bursts. The odd-numbered encrypted

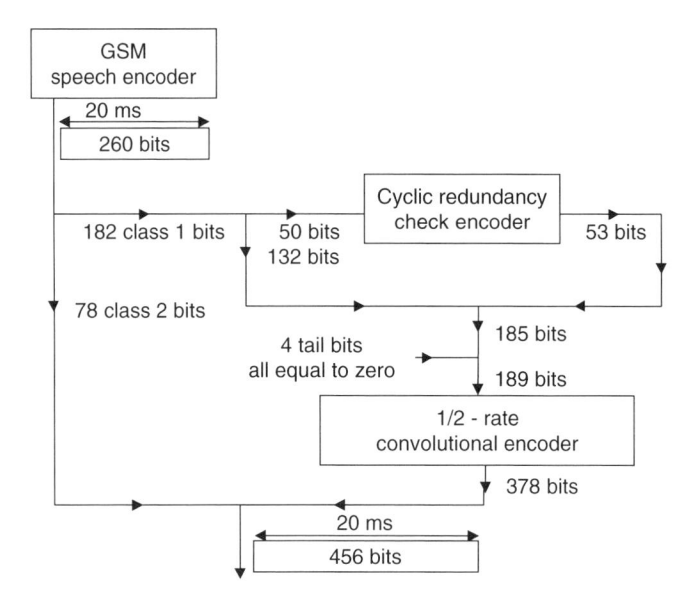

Figure 4.10 Coding of voice frames generated by the full-rate codec

bit positions carry data of one voice frame, while the even-numbered positions carry data from the subsequent frame. Considering a particular voice frame, the 228 bits transmitted in the first four time-slots share bursts with the previous frame, the 228 bits transmitted in the last four time-slots with the subsequent frame. This diagonal interleaving process, which applies for voice on a TCH/F irrespective of the voice codec used, is illustrated in Figure 4.11.

In fast fluctuating propagation conditions, the error performance of bits located far away from the training sequence may suffer from performance degradation due to inaccurate

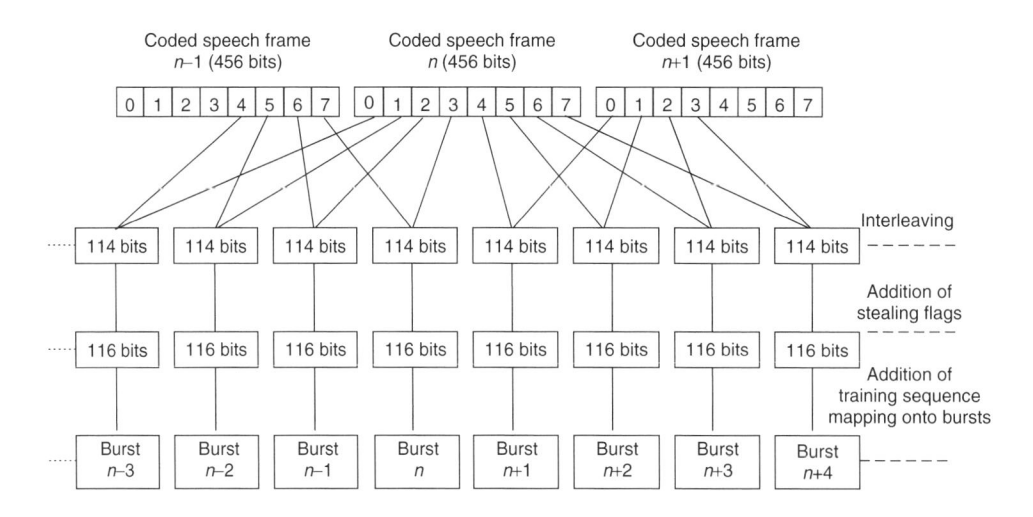

Figure 4.11 Diagonal interleaving for voice on a TCH/F

4.3 MAPPING OF LOGICAL CHANNELS ONTO PHYSICAL CHANNELS

channel estimation and, as a consequence, sub-optimal equalisation. This is why the class 1a bits are located closest to the training sequence.

It now becomes evident how an 'even' voice frame duration of 20 ms can be reconciled with an 'odd' TDMA frame duration of 120/26 ms: in a 26-multiframe, only 24 slots carry the TCH/F. Six voice frames can be carried in these 24 slots, that is 120 ms worth of speech. This matches exactly the duration of the 26-multiframe.

Due to the odd ratio of TDMA frame duration to voice frame duration and the resulting idle frames, the delay experienced on the air interface because of interleaving fluctuates slightly from voice frame to voice frame. The relevant delay is the worst-case delay, which is equal to the duration of eight TDMA frames plus one time-slot, namely 37.5 ms. To avoid jitter (i.e. delay fluctuations), all voice frames experiencing less than the worst case delay need to be 'artificially delayed', which is achieved by using a small play-out buffer. There are further delay sources related to codec and air-interface design, most notably the packetisation delay of 20 ms, since the voice codec needs to collect 20 ms worth of speech on which to perform its algorithm. Other less significant delay sources are speech and channel encoding delay (a few milliseconds) on the uplink, and processing delays for equalisation, channel decoding and speech decoding (again a few milliseconds) on the downlink. GSM 03.50 [189] reports a total delay contribution due to the above described effects of 72.1 ms for the uplink and 71.8 ms for the downlink. This obviously excludes transmission and propagation delays in the fixed network. Again according to GSM 03.50, the total roundtrip or two-way delay within an operator's network including fixed network delays should not exceed 180 ms.

The air-interface-related delay could be reduced, if rectangular interleaving over four time-slots (i.e. the same as on the SDCCH and the SACCH) were applied on the voice traffic channel instead of the diagonal interleaving over eight time-slots. However, this would come at the cost of reduced error performance. Generally, a one-way delay of 100 to 200 ms is considered acceptable for good quality voice conversation. From the above listed delay values, it can be seen that even in the case of a GSM-to-GSM call these requirements can be met, as long as the delay components outside the PLMNs can be kept low. Not surprisingly, therefore, GSM 03.50 recommends avoiding mobile-to-mobile calls with a satellite link in-between.

4.3.4.2 Voice Activity Detection and Discontinuous Transmission

Associated with voice encoding is voice activity detection (VAD). If VAD is applied and inactivity of the speaker is detected, the voice encoder will stop generating voice frames. Hand in hand with VAD goes DTX, which means stopping to transmit anything during such inactivity phases to reduce interference. Apart from saving battery power, this may lead to increased capacity, if the interference generated by a sufficiently large number of users can be averaged through application of slow frequency hopping, as discussed in more detail in Section 4.6.

Even if DTX is applied, the transmitter will not keep quiet during the whole voice inactivity phase and therefore, voice inactivity cannot be translated fully into reduced interference. This is due to four reasons. Firstly, inactivity cannot always be detected reliably in the often noisy environment, in which GSM phones are operating. Secondly, the GSM VAD will wait for an *overhang period* of roughly 100 ms before stopping to generate voice frames even after having detected inactivity. During this period, the parameters required for *comfort noise generation* can be extracted. These parameters allow the regeneration of the background noise at the receiving end (or rather the transcoding point, which is where GSM voice is transcoded to non-compressed PCM voice). This is required for the convenience of the listener, as she could get a rather awkward feeling if voice with background noise were to alternate with, contrary to expectation, completely silent phases. Worse, according to GSM 06.12 [190], DTX switching can take place rapidly, and the on–off modulation of the background noise could actually affect the intelligibility of speech seriously, especially in a car environment with high background noise levels. Thirdly, these comfort noise parameters, contained in silence descriptor (SID) frames, need not only be transmitted immediately after the hangover period, but also updated every so often, which requires the scheduling of eight bursts in every 104 bursts allocated to the respective TCH during silence phases. Finally, for radio link control purposes, the SACCH must remain active also during voice inactivity phases.

For further information, refer to the 06-series of GSM specifications dealing with speech aspects, in particular GSM 06.31 [191], which contains an overall description of DTX operation, GSM 06.32 [192] on VAD, and GSM 06.12 [190] on comfort noise.

4.3.5 Coding and Interleaving on the SACCH

Coding and interleaving of signalling frames sent on the SACCH is the same as that on the BCCH, AGCH, PCH, SDCCH, and FACCH. In all cases, 23 net octets (where 1 octet = 1 byte = 8 bit) need to be error coded. First, 40 extra bits used for error correction and detection are added to this block of 184 bits according to a shortened binary cyclic code (a so-called FIRE code). Adding four tail bits and performing rate 1/2 convolutional coding result again in 456 bits to be transmitted per signalling frame over the air interface. This time, unlike in the case of voice, these bits are interleaved over the payload portion of four full bursts; that is, there is no inter-frame interleaving. In the following, such a group of four bursts is often referred to as a *block*. This type of interleaving is also called *rectangular* interleaving, as opposed to the diagonal interleaving applied on the TCH. Since the SACCH is assigned a burst every 26th TDMA frame, SACCH frames are transmitted at a rate of one per 104 TDMA frames, or one every 480 ms, providing sufficient time diversity. With diagonal interleaving over eight bursts, an additional delay of 480 ms would have been incurred on the SACCH without significant performance gain.

Two of the 23 net octets on a SACCH frame are used for physical layer information (TA and power control), as outlined earlier. This means that the update interval for TA and power control orders is also 480 ms. As far as power control is concerned, this equates to so-called *slow power control*, which can compensate the path-loss and slow fluctuations on the radio propagation channel (i.e. shadowing), but is too slow to combat fast fading. On the BCCH, AGCH, PCH, SDCCH, and FACCH, by contrast, all 23 octets are used for higher layer signalling.

4.3.6 The Broadcast Channel and the 51-Multiframe

The 51-multiframe carries FCCH, SCH, BCCH, CCCH, and SDCCH (with its SACCH) in various combinations. Two fundamental options are distinguished.

The first option is to map all channels combined onto exactly one basic physical channel, namely TN 0 on a 'reference carrier' or BCCH carrier (denoted C0 in GSM). This is only suitable for low-traffic cells (e.g. cells with a single frequency carrier), as this combined channel offers limited PCH, AGCH and RACH capacity, as well as only four SDCCH subchannels. Together, these four SDCCH subchannels are referred to as SDCCH/4.

As a second option, the SDCCH subchannels can be mapped in groups of eight, referred to as SDCCH/8, onto one or more separate basic physical channels. In this case, a cell may feature up to four physical channels carrying a CCCH. The first one is still mapped onto TN 0, C0, which carries also the BCCH (including SCH and FCCH). Further CCCHs, where required, are mapped onto TN 2, 4, and 6 of C0. Since there must be only one SCH and FCCH in a particular cell, these additional physical channels do not carry SCH and FCCH. However, BCCH system information messages signalled on TN 0 are also signalled on these additional channels. The basis physical channel(s) used for the SDCCH/8 can be mapped onto any time-slot not used by the CCCH, or even on another carrier. An example downlink mapping of FCCH, SCH, BCCH and CCCH onto the appropriate TDMA frames on TN 0, C0, is shown for this 'non-combined' case in Figure 4.12. The uplink mapping is trivial in this case: every slot is available for the RACH. In the normal case illustrated here, there is exactly one BCCH block per 51-multiframe, allowing one system information message to be broadcast per multiframe. The BCCH may be extended to two blocks per 51-multiframe, if required.

As there can be more than one basic physical channel supporting a CCCH, mobiles need to figure out which CCCH to use, i.e. to which PCH they should listen and which RACH they must use. This can be achieved through a formula which, given the channel configuration in the cell, maps the mobile's International Mobile Subscriber Identity (IMSI) to the CCCH to be used.

The CCCH blocks available on the downlink are dynamically shared between AGCH and PCH, according to a number of parameters broadcast on the BCCH. These parameters also determine on which subset of PCH blocks mobiles pertaining to different so-called *paging groups* (again determined through the IMSI) can expect to be paged. The definition of these paging groups allows them to go into 'sleep mode' during all other PCH blocks to save battery power. This is referred to as discontinuous reception (DRX), which is the counterpart to DTX on the transmit side.

In all cases (i.e. whether on separate channels or combined with a CCCH), the SDCCH is organised in a manner such that one block of four bursts per subchannel is carried in one 51-multiframe. For the SACCH subchannel associated with each SDCCH subchannel, one block of four bursts is only set aside every second multiframe. On the SDCCH/8, this adds up to 48 bursts per 51-multiframe, leaving three frames per multiframe idle.

The CBCH (downlink only) is mapped onto the same basic physical channel as the SDCCH. If CCCH and SDCCH are combined, the CBCH takes a block per multiframe

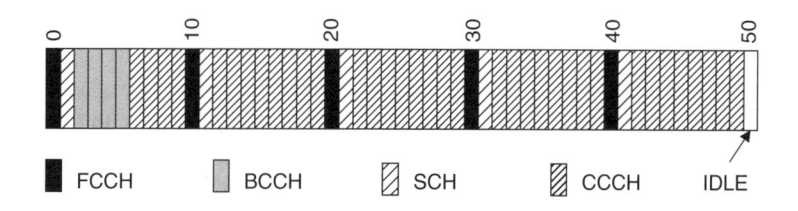

Figure 4.12 TDMA frame mapping on the downlink for the non-combined CCCH

away from the AGCH, if they are not, and thus the SDCCH is mapped onto a separate physical channel, the CBCH takes a downlink block away from the SDCCH.

One obvious question is why there needs to be a 51-multiframe for certain control channels as opposed to the 26-multiframe for traffic channels. We have mentioned earlier that mobile terminals engaged in a conversation must monitor neighbouring cells. The monitoring process entails the decoding of their BSIC every so often, to make sure that the right cells are monitored. This in turn implies that the MS must be able to listen to a frequency correction burst in one TDMA frame followed by a synchronisation burst carrying the BSIC in the next frame. The required time is only available during the idle frame in the 26-multiframe. Having an odd relation between the length of the multiframe used for the BCCH and that for TCH ensures that every so often, the FCCH and the SCH fall into this extended monitoring window, without requiring any synchronisation between base stations.

BCCH frequencies are required to transmit continuously at maximum power, enabling mobiles to perform measurements of the signal strength received from neighbouring cells. Therefore, power control and DTX cannot be applied on these downlink frequencies. Furthermore, the time-slots onto which CCCHs are mapped (i.e. TN 0 on C0, and potentially also TN 2, 4, and 6 on C0) are not allowed to hop. As far as TN 0 is concerned, this is to ease the task of initial synchronisation for the MS, as it can expect the synchronisation burst of the SCH to be sent on the same frequency as the frequency correction burst of the FCCH. Furthermore, if the BCCH were hopping, the SCH burst would have to contain a description of the BCCH hopping sequence, for which it is too short. Regarding TN 2, 4, and 6 in the case they carry common channels, if PCH, AGCH and RACH were hopping channels, their hopping sequence would have to be broadcast on the BCCH, providing little gain compared to the increased complexity [176]. The implications this restriction has on frequency planning will be discussed in Section 4.6.

4.4 The GSM RACH based on Slotted ALOHA

4.4.1 Purpose of the RACH

Users walking or driving around with a mobile phone which is switched on should *ideally* only consume precious and limited radio resources while they are engaged in a call or while sending and receiving short messages or other user data. In practice, some signalling resources are also needed for other purposes such as for the 'IMSI attach procedure', which is used to register when a phone is being switched on, and for the 'location updating procedure', to track user mobility. This was already discussed in Section 3.3. All the same, a mobile terminal will not consume any radio resources during most of the time, which in turn means that when it needs them, it has to access the system somehow and request resources.

This is exactly what the RACH, a multi-access channel, is used for. It is a common resource put aside for mobiles to request some dedicated resources by sending one or more *channel request messages* as a result of initiating the *immediate assignment procedure*. This procedure is completed successfully when the mobile station receives an *immediate assignment message*. If no suitable resources are available, the network may respond with an *immediate assignment reject message*. This is illustrated in Figure 4.13.

4.4 THE GSM RACH BASED ON SLOTTED ALOHA

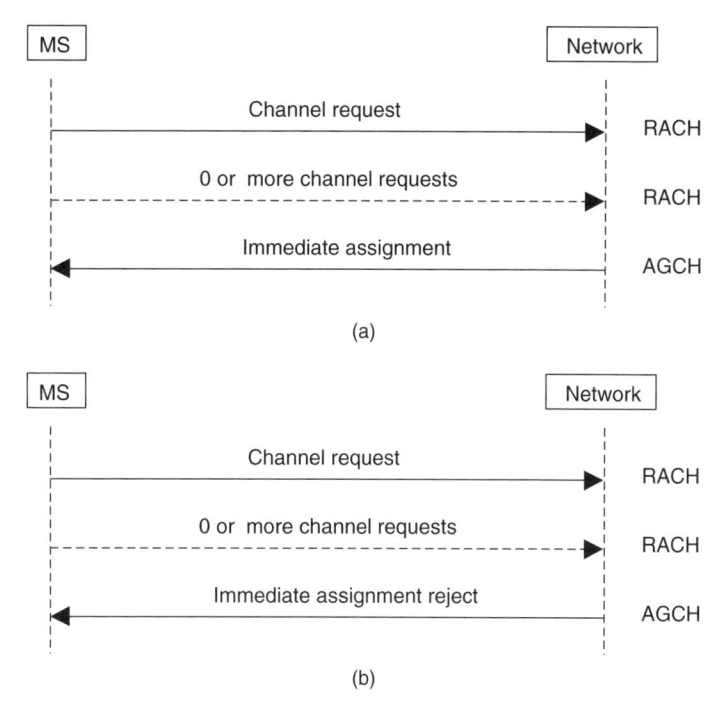

Figure 4.13 (a) Message exchange for successful immediate assignment procedure (b) Message exchange for unsuccessful immediate assignment procedure

4.4.2 RACH Resources in GSM

As pointed out in Section 4.3, the RACH is the uplink part of the CCCH. Since, as discussed in Section 3.3, the PCH is usually the bottleneck in the system, the resources available on the RACH are determined by those required for the downlink part of the CCCH. Therefore, abundant resources are normally available on the RACH. If CCCH and SDCCH are not combined, the RACH uses the full uplink capacity of a basic physical channel. Therefore, if a single such CCCH is sufficient for a given cell, 780 000 RACH time-slots are available per hour. In the worst case, namely with a combined CCCH and SDCCH, where 24 out of 51 slots in a 51-multiframe are used for SDCCH and SACCH, only 413 000 RACH slots would be available per hour. We will discuss later that this amount of RACH resources should indeed be enough for 'plain' phase 2 GSM.

4.4.3 The Channel Request Message

The channel request message is transmitted by the MS on a single time-slot pertaining to its RACH (recall that there can be more than one RACH mapped onto different time-slots, but that each MS uses only one particular RACH). The access burst format used for this purpose has already been discussed in Section 4.2, together with the rationale for the extended guard period and the synchronisation sequence, which reduce the gross payload to 36 encrypted symbols. Due to the contention-based nature of this channel and the fact that no inter-burst interleaving is applied, the error correction coding must be particularly strong. In GSM phase 2, for instance, only eight bits are available for the channel request message, while the remaining 28 add redundancy to provide robustness. Some of these eight bits carry a random reference value (the purpose of which is explained in the subsection below on contention resolution), and the remaining few bits must suffice to code the *establishment cause*. This establishment cause can indicate:

- an emergency call;
- call re-establishment following link failure (some kind of emergency handover);
- an answer to paging (containing also information on the content of the paging message and the capability of the mobile station);
- an originating call (with indication of type of call, i.e. voice or circuit-switched data, and whether a TCH/H is sufficient);
- a location updating procedure;
- other procedures for which an SDCCH is sufficient (e.g. SMS); or
- a channel request message related to GPRS, to be discussed in more detail in Sections 4.10 and 4.11.

4.4.4 The RACH Algorithm

As the name of the channel implies, the medium access protocol used on the RACH is based on random access, that is, on an ALOHA-type protocol. More precisely, since we are dealing with a time-slotted channel, it is a variant of the S-ALOHA protocol. The precise operation of the protocol is determined by three parameters, namely MAX_RETRANS, TX_INTEGER, and the S-parameter. The first two parameters are signalled in the RACH control parameter information element contained in the most important BCCH system information messages in a manner that they can be updated once roughly every 51-multiframe. The value of the S-parameter depends on TX_INTEGER and the type of CCCH (i.e. whether combined with an SDCCH/4 or not), as outlined in GSM 04.08 (or GSM 04.18 for R99), where the interested reader can also find further details on the matters dealt with in the following. Additional information can also be found in Reference [193], together with a performance assessment of the GSM random access procedure.

If a mobile terminal wants to start an immediate assignment procedure (e.g. following a paging request, or because its user has hit the 'send' button to initiate a call), it needs to establish first if it is allowed to access the cell. Every MS with an inserted SIM card containing the Subscriber Identity Module is a member of one of 10 access classes, as determined by subscription information stored in that SIM card. Furthermore, mobiles might also be a member of one or more out of five special access classes, again based on information held on the SIM card. The RACH control parameter information element contains a bitmap, which indicates for each class whether its members are allowed to access the cell. It contains also bits indicating whether emergency calls and call re-establishment are permitted. Finally, there is a cell bar bit: if set, nobody is allowed to access the cell on the RACH.

Once an MS has established that it is allowed to access the cell, it will schedule the sending of MAX_RETRANS + 1 channel request message on its RACH. This is where the details of the S-ALOHA algorithm and the parameters mentioned above come into play. The meaning of MAX_RETRANS should now be evident: it denotes the maximum number of allowed retransmissions of a channel request message following the first transmission attempt. This parameter can assume the values 1, 2, 4 and 7 (coded in two bits).

The time-slot in which the first channel request message is sent is drawn at random from the eight or TX_INTEGER slots (whichever is lower) pertaining to the RACH which follow the initiation of the immediate assignment procedure. TX_INTEGER can assume 16 different values from 3 to 50.

The number of RACH time-slots *between* two successive channel request messages (excluding the slots on which the messages are sent) is selected by drawing a random value for each retransmission with uniform probability distribution in the set $\{S, S + 1, \ldots, S + TX\_INTEGER - 1\}$. In other words, the retransmission interval is the sum of a deterministic component S and a random component. This is illustrated in Figure 4.14, showing only time-slots belonging to the considered RACH (i.e. one time-slot per TDMA frame in the 'non-combined case'). In this example, $TX\_INTEGER < 8$.

If the base station receives a channel request message correctly, and a channel of the requested type (e.g. an SDCCH subchannel) is available, it will respond with an immediate assignment message. This message contains the description of the dedicated channel to be used — the operator may choose to assign an SDCCH, even if a TCH was requested, and only assign a TCH after completion of some call set-up-related signalling exchange. It also contains the initial timing advance value to be used, which was calculated by the network based on the timing of the received channel request message. Finally, it repeats the content of the channel request message, and indicates the frame number in which the message was received. Similarly, if no channel of the requested type is available, the BTS will send an immediate assignment reject message (which includes again the original channel request message and the frame number). The MS, in turn, will stop sending channel request messages as soon as it receives such a message with the right frame number, establishment cause and random reference (if it has not already stopped because it sent MAX_RETRANS + 1 requests).

Assume for now that if a channel request message is received successfully by the base station, it sends the immediate assignment (reject) message before the MS schedules a further channel request message. If we ignore the S-parameter, in essence, the RACH algorithm in GSM is S-ALOHA with uniform retransmission intervals as determined by TX_INTEGER. In Chapter 3, on the other hand, we were mostly focusing on algorithms with a given transmission permission probability p indicated on a per slot basis, which

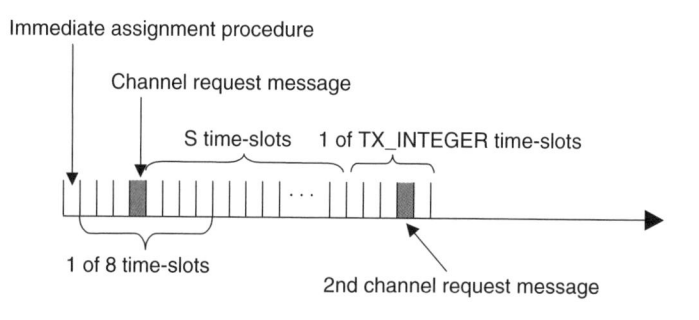

Figure 4.14 Scheduling of channel request messages

requires the execution of a Bernoulli experiment in each slot and results in a geometric distribution for the retransmission interval.

According to Reference [194], the average delay behaviour of the protocol is quite insensitive to the exact distribution of the retransmission interval. Therefore, the protocol performance should be very similar for these two cases, if $p = 1/\text{TX\_INTEGER}$, such that the same average retransmission interval results. This also means that we will have to expect the same bi-stable behaviour as reported in Reference [125], which is intuitively clear. Should it happen that a few mobiles initiate the immediate assignment procedure simultaneously, if TX_INTEGER is not large enough, there will be an increased likelihood of repeated collisions, exactly in the same manner as with a p chosen too large. The protocol may then get caught in an operating point with low throughput and high delay.

Let us now discuss the extension of the retransmission interval by a constant number of RACH slots, as determined by the S-parameter. There are two sets of five different values for this parameter, one set for the case where the CCCH is combined with the SDCCH/4, and one for the non-combined case. On the former, S can assume values from 41 to 114, on the latter from 55 to 217, in both cases quite substantial compared to TX_INTEGER, leading to an additional retransmission delay of at least 250 ms in the non-combined case.

At first glance, this increased delay is incurred without gaining stability, since S assumes the same deterministic value for all mobiles, thus unlike TX_INTEGER, it does not help in spreading channel request messages of mobiles previously involved in collisions over different time-slots to reduce the collision probability. The question is therefore why this S-parameter was introduced in the first place. Well, the base station may not be able to send an immediate assignment (reject) message as soon as it receives a channel request message, because of limited occurrence of available AGCH blocks (on which to send this message) on the downlink part of the CCCH. Furthermore, it may also need some time to process the request. If the S-parameter did not exist, an MS would potentially retransmit the channel request message MAX_RETRANS times even though the first request was received successfully by the base station. This would cause unnecessary load on the RACH and drain further processing resources from the BS for the processing of these unnecessary requests. Even worse, the BS cannot tell whether these request messages were sent by one or multiple mobiles, since they do not contain mobile station identifiers. It may therefore send multiple immediate assignment messages to the same MS and reserve dedicated resources with each of these messages, thus wasting SDCCH or even TCH resources. An additional reason listed in Reference [193] is the possibility to queue a request during this fixed part of the retransmission interval while waiting for a dedicated channel to free up. It should now be clear that the existence of the S-parameter is well justified. All the same, the minimum value it can assume seems to be unnecessarily large.

In summary, the GSM RACH algorithm conforms to a relatively plain S-ALOHA protocol with uniformly distributed retransmission intervals. To be precise, also the initial transmission does not necessarily occur on the first available RACH slot, but is rather spread over up to eight slots. The rather loose update interval of the RACH control parameters (roughly once every 51-multiframe) means that efficient stabilisation of the protocol is not possible. It is not even clear from GSM 04.08 whether an MS, after scheduling the first transmission attempt, is obliged to take updates of the RACH control parameters into account for the scheduling of subsequent retransmission attempts. Furthermore, the range of possible parameter values will cause high access delays, as soon as retransmissions are required, that is, if the initial channel request message is corrupted due to bad propagation conditions or suffered a lethal collision with request messages sent at the same time by other mobiles. Should the protocol suffer stability problems, that is, get caught in an operating point with low throughput and high delay, the protocol needs to be restarted by denying (possibly gradually) all access classes access to the cell, until the RACH is cleared again.

Fortunately, in GSM phase 2, collisions should be rare, since the RACH load is extremely low, as discussed in more detail below. Packet capture, which is possible owing to the lack of power control on the RACH, the strong error protection, and the fact that channel request messages may only partially overlap due to the extended guard periods and lack of timing advance, will further reduce the likelihood of stability problems. Finally, the finite number of permitted retransmissions will prevent mobiles from remaining backlogged over extended periods of time. Note also that for procedures such as setting up a circuit-switched call, transmission of short messages, etc., access delays in the order of a few hundred milliseconds, even a few seconds, should not affect the perceived quality of service significantly, and can therefore be tolerated.

4.4.5 Contention Resolution in GSM

A channel request message does not identify the MS sending this request. The eight-bit message is composed of an establishment cause and a random reference value. The most common establishment causes for GSM phase 2 are coded using three bits, leaving five bits for the random reference. Establishment causes which were added for GSM phase 2+ share leftover code-points, resulting for instance in GPRS-related establishment causes using at least five bits (see again GSM 04.08 for details).

To identify the MS which sent a particular channel request message, the base station will mirror the content of this message back in the immediate assignment (reject) message, together with the number of the frame in which the request was received. The slot number is redundant, since both the request and the assignment message must be sent on the same CCCH. On a perfect collision channel, an assignment message specifying in which RACH slot the request message was received would identify unambiguously the mobile terminal to which the assignment message is addressed, since it must have been the only terminal using that particular RACH slot. However, in a GSM environment, where a request message may be captured by the base station even in the presence of other request messages sent on the same time-slot, this is not the case. This is exactly why a random reference needs to be included in the request message: to distinguish between two mobiles having sent a request with the same establishment cause using the same RACH slot.

Due to the limited number of bits available for the random reference, there is a non-negligible probability that two mobiles using the same RACH slot with the same establishment cause end up picking the same random reference value. Provided that one of the two channel request messages is captured by the BTS, and that both mobiles manage to receive the resulting immediate assignment message, they will start transmitting on the same dedicated channel. This situation needs to be resolved, which is referred to as *contention resolution* in GSM. It should be noted that this contention resolution is quite different from the collision resolution algorithms discussed in Section 3.5 associated with the MAC layer. The GSM contention resolution occurs after successful execution of the 'MAC algorithm' and, since contention resolution occurs on a dedicated rather than a shared channel, it is handled by the DLL.

The solution, which is described in detail in GSM 04.06 [195], is the following: the layer 3 service request (or location updating request) message sent by the MS in the first DLL frame is mirrored back by the network in its response DLL frame. The mobile terminal, having stored the content of this message, will compare it with the received message, and leave the channel if there is a mismatch. This approach must work, since the service request message contains an unambiguous mobile identity, either the Temporary Mobile Subscriber Identity (TMSI) or the IMSI. If a collision occurred on the RACH with capture of the stronger of two or more identical channel request messages, two things may happen. The network may be able to capture also one of the DLL frames (possibly after a few retransmissions), and mirror the layer 3 content back, in which case only the 'captured mobile' will survive on the dedicated channel, as desired. Although unlikely, this could potentially even be a different MS than that 'captured on the RACH'. The other mobiles would have to initiate a new immediate assignment procedure. Alternatively, the signalling exchange on the dedicated channel may fail altogether, requiring all mobiles involved in the collision to start over again.

4.4.6 RACH Efficiency and Load Considerations

Usage of RACH resources in GSM is inefficient. This is not so much due to the limited theoretical protocol throughput of 37%, which compares to up to 100% on a TCH. True, with capture, the throughput could be higher, but since we are dealing with a non-stabilised version of S-ALOHA, 'viable' throughput ratios may be significantly lower than the theoretical maximum. The main inefficiency comes from the fact that a RACH burst carries only eight information bits, whereas a burst on a TCH/FS, deducting FEC overhead, carries 65 net voice bits. This is due to factors specific to cellular communication systems, namely the maximum propagation distance to be catered for, requiring extra guard periods, and the adverse propagation conditions, requiring strong error protection.

The minimum amount of available RACH resources provided by the uplink part of a CCCH combined with an SDCCH/4 is 413 000 slots per hour. There is no immediately obvious alternative use for these slots. Often, the capacity requirements for paging and access grant traffic are such that one or several 'non-combined' CCCHs are needed, offering 780 000 RACH slots per hour and CCCH. Through considerations on the expected RACH traffic in a GSM system offering only phase 2 services, we will demonstrate that there are more than enough RACH resources, such that the inefficient use of these resources is not a cause for concern. The situation may be significantly different once GPRS is introduced, though, as will be discussed in more detail in subsequent sections of this chapter.

In GSM phase 2, the load on the RACH is mainly made up of three components:

(1) channel requests to set up a mobile originated or mobile terminated call;

(2) channel requests for transmission or reception of short messages; and

(3) channel requests for location updating procedures.

Apart from the number of mobile terminals camping on the cells of a particular network (that is, listening to their broadcast channel), the total load depends on many factors, which can vary from network to network. We will perform some 'thought experiments' in the following based on publicly available figures, to determine roughly the contribution of the first two components, using the UK as an example. The contribution of the third component is more difficult to estimate.

The number of cells and the number of carriers in each cell must be planned by the network operator in such a way that they can carry peak traffic levels without having to block more than a few per cent of the service requests of the network customers. Judging from the calculations provided in Section 4.6 dealing with the utilisation of dedicated channels, 10 000 call set-up attempts per hour originating in a particular cell should be a generous estimation, to be exceeded only in the busiest of cells (if at all). In such cells, we would expect to have at least one 'non-combined CCCH' (most likely more), resulting in an input traffic (normalised to the number of available RACH slots) of barely more than 1% (which would be 7800 attempts per hour for a single 'non-combined RACH').

According to the GSM Association [70], 750 million messages were sent in a single month, namely December 2000, in the UK alone. There are four GSM network operators in the UK. As a very conservative (and for our purposes worst-case) estimate, assume that they have deployed together 10 000 cell-sites with omni-directional base stations. Based on numbers published by some operators, the current total is certainly higher, since more than 15 000 sites were in operation already by the end of 1998, and a lot of these sites feature three sectors, each with its own RACH. The SMS traffic would have to grow to three billion messages per month to exceed on average 10 000 messages per base station and day based on the worst-case assumption. Even if the temporal and spatial distribution of the messages sent are extremely uneven, the normalised RACH input traffic due to SMS during busy hours is unlikely to exceed 1% for any given base station. True, it is possible that three billion messages will seem rather modest by the time you read this, but together with the traffic, the number of base stations will have grown as well.

The frequency of channel request messages for location updates depends (apart from the subscriber number) mainly on two factors. The first is how often (if at all) the network requests periodic location updates to be sent (the cells onto which the author's mobile phone camped, while writing this text, requested one such update every hour). The second relates to mobility patterns of the users and the size of location areas, since the location must be updated when crossing location area borders.

Consider first the RACH load contribution due to periodic location updating. Assume an operator with 10 million subscribers, all having their handsets switched on permanently, and 2500 omni-directional base stations. For the UK, this is again a very conservative assumption, not so much on the subscriber side (at the time of writing, three UK operators already exceeded the 10-million-subscriber mark), but on the base station side[2]. Under these assumptions, periodic location updating once every hour would result on average in 4000 immediate assignment procedures per hour and base station, resulting in a normalised RACH input traffic of clearly less than 1%. Allowing again for a certain variance in the distribution of users over base stations, certain base stations could experience a normalised RACH traffic due to periodic location updates exceeding 1%.

The network operator will choose a location area size that provides the best trade-off between frequency of mobility induced location updates (the lower, the larger the area) and load on the paging channel (the lower, the smaller the area). This trade-off depends

---

[2] One GSM 1800 operator claimed to have 5300 sites in operation in April 1999 and planned to reach 10 000 in the year 2001.

on numerous parameters, such as user mobility and frequency of incoming calls per user, which would all have to be modelled first, to be able to estimate the RACH load due to mobility-induced location updating procedures. Interestingly, while writing this text, the author observed that he must have been living on the border of two cells, which were received at roughly equal signal strength, causing frequent cell reselections even though his handset was lying still on a table. These cells happened to be in different location areas, triggering a location updating procedure whenever a reselection occurred. Such real-life effects make the modelling even more difficult. It would be convenient to have some measured data from real networks. Typically, though, network operators are rather reluctant to give away such data (if they have it in the required shape in the first place), and the authors did not attempt to get hold of it.

So, where does this leave us? Always under the assumption of a single 'non-combined CCCH' per cell, based on worst-case estimates, we ended up with a normalised input traffic for the RACH of not more than 3%. However, this is before accounting for mobility-induced location updating procedures for which we cannot provide estimates. What we can say is that there could be many more than periodic location updates, and the normalised RACH load would still remain in the single per cent figures, at which level collisions on RACH slots are very rare. We can therefore be fairly confident about two things:

(1) stability should not be a problem on the RACH for a GSM phase 2 system;

(2) there is even spare capacity for additional RACH load generated by new services, which are being introduced as part of phase 2+ enhancements.

Indeed, the latter was a design assumption made for GPRS, as will be discussed in more detail in Section 4.7.

4.5 HSCSD and ECSD

4.5.1 How to Increase Data-rates

It would be desirable to support higher data-rates in GSM in a seamless manner; that is, building as much as possible on the existing infrastructure and the existing air interface. In particular, it would be desirable to keep the carrier spacing of 200 kHz and the GSM slot/frame structure. Within these constraints, there are three obvious ways of increasing user data-rates. These are listed together with the price at which this increase comes.

(1) One could increase the code-rate, that is reduce the FEC coding redundancy, which would allow for a theoretical user data-rate per slot of up to 22.8 kbit/s. However, with decreased redundancy comes decreased quality, or put the other way round, to maintain a given quality (in terms of error rates), the signal-to-interference-plus-noise ratio would have to be increased. This in turn means that less channels can be offered per cell with a given spectrum allocation.

(2) Basic physical channels could be pooled together; that is, a single user would transmit and receive on more than one time-slot per TDMA frame. The immediate effect seems to be the following: as users occupy n physical channels, the number of users that a cell can sustain simultaneously shrinks also by a factor of n. In reality,

however, the price to be paid is higher, due to the reduction in trunking efficiency. This is discussed in detail in the next section.

(3) Instead of GMSK, where one modulated symbol corresponds to one bit, higher order modulation schemes could be used, where every modulated symbol 'carries' multiple bits. Such schemes are less robust than GMSK. As with increased code-rates for error coding, higher SINR would be required to maintain a given quality level.

Along the GSM evolution path, all these methods have been introduced. Somewhere between R96 and R97 (depending on which series of GSM specifications is being looked at), the TCH/F14.4 was introduced, increasing the user data-rate available on a single basic channel from 9.6 kbit/s to 14.4 kbit/s. For the TCH/F14.4, the same half-rate convolutional encoder is used as for the TCH/F9.6, but more encoded bits are punctured (i.e. deleted) to reduce the redundancy and increase the net user data-rates. R96 brought us also HSCSD, allowing up to eight time-slots to be allocated to a single user. GPRS, the main new feature in R97, provides also multi-slot operation and multiple coding schemes, including a 'non-coding' scheme, to maximise the user data-rate when good propagation conditions and little interference are experienced. Finally, owing to the introduction of 8PSK modulation in release 1999, user data-rates can in theory be trebled. From an air interface perspective, all these features can be combined, i.e.:

- multiple TCH/F14.4 can be aggregated;
- for 8PSK, using the same FEC code-rate as on the TCH/F9.6 and TCH/F14.4, the E-TCH/F28.8 and E-TCH/F43.2, respectively, are defined (additionally, an E-TCH/F32.0 is also defined, as listed in Table 4.1, which uses a different coding-scheme); and
- with the exception of the E-TCH/F32.0, these E-TCH can also be aggregated.

The theoretical upper limit for the speed of circuit-switched data services would therefore be 8×43.2 kbit/s $= 345.6$ kbit/s. However, the GSM core network was not designed for these data-rates, so these figures are for the time being indeed only of a theoretical nature.

4.5.2 Basic Principles of HSCSD

For R98 and earlier releases, HSCSD stage 1 and stage 2 descriptions (relating to service aspects and network aspects respectively) can be found in the ETSI specifications GSM 02.34 [75] and GSM 03.34 [196]. From R99 onwards, these descriptions are covered in 3GPP technical specifications (TS) 22.034 [197] and 23.034 [198]. As usual, air-interface-related matters are to be found in the GSM 05 series.

The basic design principle for the chosen HSCSD solution was to keep it simple and to introduce the minimum number of modifications to existing specifications, such that this service can be introduced through a pure software upgrade of GSM network elements (the handsets are a different matter). As a result, the existing physical channels, burst formats, etc., are reused for HSCSD. Proposals such as new burst formats extending over two time-slots, which would result in a reduction of the relative overhead by using a single training sequence and eliminating the need for guard periods between these two slots, were rejected.

HSCSD allows the aggregation of up to eight time-slots or basic (full-rate) physical channels for a single user. This user allocation is referred to as *HSCSD configuration*, a special case of the generic *multi-slot allocation*. The basic physical channels are exactly the same as those in GSM with one exception: an HSCSD configuration can include unidirectional downlink-only channels, denoted TCH/FD, provided that at least one channel in the configuration is a conventional bi-directional channel. Configurations that include unidirectional channels are *asymmetric*, else they are *symmetric*.

Additional restrictions regarding the channels included in an HSCSD configuration are that they need to be on the same carrier, and that they are of the same type, i.e. the same coding scheme is applied. The TCH types that may be used in a multi-slot configuration are the TCH/F14.4, the TCH/F9.6, and the TCH/F4.8. If frequency hopping is applied, the channels need to hop together, which means that they must use the same frequency hopping sequence, so that only one hop per link direction and TDMA frame is required.

Some interfaces in the circuit-switched core network of GSM support only up to 64 kbit/s per circuit, which is why the maximum Air-Interface User-Rate (AIUR) is currently limited to 57.6 kbit/s. Therefore, the practical upper limit for multi-slot allocation seems to be, at least for the time being, an allocation of no more than four TCH/F14.4 (although GSM 04.21 [199] lists among the possible combinations also five TCH/F14.4).

HSCSD can be used for transparent and non-transparent circuit-switched data services. According to GSM 03.10 [200], a *transparent* service is characterised by constant throughput, constant transit delay and variable error rates. A *non-transparent* service, on the other hand, is characterised by improved error rates owing to the application of an ARQ mechanism, which in turn results in variable transit delay and throughput. The radio link protocol (RLP) providing this ARQ mechanisms is defined in GSM 04.22 [201] (TS 24.022 [202] for R99). The AIUR must remain constant during the entire call when a transparent service is requested, while it may fluctuate for non-transparent calls. This has implications on handover and resource utilisation, as discussed below. Note also that asymmetric HSCSD configurations can only be used for the non-transparent service.

Apart from restrictions relating to frequency hopping and handovers, from a physical-layer point of view, transmission occurs on independent TCHs. Accordingly, each TCH has its own SACCH and is power controlled individually. However, power control information on downlink unidirectional channels is ignored.

4.5.3 Handover in HSCSD

HSCSD offers full GSM mobility. Should a handover be required, then all traffic channels being part of a HSCSD configuration must be handed over simultaneously. Therefore, a single FACCH associated with only one traffic channel is sufficient for an HSCSD configuration. This traffic channel is the *main* channel in the configuration, and must be a bi-directional TCH.

In the case of the transparent service, where the AIUR must remain constant, the target cell must provide the necessary amount of channels to sustain this user-rate. Otherwise the handover cannot be executed, and if the serving cell cannot sustain the call (e.g. because of deteriorating transmission and reception quality) and no other suitable handover target cell can be found, the call will have to be dropped. This has implications on resource utilisation. Already from a single-cell point of view, aggregation of traffic channels will result in reduced trunking efficiency, as discussed in detail in Section 4.6. However, in a cellular environment with user mobility, we must look beyond the borders of single cells. If a user wants to set up a call in one cell, using for instance four traffic channels, the network operator should ideally not only check whether the required resources are available in that cell, but also whether the call could be sustained if the user were to move to other cells. This is also true for basic GSM single-channel operation, and may require the operator to put aside a certain fraction of channels in each cell for calls to be handed in (as opposed to calls being set up in the cell via RACH). However, the fraction of resources, which needs to be reserved to guarantee a certain quality of service in terms of call dropping rate, is larger for HSCSD than for basic GSM. This is because if four GSM calls on full-rate TCHs each are set up in a cell, the likelihood of all four users moving simultaneously from this cell to the same neighbouring cell is low — unless we are dealing with scenarios such as multiple users moving together, e.g. in a train.

Excluding such special cases, betting on averages, we might get away with putting only one channel aside in each of the neighbouring cells for a potential handover. Not so with a single HSCSD call on four traffic channels.

In the case of a non-transparent service, the AIUR may be changed during a call, provided that the maximum rate specified at call set-up is never exceeded. The user may request an in-call change in the maximum number of TCHs and the user-rate, and so can the network. The latter is useful for handover purposes, for which the network may downgrade the user-rate, such that it is enough for a handover target cell to have only one traffic channel available at the time a handover is required to prevent the call from being dropped. Should further channels free up later, the user-rate can again be upgraded. This will reduce the call dropping ratio at a fixed resource utilisation or, alternatively, improve the resource utilisation at a fixed call dropping ratio.

4.5.4 HSCSD Multi-slot Configurations and MS Classes

We have seen to what great length the designers of GSM went to enable half-duplex operation in mobile handsets. The question that arises now is whether half-duplex operation is still possible if more than one time-slot is allocated to a single MS. The answer is yes, but only up to a given number of slots, and even then only provided that these slots are allocated in specific configurations. What the constraints on the number of slots and the slot configuration are, depends on the capability of the handsets.

Consider Figure 4.15. It shows that a handset can transmit and receive on two time-slots in half-duplex mode, provided that they are adjacent. However, the time available for switching between receive and transmit mode and for adjacent cell monitoring is significantly reduced, requiring either faster synthesisers than those used in plain GSM phones, or multiple synthesisers. Asymmetric multi-slot configurations were introduced specifically to increase downlink data-rates while still allowing for half-duplex operation, assuming that downlink transmission speed is in general more important to mobile users than uplink speed. For instance, while it is impossible for a half-duplex MS to receive and transmit on four time-slots each, it is in theory possible to receive on four time-slots and transmit on one time-slot in half-duplex mode.

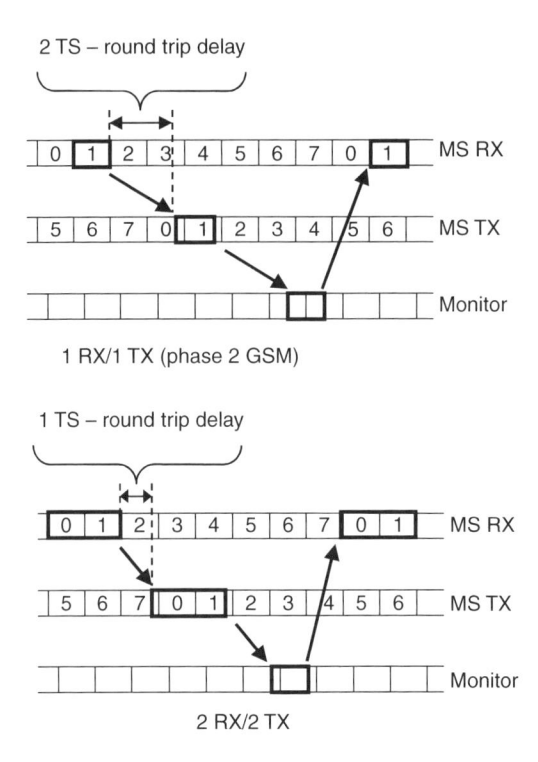

Figure 4.15 Single-slot and double-slot operation in half-duplex mode (TS = time-slot)

To allocate resources while respecting the handset constraints, the network needs to know the handset capabilities in terms of the maximum number of transmit and receive time-slots it can handle, whether these need to be contiguous, etc. For this purpose, annex B in GSM 05.02 defines 29 multi-slot classes, 18 of which are recognised for HSCSD. Of those, 12 classes are for half-duplex operation ('type 1' classes), while the other six require a handset to be capable of transmitting and receiving simultaneously ('type 2').

Along with other parameters, for each class, the maximum number of receive slots (rx), transmit slots (tx), and the sum of receive and transmit slots is indicated. It is $rx + tx \geq$ sum, that is, class 3 for instance indicates two transmit and receive slots each, but the sum is three, so if two transmit slots are used, reception is only possible on one slot per frame, and vice versa. For a 'type 1' MS, receive and transmit slots must be assigned within a window of size rx and tx, respectively, which means that contiguous time-slots need to be assigned if the full receive or transmit capability of an MS is exploited for an HSCSD configuration. This in turn means that it is not enough for a cell to have the required *amount* of resources available for an HSCSD call to be set up or handed in. These resources must also be available in the *right configuration*, which will often require intracell handovers to be performed first to reshuffle the channels allocated to other users. Since this will result in a mute period for these other users, this clearly has an impact on the quality of service they experience. The number of required handovers should therefore be minimised, which in turn has repercussions on the strategy to be used regarding initial channel allocation.

4.5.5 Enhanced Circuit-Switched Data (ECSD)

For circuit-switched services, the EDGE capability (i.e. 8PSK modulation) can be used to enhance data-rates, both in single-slot and multi-slot configurations. Two different E-TCH types can be used in multi-slot configurations, namely the E-TCH/F28.8 and the E-TCH/F43.2.

For basic HSCSD without EDGE capability, all traffic channels in an HSCSD configuration need to use the same coding scheme (in the case of bi-directional channels both on the uplink and the downlink). For enhanced circuit-switched data (ECSD), this restriction does not apply anymore. According to TS 22.034 [197], channel coding asymmetry may be set up by the network in three cases:

- if the MS only supports enhanced modulation on the downlink (8PSK receive capability only is cheaper to implement than both transmit and receive capability);
- if the MS supports enhanced modulation on both links, but the user indicates preference for uplink or downlink biased channel coding asymmetry;
- if the MS supports enhanced modulation on both links, and the user indicates preference for channel coding symmetry, but the link conditions justifies different channel coding on up- and downlink.

This asymmetry is only possible with non-transparent services and only with the TCH/F14.4, the E-TCH/F28.8, and the E-TCH/F43.2. In HSCSD configurations with channel coding asymmetry, one or more TCH/F14.4 with GMSK modulation can be used in the unbiased link direction, while in the biased direction, either of the two mentioned E-TCH/F can be used (but only one type at any given time).

Note that both the slow and the fast associated control channels of an E-TCH use GMSK rather than 8PSK modulation. In fact, the SACCH is the normal SACCH known from GSM phase 2, while the E-FACCH is a slightly modified normal FACCH with a different interleaving depth. As with 'plain' HSCSD, the E-FACCH exists only on the main channel of a multi-slot configuration.

Additionally, the E-TCH requires a new associated control channel, the E-IACCH, or In-band Associated Control CHannel, for fast in-band signalling enabling faster power control than that possible when power control commands are delivered on the SACCH. This channel is carried on the 8PSK bursts of the E-TCH, making use of the symbols that would normally carry stealing flags. The stealing flags are not required when applying 8PSK modulation, since the E-FACCH distinguishes itself from the E-TCH by using GMSK instead of 8PSK modulation, which can be distinguished blindly by the receiver. If fast power control is applied, the E-IACCH is used by the MS in the uplink direction to signal the downlink reception quality of this E-TCH, and by the BTS in the downlink direction to order power levels to be used by the MS for this channel. The power control commands and measurement reports are interleaved over four TDMA frames, allowing for six update commands per 26-multiframe rather than only one every 104 TDMA frames as with SACCH-based power control.

Obviously, in a multi-slot configuration with asymmetric uplink biased channel coding, where the in-band channels do not exist on the downlink, fast uplink power control is not possible. The E-IACCH is still used on the uplink for fast measurement reporting.

4.6 Resource Utilisation and Frequency Reuse

4.6.1 When are Resources Used and for What?

4.6.1.1 Handset Switch-on and Idle Mode

When a GSM handset is being switched on, it (more precisely, its physical layer entity) will first enter search mode, scan all possible GSM carriers on which a BCCH could be located, select the cell with the strongest BCCH it detects, and 'camp' on this BCCH. It will then extract the necessary information from the BCCH to access the system on the RACH and enter *dedicated mode*. In the channel request message sent on the RACH, it will request an SDCCH to register the MS as active in the system. This may either be performed through an *IMSI attach procedure* (if the MS was last switched on in the same location area) or else a *location updating procedure*. Once this registration is successfully completed, the MS will release the SDCCH, and enter idle mode. From the nature of this activity (i.e. the fact that it is essentially only performed when handsets are switched on), it is clear that this procedure will only consume a negligible amount of air-interface resources available in a GSM system.

In *idle mode*, during which no dedicated resources are consumed, the handset will listen to the BCCH and the PCH (that is, in DRX mode only to a subset of PCH blocks according to the mobile's paging group) on the currently selected cell. It will also monitor neighbouring BCCHs and select a new cell, if the BCCH of that cell is received at a stronger signal level than the one on which it is currently camping. To avoid a ping-pong effect (i.e. switching endlessly back and forth) at location area borders, this cell reselection occurs only if the difference in reception level between the potential target cell and the current cell exceeds a certain threshold value. In other words, a hysteresis is applied. The idle mode behaviour of GSM handsets is specified in detail in GSM 03.22 [203] (TS 23.122 [204] for R99).

4.6.1.2 Location Updating Procedure

When the selected new cell is in a different location area than the old cell, the handset has to perform a location updating procedure, to allow the network to track user movement. Similarly, the network may request the handset to perform every so often a periodic location updating procedure, just to check whether the handset is still operational and in reach of the network, to avoid unnecessary paging of handsets with empty batteries, or those out of coverage. Additionally, as outlined above, a location updating procedure might also be used for initial registration in the network.

To perform this procedure, the handset needs to enter dedicated mode, that is, access the system on its RACH, indicating location updating as establishment cause, and then wait for an immediate assignment sent by the BTS on the AGCH, if the random access was successful and an SDCCH is available. Finally, in dedicated mode, on the allocated SDCCH subchannel, the required signalling messages to perform a location updating procedure can be sent. For releases up to R98, this message exchange is detailed in GSM 04.08 [178], which contains the complete specification of the mobile radio interface layer 3 (or RIL3, as indicated earlier), a sizeable document indeed. The release 1999 version of this document, however, is only an empty shell pointing to three other specifications, into which the original document was split. The RR part of RIL3 is contained in GSM 04.18 [188] (still more than 300 pages!), while MM and CM (the latter entailing both circuit-switched call-control and session management for packet-data services) are contained in 3GPP TS 24.008 [205], as they are common to GSM and UMTS. Finally, another 3GPP specification, namely TS 23.108 [206], provides examples of structured procedures, such as the complete message exchange required to perform the location updating procedure (a short document, for a change).

From GSM 04.08 or TS 23.108, it can be seen that four messages need to be sent in each direction on the SDCCH for a typical location updating procedure, however, additional messages may optionally be sent, increasing the total to six messages per direction. Each of these messages should normally fit into the 23-octet payload available on one SDCCH block. For each SDCCH subchannel, whether mapped onto an SDCCH/8 or on an SDCCH/4 (i.e. combined with a CCCH), one block per link direction and per 51-multiframe is scheduled in a manner that downlink messages can be replied to on the uplink in the same multiframe period. Even when allowing for a few retransmissions due to messages received in error, a two-second allocation of the SDCCH will typically suffice to complete a location updating procedure.

This irregular transmission of bursts creates a sound pattern, which should be all too familiar to every GSM user having put her handset near to a computer, telephone, or other equipment susceptible to interference, while sending or receiving a short message, or while her phone performed a location updating procedure. It is a superposition of one 'click' every 51-multiframe (i.e. 235 ms) due to sending of SDCCH blocks, and another every two 51-multiframes due to SACCH blocks. The gap between the SACCH block and every second SDCCH block it follows depends on whether we are dealing with an SDCCH/4 or an SDCCH/8, and on the subchannel number which was allocated (see GSM 05.02 for details), so the exact rhythmic pattern varies from case to case.

4.6.1.3 Setting up Voice or Data Calls, Short Message Transfer

If the network wants to deliver a short message or a mobile terminated (MT) call to a registered mobile user (i.e. an incoming call from the considered mobile user's perspective), it first needs to page his handset on the PCH. Paging messages need to be sent in every cell pertaining to the location area last stored by the network for that handset. If the intended user's handset receives the message, it will try to access the system on its RACH, indicating 'paging response' as establishment cause. If a registered mobile user initiates a so-called mobile originated (MO) call or wants to send a short message, the paging procedure is obviously omitted, and the establishment cause indicated on the RACH is different.

Short messages are delivered on an SDCCH (unless the user is engaged in a call, in which case the SACCH is used, since SDCCH and TCH are mutually exclusive). Therefore, the sequence of events is the same as with a location updating procedure (save for the additional paging required for the mobile terminated case). However, for a short message of 160 characters, a few more control blocks have to be exchanged than for a location updating procedure, but less than three seconds on the SDCCH will normally do all the same.

If a voice or data call is to be set up, the operator has essentially three options (only two for MT calls). The type of TCH requested in the channel request message sent on the RACH can be allocated directly in the immediate assignment procedure. This option is referred to as *very early assignment* and seems only to be available for mobile originated calls. However, if the ensuing call set-up signalling results in the call to be aborted (e.g.

because of lack of authorisation of the user, whether that is because he is fraudulent, or because he attempts to make a type of call not covered by his subscription, such as an international call), then precious TCH resources have been wasted. Alternatively, the call set-up signalling could be exchanged on the SDCCH, and the TCH is only allocated when the called party picks up the handset, hits the 'yes' button, or whatever the appropriate action required to be taken by the called party *(late assignment)*. In this case, if it is an MO call, the ringing tone of the counter-party is generated synthetically at the handset. This is the most resource-efficient option, but may result in unnecessary 'post-pick-up' delay, potentially swallowing the first few syllables of the conversation, if it is a voice call. A minor detail is that the ringing tone generated by a GSM 900 or 1800 handset is normally chosen to continental European taste (i.e. tones of equal length at regular intervals, as specified in GSM 02.40 [207]). This could confuse for instance a British caller, who is accustomed to hearing ringing tones in pairs (i.e. two tones close together, separated from the next pair by a longer break).

The third option, referred to as *early assignment*, which offers a compromise and seems to be the most common option chosen by operators, is to allocate an SDCCH in the immediate assignment message and to start the call set-up signalling exchange on that SDCCH. The TCH is then allocated *before* user alerting, i.e. before ringing tones are generated. Obviously, if the calling party lets the phone ring for a long time without the called party answering the phone, then early assignment is not much more efficient than very early assignment.

4.6.1.4 Dedicated Mode Behaviour, Handover

Both the SDCCH and the TCH are associated with an SACCH, which is required mainly for radio link control, i.e. timing advance commands, power control commands, measurement commands and reports. The status (and possible loss) of the radio link is determined by the quality of received SACCH frames and the number of lost SACCH frames. The available SACCH capacity must be used continuously in both directions while in dedicated mode (that is, also during voice silent intervals in DTX mode), e.g. to keep track of the TA and to make the counting of lost SACCH frames easy. Based on the measurement reports sent on the SACCH, the network can determine when a handover is required.

From an air-interface perspective, there are two fundamental types of handovers, namely intracell and intercell handover. An *intracell handover* may be required when the allocated channel suffers particularly bad interference, or because the channels in a cell must be reshuffled, for instance to accommodate an HSCSD call requiring multiple contiguous time-slots. An *intercell handover* is typically induced by movement, the extreme case being that the MS moves completely out of the coverage area of the serving or source cell. However, a handover between cells is also advisable when the reception quality in the source cell is sufficient to satisfy the quality of service requirements, but that in the potential target cell is better. In such a case, the execution of a handover allows transmission power levels to be reduced, which in turn reduces the total co-channel interference level generated in the system.

As with cell reselection in idle mode, a hysteresis is applied to avoid ping-ponging between two cells, that is, the reception quality in the potential target cell must exceed that in the source cell by a certain margin.

The need for a handover is signalled by the network on the FACCH in the shape of a handover command, which specifies the new cell and the location of the channel to be

4.6 RESOURCE UTILISATION AND FREQUENCY REUSE

used in that cell. As mentioned earlier, the FACCH steals into the TCH (or SDCCH). The MS will then access the new channel transmitting a series of handover access messages, which use the access burst format rather than the normal burst format to establish the required timing advance. This is followed by a 'handover complete' message sent on the new FACCH, and only then, payload transmission (e.g. of voice frames) can be resumed, i.e. the physical channel can be used again as TCH or SDCCH. In general, for intercell handover, when some kind of synchronisation between cells is available and the MS can calculate the TA to be used on the new channel, only four access bursts are sent to confirm the TA value, and sending of these access bursts may even be optional. On the other hand, when there is no synchronisation, the MS must continue to send access bursts until it receives a message from the network containing an ordered TA value, calculated by the network on the basis of these access bursts, before reverting to user data transmission.

If the connection cannot be established in the new cell, the terminal may try to return to the old cell, indicating a handover failure. If this does not work either, or if the connection is suddenly lost before a handover could even be initiated, a call re-establishment procedure may be initiated by the MS, if so permitted by the network. This requires the sending of a channel request message on the RACH, with a special establishment cause indicating call re-establishment.

Note that a handover may also occur during the call set-up signalling exchange on the SDCCH.

4.6.2 How to Assess Resource Utilisation

The achievable resource utilisation in a 'circuit-switched environment' depends on various factors, among them the available amount of resources and the required Quality of Service (QoS) in terms of call blocking probability, call dropping probability, frame erasure rate and various bit-error-rate measures. For call blocking and dropping probability, the term Grade of Service (GoS) is also often used instead of QoS. To assess the resource utilisation in a GSM system accurately, one would have to perform a detailed modelling of the use of the different common and dedicated channel resources in a cellular environment. Furthermore, the mobility behaviour of terminals and the propagation conditions would also have to be modelled, as the former has an impact on call dropping probability, while both effects combined together affect frame erasure rates and bit error rates.

Given the spectrum available to an operator, the number of hierarchical cell layers deployed in a network, and the frequency reuse factor, the number of carriers available per base station, and thus the number of basic physical channels can be determined. These physical channels carry both common and dedicated channels. The first step is to determine the required amount of common channels, which, as discussed earlier, depends mainly on the capacity required for the PCH. It is not trivial to assess the latter, since the required capacity depends among other things on the size of location areas, which in turn has an impact on the signalling load on the SDCCH. Rather than investing a significant amount of effort into these matters, we simply state that a single-carrier cell requires typically one basic physical channel for a CCCH combined with an SDCCH/4, while for a two-carrier cell, two basic physical channels, one for the CCCH and one for the SDCCH/8, should do. As more carriers are added, additional CCCHs and SDCCH/8s may become necessary. In any case, the lion share of resources available in a cell will be used for traffic channels, and erring slightly on the amount of resources required for CCCH and SDCCH will not have a significant impact on the accuracy of the assessment of the total resource utilisation.

From the discussion provided in Section 4.4, it should be evident that the RACH load is normally negligible, so RACH inflicted call blocking (i.e. having to refuse a call because no RACH resources are available) can safely be ignored. Also, given the comparatively moderate amount of resources required on the SDCCHs for location updating procedures, SMS, and call set-up signalling, it is clearly in the interest of an operator to dimension SDCCH resources in such a way that SDCCH-inflicted call blocking does not occur. After all, it would be a pity if entire basic physical channels would have to be left idle because of too few SDCCH subchannels being available, each of which occupies only an eighth of a basic physical channel.

In summary, resource utilisation is to a large extent determined by the traffic channels, that is by blocking inflicted due to lack of them (for blocking-limited system) or the transmission quality experienced on them (for interference-limited systems). As long as we are making an allowance for a few physical channels being required for common channels and SDCCHs, it is entirely justifiable to focus on the traffic channels. This is what we will do in Subsection 4.6.4 for investigations on resource utilisation in a blocking-limited GSM system. Similarly, in the literature on interference-limited GSM reviewed in Subsection 4.6.5, the respective authors were all concerned with TCH performance.

The investigations performed for Subsection 4.6.4 are restricted to resource utilisation in a single test cell, and they do not distinguish between traffic channels required for calls to be handed in and those for calls to be established in the test cell (whether these be MO or MT calls). The blocking probability is simply established as a function of the total number of bids or requests made for traffic channels in this cell. The consequences of a hand-in request being blocked depend on propagation conditions and load conditions in neighbouring cells. If the call can be reconnected to the old cell or handed over to another cell, which provides sufficient signal quality, the quality of service is not or only to a limited extent affected. However, if the call needs to be dropped due to unavailability of resources in the test cell, the quality of service is seriously affected. In fact, call dropping is often considered to be much more annoying for users than call blocking, thus the target call dropping rate is usually lower than the call blocking rate. To minimise call dropping, typically some resources in each cell need to be reserved for the exclusive use of calls to be handed in. Depending on the mobility pattern of users, this may affect the total resource utilisation significantly. Particularly when dealing with multi-slot calls, resource utilisation is expected to drop substantially, as already discussed in Section 4.5.

4.6.3 Some Theoretical Considerations — The Erlang B Formula

Assume a scenario where a base station offers N_{tc} traffic channels, and mobiles request a service which requires exactly one, but any one of these channels. The service time is exponentially distributed with mean duration $1/\mu_s$ seconds. The arrival of service requests is Poisson distributed (assuming an infinite number of mobiles) with an arrival rate λ_c calls per second, that is with $3600 \cdot \lambda_c$ calls per hour. In this case the probability of blocking can be calculated using the so-called *Erlang B formula* (see e.g. Reference [104]), namely

$$P_b = \frac{\frac{(\lambda_c/\mu_s)^{N_{tc}}}{N_{tc}!}}{\sum_{k=0}^{N_{tc}} \frac{(\lambda_c/\mu_s)^k}{k!}}. \tag{4.1}$$

To be precise, according to Reference [104], it can be shown that the Erlang B formula holds for arbitrary service time distributions, as long as the mean of the respective distribution is $1/\mu_s$.

The offered traffic in *Erlang* (or E), G_E, is λ_c/μ_s and the resource utilisation normalised per offered channel is

$$\eta_r = \frac{\lambda_c}{\mu_s \cdot N_{tc}}(1 - P_b). \tag{4.2}$$

To analyse accurately the blocking performance in a GSM system, we would not only have to account for blocking due to lack of traffic channels, but also have to consider blocking due to lack of signalling resources and the impact of potential congestion on the RACH. However, as outlined above, we can safely assume that RACH-congestion does not contribute to blocking in a GSM system offering only circuit-switched services (repeated transmission failures on the RACH due to radio conditions is another matter). If we assume furthermore that the amount of signalling resources is dimensioned generously, and provided that the above assumptions on traffic model and service times hold, Equation (4.1) can be used to calculate blocking in a GSM system with homogeneous traffic. With Equation (4.2), the *gross* resource utilisation on the traffic channels given a certain blocking limit or grade of service can be assessed (gross because inactivity periods due to voice activity detection are not accounted for).

4.6.4 Resource Utilisation in Blocking-limited GSM

4.6.4.1 General Assumptions

A single test cell is considered with target call blocking levels of 2% and 5%. As outlined above, resource utilisation assessed in this manner is expected to be optimistic, since it ignores call dropping due to mobility events and failed handovers. Given the number of frequency carriers available in the test cell, the number of basic physical channels which are assumed to be used as TCH/F (or pairs of TCH/Hs) is listed in Table 4.3. The remaining physical channels are assumed to be required for common channels and SDCCHs.

The system is considered to be fully blocking-limited, that is, the frequency reuse factor is assumed to be large enough for the quality requirements in terms of voice

Table 4.3 Number of TCH/F available given number of carriers

Carriers	1	2	3	4	5
TCH/F	7	14	22	29	36

frame erasure rates and bit-error-rates on the signalling channels to be met, even at full resource utilisation. Judging from References [79] and [114], with ideal hexagonal cells, this condition seems to be met for instance at a 4/12 reuse-factor, meaning that the site-reuse-factor is four, where each site covers three sectors operating at different frequencies. Equating a sector to a cell, this effectively means a reuse-factor of 12. Considering the GSM 900 band, and assuming two operators in this band in a given country, then up to five carriers would be available per sector in a system with a single layer of cells, consistent with the range of values listed in Table 4.3. Under the same assumptions, typical GSM 1800 operators could potentially deploy more carriers per sector, though. Furthermore, as discussed in more detail in the next subsection, the system may also be blocking limited at a 3/9 reuse factor, allowing for six or more carriers per cell at an operator allocation of 12.5 MHz. On the other hand, as systems mature, multiple hierarchical layers of cells are often deployed, requiring the total spectrum available to an operator to be divided between these layers, which again reduces the number of carriers that can be deployed in a particular cell.

4.6.4.2 Resource Utilisation in Homogeneous Traffic Conditions

Figure 4.16 shows the blocking probability, assuming homogeneous full-rate traffic (i.e. every user requires exactly one full-rate channel), as a function of the number of Erlangs, for one to five carriers per cell. Table 4.4 summarises the important results, namely the number of Erlangs that can be supported at blocking probability levels of 2% and 5% respectively, and thereby the resulting gross resource utilisation according to Equation (4.2).

From Table 4.4, it is immediately evident that the gross resource utilisation is extremely low in a single-carrier cell, but improves as the number of carriers per cell increases. This illustrates the effect of *trunking efficiency*: the more resources there are available, the

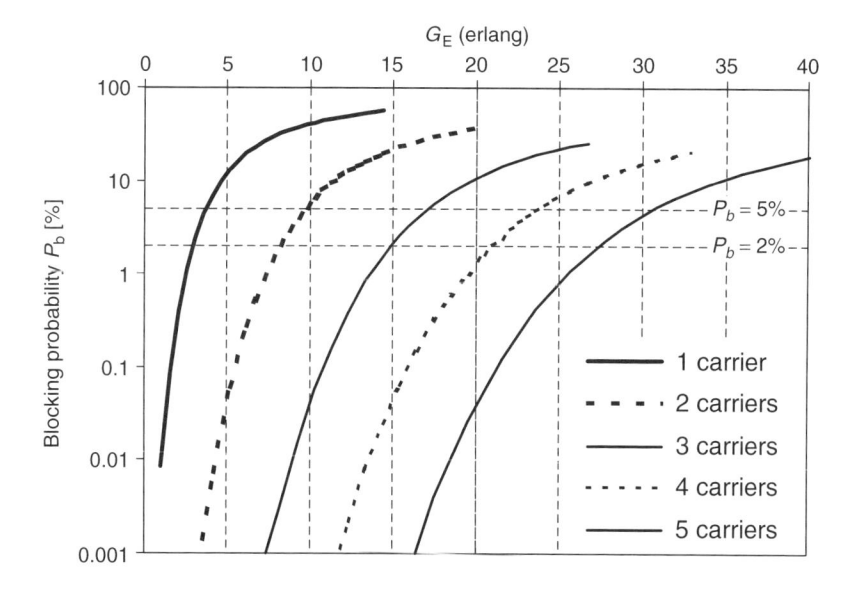

Figure 4.16 Blocking probability for homogeneous full-rate traffic

4.6 RESOURCE UTILISATION AND FREQUENCY REUSE

Table 4.4 Sustained offered traffic and resource utilisation for TCH/F

Carriers	N_{tc} (TCH/F)	G_E [Erlang] $P_b = 2\%$	G_E [Erlang] $P_b = 5\%$	η_r $P_b = 2\%$	η_r $P_b = 5\%$
1	7	2.93	3.73	0.410	0.506
2	14	8.19	9.72	0.573	0.660
3	22	14.89	17.12	0.663	0.740
4	29	21.04	23.82	0.711	0.781
5	36	27.34	30.65	0.744	0.809

better use of these resources can be made. A detailed explanation of why this is so can be found in Section 7.2 in the context of discussions on multiplexing efficiency, as the effects which determine trunking efficiency and multiplexing efficiency are very similar.

For a five-carrier cell, gross resource utilisation is not so bad, at $P_b = 2\%$, it is almost 75%, and at $P_b = 5\%$ even higher. However, if all the traffic considered were to consist of voice calls and voice activity detection were applied, then the net resource utilisation would halve (assuming 50% voice activity). In a single-carrier cell, this would equate to a meagre net resource utilisation of 20% to 25%, or in other words, on average, more than three-quarters of the available resources in the cell would remain idle.

Equivalent results as those produced for full-rate traffic channels can easily be produced for half-rate channels as well. Keeping in mind that N_{tc} simply doubles, the full-rate results provided can be translated into half-rate results; for instance, the two-carrier curve is now relevant for a single-carrier cell, and the four-carrier curve roughly for a two-carrier cell. For a four-carrier cell that can serve 58 half-rate users simultaneously, η_r assumes a value of 0.86 at $P_b = 5\%$.

Similarly, results can also be produced for HSCSD calls. Consider for instance a multi-slot class 5 MS, which can transmit and receive on two time-slots simultaneously, provided that these slots are adjacent. In the following, it is always assumed that such a mobile terminal will actually request two time-slots (in theory, it could content itself with a single time-slot). On a single-carrier cell, in that case, three such calls could be served simultaneously, on a two-carrier cell seven calls (provided that the SDCCH/8 is assigned to TN 1 on C0, i.e. the time-slot adjacent to the CCCH), 11 calls on three carriers and 14 on four. The results for two and four carriers can readily be gleaned from entries in Table 4.4 for appropriate N_{tc} values. With three carries, η_r is 0.52 and 0.61 for $P_b = 2\%$ and 5% respectively, and the one-carrier case is too sad an example to quote results.

For the RACH load estimation performed in Section 4.4, we considered 10 000 call set-up attempts per hour and claimed that this was a generous figure for a cell with a single 'non-combined CCCH'. We can now substantiate this claim. From the number of Erlangs listed in Table 4.4, one could calculate the total number of traffic channel bids per second as $\lambda_c = \mu_s \cdot G_E$, if the average TCH holding time $1/\mu_s$ were known. The TCH holding time is a fraction of the cell dwell time, which in turn depends on several factors, such as cell size and mobility patterns of the considered users [208]. For one numerical example in Reference [209], a cell dwell time of 30 seconds was used. Without elaborating further on realistic values in deployed cellular communication systems, we just note that at $G_E = 30.65$ E (the maximum value reported in Table 4.4), $1/\mu_s$ could be as low as 11 seconds without exceeding 10 000 TCH bids per hour. Since not all of these bids are actually due to calls originating in the test cell requiring RACH resources (calls handed in from other cells do not need to access the RACH), $1/\mu_s$ could be even lower without exceeding 10 000 set-up attempts per hour.

4.6.4.3 Resource Utilisation in Heterogeneous Traffic Scenarios

Heterogeneous traffic is much more difficult to deal with than homogeneous traffic, both from a practical or operator perspective, and from a theoretical perspective. The blocking probability depends in this case not only on the number of users of each category to be served, but also on the current resource constellation. Consider for instance a single-carrier cell with CCCH and SDCCH/4 combined, which serves both full-rate and half-rate users. If, at a certain moment in time, five full-rate calls and two half-rate calls were served and an additional full-rate call were to arrive, one would expect that this new call could be accommodated. However, if the two half-rate calls happen to be assigned to different time-slots, then this new call would have to be blocked. This is illustrated in Figure 4.17.

One might think that it would be a silly thing to assign half-rate calls to separate time-slots, that the obvious thing to do would be to pair two half-rate calls on a single time-slot. True, but this ignores the dynamics of the system. It might be that this cell served initially four half-rate calls on two time-slots, but that in the meantime, two of these calls have left the cell, unfortunately two calls which happened to be assigned to different time-slots, leaving behind two 'holes', i.e. unpaired half-rate calls. The obvious answer to this problem is to perform an intracell handover to 'repack' the channels, that is, to pair these two calls. Two extreme approaches are conceivable, namely to repack whenever holes open up, or only to repack when required, e.g. upon arrival of a full-rate call, if this call cannot be served without prior repacking.

Since a handover causes a short mute period, which affects the quality of service of the user subjected to it, unnecessary handovers should be avoided. This means that the first approach is not a good idea, it might be that the resources wasted due to holes are not required during the lifetime of the unpaired calls, or that holes could be filled up with newly arriving half-rate calls, without requiring any handover. With regards to the second approach, the question arises whether intracell handovers could be performed while a new caller is engaged in call set-up signalling, such that the required resource is freed

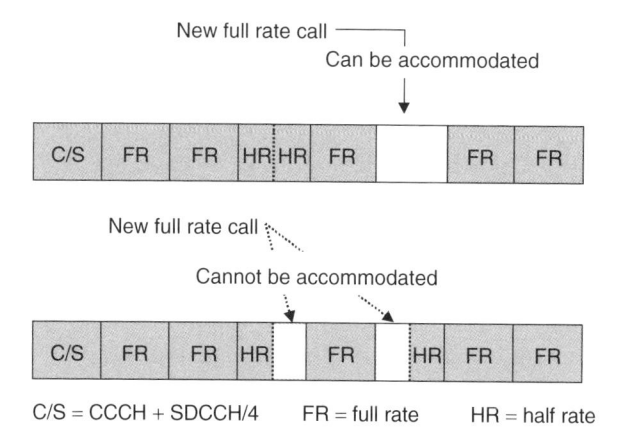

Figure 4.17 With heterogeneous traffic, call blocking depends on the resource constellation

4.6 RESOURCE UTILISATION AND FREQUENCY REUSE

up just in time to serve this call. This is potentially possible, although it might require a call queuing facility, which is optional in GSM. If it were possible and would not have a significant impact on the call set-up delay of the new call, it would certainly be the preferred approach.

Channel allocation schemes for heterogeneous half-rate and full-rate traffic in GSM are analysed in detail in References [210] and [211]. An analytical assessment of the resource utilisation is significantly more complex than for homogeneous traffic. Adding multi-slot calls into the equation would make an analysis even more difficult.

Here, we content ourselves with an assessment of the resource utilisation in heterogeneous scenarios through simulations, using a very plain initial-channel-assignment scheme and ignoring channel-repacking schemes altogether. 'Plain' in this context does not prevent from applying some common sense: upon arrival of a new half-rate call, a first search of suitable time-slots is restricted to those that already carry a single half-rate call. The simulation tool used for this purpose has been validated by comparing theoretical with simulation results for homogeneous traffic conditions.

Figure 4.18 shows the blocking probability as a function of the total number of Erlangs for a two-carrier cell, separately for each type of call considered, and averaged over all calls. The offered traffic is composed in such a manner that full-rate calls, half-rate calls, and HSCSD multi-slot class 5 calls contribute each a third to the call arrivals. The Erlangs are calculated irrespective of the resource requirements of each call, i.e. as λ_c/μ_s, where λ_c is the total call arrival rate and where $1/\mu_s$, the average service time of the calls in the test cell, is assumed to be the same for all types of calls.

The resource utilisation is determined at the target blocking probability level for the service suffering most from blocking, which is, not surprisingly, the HSCSD service. Table 4.5 summarises all results obtained for the two-carrier cell, that is, both results for homogeneous traffic conditions already discussed earlier, and results for heterogeneous traffic conditions. To assess the impact of the traffic heterogeneity, the resource utilisation

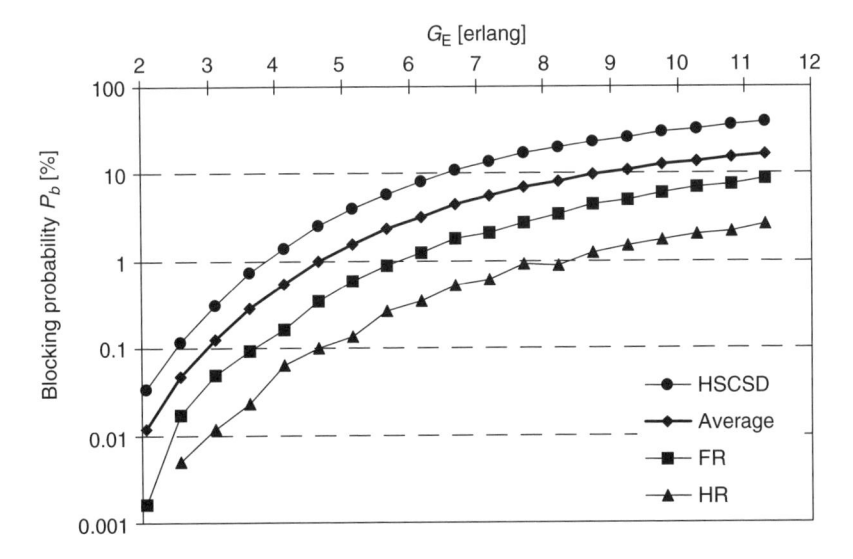

Figure 4.18 Blocking performance in a two-carrier cell, equal share of HSCSD class 5, full-rate, and half-rate calls

150   4 MULTIPLE ACCESS IN GSM AND (E)GPRS

Table 4.5 Comparison of result for the two-carrier case

Traffic composition	Method	N_{tc}	$P_b = 2\%$		$P_b = 5\%$	
			G_E [E]	η_r	G_E [E]	η_r
HSCSD class 5 only	Theory	7	2.93	0.40	3.73	0.51
50% FR, 50% HSCSD	Simulation	9.33	3.49	0.37	4.32	0.44
Equivalent homogeneous	Theory	9.33	4.58	0.48	5.64	0.57
33% FR,HR,HSCSD, each	Simulation	12	4.42	0.36	5.45	0.43
Equivalent homogeneous	Theory	12	6.61	0.54	7.94	0.63
FR voice only	Theory	14	8.19	0.57	9.72	0.66
50% HR, 50% FR	Simulation	18.67	10.48	0.55	12.33	0.63
Equivalent homogeneous	Theory	18.67	12.05	0.63	13.65	0.69
HR voice only	Theory	28	20.14	0.69	22.86	0.78

achieved with a given traffic composition is compared to that achieved with an 'equivalent' homogeneous scenario. N_{tc} for this equivalent scenario is calculated as the number of full-rate traffic channels available in the cell divided by the average number of full-rate channels required per call in the given heterogeneous traffic scenario. For the example just discussed, the average number of full-rate channels required is $0.33 \cdot 1 + 0.33 \cdot 0.5 + 0.33 \cdot 2 = 1.17$, hence the equivalent $N_{tc} = 14/1.17 = 12$. For other heterogeneous traffic scenarios also shown in the table, for which the equivalent N_{tc} for homogeneous traffic is non-integer, the theoretical results are obtained through interpolation of the respective results for the nearest two integers.

The table illustrates the negative effect of traffic heterogeneity on the sustained traffic level G_E and on resource utilisation, which is because of the constraints regarding permitted resource allocation: an HSCSD call, for instance, is blocked even if there are two time-slots available, unless these time-slots happen to be adjacent. Similarly, half-rate calls save only resources compared to full-rate calls if most of them can be paired on full-rate channels. This means that a substantial increase in terms of sustained traffic level G_E can be achieved through half-rate terminals only if the fraction of half-rate users is substantial. With a 50/50 split between half-rate and full-rate calls, the resource utilisation is lower than that in the homogeneous full-rate case, and the increase in Erlangs compared to the full-rate case is only about two-thirds of that achieved in the equivalent homogeneous scenario with $N_{tc} = 18.67$.

Table 4.6 shows similar results for the four-carrier case. Here, two additional scenarios for mixed full-rate, half-rate and HSCSD traffic are considered. In the second scenario, the traffic composition is altered in a manner that the average resource requirement per call is exactly one full-rate channel. The considered traffic composition consists of 50% full-rate traffic, 33.33% half-rate traffic, and 16.66% HSCSD multi-slot class 5 traffic. The third scenario includes calls generated with a multi-slot class 9 MS which can receive on three time-slots and transmit on two of them. Again, these time-slots need to be adjacent. It is assumed that these 'class 9 calls' require all three time-slots in the downlink direction, and thus three full-rate channels (the third uplink slot remains idle). The traffic composition is again such that the average resource requirement per call is exactly one full-rate channel, that is, 33.33% of the calls are full-rate calls, 50% half-rate calls, 8.33% HSCSD class 5 calls, and the remaining 8.33% HSCSD class 9 calls.

4.6 RESOURCE UTILISATION AND FREQUENCY REUSE

Table 4.6 Comparison of result for the four-carrier case

Traffic composition	Method	N_{tc}	$P_b = 2\%$ G_E [E]	η_r	$P_b = 5\%$ G_E [E]	η_r
HSCSD class 5 only	Theory	14	8.19	0.57	9.72	0.66
50% FR, 50% HSCSD	Simulation	19.33	10.69	0.54	12.33	0.61
equivalent homogeneous	Theory	19.33	12.61	0.64	14.62	0.72
33% FR,HR,HSCSD, each	Simulation	24.86	14.18	0.56	16.24	0.62
equivalent homogeneous	Theory	24.86	17.38	0.69	19.85	0.76
FR voice only	Theory	29	21.04	0.71	23.82	0.78
FR,HR,HSCSD, scenario 2	Simulation	29	16.03	0.54	18.04	0.59
FR,HR,HSCSD, scenario 3	Simulation	29	13.77	0.47	15.98	0.52
50% HR, 50% FR	Simulation	38.67	27.34	0.69	30.42	0.75
equivalent homogeneous	Theory	38.67	29.76	0.75	33.28	0.82
HR voice only	Theory	58	47.75	0.81	52.55	0.86

4.6.4.4 Conclusions

Our investigations on resource utilisation in a blocking-limited system illustrated the importance of trunking efficiency. Gross resource utilisation in a single-carrier cell is extremely low, for homogeneous full-rate traffic for instance between 40% and 50%. As the number of carriers, and thus the number of physical channels available increases, gross resource utilisation improves as well. For homogeneous full-rate traffic, satisfactory levels can be achieved in a four-carrier cell, for half-rate traffic even better ones, exceeding 80% resource utilisation.

However, there are several problems. Firstly, the picture looks different when heterogeneous traffic scenarios are considered, particularly when multi-slot calls are involved. For instance, certain traffic scenarios looked at for a four-carrier cell lowered the utilisation to below 60%. Secondly, as outlined before, these investigations, which were limited to a single cell and ignored mobility, produced optimistic results, so the resource utilisation in a cellular environment will be lower. Thirdly, even when gross utilisation levels are acceptable, net levels may be significantly lower. For instance, a voice channel may only be utilised to about 50% per link direction, if DTX is applied. Similarly, data channels will often carry data only in one direction at any given time.

Clearly, in a purely blocking-limited system, which could in theory sustain 100% resource utilisation, any utilisation level below that equates to wasted capacity. Three ways of improving resource utilisation will be discussed in the remainder of this book:

- The first is to use this spare capacity for packet-switched services in a manner that blocking levels for circuit-switched services are not increased. 'Basic' GPRS (according to R97) allows, with some constraints, to do exactly this, by using this spare capacity temporarily for services, which are not delay-critical. This is primarily improving the gross resource utilisation. Idle periods due to DTX on TCHs serving voice calls cannot be exploited.

- The second is to allow for statistical multiplexing of real-time and non-real-time services in an 'all-packet system', using mechanisms such as PRMA, which would allow the net resource utilisation to be increased. Statistical multiplexing of voice is discussed extensively in the following chapters, the specific application to GSM is dealt with in Chapter 11.

- The third is to operate GSM as an interference-limited rather than a blocking-limited system. In this case, low resource utilisation does not necessarily mean low capacity. Rather, it leads to reduced interference, which allows for tighter frequency reuse, and thereby potentially more efficient use of air-interface resources. This is discussed in the next subsection.

4.6.5 Resource Utilisation in Interference-limited GSM

In a blocking-limited system, the acceptable grade of service is determined by a certain blocking probability, which must not be exceeded. This is sometimes referred to as *hard-blocking* level (there is a hard limit of available channels), in contrast to the *soft-blocking* level which is defined for interference-limited systems — exceeding it leads to a gradual rather than sudden quality deterioration. In the 'CDMA world', soft-blocking is also referred to as soft-capacity.

According to GSM 03.30 on radio network planning [181], the threshold carrier-to-interference ratio (CIR) which provides sufficient performance is 9 dB. This includes a 2 dB implementation margin over simulated residual BER threshold results. The system performance is generally accepted to be good enough if the experienced CIR is below the 9 dB threshold for no more than 10% of the time (e.g. Reference [79]). This situation defines the soft-blocking level; calls have to be blocked if their admission to the system leads to a violation of this requirement. However, this quality measure does not specify whether it is acceptable if 10% of the users experience a CIR which is consistently below 9 dB or whether each individual connection must not suffer a CIR below 9 dB for at most 10% of its duration. In the former case, this 10% would essentially be deprived of a decent service; that is, there is only appropriate service coverage for 90% of the users. If a certain blocking probability P_b is tolerated in a blocking-limited system, it would appear to be fair to request for an interference-limited system that no more than P_b per cent of the connections may experience a CIR, which falls below 9 dB for more than 10% of their duration. However, it is generally assumed that the interference-averaging feature provided by random hopping is a requirement for operating an interference-limited system. In particular, to translate the reduced average interference due to DTX into increased capacity, random hopping is required. Therefore, requesting to meet the 9 dB threshold for 90% of the time over *all* connections, although less stringent a requirement, may be considered just acceptable since, owing to SFH, the CIR samples falling below the threshold should be reasonably well distributed over all the connections.

From References [79] and [114], we deduce that if ideal hexagonal cells are considered, both a 4/12 and a 3/9 reuse pattern will result in a blocking-limited system, that is, at the load-limit determined by an accepted blocking level of, e.g. 2%, the CIR falls below 9 dB for well less than 10% of the time. These results were obtained for a GSM system applying DTX (at an activity factor of 0.5), SFH and slow power control. The results reported in Reference [184], where power control was not applied, seem to confirm this, although depending on the antennas deployed, it was found that a 3/9 system was only just blocking-limited[3].

---

[3] There seem to be some discrepancies regarding 4/12 reuse without SFH and DTX, where the soft-blocking level was found to be 0.7% at 2% hard blocking in Reference [79], while according to Reference [184], the load

4.6 RESOURCE UTILISATION AND FREQUENCY REUSE

The question is now: can we further tighten reuse factors and operate a fully interference limited system, where the capacity is essentially determined by the average interference level experienced by the users? On the BCCH frequencies, neither DTX nor power control can be applied, which limits the scope for tightening the reuse factor, unless special measures are taken such as those proposed for EDGE COMPACT (see Section 4.12). It would in theory be possible to use the same frequency carrier in certain cells as a BCCH frequency, and in other cells as a 'normal frequency', allowing for hopping, DTX and power control without restrictions in the latter case. However, it was found in References [185] and in [212] that it is better to split the total available spectrum into two separate pools of frequency carriers, one dedicated to BCCH frequencies and the other to non-BCCH frequencies. The BCCH frequencies are operated in blocking-limited mode, e.g. at a 4/12 reuse, and the scope for tightened frequency reuse is constrained to the remaining frequencies. This is exactly the scenario considered in References [79] and [184]. Hence the considerations that follow apply only to non-BCCH frequencies.

In References [79] and [80], for the non-BCCH frequencies, 1/3 and 1/1 reuse schemes are considered. 1/1 reuse implies uniform frequency reuse with omni-directional antennas. With 1/3 reuse, each base-station site covers three sectors or cells, as depicted in Figure 4.19. Such tight reuse factors imply interference-limited operation: the load tolerated due to soft-blocking is much lower than that due to hard-blocking. This is also referred to as *fractional loading*, since less *transceivers* (TRX, each one having

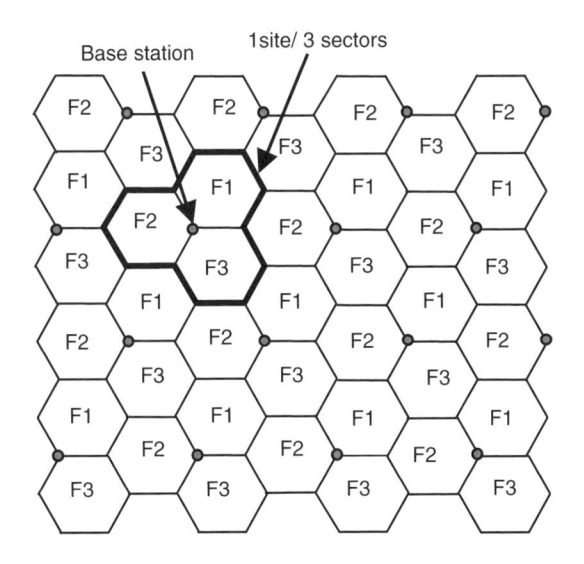

Figure 4.19 Ideal cell layout with 1/3 frequency reuse

---

tolerated in terms of soft-blocking is not much higher than that in terms of hard-blocking. One would therefore expect the soft-blocking level to be close to 10% at the traffic load determined by the hard-blocking level, which, by the way, should be lower in Reference [184] than in Reference [79] due to the smaller spectrum allocation considered in Reference [184], resulting in lower trunking efficiency. The discrepancies may be due to different approaches in how the soft-blocking level was determined, the lack of power control in the scenario considered in Reference [184] (thus a scenario relevant for the BCCH frequencies), and the different antenna types considered.

a capacity of eight basic physical channels) are used simultaneously in a cell than the number of carriers over which is hopped. Whether interference-limited operation can provide higher capacity than blocking-limited operation then depends on the trade-off between the additional number of carriers which become available per sector owing to the tighter reuse factor, and the tolerated fractional load level.

In Reference [79], results from a study for a system with power control, DTX, and SFH are presented, assuming that 72 carriers are available as pure TCH frequencies. It is reported that the supported traffic load in terms of Erlangs per km^2 (with the cell area being 10.4 km^2) and time-slot is 0.23 for 4/12 reuse, 0.35 for 3/9 reuse, 1.0 for 1/3 reuse and 0.5 for 1/1 reuse respectively. In other words, the spectrum efficiency in the interference-limited 1/3 system is 186% higher than that in the blocking-limited 3/9 system. However, these results are based on a single time-slot per carrier (to simplify simulations). With eight time-slots, the efficiency of the blocking limited system is substantially better owing to increased trunking efficiency. For instance, at a 4/12 reuse, there are six carriers available per sector, hence 48 TCHs. At $P_b = 2\%$, from Equation (4.1) we obtain 38.39 E. Divide this by the sector area of 10.4 km^2 and the eight time-slots per carrier, to obtain the result of 0.46 E/km^2 and slot, as quoted in Reference [79]. Carrying out similar calculations for the 3/9 case, for which no equivalent results were provided in Reference [79], we obtain a load limit of 0.64 E/km^2, which means that the gain achievable through 1/3 reuse is reduced to 56%.

A further problem with the results reported in Reference [79] is that they are based on average interference (as are those in References [80] and [184]), that is, they assume 'perfect interference averaging' owing to the frequency hopping. This is always an ideal assumption, but provided that the population of users over which the interference is averaged is large, which in turn depends on the number of frequencies over which is hopped, then 'near-perfect averaging' can be achieved and the results are only slightly optimistic in this respect. With the generous frequency allocation considered in Reference [79], the relevant user population is indeed large both for 1/3 reuse and 1/1 reuse, where hopping is performed over 24 and 72 frequencies respectively. Incidentally, though, GSM allows hopping over at most 64 frequencies.

According to References [80] and [113], as a result of the interference diversity gain, which can be obtained with random hopping, the 9 dB CIR threshold is too conservative. With a reduced CIR threshold of 6.5 dB (justified by the gain of 2.5 dB already discussed in Subsection 4.2.3), it is claimed that the efficiency of the interference-limited system configurations can be increased proportionately, i.e. by a factor of 1.8. This would take the efficiency advantage of the 1/3 configuration as compared to the 3/9 configuration again beyond 180%. However, as pointed out in References [80] and [184], there is no unique relationship between mean CIR and error rate in a random hopping environment, hence scaling the interference-limited results obtained in Reference [79] by a factor of 1.8 to account for the interference diversity gain may not yield accurate results. More importantly, the interference diversity gain decreases as the relative load increases, judging from Figure 6 in Reference [80], from 2.5 dB at a load of 25% to around 0.5 dB at 75%, if a target frame erasure rate of 2% is being considered. It seems therefore inconsistent to use a CIR threshold based on 25% system load to calculate a system capacity which implies a system load of more than 77%.

In Reference [185], real cell coverage areas based on the output of frequency planning tools were considered rather than ideal hexagonal cells, and both the effects of co-channel and adjacent channel interference were studied. As in Reference [184], the soft-blocking level was determined by simulated error performance rather than CIR threshold. It was found that 1/3 reuse with random hopping performed not particularly well compared to optimised frequency planning strategies with less tight reuse factors and cyclic hopping. On the other hand, provided that the sectors were synchronised and MAIO assignment was optimised to minimise co- and adjacent channel interference, 1/1 reuse with random hopping was found to outperform configurations with large reuse factors.

In summary, in terms of spectral efficiency, it appears that interference-limited system configurations can provide a performance advantage compared to blocking-limited configurations, but the choice of the right configuration and the quantification of this advantage are a tricky business, and depend on many parameters. Moreover, it should be taken into account that the voice quality will typically be lower in an interference-limited system than in a blocking-limited system, where excess CIR will lead to lower frame erasure rates (but also to wasted capacity). Furthermore, since it is better to have dedicated bands for BCCH frequencies and TCH-only frequencies rather than a single common band, interference-limited operation is not possible on the total spectrum available to an operator. Exploring ways of increasing resource utilisation in blocking-limited GSM, e.g. through GPRS, is therefore highly desirable.

What remains to be discussed before moving on to GPRS is the impact of frequency hopping on base stations. In Reference [184], a distinction is made between *baseband* frequency hopping and *synthesiser hopping*, the latter also referred to as wideband frequency hopping. Baseband hopping implies that the number of frequencies over which is hopped is equal to the number of TRX available in a cell. This in turn means that each TRX can remain tuned to a particular carrier frequency. However, in the case of an interference-limited, thus fractionally loaded cell, this may require deploying many more TRX than the number required to serve the cell traffic. For instance, in the 1/3 configuration considered in Reference [79], the soft-blocking level is reached at a system load of 43%, i.e. the gross hardware resource utilisation (ignoring DTX) would be 43%, if there were as many TRX as there are frequencies being hopped over. Clearly, it would be beneficial to deploy only as many TRX as required to serve the cell traffic at the soft-blocking level to avoid wasteful use of hardware resources. This is possible, but implies wideband frequency hopping or synthesiser hopping, i.e. the capability of a TRX to switch frequencies quickly. This is more demanding than the equivalent task at the mobile station, since a base station may be transmitting (and receiving) on every slot in a TDMA frame.

4.7 Introduction to GPRS

4.7.1 The Purpose of GPRS: Support of Non-real-time Packet-data Services

GSM was designed as a circuit-switched system. Both the fixed network and the air interface provide resources in the shape of circuits, that is, in the latter case by allocating dedicated channels. GPRS, by contrast, allows the sending and receiving of data in an end-to-end packet transfer mode, without utilising any network resources in circuit-switched mode. As a result, packet-mode data applications can be supported by GPRS much more efficiently than by a phase 2 GSM system. In particular, applications that exhibit one or more of the following characteristics are identified in GSM 02.60 [213], the so-called 'stage 1' service description of GPRS, as target applications:

(1) intermittent, non-periodic (i.e. bursty) data transmissions, where the time between successive transmissions greatly exceeds the average transfer delay;

(2) frequent transmissions of small volumes of data, for example transactions consisting of less than 500 octets of data occurring at a rate of up to several transactions per minute;

(3) infrequent transmission of larger volumes of data, for example transactions consisting of several kilobytes of data occurring at a rate of up to several transactions per hour.

Although not explicitly stated, from the above characteristics and the delay classes specified in GSM 02.60 in terms of total transfer delay of fixed-length packets, as shown in Table 4.7, it is evident that GPRS was not designed to carry real-time or conversational services. This is also clear from GSM 03.60 [76], the 'stage 2' service description of GPRS (stage 2 being more detailed than stage 1), which states that 'GPRS is designed for fast reservation to begin transmission of packets, typically 0.5 to 1 seconds' — clearly not fast enough to support packet-voice in a manner similar to PRMA. Interestingly, though, both GSM 02.60 and its 'UMTS spin-off', 3GPP TS 22.060 list telnet as an example of a real-time or conversational service supported by GPRS, so we have to spend some effort in defining 'real-time' and 'conversational'. Concurring with TS 22.060, 3GPP TS 22.105 on services and service capabilities [214] lists telnet as an error intolerant, conversational service (as opposed to voice, which is an error tolerant conversational service). By contrast, 3GPP TS 23.107 on the UMTS QoS architecture [215] lists four QoS classes, namely conversational, streaming, interactive, and background in descending order of delay-sensitiveness (i.e. conversational is the most delay-sensitive), and subsumes telnet under interactive services. This classification seems to make more sense, and for the purposes of this book, real-time services are services such as voice and video conferencing, which have much more stringent delay requirements than for instance telnet. The above-mentioned targeted reservation speed of 0.5 to 1 seconds is insufficient for such services. In short, we can state that GPRS (at least in its first iterations) supports only non-real-time services.

One immediate consequence of the limitation to non-real-time services is that proper handover need not be supported in GPRS. Cell selection and reselection, similar to what

Table 4.7 GPRS delay classes

Delay Class	Total transfer delay			
	Packet size: 128 octets		Packet size: 1024 octets	
	Mean transfer Delay	*95 percentile Delay*	*Mean transfer Delay*	*95 percentile Delay*
1. (Predictive)	<0.5 s	<1.5 s	<2 s	<7 s
2. (Predictive)	<5 s	<25 s	<15 s	<75 s
3. (Predictive)	<50 s	<250 s	<75 s	<375 s
4. (Best Effort)	Unspecified		Unspecified	

is used in GSM idle mode, will therefore do in GPRS, whether in idle mode or in the GPRS-equivalent of 'dedicated' mode.

4.7.2 Air-Interface Proposals for GPRS

ETSI set up two ad hoc groups to deal with GPRS standardisation, namely the SMG2 GPRS ad hoc group responsible for air-interface-related matters, and the SMG3 GPRS ad hoc group responsible for the design of the fixed network entities. The bulk of the work in these groups was carried out between 1995 and 1997.

The main question that arose for SMG2 GPRS ad hoc was how packet-data services could be supported over the air and what changes would have to be made to the GSM air interface, which was designed with circuit-switched services in mind. It was pretty clear from the outset that basic GSM parameters, such as carrier spacing and frame/time-slot structure, would have to remain unaltered.

Eventually, four different proposals were presented to the group, all based on some kind of reservation ALOHA scheme. The most conservative one was to use existing GSM mechanisms to perform 'fast circuit-switching'. In particular, given the spare capacity on the RACH, as discussed extensively earlier, it was proposed to use the existing immediate assignment procedure on the GSM RACH to request traffic channels for GPRS, obviously now only for the duration of the transfer of single packets rather than an entire data session. With this approach, no dedicated air-interface resources would be required for GPRS; instead, they would be shared on a fully dynamic basis between GSM and GPRS. The beauty of this proposal is that it would allow for increased total resource utilisation without affecting the blocking probability experienced by circuit-switched services. Obviously, this requires a means to pre-empt packet-data users in a manner that traffic channels can be freed up for circuit-switched users, while they are queued at the random access stage or in the process of exchanging signalling messages on the SDCCH.

There are several downsides to this approach. As far as the GSM immediate assignment procedure is concerned, due to limited code-points (or code-words) available, almost no GPRS specific information (which could be useful for speeding up the assignment procedure) can be conveyed in the channel request message. Additionally, for the purposes of a packet-data service, the RACH delay performance is rather mediocre. When it comes to resource utilisation on the traffic channels, packet transfer occurs typically in only one direction, while the other link direction is used sparsely for acknowledgements. Setting up a bi-directional traffic channel may be advantageous in terms of delay performance, since plenty of resources are available for acknowledgements or other control signalling, but would be wasteful in terms of resource utilisation.

The most radical proposal was a new multiple access scheme referred to as variable-rate reservation-access (VRRA). This proposal, described for instance in Reference [216], has quite some similarities with PRMA++ [46]. Each basic physical channel assigned to GPRS carries four information slots followed by a control slot, both on the uplink and on the downlink. On the downlink, the control slots are used to signal paging, access grant and channel status information messages together with acknowledgements for uplink data transfer, while the uplink control slots seem to be intended for acknowledgements in the case of downlink data transfer. Uplink and downlink resources can be allocated independently of each other, i.e. uplink information slots are paired with downlink control slots on the same basic physical channel and vice versa. The existing normal burst and access burst formats are used; the latter for random access attempts on idle uplink information slots (as identified through signalling messages sent on the downlink control slots). The coding and interleaving for the information slots could be based on the SDCCH coding (except for random access attempts), while information on the control slots is conveyed in single bursts, i.e. no interleaving is applied and a new coding scheme would have to be used. If more than one basic physical channel is assigned to GPRS, the control slots on the different channels could either be scheduled all in the same TDMA frame (referred to as rectangular configuration), or staggered (diagonal configuration).

VRRA as initially proposed seems to require dedicated GPRS resources, i.e. every cell offering the service would require at least one time-slot allocated permanently to GPRS, which is wasteful in cells with little GPRS traffic. The main reason why this proposal dropped out of the race eventually was however different: some operators declared in 1995 that they would not introduce GPRS, if it implied hardware changes at the base stations. It was perceived that the proposed channel structure and coding schemes in VRRA would need such changes.

Like VRRA, the two intermediate proposals allow for independent assignment of uplink and downlink resources to mobiles. Since GSM uses frequency division for duplexing and operates only in paired bands, the total amount of resources available for GPRS in the two link directions is necessarily symmetric, though. For packet data transfer, both proposals require the setting up of dedicated physical channels, referred to as Packet Data CHannels (PDCH). The major distinguishing feature of the third proposal is the introduction of so-called Uplink State Flags (USF), which are piggybacked onto user and control data on the downlink PDCH. These flags allow several Packet Data Traffic CHannels (PDTCH) and the Packet Random Access CHannel (PRACH) to be multiplexed dynamically onto a single PDCH. The fourth proposal, by contrast, does not rely on USF, but rather on bitmaps indicating precisely which resources can be used for how long. Although the starting point for the GPRS solution chosen eventually was the USF proposal, in essence, both these proposals have been adopted, the one based on USF referred to as *dynamic allocation* MAC mode, the other one as *fixed allocation* MAC mode. They will be discussed in detail in subsequent sections. Additionally, some concepts of the first proposal have also been adopted, namely that no dedicated resources need to be allocated permanently in a cell supporting GPRS and as a consequence, that a GPRS packet transfer can also be initiated by accessing the GSM RACH. This leads us directly to the basic GPRS principles.

4.7.3 Basic GPRS Principles

As regards the air interface, the two fundamental radio resource management principles for GPRS are the *master–slave principle* and the *capacity on demand principle* [54]. The former points at the two different classes of PDCHs, namely the Master PDCH (MPDCH) and the Slave PDCH (SPDCH). The SPDCH is used for data transfer and for dedicated signalling. The MPDCH accommodates also packet common control channels which carry all the necessary control signalling required to initiate a packet transfer. The latter implies that GPRS does not require permanently allocated PDCHs in a cell. In cells without a PDCH, GPRS mobiles camp on the CCCH. They can be paged on the PCH for mobile terminated packet transfer, and access the RACH for both mobile originated and terminated packet transfer. Only when there are packets to be transferred and resources are available, is either an MPDCH or an SPDCH set up.

4.7 INTRODUCTION TO GPRS

If a master PDCH is set up, then idle GPRS mobiles will camp on the relevant control channels mapped onto this physical channel. An MPDCH would typically be set up only if it were intended to assign some resources to GPRS on a semi-permanent basis. Slave PDCHs, on the other hand, can be released quickly, hence pre-empted for circuit-switched services, should resources in a cell run out. It is this feature which enables temporarily unused traffic channels to be used for GPRS data transfer, to improve resource utilisation, without increasing the blocking probability of circuit-switched services.

The capacity on demand principle is particularly important from the point of view of resource utilisation in blocking-limited systems. For the two-carrier full-rate example listed in Table 4.5, at $P_b = 2\%$ (i.e. $G_E = 8.19$ E), the cumulative occupancy of traffic channels for circuit-switched services is shown in Figure 4.20. The shaded area is the effectively utilised capacity, while the remaining area shows the unused capacity, which could potentially be used by GPRS. The extent to which this is possible depends on various factors, such as GSM and GPRS traffic patterns, what type of GSM call queuing facility exists, what call set-up delay is tolerated, and how quickly PDCHs, in particular SPDCHs, can be released. Furthermore, it also depends on the QoS required by the GPRS subscribers. While it is potentially possible to serve low-priority GPRS users by exploiting only this latent capacity, traffic generated by high-priority GPRS users will affect the blocking probability of circuit-switched services, since to satisfy their QoS requirements, they cannot be pre-empted arbitrarily. In cells with several carriers and a substantial fraction of GPRS traffic, one would therefore expect one or a few basic physical channels (at least one of them an MPDCH) to be set aside for GPRS on a semi-permanent basis to provide adequate QoS for GPRS users. Obviously, this does not prevent from adding other SPDCHs dynamically when appropriate resources are available.

GPRS allows several mobile terminals to be multiplexed onto a single basic physical channel (or PDCH), but it also offers multi-slot capability, allowing a single MS to use up to eight PDCHs. This allows a wide range of data-rates to be supported from very

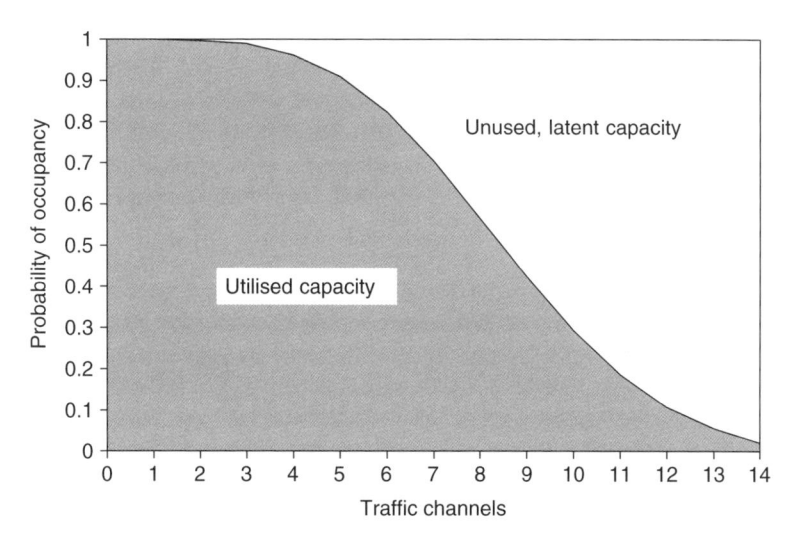

Figure 4.20 Cumulative probability of traffic-channel occupancy for a cell with 14 traffic channels at $P_b = 2\%$

low rates to more than 100 kbit/s, as discussed in more detail in Section 4.9. As pointed out before, up- and downlink resources are allocated separately, allowing for asymmetric resource allocation, although the global resource available in a cell for GPRS is symmetric. Finally, and crucially, resources are only assigned to individual users when they need to transmit data.

4.7.4 GPRS System Architecture

GPRS introduces two new network nodes in the GSM PLMN, namely the Serving GPRS Support Node (SGSN), and the Gateway GPRS Support Node (GGSN). A detailed description of the architecture can be found in GSM 03.60 [76], while in the following, only a very rough overview is provided.

The SGSN is at the same hierarchical level as an MSC. It keeps track of individual mobiles' location and performs security functions and access control. It is connected to the base station system via frame relay. The GGSN provides interworking with external Packet Data Networks (PDN), such as the Internet. It is connected with SGSNs via an IP-based GPRS backbone network. Mobile terminated IP datagrams, for instance, are delivered via IP routing mechanisms to the GGSN (the IP address allocated to the MS effectively 'points' to the GGSN). Provided that an active PDP context exists for this address (see below for details), the GGSN delivers them onwards to the relevant SGSN by encapsulating them with a GPRS tunnelling protocol (GTP) header, a UDP and an IP header (with the IP destination address pointing to the SGSN). Finally, the SGSN delivers them to the right BSS via frame relay, which requires the cell selected by the MS to be known. If the latter is not the case, then the SGSN must first page the MS in the last known routing area (RA, which is the GPRS equivalent to the location area for circuit-switched services).

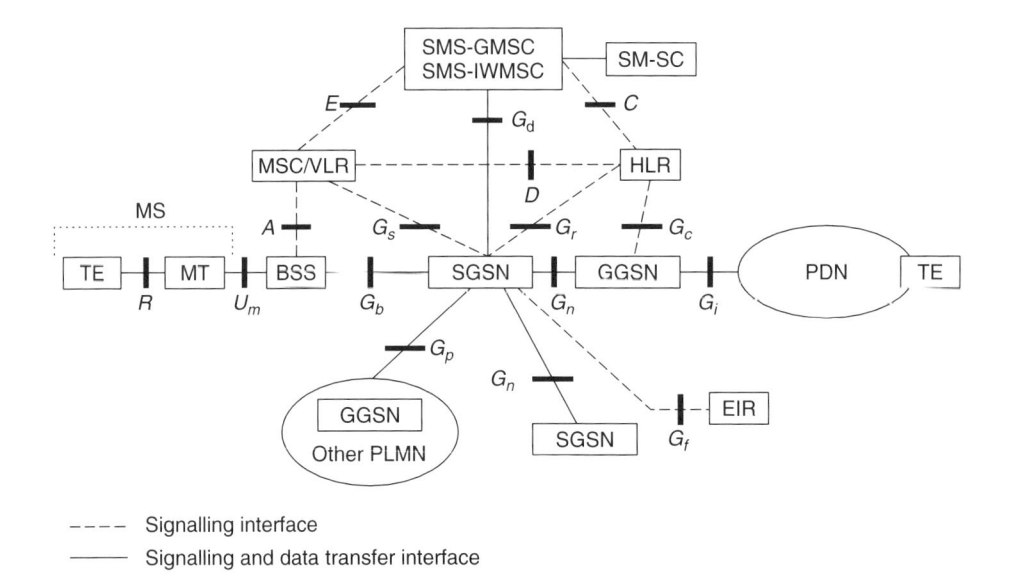

Figure 4.21 GPRS logical architecture

4.7 INTRODUCTION TO GPRS

The logical GPRS architecture is shown in Figure 4.21. This is essentially the same as Figure 4.1, but the A- and the E-interface are here shown as signalling only, since GPRS user data is delivered via SGSN and GGSN. The G_s interface is optional. When present and when the MSC/VLR is enhanced appropriately, it allows for more efficient co-ordination of GPRS and non-GPRS services and functionality. Paging for circuit-switched calls for example can be performed more efficiently via the SGSN, and GPRS and non-GPRS location updates can be combined. The G_c interface is also optional.

Care has been taken in the design of GPRS to maintain a strict separation between the radio subsystem and the network subsystem, allowing the network subsystem to be reused with other radio access technologies, e.g. UTRA. However, this has only partially materialised. While the UTRAN is indeed connected to a packet-switched core network composed of SGSNs and GGSNs, this network subsystem is an evolution of GPRS, with the UMTS SGSN quite different in functionality from the GPRS SGSN.

GPRS does not mandate changes to an installed MSC base. It does also not require hardware changes to the BTS (software changes are obviously required). The BSC needs to be enhanced to support the new G_b interface. The HLR is enhanced with GPRS subscriber information, and the SMS-GMSCs and SMS-IWMSCs are upgraded to support SMS transmission via the SGSN.

4.7.5 GPRS Protocol Stacks

Separate protocol stacks exist in GPRS for the transmission plane (dealing with user data transfer) and the signalling plane. The transmission-plane protocol-stack is shown in Figure 4.22, while the stack for the signalling plane is shown in Figure 4.23, in the latter case only between MS and SGSN. In the following sections, the main focus will be on the air-interface specific layers over the U_m interface, i.e. the GPRS physical layer (PHY), the MAC and the RLC, in accordance with the scope of GSM 03.64. The reader may find more information on the architecture and the other layers in GSM 03.60. For completeness, we enumerate those layers not dealt with in detail in the following sections:

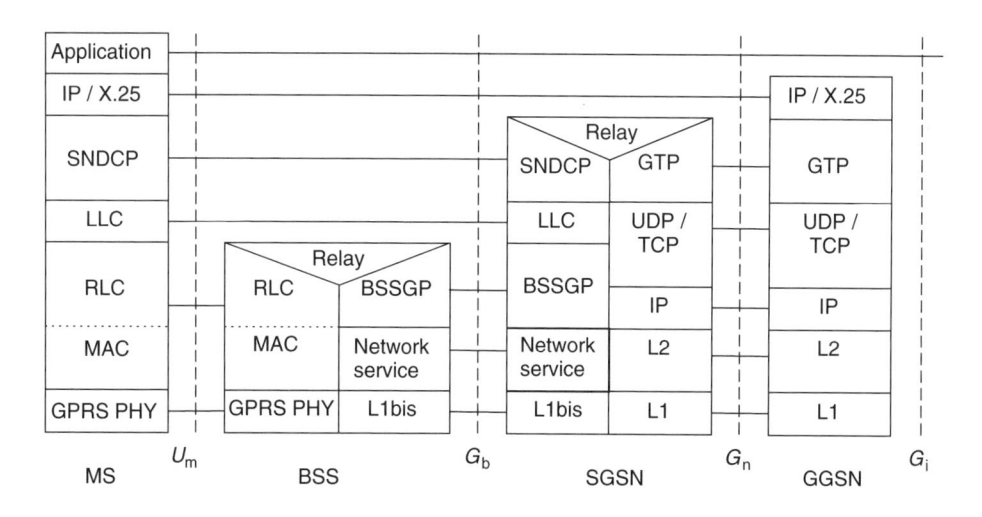

Figure 4.22 Transmission-plane protocol-stack

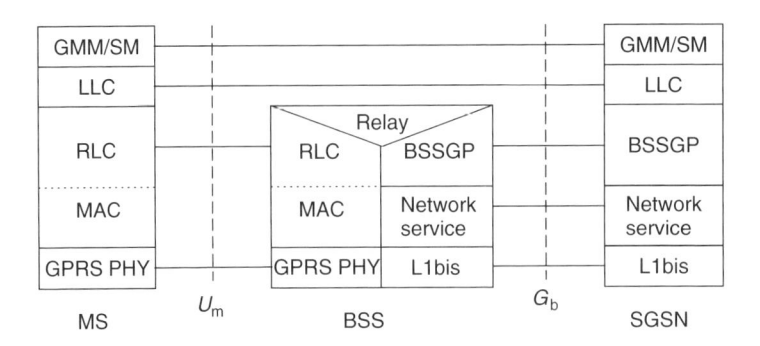

Figure 4.23 Signalling-plane protocol-stack between MS and SGSN

- The logical link control (LLC) specified in GSM 04.64 [217], provides a highly reliable ciphered logical link. To achieve reliability, it includes ARQ functionality in its so-called acknowledged mode. Since RLC performs ARQ as well and for applications running over TCP/IP, TCP does it as well, some readers may wonder whether this was really the best possible design. The people standardising UMTS wondered as well, and ended up with a somewhat different protocol stack which does not feature the LLC anymore.

- The SNDCP is the subnetwork dependent convergence protocol. It maps network-level characteristics onto the characteristics of the underlying layers, e.g. by performing segmentation of a network protocol data unit (PDU)[4] into multiple LLC PDUs, and re-assembly of these LLC PDUs into a single network PDU. There is a single SNDCP entity at the MS and one at the SGSN, which provide multiplexing of data coming from different sources, and potentially using different packet data protocols (the most common being IP). The SNDCP also compresses redundant protocol and user information, e.g. by performing TCP/IP header compression. The SNDCP is specified in GSM 04.65 [218].

- BSSGP stands for BSS GPRS Protocol, which is specified in GSM 08.18 [219]. This layer conveys routing- and QoS-related information between BSS and SGSN. Unlike its equivalent over the air interface, namely the RLC, BSSGP does not perform error correction.

- The network service layer specified in GSM 08.16 [220] transports BSSGP PDUs. It is based on the frame relay connection between BSS and SGSN.

- GMM/SM in the signalling protocol stack stands for GPRS mobility management and session management. Subsection 4.7.7 describes its basic principles briefly.

In the above figures, it is not specified whether the MAC and the RLC are terminated at the BTS or the BSC. In fact, this depends very much on the implementation, in particular the placement of the so-called Packet Control Unit (PCU), which deals for instance with channel access control and with RLC-related ARQ functions. The PCU can be placed at

---

[4] A PDU of protocol X is the unit of data specified at the X protocol layer consisting of X protocol control information and possibly X protocol layer user data [213].

the BTS or remote to it, in the latter case typically at the BSC, but potentially even at the SGSN (the G_b interface would then be located between PCU and SGSN).

4.7.6 MS Classes

Three modes of operation are supported for GPRS mobiles. *Class A* mobiles can support GPRS and other GSM services simultaneously. *Class B* mobiles monitor control channels for GPRS and other GSM services simultaneously, but can only support one set of services at one time. Finally, *class C* mobiles cannot even monitor GPRS and other GSM control channels simultaneously, thus they operate either as a normal GSM mobile terminal, or as a GPRS-only MS, as chosen by the user.

On top of these modes of operation, GPRS mobiles are also characterised by the multi-slot class they belong to, which defines the capability in terms of multi-slot operation, very much as in the case of the HSCSD service. In fact, multi-slot classes 1 to 18 are used both for HSCSD and GPRS, the first 12 being 'type 1' classes without simultaneous receive/transmit capability suitable for 'conventional' half-duplex operation. Additionally, specifically for GPRS, classes 19 to 29 were introduced, which are also 'type 1' classes, but are designed for a GPRS-specific way of 'half-duplex' operation, where uplink and downlink transfer of user data must not occur simultaneously and neighbour cell measurements do not need to be performed in every TDMA frame. In other words, in certain TDMA frames, the MS either only transmits or only receives. With this mode of operation, aggregation of a large number of slots is possible without requiring simultaneous receive/transmit capability. Indeed, all 11 new classes feature at least six receive slots, whereas the other 'type 1' classes feature at most four. For the detailed specification of all 29 multi-slot classes, refer to appendix B of GSM 05.02.

4.7.7 Mobility Management and Session Management

From a mobility management perspective, ignoring a few transitional states, a GSM MS is either in idle state or in dedicated state. In the former, the network knows the mobile location on the basis of location areas and no dedicated resources are consumed. In the latter, the MS is transmitting and receiving on dedicated resources, thus by necessity, the network knows the location of the MS on a cell basis.

In GPRS, two separate sets of states are distinguished, namely mobility management states for GPRS mobility management and packet data protocol (PDP) states for session management. From a GMM perspective, the MS can either be in idle, standby, or ready state. In *idle state*, the network does not know the mobile location at all. In *standby state*, it (or rather, the serving SGSN) knows the location on the basis of routing areas. In *ready state*, finally, the network knows the location on a cell basis. Therefore, the standby state in GPRS corresponds roughly to the idle state in GSM.

The PDP states indicate the status of a particular PDP *context* (i.e. information related to the subscription of a particular service, such as the PDP address to be used, and the subscribed QoS profile). A GPRS subscriber can subscribe to several services and, therefore, this subscriber may be associated with several PDP states simultaneously (but his handset only with one MM state — one reason for introducing these two sets of states). In *PDP inactive state*, the PDP context has no routing information, meaning that packets

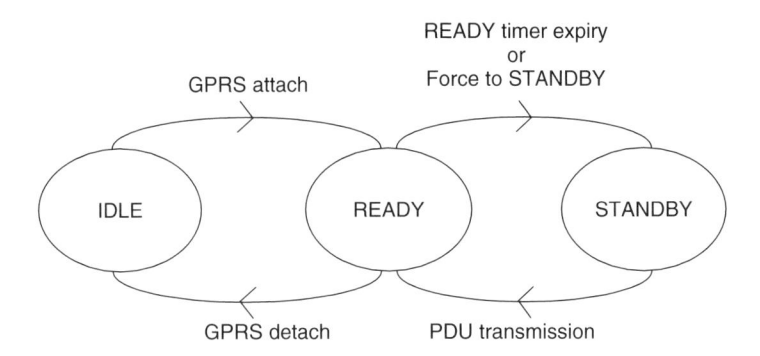

Figure 4.24 Mobility management states at the mobile station

arriving at the GGSN cannot be delivered onwards to the SGSN serving the target MS. In *PDP active state*, the required routing information is available and packets can be delivered to the SGSN and from there onwards to the MS, provided that it is in MM ready state. If in MM standby state, then it needs to be paged first, as explained in Subsection 4.7.4, before delivery from the SGSN onwards is possible. The MS can only enter the active PDP state, if it is either in MM standby or MM ready state.

From this discussion, it should be clear that it does not make sense for a switched on GPRS MS to stay in MM idle state. It will typically enter MM ready state after being switched on and, if no data transfer takes place for a while, fall back to MM standby state and stay there. It is therefore always reachable by the network (allowing for instance services to be pushed to the MS), a feature which is also referred to as *always on*.

In accordance with the two sets of states, there are two registration mechanisms. The *GPRS attach* procedure is used to establish an MM context, i.e. a logical link between the MS and the SGSN. It results in a transition from MM idle to MM ready state, from which a further transition to standby state may occur, for instance if the 'ready timer' expires (in neither direction is a direct transition from idle to standby state possible). This is illustrated in Figure 4.24. Once attached, the MS is available for SMS over GPRS, paging via SGSN, and notification of incoming data. The *PDP context activation* procedure, as the name reveals, is used to move from PDP inactive to PDP active state. It causes the SGSN to establish a context with the GGSN to be selected, and the GGSN to activate the assigned PDP address (e.g. an IP address), through which the GPRS subscriber can be reached by the outside world (the Internet, for instance) for user data transfer. The GGSN is selected based on the so-called Access Point Name (APN) contained in the user subscription information for the relevant PDP context. Note that the GGSN does not change while a session is active. The SGSN, on the other hand, as the name suggests, may change during a session as a result of movement.

4.8 GPRS Physical and Logical Channels

4.8.1 The GPRS Logical Channels

The following logical channels, also referred to as *packet data logical channels* in GSM 03.64, were introduced in R97 specifically for GPRS:

- the Packet Common Control CHannel (PCCCH), consisting of:
  — the Packet Random Access CHannel (PRACH) in uplink direction;
  — the Packet Paging CHannel (PPCH), the Packet Access Grant CHannel (PAGCH), and the Packet Notification CHannel (PNCH) in downlink direction;
- the Packet Broadcast Control CHannel (PBCCH), only downlink;
- the Packet Data Traffic CHannel (PDTCH), which is allocated independently in uplink and downlink direction, to allow for asymmetric resource allocation; and
- the following two packet dedicated control channels, which are both bi-directional: the Packet Associated Control CHannel (PACCH) and the Packet Timing Advance Control CHannel (PTCCH).

It is easy to see that most of these channels are more or less the exact equivalent of the logical channels already known from circuit-switched GSM operation, with well known acronyms such as RACH being prefixed by a 'P' for 'packet' (or a 'PD' for 'packet data' in the case of the PDTCH). Indeed, the PPCH cannot only be used for packet-data related paging, but also to page a class A or class B mobile station for a circuit-switched call. There are some differences, though. While the NCH is used for group calls, the PNCH is used for the 'point-to-multipoint multicast' service, which is not yet defined in the first phase of GPRS. The PACCH combines some of the functionality catered for by the SACCH, the FACCH, and the SDCCH in the 'circuit-switched world'. Finally, the PTCCH is new.

In GSM dedicated mode, at least one SACCH is assigned to an MS permanently, providing regularly occurring signalling capability for link control, which includes the ability to track and control the timing advance. For bursty packet-switched services, we would preferably like to avoid such a channel which consumes radio resources even while there is no user data to be sent. This is why PACCH blocks, which convey for instance resource assignment messages, acknowledgements and power control information, are only scheduled as and when required. However, some regularly occurring signalling resource is all the same required during the so-called packet transfer mode to track and control timing advance, which is where the PTCCH comes into play. Thanks to the exclusive focus on timing advance, the same amount of physical resources required for one SACCH associated to a single full-rate TCH is sufficient to control the timing advance of 16 GPRS mobiles.

4.8.2 Mapping of Logical Channels onto Physical Channels

The Packet Data CHannels (PDCHs) were already introduced earlier. These are the basic physical channels carrying the GPRS logical channels. A master PDCH can carry all GPRS logical channels listed above. There can be from 0 to 16 MPDCHs in a cell, however, only one of them may carry the PBCCH. The slave PDCH carries only PDTCH, PTCCH, and PACCH. Note that a PDTCH is allocated to a specific MS, but owing to the multiplexing capability, a PDCH can serve several mobiles, thus carry several PDTCHs. Recall also that the PDTCH is allocated independently in uplink and downlink direction,

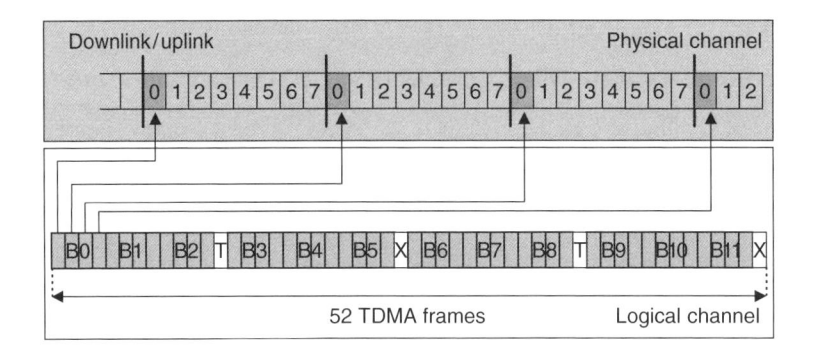

Figure 4.25 PDCH multiframe structure (X = idle frame, T = frame used for PTCCH)

which means that different sets of mobiles may be multiplexed in uplink and downlink direction respectively.

A PDCH multiframe consists of 52 TDMA frames, divided into 12 *radio* blocks (B0 to B11) of four frames each, which in turn carry RLC/MAC blocks, furthermore two idle frames and two PTCCH frames, as shown in Figure 4.25. As any other basic physical channel in GSM, a single PDCH is mapped onto one time-slot per TDMA frame (as shown in the figure), and uplink and downlink time-slots are staggered by three slot periods. Frequency hopping may be applied on both SPDCH and MPDCH (the latter even when carrying the PBCCH). The 52-multiframe structure is used both for master PDCHs carrying a PCCCH and slave PDCHs not carrying a PCCCH. Because of this choice, a GPRS MS can decode neighbour-cell SCHs in the usual way known from GSM, irrespective of whether its allocated PDTCHs are carried on MPDCHs or SPDCHs.

As pointed out earlier, there is no requirement for an MPDCH to be allocated permanently in a cell supporting GPRS. Initial synchronisation is aided by FCCH and SCH (there is no packet-equivalent of these channels), and a new system information message type on the BCCH will indicate whether MPDCHs exist in the cell. If this is the case, the message contains all parameters characterising the MPDCH carrying the PBCCH (including parameters relating to frequency hopping, where applied). A GPRS MS will then first listen to the PBCCH to extract all GPRS relevant system information, including number and location of other MPDCHs. There is a similar algorithm as in GSM to determine the PCCCH to which an MS belongs, i.e. on which it can expect to be paged and which it can use to place channel request messages. Unlike GSM, however, the PBCCH is not duplicated on every MPDCH, because a duplication would take away resources unnecessarily, which could otherwise be used, for example, by downlink PDTCHs sharing the same PDCH. Hence an MS may have to listen to the PBCCH and to its PCCCH on different time-slots.

Due to the choice of the 52-multiframe for PDCHs, an MS cannot possibly listen to its PCH mapped onto a 51-multiframe while involved in GPRS-related activities, without seriously impacting the scheduling of PDTCH and PACCH blocks on the PDCHs it is assigned to. It is therefore necessary that class A and B mobiles be paged on their PPCH for circuit-switched services rather than their PCH.

The mapping of the different logical channels onto the physical channels (i.e. MPDCH and SPDCH) is much more flexible in GPRS than it is in GSM. With the exception of

PTCCH and PBCCH, this mapping can change from radio block to radio block in a very dynamic fashion. Recall that dedicated channels of several users can be multiplexed onto a single PDCH, hence means are not only required to distinguish between different types of logical channels, but also to distinguish between channels of a single type allocated to different users.

4.8.2.1 Downlink Mapping

On the downlink, the radio block headers (or rather, the MAC headers) identify the type of logical channel. Additionally, for PDTCH and PACCH, these radio blocks carry also an MS identity in the shape of a Temporary Flow Identity (TFI) associated with a Temporary Block Flow (TBF), which is unambiguous in the considered context. Although in general very flexible, there are certain rules and constraints regarding the scheduling of downlink blocks. For instance, on the MPDCH carrying the PBCCH, PBCCH blocks must be scheduled in a regular fashion to facilitate MS operation. Also, there are similar rules for the PPCH as for the PCH in GSM, to allow for battery-efficient operation, e.g. DRX mode.

4.8.2.2 Uplink Mapping — The Uplink State Flag

On the uplink, apart from the PTCCH, three different types of logical channels may occur, the PRACH on one side (only on an MPDCH, though), PDTCH and PACCH blocks assigned to a specific user on the other. The uplink state flag has an important role to play in this context. There are eight different USF values. The value relevant for a particular uplink block is signalled in the preceding downlink block on the same physical channel, see Figure 4.26. On the MPDCH, one value is reserved to indicate the occurrence of a PRACH block, while the other seven values can be used to multiplex PACCH and PDTCH blocks of seven different users onto this MPDCH. On the SPDCH, all eight values are available for user multiplexing. Clearly, therefore, an uplink assignment message signalled by the network must not only contain the usual description of the relevant physical channel(s), but also the USF value to which the MS is assigned (individually for each channel, if multiple channels are assigned). It is then up to the MS to schedule either a PACCH block or a PDTCH block in an uplink radio block with its USF value.

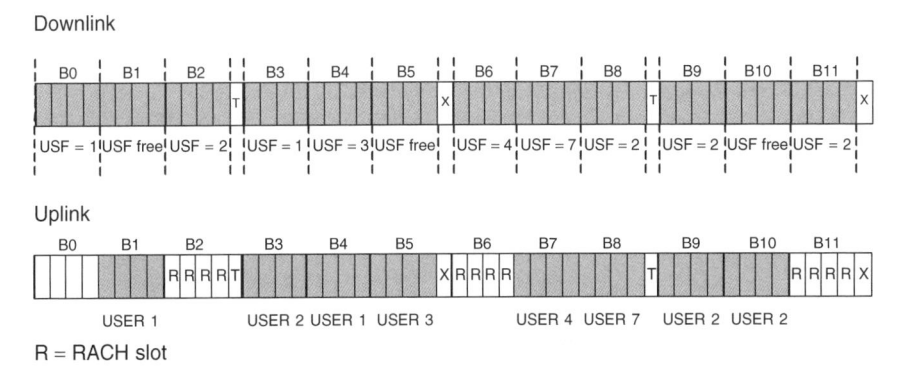

Figure 4.26 Uplink multiplexing with USF on an MPDCH (B0 = Block 0, R = PRACH, T = PTCCH, X = idle slot/frame)

As an alternative approach to uplink multiplexing, instead of this USF-based so-called *dynamic allocation*, the network can decide to use a *fixed allocation* scheme. With this scheme, rather than signalling the relevant USF values, the assignment messages sent by the network specify exactly which radio blocks are to be used by the MS; possibly different radio blocks on different time-slots, if multiple slots are assigned. Mobile terminals obtaining a fixed allocation do not need to monitor the USF. If fixed and dynamic uplink allocations are mixed on a single physical channel, the USF for radio blocks assigned using fixed allocation must be set to a value that is not used for any of the mobiles holding a dynamic allocation, to avoid collisions between these two mobile categories. The different resource allocation modes are discussed in more detail in Section 4.10.

On an MPDCH, there are two ways to identify which uplink radio blocks are assigned to the PRACH:

(1) with an uplink state flag marked as 'free' for this block (i.e. fully dynamic allocation);

(2) by specifying (through signalling on the PBCCH) a subset of the 12 radio blocks per 52-multiframe, which are exclusively assigned to the PRACH. Although the PRACH allocation can be modified over time, this approach is more static in nature due to the limited occurrence of PBCCH blocks.

One reason why the second approach was introduced was to reduce the processing burden at the MS arising from having to monitor the USF and to attempt a random access in the uplink block following a downlink block with a flag marked as free. However, even when the second approach is used, the network operator may allocate additional PRACH resources in a dynamic fashion by setting the USF for additional blocks to free. Note that the USF values relating to the blocks assigned to the PRACH through PBCCH signalling must also be set to 'free'. A handset can therefore either listen to the USF and access any radio block assigned to the PRACH, or rely solely on the PRACH parameters signalled on the PBCCH, which means that it can only 'see' and access those PRACH blocks allocated on a semi-static basis.

4.8.2.3 Mapping of the PTCCH

As illustrated in Figure 4.25, the PTCCH is mapped onto two slots in every 52-multiframe. The PTCCH resources on a single PDCH suffice to track and control the TA of up to 16 different mobiles. For this purpose, there are 16 different uplink PTCCH subchannels (PTCCH/U), each consisting of a single time-slot every eighth 52-multiframe, in which the respective MS sends an access burst. On the downlink, by contrast, TA messages for four mobiles are multiplexed onto one radio block, meaning that a PTCCH/D is paired with four PTCCH/U. The exact mapping, i.e. the assignment of the individual subchannels, is determined by the Timing Advance Index (TAI) signalled to the MS together with other parameters contained in the so-called packet uplink (or downlink) assignment message. For more details on the timing advance procedure, refer to Subsection 4.9.6.

4.8.3 Radio Resource Operating Modes

In GSM, the RR idle state and the RR dedicated state are more or less equivalent to the MM states of the same names, which were briefly discussed in Subsection 4.7.7 (this is

4.8 GPRS PHYSICAL AND LOGICAL CHANNELS

why we can assume that a GSM MS in MM dedicated state consumes dedicated radio resources). In GPRS, the situation is a little bit more complicated. There are three so-called *radio resource operating modes*, namely packet idle mode, packet transfer mode, and dual transfer mode. For the time being, we focus on the first two, since the last one, which is discussed separately in the next subsection, is not supported by all GPRS mobiles.

In *packet idle mode*, an MS listens to the PBCCH (or BCCH, if no PBCCH is available in the selected cell), and to the paging subchannels on the PPCH (or PCH) corresponding to the paging group of the MS. If data is to be transmitted, the MS must transit from the packet idle mode to the packet transfer mode. In *packet transfer mode*, the MS may transmit or receive data on one or several physical channels. Having identified the GPRS multiple access protocol as belonging to the family of R-ALOHA protocols in Section 3.5, we can now say that the packet transfer mode corresponds to the R-ALOHA reservation phase. Correspondingly, the transition from idle to transfer mode may (but does not always) involve a random access procedure. It must be noted, however, that the behaviour in 'reservation mode' is quite complex. This will be discussed in detail in Section 4.10.

Regarding the correspondence between MM states and RR operating modes, note first that the MM idle state is irrelevant, i.e. does not correspond to any RR mode. In MM standby state, the MS is necessarily in RR packet idle mode, while packet transfer mode implies MM ready state, since to assign PDTCH resources, the MS location must be known at the cell level. Importantly, however, the MS can be in RR packet idle mode while in MM ready state. This is illustrated in Table 4.8.

It was pointed out earlier that due to the nature of the services to be provided by GPRS, a real-time handover capability was not required, cell (re)selection would do instead. Indeed, when selecting a new cell, a GPRS MS involved in packet transfer leaves the packet transfer mode, enters the packet idle mode, switches to the new cell, reads the system information in that cell, and only then may attempt to re-enter packet transfer mode.

When a GPRS MS in MM ready state, but currently not involved in a packet transfer, selects a new cell, it must enter packet transfer mode briefly after having switched to the new cell to transmit an LLC frame serving as a cell update. This allows the network to keep track of the MS position at the cell level, as required in MM ready state.

4.8.4 The Half-Rate PDCH and Dual Transfer Mode

Initially, it was planned that GPRS would only use full-rate physical channels. This was consistent with the early focus on solutions such as VRRA, which would only make sense if dedicated physical resources were available for GPRS. However, back in August 1996, when it was clear that GPRS would be built on the capacity on demand principle, and according to the master–slave concept, the author submitted a document to the SMG2 GPRS *ad hoc* group making a case for the introduction of half-rate slave PDCHs. These

Table 4.8 Relationship of RR modes and MM states in GPRS

RR mode	Packet transfer mode	Packet idle mode	Packet idle mode
MM state	Ready	Ready	Standby

would only be used to fill up 'holes', that is the unused portion of a basic physical channel carrying only one half-rate (voice or data) call. Such 'holes' can occur due to an odd number of half-rate calls being served in a particular cell, or due to departing half-rate calls, as discussed in detail in Subsection 4.6.4. Since these half-rate SPDCHs would only exist during the 'lifetime of a hole', they would have to be used either for low-priority services, or to increase temporarily the resources allocated to multi-slot capable mobile stations already in packet transfer mode. The latter would be provided that the hole opens up on a slot which is 'within reach' of potential target GPRS mobile stations, i.e. could be used within the constraints of the multi-slot class they belong to, given their current slot allocation.

Due to the earlier decision of the SMG2 GPRS *ad hoc* group to focus on full-rate PDCHs, and since it was unclear whether the half-rate voice codec would ever become popular (indeed, it has not so far), the proposal was not accepted at the time. However, for release 1999 of the GPRS specifications, the Dual Transfer Mode (DTM) was introduced, which can make use of a half-rate SPDCH, albeit in a somewhat different fashion and for a different purpose than originally proposed by the author. Dual transfer mode describes the capability of a mobile terminal to be engaged simultaneously in a circuit-switched call and in GPRS packet transfer in a co-ordinated fashion (conventional class A mobile stations may be involved in these two activities independently from each other). This may happen on a multi-slot configuration or using a single time-slot. In the latter case, the call takes place on a TCH/H, while the other half of the same basic physical channel becomes a half-rate SPDCH for the exclusive use of this MS, that is, the SPDCH/H carries exactly one PDTCH/H with its PACCH/H. If a multi-slot configuration is used, the total resource allocated to the DTM MS consists of a single traffic channel (potentially a half-rate channel) and one or more full-rate PDTCHs, i.e. PDTCH/Fs. In this case, unless handset limitations mandate the use of exclusive allocation, fixed, dynamic, or extended dynamic allocation (a variation of dynamic allocation) is used on the PDTCHs. In the remainder of this chapter, 'PDTCH' implies 'PDTCH/F', unless otherwise indicated.

With this information in hand, we can briefly reconsider the different MS classes mentioned in Subsection 4.7.6. The simplest case is a class C MS, which can only be either GSM attached (and GPRS *de*tached) or GPRS attached (and GSM *de*tached), thus in the former case switch between idle and dedicated mode, while in the latter case between packet idle and packet transfer mode. A class B MS can be in idle mode and packet idle mode simultaneously, but it can only be either in dedicated or in packet transfer mode at any given time. An MS supporting DTM is a special case of a class A MS. The dual transfer mode can only be reached via GSM dedicated mode. Finally, the RR state machine of a conventional class A mobile station not supporting DTM features all four combinations of GSM and GPRS RR states, with state transitions occurring as a result of the respective GSM or GPRS operation. Since this information on RR operating modes was added to the GSM specifications shortly before this text was written, it might be subject to further modifications. For full details, the reader is therefore referred to the newest version of GSM 03.64.

4.9 The GPRS Physical Layer

In GSM 03.64, the GPRS physical layer is split into two sub-layers, the *physical RF layer*, dealing with basic transmission and reception functionality such as modulation and demodulation, and the *physical link layer* above that, responsible for forward error correction coding and interleaving, as well as providing functionality for cell (re)selection, power control, etc. Whether necessary or not, this split introduced specifically for GPRS is convenient insofar as it restricts GPRS-specific modifications and enhancements to the link layer, while the RF layer is common to all GSM services including GPRS. Since the RF layer was already discussed in earlier sections of this chapter, the subject matter in this section is the physical link layer. EGPRS specific extensions are dealt with in Section 4.12.

4.9.1 Services offered and Functions performed by the Physical Link Layer

The basic purpose of the physical link layer is to convey higher layer information (e.g. MAC/RLC information) across the GSM radio interface. Furthermore, it is responsible for:

- channel (or FEC) coding and interleaving;
- synchronisation (including determination of the timing advance);
- monitoring and evaluation procedures relating to radio link signal quality;
- cell (re)selection procedures;
- transmitter power control procedures; and
- battery power conservation procedures, such as DRX procedures.

Some of these matters are discussed in more detail in the following.

4.9.2 The Radio Block Structure

A GPRS radio block consists of four normal bursts and is used to convey RLC/MAC blocks. An RLC/MAC block serving the purpose of *data transfer* consists of a one-octet MAC header and an RLC data block composed of a variable length RLC header, an RLC data field containing octets from one or more LLC PDUs, and a Block Check Sequence (BCS).

An RLC/MAC block used for *control message transfer* features also a one-octet MAC header (sometimes extended by one or two more octets) and a BCS, but instead of an RLC header and an RLC data field, it carries the RLC/MAC control message (or parts of it, if one block is not sufficient).

The relevant MAC and RLC header content will be discussed in Section 4.10, however, one aspect is important here, namely that downlink MAC headers contain three USF bits, since these receive special treatment in terms of error correction coding.

4.9.3 Channel Coding Schemes

The basic channel coding scheme or FEC coding scheme used for GPRS is the same as that applied on the GSM SACCH, i.e. half-rate convolutional coding with a 40 bit

FIRE code used as BCS. This coding scheme, referred to as CS-1, is used for all control messages which are sent using the normal burst format (as opposed to the access burst format used for example on the PRACH and on the PTCCH/U). CS-1 can also be used on the PDTCH. Additionally, three more coding schemes featuring higher code-rates are defined for use on the PDTCH only, namely CS-2, CS-3, and CS-4, in the order of decreasing redundancy. CS-2 and CS-3 are based on half-rate convolutional coding, but redundancy is decreased by puncturing (that is, deleting) some of the bits generated by the channel encoder. In CS-4, no coding is applied except for the USF. A 16-bit CRC sequence is used as a BCS for CS-2 to CS-4.

Correct interpretation of the USF values by the mobile terminals is crucial for proper operation of the GPRS multiple access protocol. Consequently, the USF needs to be coded in a very robust manner. Owing to the extended BCS, CS-1 is considered to be robust enough, so the USF does not need to receive special treatment, when this coding scheme is used. Instead, it is convolutionally coded together with the rest of the radio block and, as a result, it needs to be decoded by the mobile terminals as part of the data. In CS-4, by contrast, a block-code with a length of 12 bits is used to code the three USF bits, while the rest of the data is not coded at all. In CS-2 and CS-3, the USF is first 'pre-coded' to six bits in such a manner that the same 12-bit words result at the output of the convolutional encoder as with the block code used in CS-4. Unlike other data bits, the USF bits are not punctured in CS-2 and CS-3. As a result, for CS-2 to CS-4, the receiver can treat the USF bits in exactly the same manner: they can be decoded separately as a block code.

As is typical for GSM, every GPRS MS must support all four coding schemes, while for a GPRS-enabled network, all schemes but CS-1 are optional. In practise, early GPRS infrastructure will typically support CS-1 and CS-2, with CS-3 and CS-4 capability potentially added later, since they require hardware upgrades at the BTS.

Table 4.9 taken from GSM 03.64 provides an overview of the main parameters of the four coding schemes. The radio blocks transmitted using one of these four schemes are interleaved over four normal bursts, hence 456 raw bits are available to carry all necessary information. The stealing bits, not included in these 456 bits, are used to signal the applied coding scheme.

4.9.4 Theoretical GPRS Data-Rates

The last column in Table 4.9 shows the data-rate achievable per time-slot, excluding the overhead due to FEC, BCS, and USF (that is, accounting for the number of bits shown in the fifth column). These data-rates are calculated on the basis of 12 radio blocks per

Table 4.9 Coding parameters for the GPRS coding schemes

Scheme	Code rate	USF	Pre-coded USF	Radio Block excl. USF and BCS	BCS	Tail bits	Coded bits	Punctured bits	Data rate [kbit/s]
CS-1	1/2	3	3	181	40	4	456	0	9.05
CS-2	≈2/3	3	6	268	16	4	588	132	13.4
CS-3	≈3/4	3	6	312	16	4	676	220	15.6
CS-4	1	3	12	428	16	—	456	—	21.4

52-multiframe, or on average one every 20 ms, hence they account correctly for overhead due to PTCCH and idle frames. The indicated values seem to imply that GPRS can support data-rates of up to 171.2 kbit/s in each link direction, if all eight time-slots of a GSM carrier are assigned to a single GPRS MS for exclusive use. However, these results have to be interpreted with great care. Ignoring for the time being physical layer issues (such as the impact of interference), there is a substantial amount of protocol overhead, which will reduce the net throughput that can be achieved under ideal circumstances.

First, while these figures do account for the three USF bits, they ignore the other five bits overhead contained in the MAC header. Second, RLC blocks (i.e. the RLC data carried in a radio block) must carry an integer number of octets, resulting in a few unusable spare bits for CS-2 to CS-4, namely seven, three, and seven respectively. Third, radio blocks used for data transfer carry an RLC header, which is at least two octets long. At this point in time, the net data-rates as seen by the LLC have already reduced to 8 kbit/s, 12 kbit/s, 14.4 kbit/s, and 20 kbit/s per time-slot respectively, for the four coding schemes.

To provide a fair appreciation of the net data-rates as seen by the IP layer, we would also have to account for the optional RLC header octets, the overhead generated by the LLC and the SNDCP, and the signalling overhead carried on the PACCH, which takes resources away from the PDTCH. This is far from easy to establish. The occurrence of the PACCH, for instance, is irregular. Ignoring this signalling overhead, the biggest problem is that the additional RLC, LLC, and SNDCP overhead *per radio block* depends on the length of the LLC PDU, which is variable. To get a feeling for the overhead we can expect, a digression into SNDCP and LLC matters cannot be avoided.

An SNDCP PDU contains data from a single network PDU (e.g. an IP packet). If the network PDU exceeds the maximum size of the SNDCP PDU, then it needs to be segmented and carried on multiple SNDCP PDUs. The SNDCP header overhead is typically three or four octets for the first segment of a network PDU, and one or three octets for subsequent segments, depending on whether the acknowledged or the unacknowledged LLC transmission mode is applied. The SNDCP PDU is then mapped onto an LLC PDU or frame, which results in additional overhead, namely an LLC frame header (variable in length, but typically four octets for acknowledged mode and three for unacknowledged mode [221]), and an LLC Frame Check Sequence (FCS) spanning three octets. The maximum theoretical LLC PDU size is 1600 octets, however, the practical maximum value available for SNDCP PDUs is around 1500 octets, and can be lower depending on the LLC transmission mode selected.

Depending on its length, an LLC PDU is transmitted on one or many RLC/MAC blocks, each fitting into one radio block. A maximum size LLC PDU would have to be carried over many radio blocks and would add significantly less than one additional octet of overhead per block. Ignoring signalling overhead required for entering into packet transfer mode, acknowledgements, etc., the ideal net throughput rates which could be achieved would therefore come close to the above-listed revised data-rates, i.e. 8 kbit/s, 12 kbit/s, 14.4 kbit/s, and 20 kbit/s per time-slot. On the other hand, a radio block can also carry several short LLC PDUs. Potentially, a single radio block might carry a segment of an LLC PDU at the beginning of the RLC data field, an integer number of complete LLC PDUs in the middle, and another segment of an LLC PDU at the end of the RLC data field. Apart from the excessive relative SNDCP and LLC overhead per PDU, this will require additional optional RLC header octets for LLC PDU delimitation. Figure 4.27 illustrates

LLC PDU

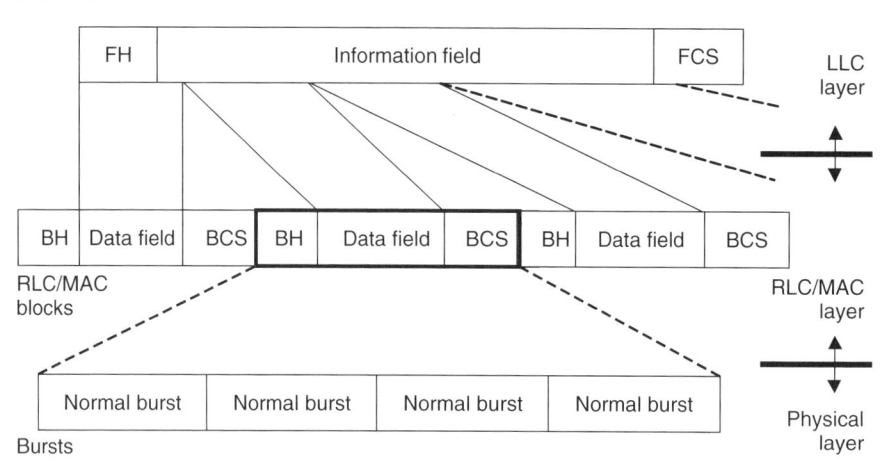

```
┌────┬──────────────────────────────┬──────┐
│ FH │      Information field       │ FCS  │          LLC
└────┴──────────────────────────────┴──────┘         layer
```

```
┌──┬──────────┬───┬──┬──────────┬───┬──┬──────────┬───┐
│BH│Data field│BCS│BH│Data field│BCS│BH│Data field│BCS│
└──┴──────────┴───┴──┴──────────┴───┴──┴──────────┴───┘
RLC/MAC                                        RLC/MAC
blocks                                          layer
```

```
┌─────────────┬─────────────┬─────────────┬─────────────┐
│Normal burst │Normal burst │Normal burst │Normal burst │
└─────────────┴─────────────┴─────────────┴─────────────┘
Bursts                                          Physical
                                                 layer
```

FH = frame header
FCS = frame check sequence
BH = block header (MAC and RLC header)
BCS = block check sequence

Figure 4.27 Mapping of LLC PDU onto RLC/MAC blocks and then bursts

how LLC PDUs are mapped onto RLC/MAC blocks and then transmitted on normal bursts, in the example shown requiring multiple RLC/MAC blocks for the transmission of a single LLC PDU.

Consider now what is a rather unsuitable application for the original GPRS design (as per release 1997), namely a voice over IP service. Due to delay constraints, the network PDU size is extremely small, it will typically not contain more than a few 10 ms worth of speech. Assume that an IP packet contains 20 ms worth of speech coded at 8 kbit/s, i.e. a voice payload of 20 octets, which we would ideally want to transmit on one radio block to avoid multi-slot transmission. Suppose further that a header compression scheme can be applied by the SNDCP, which manages to compress the header overhead generated by the IP stack (typically an IP/UDP/RTP header measuring at least 40 octets in the case of IPv4) to two octets. This is possible with new header compression schemes developed specifically for application over low-bandwidth radio links [222], which are, however, not supported by the SNDCP. Assume next that the overhead introduced by the SNDCP (4 octets) and the LLC (3-octet header in unacknowledged mode + 3 octets FCS) amounts to 10 octets. Add an additional optional RLC octet to indicate the length of the LLC PDU. We have now managed to fill all but three octets of an entire CS-3 radio block, which was supposed to give us a net data-rate of 14.4 kbit/s, to transmit voice coded at 8 kbit/s! If the LLC PDU were to fill precisely all data field octets in the RLC/MAC block, the optional RLC octet would not be required, hence a voice codec rate of 9.6 kbit/s could at best be sustained with CS-3. Fortunately, GPRS as specified in release 1997 was never meant to carry voice, so this extreme example is not very relevant in practice. However, there are now endeavours to enhance GPRS in a manner that voice can be supported efficiently, as discussed in more detail in Chapter 11. The

4.9 THE GPRS PHYSICAL LAYER

above calculations show that apart from having to solve the delay problem (since GPRS is too slow for voice), the protocol stacks have to be reworked to contain the protocol overhead resulting from this application.

4.9.5 'Real' GPRS Data-rates and Link Adaptation

So far, we have ignored some real-world effects such as the impact that transmission errors will have on the achievable throughput rates. The error-rate of interest here is the block error rate (BLER) experienced on the radio interface (fixed network errors should be negligible by comparison). It will depend on the experienced SIR (or alternatively, for a coverage-limited system, on the SNR) and on the chosen coding scheme. Figure 4.28 illustrates the performance of the different coding schemes, assuming a single interferer, and using the 'TU 3' channel model defined in GSM 05.05 [223] with ideal frequency hopping. 'TU' stands for 'typical urban', and the number refers to the assumed MS speed, i.e. 3 km/h. The results are taken from Reference [224]. They were produced at a time when the coding schemes were not yet fully specified, thus differ slightly in a few parameters from the finally specified schemes. However, the parameters relevant for the BLER performance remained virtually identical.

If blocks are in error, they need to be retransmitted, reducing the achievable throughput accordingly. Consider a simple model, which ignores details of the RLC operation, and does not account for the overhead required to request retransmissions. We can therefore simply state that the fraction of blocks in error will reduce the real throughput S_{real} according to

$$S_{\text{real}} = S_{\text{ideal}} \cdot [1 - \text{BLER}(\text{CIR})], \qquad (4.3)$$

where S_{ideal} is the throughput that can be achieved in ideal conditions, i.e. ignoring transmission errors. Considering long LLC PDUs, thus ignoring LLC and SNDCP overhead, we set S_{ideal} to 8 kbit/s for CS-1, 12 kbit/s, 14.4 kbit/s, and 20 kbit/s respectively, for CS-2 to CS-4, as discussed in the previous subsection. The resulting real throughput per

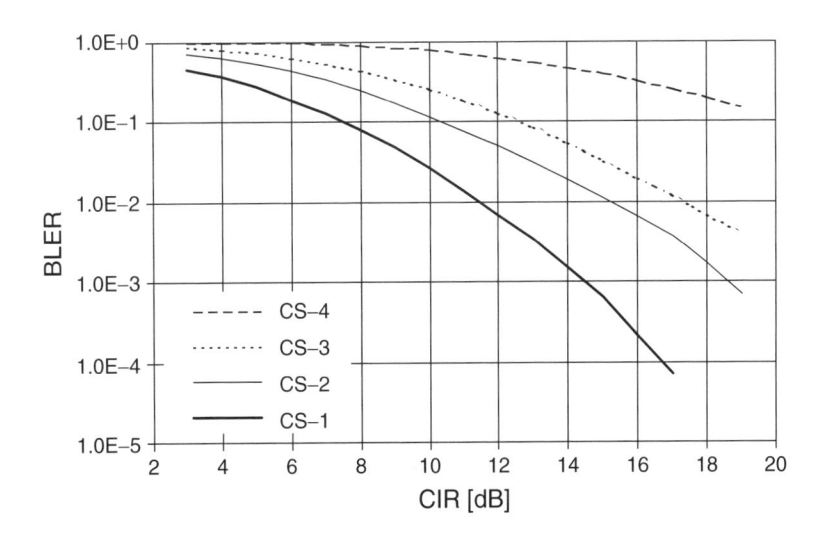

Figure 4.28 BLER for the four coding schemes

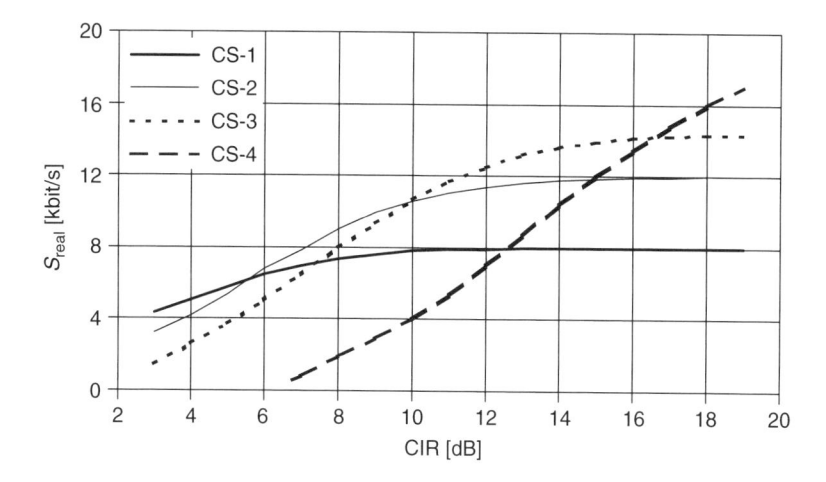

Figure 4.29 Throughput as a function of CIR for the four coding schemes

time-slot is illustrated in Figure 4.29. These results are very similar to those shown in Reference [225], Figure 2, the major difference being that in Reference [225], some of the additional overhead we accounted for was ignored, and instead, S_{ideal} was set according to the values reported in the last column of Table 4.9.

This figure illustrates well why additional coding schemes were introduced in the first place. CS-1 is the most robust scheme and performs best at a low CIR, but S_{real} saturates at an experienced CIR of about 10 dB. Higher CIRs cannot be translated into increased throughput. To exploit this 'excess' CIR, less robust coding schemes can be used. This is where *link adaptation* comes into play. An ideal adaptation algorithm would switch between the coding schemes in a manner that in every moment in time, the coding scheme providing highest throughput is used. Based on Figure 4.29, the switching point between CS-1 and CS-2 is at 5.5 dB, that between CS-2 and CS-3 at 10 dB, and finally between CS-3 and CS-4 at 17 dB. We could be tempted to add an additional curve to this figure, which follows the best throughput segments, and claim that this curve represented the net throughput (for long LLC PDUs) that can be achieved per time-slot at a given CIR.

Alas, it is more complicated than that. The sending entity needs to perform link adaptation based on some measurements. Due to the GSM duplex spacing, the CIRs may only be correlated loosely between up- and downlink (as a result of certain propagation phenomena, in particular fast fading, being fairly independent on the two links). This applies to GPRS even more, since resources on the two link directions are allocated independently from each other. As a consequence, these measurements have to be performed in the right link direction, such that for uplink adaptation, the MS depends on some feedback provided by the BTS, and vice versa. Also, measurements have to be averaged over a certain period of time, to avoid errors introduced by channel noise. Hence this is by no means a quick process. In Reference [225], for single-slot operation, it is suggested that the receiving side sends acknowledgements every 16 radio blocks (to avoid RLC protocol stalling, see Subsection 4.10.2), which determines the measurement interval and the speed of link adaptation. The link adaptation algorithm chooses the coding scheme based on the BLER it can calculate from the number of positively and negatively acknowledged radio blocks.

There is an additional problem, namely excessive BLER at some of the switching points established through Figure 4.29. For instance, with CS-2 at 6 dB, the BLER is in excess of 40%. This may not only have an impact on RLC and LLC operation, but also on that of higher layers, since high error rates mean a lot of retransmission, leading to delayed transfer of LLC PDUs. As an example, TCP/IP connections are known to suffer from this [230]. It would therefore be better to use CS-1 at 6 dB, although S_{real} is slightly lower than with CS-2. In essence, therefore, the optimum switching points may well depend on higher layer protocols associated with specific applications to be supported over GPRS, but this is information that is typically not available to the GPRS network.

So, what 'real' data-rates can be achieved with GPRS? It should be evident by now that it is very difficult to provide a conclusive answer, since it depends on so many parameters. The results shown in Figure 4.29 show the data-rates that can be achieved per time-slot as a function of the CIR for a specific channel model, when the overhead due to FEC coding, MAC header, mandatory RLC header fields, PTCCH and idle frames is accounted for. They show for example, that the 'full' 20 kbit per second and slot are only available to mobiles, which experience good propagation conditions and very little interference, since a CIR in excess of 20 dB would be required. With loose frequency reuse factors, a substantial proportion of the MS population may indeed experience such conditions. As frequency planning becomes more aggressive, however, the fraction of terminals experiencing a CIR of 20 dB or more becomes smaller.

To assess the average data-rates over the entire MS population, one would have to collect statistics on the experienced CIR, and then average the rates shown in Figure 4.29 over the relevant distributions[5]. From results reported in Reference [225] for a specific live network, it would appear that the average throughput per time-slot can indeed exceed 90% of the ideal throughput obtained through CS-4, say 18 kbit/s. From the same reference, we may assume that the PACCH overhead due to signalling of acknowledgements is 1/16 for single-slot operation. The figure may be different for multi-slot operation, and additional PACCH blocks may be required for other signalling messages. One would then also have to introduce some loss factors to account for imperfect link adaptation. As a rough guess, to be interpreted with caution and depending very much on the radio planning applied in a specific network, somewhere between 15 and 17 kbit/s per time-slot may be achieved on average, before accounting for overhead due to LLC and higher layers.

To complete the picture, note that early GPRS handsets support typically operation on one uplink slot and two downlink slots, with handsets supporting four downlink slots to come soon after. Recall also that early GPRS infrastructure may not support CS-3 and CS-4. At the time of writing, it is not possible to say whether we can realistically expect handsets with eight-slot capability. This may be rather challenging in terms of handset design and battery life, and equivalent data-rates may be provided more cost efficiently through UMTS handsets, for example.

4.9.6 The Timing Advance Procedure

The timing advance procedure is required to determine the correct TA value to be used for the uplink transmission of radio blocks. It consists of an initial timing advance estimation

---

[5] To obtain accurate results, a careful mapping between CIR statistics and block error rates is required for reasons discussed in References [227–229] for example.

and, while in packet transfer mode, continuous timing advance updating. The timing advance procedure is neither relevant for packet idle mode (since no uplink radio blocks are transmitted), nor for dual transfer mode, where the method known from GSM dedicated mode is applied.

The initial timing advance estimation is obtained by the network through evaluation of the received packet channel request message on the RACH or PRACH (exactly as in circuit-switched GSM). It is then transmitted in the downlink signalling block sent on a PAGCH, typically a packet downlink assignment or a packet uplink assignment message, and used by the mobile station for subsequent uplink transmission. This assignment message will also contain the PTCCH and the TAI determining the subchannel on this PTCCH to be used for the continuous timing advance updating procedure. On the PTCCH/U, the MS will send access bursts on the time-slots identified by its TAI value. On the PTCCH/D, the network will send TA messages in the usual radio block format (i.e. interleaved over four normal bursts), each message containing updated TA values for four mobiles calculated on the basis of the received access bursts. For uplink radio block transmission, the mobiles will then always use the latest TA values ordered by the network.

From the PTCCH mapping described in Subsection 4.8.2, it is evident that the TA values are updated once every eight 52-multiframes, or once every 1.92 s. According to GSM 05.10 [226], the error tolerance for the timing advance value is ± 1 symbol, or ± 3.69 µs, one symbol thus corresponding to a two-way distance at the speed of light of roughly 1100 meters. The PTCCH update interval should therefore be tight enough to cater even for very fast moving mobiles. However, to increase the accuracy and to be able to average the measurements over several bursts, the network may also monitor the delay of the normal bursts and access bursts sent by the MS on PDTCH and PACCH. If it should find that the TA needs earlier updating than would be possible via the appropriate PTCCH subchannel, it can immediately send a new TA value on the PACCH, using the packet power control/timing advance message format. This is possible since the MS needs to monitor continuously all downlink physical channels on which the network may send PACCH blocks addressed to that MS.

So why was the PTCCH introduced, why could one not rely entirely on the PACCH for TA purposes, in a similar manner as circuit-switched GSM relies on the SACCH? Because an MS involved in a downlink packet transfer, but not simultaneously in an uplink transfer, will only need uplink PACCH resources from time to time to signal acknowledgements. Regular PACCH blocks would have to be scheduled to allow for the tracking of the TA, which is inefficient. Using the PTCCH/U is more efficient. Furthermore, on the downlink, a single radio block can carry TA values for four mobiles on the PTCCH/D, while the power control/timing advance message would have to be addressed to a single MS.

There are a few special cases, which will become clearer once the GPRS MAC/RLC protocols are described in more detail in the next section.

- In response to the packet channel request message, the network may send a packet queuing notification, if no suitable radio resources are available to be assigned to the MS. By the time resources become available and an assignment message is sent, the initial TA estimate may be outdated.

- If the MS is in MM ready state, packet downlink assignment messages may be sent without prior paging, and thus no packet channel request message is sent by the MS which could be used for the initial TA estimate.

In these cases, an initial TA estimate must be obtained through other means. One option is polling by the BTS, which causes the MS to send a packet control acknowledgement message on its PACCH/PDTCH using four access bursts. Alternatively, it can be obtained through the continuous TA updating procedure on the PTCCH (that is, the MS has to wait with uplink PACCH and PDTCH transmission until it first receives a TA update from the PTCCH/D based on the access bursts sent on the PTCCH/U).

4.9.7 Cell Reselection

In GSM, cell reselection is only performed during idle mode, while in dedicated mode, the handover procedure is used to change the serving cell. In GPRS, by contrast, no handover procedure is supported, and cell reselection applies not only to packet idle mode, but also to packet transfer mode. To be precise, when selecting a new cell, the MS must leave the packet transfer mode, switch to the new cell while in packet idle mode, read the system information on the (P)BCCH in that cell, and can only then resume the packet transfer in the new cell. However, a class A MS in dedicated mode or DTM mode will rely on GSM cell reselection and handover procedures.

Additional cell-reselection criteria are provided to complement the existing GSM reselection criteria, for instance, to specify individual hysteresis values for each pair of cells. Furthermore, it is possible for the network to order the mobile station to suspend its normal cell reselection process, and to accept decisions made by the network, in which case *network-controlled* cell reselection is performed. These decisions may be based on additional measurement reports which the network ordered the MS to send.

Mobile terminals belonging to multi-slot classes 19 to 29, which operate in a GPRS-specific half-duplex mode, may not be able to perform neighbouring cell measurements within the TDMA frame, if the downlink PDTCHs assigned to these terminals exceed a certain number. In this case, the network must provide measurement windows, during which it does not deliver PDTCH blocks to such mobile terminals, to ensure that they can perform the required number of measurements. Furthermore, the resource allocation algorithm must also take into account the need for extended measurement windows to allow for BSIC detection. This is all implicitly provided for in GSM (through 'conventional' half-duplex operation and idle frames in the 26-multiframe structure), but due to multi-slot transmission/reception and the flexibility of the GPRS MAC/RLC, this cannot be taken for granted in GPRS. On the plus side, this allows individual equipment manufacturers to excel in implementing the ultimate resource allocation algorithm, on the minus side, it leaves a lot of room for errors.

4.9.8 Power Control

Both open-loop and closed-loop transmission power control are possible on the uplink. Open-loop power control is based on the received downlink signal strength, which must be derived from radio blocks which are not downlink power controlled, e.g. BCCH or certain PCCCH blocks. For details, refer to GSM 05.08 [231].

Uplink power control is not applicable on the RACH or PRACH, where the maximum output power allowed in the respective cell is used.

Downlink power control is optional. When applied, it is based on the channel quality reports provided by the mobiles (e.g. in certain acknowledgement messages), which contain measures such as the quality of the received signal and interference levels. Packet system information messages indicate whether downlink power control is to be applied or not. On PDCHs which deliver PBCCH or PCCCH blocks, the BTS should use constant output power. On other PDCHs, where the USF is used for uplink multiplexing, the BTS output power must be regulated in a manner that it is not only sufficient for the MS for which a specific RLC/MAC block is intended, but also for other mobiles which have to read the USF on this block. In essence, therefore, using USF-based dynamic allocation rather than the fixed allocation scheme may have to be paid for by increased average downlink power level (and thus downlink interference).

4.10 The GPRS RLC/MAC

The full name of the 'radio interface layer 2' in GPRS is 'medium access control and radio link control layer'. The MAC is the lower sub-layer, the RLC the higher sub-layer. We are primarily (but not exclusively) interested in the MAC sub-layer in the following.

4.10.1 Services offered and Functions performed by MAC and RLC

The main purpose of the MAC entity is to enable multiple mobiles the sharing of a common transmission medium. In a given cell, this common transmission medium available for GPRS may consist of several physical channels (i.e. PDCHs). The MAC entity may allow a mobile station to use several physical channels in parallel, that is, use several timeslots within the TDMA frame. For mobile terminated data traffic, the MAC entity provides the procedures for queuing and scheduling of access attempts. For mobile originated data traffic, the MAC provides arbitration between multiple mobiles attempting to access the shared medium simultaneously.

To achieve all this, the MAC performs the following functions.

- It provides efficient multiplexing of data and control signalling on both uplink and downlink, as controlled by the network. On the downlink, multiplexing is controlled by a scheduling mechanism. On the uplink, multiplexing is controlled by medium allocation (in response to service requests) to individual users.

- For mobile originated channel access, it provides contention resolution between channel access attempts, including collision detection and recovery.

- For mobile terminated channel access, it schedules access attempts, which may include queuing of packet accesses.

- Priority handling.

The GPRS RLC sub-layer is responsible for the following.

- Interface primitives allowing the transfer of LLC PDUs between the LLC layer and the MAC sub-layer.

- Segmentation of LLC PDUs into RLC data blocks at the transmit side and re-assembly at the receive side.
- Procedures for backward error correction enabling the selective retransmission (as defined by a bitmap) of unsuccessfully delivered RLC data blocks.
- Link adaptation, that is, the transmission of code words according to the channel conditions.

4.10.2 The RLC Sub-layer

The RCL/MAC function provides two modes of operation, namely *unacknowledged* operation and *acknowledged* operation. In the former, no retransmissions are performed at the RLC/MAC layer, in the latter, *backward* error correction is applied, that is, RLC blocks detected to be in error (through evaluation of the block check sequence) are retransmitted. This retransmission feature is provided by the RLC sub-layer. It is controlled by a selective *type I hybrid ARQ mechanism* (hybrid, because it is combined with *forward* error correction coding), which works as follows.

The sending side transmits blocks within a sending window and the receiving side responds with packet uplink or downlink ACK/NACK messages providing positive (ACK) and negative (NACK) acknowledgements, as required. These acknowledgements indicate up to which point *all* RLC data blocks were received correctly, by signalling the appropriate *Block Sequence Number* (BSN). On top of that, starting from this sequence number, a bitmap identifies for each subsequent block, whether it has been received correctly or in error. The sending window moves in accordance with the block sequence number contained in the acknowledgement, which identifies the current starting point of the sending window.

Compared to conventional sliding window protocols, the added selective bitmap feature saves the transmitting side from retransmitting correctly received blocks beyond this current starting point. Ideally, the sending side would only retransmit blocks, which were explicitly identified by the receiving side as received in error, in order not to waste capacity by retransmitting blocks which might already have been received correctly. However, depending on the frequency of acknowledgement messages sent by the receiving side, and the size of the sending window, the sending side might not be able to move the sending window quickly enough. It may then end up retransmitting blocks all the same before they were identified explicitly by the receiving side as in error. When this happens, the RLC protocol is said to be *stalling*. In GPRS release 1997, the window size is 64 RLC/MAC blocks.

4.10.3 Basic Features of the GPRS MAC

For mobile stations in packet idle mode, the MAC features procedures for the reception of information signalled on the PBCCH and the PCCCH. Information contained in packet system information messages is required for the mobile stations to understand how to access the cell and to perform autonomous cell reselection. In packet idle mode, mobile stations also monitor the relevant paging subchannels on their PCCCH. If none is present in the cell, they monitor the relevant paging subchannels on their CCCH.

For mobile stations in packet transfer mode, the MAC procedures support the provision of *temporary block flows* that allow the unidirectional transfer of signalling and user data (in the shape of LLC PDUs) within a cell between the network and a mobile station. Each TBF is identified by a *temporary flow identity* assigned by the network. TBFs will be discussed in more detail below.

One of the interesting aspects of the GPRS MAC is the transition from packet idle to packet transfer mode. This will often, but not always, involve a random access procedure based on an S-ALOHA algorithm, which, if successful, will lead to a reservation of resources to be used in packet transfer mode. In other words, we are dealing here with a member of the R-ALOHA family of protocols. It is therefore appropriate to review the GPRS MAC briefly in terms of the R-ALOHA design options identified in Section 3.7 before digging into more details.

4.10.3.1 R-ALOHA Design Choices

The following design choices have been made for the six R-ALOHA design options identified in Section 3.7.

(1) The split of channel resources between C-slots and I-slots (here PRACH and PDTCH/PACCH respectively) is hybrid, as already identified in Section 3.7. There are two rather static aspects, namely the split of total GPRS resources in a cell into MPDCHs and SPDCHs, only the former being able to carry PRACH slots, and the semi-static reservation of a certain number of radio blocks per 52-multiframe on the MPDCH for the PRACH (as signalled on the PBCCH). Additional slots can be assigned to the PRACH in a dynamic fashion through USF signalling. The split can be made fully dynamic by using only MPDCHs for GPRS and on each MPDCH reserving zero blocks for the PRACH through PBCCH signalling. In this way, PRACH slots are identified exclusively through USF signalling.

(2) Resource assignment signalling must be explicit. GPRS supports a variety of packet services, which may have substantially different resource requirements. Furthermore, the amount of resources, which can be assigned to a specific GPRS MS, will also depend on this mobile station's multi-slot class.

(3) The duration of the reservation phase will be discussed in more detail below. It depends, among other things, also on the chosen medium access mode. The medium access modes supported in GPRS are fixed allocation, dynamic allocation (both already introduced in Subsection 4.8.2), extended dynamic allocation, and (for DTM only) exclusive allocation (see Subsection 4.8.4).

(4) There are several means to signal modified resource requirements or extension requests. Typically, these are sent on PACCH blocks, as will be discussed in more detail below.

(5) Access control is centralised, a detailed discussion is provided in Section 4.11.

(6) Given the very limited payload of access bursts, it is clear that they contain only signalling, no user data. In fact, if the GSM RACH is used for GPRS, even the signalling content of the access bursts is very limited, since only two establishment cause values are reserved for GPRS. On the PRACH, the situation is somewhat better. With the conventional coding scheme, all eight payload bits are available for GPRS purposes, and an extended access burst format provides 11 payload bits.

However, this is still very limited, for instance, it does not even allow for the inclusion of an unambiguous mobile identity.

To understand better how the GPRS MAC operates, we need to discuss in more detail the multiplexing principles, what signalling messages are defined, their structure and how they are exchanged between the network and the MS to initiate a packet transfer (in either direction) and during the transfer.

4.10.4 Multiplexing Principles

A key component for GPRS multiplexing is the concept of temporary block flows. A TBF supports *unidirectional* transfer of LLC PDUs either from the MS to the network or in the other direction. As GPRS is meant to provide concurrent and independent packet transfer in uplink and downlink direction, it is clear that concurrent TBFs may be established in opposite directions. PDTCH resources to be used for a TBF in the appropriate link direction may be allocated on one or several PDCHs (within the limitations of the MS multi-slot class, and with similar general constraints as in HSCSD, i.e. they must be on the same carrier frequency or hop together). For signalling purposes, such as sending of acknowledgements on the PACCH, some resources will also have to be assigned in the opposite direction.

Each TBF is identified by a 5-bit *temporary flow identity* assigned by the network, which is included in every RLC/MAC block associated with that TBF. The TFIs must be chosen such that the TFI value is unique among concurrent TBFs in the same direction on all PDCHs used for this TBF. That is, the same TFI value in use on a given PDCH in the downlink direction can be reused:

(1) for another TBF in the uplink direction on the same PDCH or other PDCHs; and

(2) for another TBF in the downlink direction, as long as the two sets of PDCHs used for these two downlink TBFs are disjunctive.

By sticking to this rule, for RLC/MAC blocks carrying user data, the TFI together with the link direction identifies unambiguously a TBF. For RLC/MAC control blocks, which may be sent in both directions (e.g. acknowledgements sent on the PACCH in the opposite direction to the PDTCHs), on top of TFI and link direction, also the message type (six bits) is required to identify the relevant TBF unambiguously. For instance, if a packet uplink acknowledgement message is sent on the downlink, it is clear that the TFI must relate to an uplink TBF.

Obviously, when a TBF is being set up, before being able to assign a TFI, the mobile station must identify itself with another suitable, unambiguous identity (very much as in the case of plain GSM, see Section 4.4). How this is done will be discussed in more detail in subsequent subsections, which deal with mobile originated and mobile terminated packet transfer.

We have already discussed in Section 4.8 how USFs are used for uplink multiplexing. With eight values available, at most eight uplink TBFs can be supported on one SPDCH, if dynamic allocation is used throughout. With fixed allocation, the number of mobiles that could be multiplexed onto a single PDCH may be higher, it is (among other factors) constrained by the number of available TFI values. Although we know that there are 32 in total, given various options of the GPRS resource allocation scheme on one side, and handset constraints (multi-slot classes) on the other, it is quite tricky to figure out exactly how many mobiles could be multiplexed onto the uplink part of a single PDCH. The constraints on the downlink are similar to those with fixed allocation on the uplink, irrespective of what type of allocation is used.

To complete the picture on medium access modes, on top of fixed and dynamic allocation supported by all GPRS mobiles, *extended dynamic allocation*, a variation of dynamic allocation when radio resources on multiple PDCHs are allocated to a single TBF, must also be supported by mobiles pertaining to certain multi-slot classes. As usual, what is mandatory for mobiles is optional for networks: they must support only either fixed or dynamic allocation, the other one is optional, as is extended dynamic allocation. Finally, *exclusive allocation*, introduced in release 99, must be supported by DTM-capable mobiles, while it is optional (surprise!) for release 99 networks (that is, conditional on whether the network wants to support DTM or not).

We will discuss the differences between the first three medium access modes in the subsection dealing with uplink packet transfer. Exclusive allocation was already discussed in Subsection 4.8.4, it is comparatively trivial, tightly coupled with the dedicated mode behaviour of the DTM MS on the TCH, and will not be elaborated upon any further.

4.10.5 RLC/MAC Block Structure

An RLC/MAC block is the protocol data unit exchanged between RLC/MAC entities. It is carried on a radio block, which in turn consists of a sequence of four normal bursts. An RLC/MAC block consists of a MAC header, and:

- either an *RLC/MAC control block*; or

- for packet data transfer on the PDTCH, an *RLC data block*, which in turn is composed of an RLC header, an RLC data unit and spare bits (the latter because the data unit consists of an integer number of octets, see Subsection 4.9.4).

This is illustrated in Figure 4.30.

The normal MAC header is 8 bits or one octet long. The first 2 bits indicate the payload type, namely whether it is a data or a control block and, only relevant for downlink control blocks, whether there is an optional MAC header octet.

In the downlink direction, the next two bits contain the *Relative Reserved Block Period* (RRBP) field, which is followed by the *Supplementary/Polling* (S/P) bit. These three bits

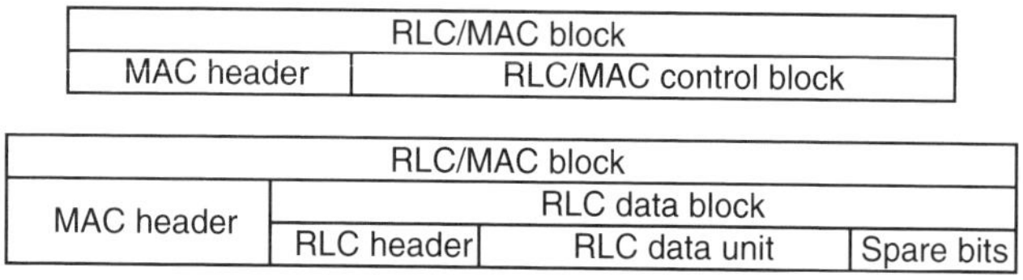

RLC/MAC block	
MAC header	RLC/MAC control block

RLC/MAC block			
MAC header	RLC data block		
^^	RLC header	RLC data unit	Spare bits

Figure 4.30 RLC/MAC block structure for control and data blocks

implement a polling mechanism as follows: by setting the S/P bit, the network can poll the MS to send a control block in a reserved uplink radio block on the same physical channel; which radio block is specified relative to the downlink polling block by the RRBP bits. If the S/P bit is not set, the RRBP bits are meaningless. The final 3 bits signal the well-known USF values. The first optional MAC header octet is required for downlink RLC/MAC control messages which do not fit into a single RLC/MAC control block, and thus need to be segmented into two such blocks. The second optional MAC header octet contains a 'power reduction' field relevant for power control, a TFI field and a bit indicating the link direction of the TBF identified by the TFI (note though that TFI fields, where required, are typically included in the message part of downlink control messages).

In the uplink direction, the MAC header fields are not the same for control and data blocks. For the former, apart from the 'payload type' bits, there is only one relevant bit, the 'retry' bit, which is set if the (packet) channel request message had to be sent more than once during the most recent channel access. For the latter, on top of these fields, there is a 'countdown value' field spanning 4 bits, which is relevant for the release of resources, and a 'stall indicator' bit, which indicates whether the MS RLC transmit window is stalled (that is, it cannot advance), or not stalled.

The minimum RLC header size in RLC data blocks is two octets. In the downlink direction, the first octet features two 'power reduction' bits, five TFI bits identifying the TBF, and one 'final block indicator' bit, which indicates, if set, that the downlink RLC data block is the last such block of this downlink TBF. The second octet contains a 7-bit block sequence number required for RLC operation and an extension bit used to indicate the presence of an optional octet in the RLC header. In the uplink direction, the 'power reduction' field is replaced by two spare bits, whereas the 'final block indicator' bit is replaced by a 'TLLI indicator' bit. If set, it means that an optional TLLI field with a *Temporary Logical Link Identity* is contained within the RLC data block, which identifies the communication source unambiguously and can therefore be used for contention resolution. Everything else remains the same. Additional RLC header octets containing a 'length indicator' field, a 'more bit' and an extension bit are used for LLC frame delimitation. In most cases, one octet is required per LLC frame. If this frame is segmented and carried on multiple RLC data blocks, then only the last segment is identified by a 'length indicator' field in the corresponding RLC data block.

4.10.6 RLC/MAC Control Messages

In the following, the RLC/MAC control messages defined for GPRS are listed and discussed briefly. The first six bits in an RLC/MAC control block (to be precise, in the first one if the message spans two blocks) identify the message type. For each message, the type of logical channel it may be sent on and the link direction are indicated.

Five RLC/MAC control messages are related to uplink TBF establishment as follows.

- The *packet channel request* message, similar to the channel request message in GSM, sent by the MS on the PRACH (hence on the uplink) using one of the two access burst formats (the network signals on system information messages which one to use).

- The *packet resource request* message, sent on the uplink PACCH to request a change in the assigned uplink resources.

- The *packet uplink assignment* message, sent on the downlink PCCCH (i.e. the PAGCH) as a response to a packet channel request message, or on the downlink PACCH, e.g. as a response to the packet resource request message. It specifies resources (time-slots, frequency parameters for frequency hopping, the USF value on each time-slot assigned where applicable, etc.) to be used either to send a packet resource request message or to send data packets to be transferred using this TBF. Additionally, it indicates the TA and the channel coding scheme to be used, and may signal the assigned TFI value, power control parameters and what downlink measurements need to be performed by the MS.

- The *packet queuing notification*, sent on the downlink PCCCH as a response to a packet channel request message, if for whatever reason (most likely unavailability of suitable resources) no packet uplink assignment message can be sent, and the request must be placed in a queue. For later referencing, this message assigns a temporary queuing identity to the requesting MS.

- Finally, the *packet access reject* message, sent by the network on the PCCCH or PACCH in response to a packet channel request message or packet resource request message to indicate that it has rejected the mobile station's packet access request.

The message used to establish a downlink TBF is the:

- *packet downlink assignment* message, which is sent by the network on the PCCCH or PACCH to specify the resources (time-slots, optionally starting time and frequency parameters for frequency hopping), on which the data packets associated with this TBF are sent to the mobile station. Additionally, it indicates the TA to be used, and may indicate the TFI, how often and on which time-slots downlink measurements need to be made, and power control parameters, if power control is to be applied.

The message used by the network to signal the release of a TBF, as identified by the link direction and the TFI value contained in the message, is:

- the *packet TBF release* message, which is sent on the downlink PACCH.

The message used to page a mobile station, either for TBF establishment (for downlink packet data) or GSM RR connection establishment is:

- the *packet paging request* message, which is sent by the network either on the PCCCH, or for an RR connection establishment, while an MS is in packet transfer mode, also on the PACCH. Depending on the method used to identify the mobile stations (IMSI, TMSI, or packet-TMSI), a single message fitting into one radio block can page up to four mobile stations.

The following two messages are required for RLC operation.

- The *packet uplink ACK/NACK* message sent by the network on the downlink PACCH, containing the 7-bit sequence number of the last RLC data block up to which all uplink RLC data blocks were received correctly by the network, plus a 64-bit selective bitmap for subsequent blocks, as outlined in Subsection 4.10.2. It may also be used to update the timing advance and power control parameters and to assign new uplink resources in the case of fixed allocation.

- The *packet downlink ACK/NACK* message sent by the MS on the uplink PACCH, containing again the sequence number up to which all downlink RLC data blocks were received correctly by the MS plus the selective bitmap. It is also used to report the downlink channel quality and may be used to initiate an uplink TBF.

The signalling of system information relevant for GPRS occurs on the following.

- *Packet System Information* (PSI) messages. Seven different message types were defined for GPRS R97. Six of them are sent on the PBCCH where it is available, they can also be sent on downlink PACCHs. When a PBCCH is present, the regular sending of four of these message types is mandatory, the other two are optional. The mandatory messages contain information on number, location and structure of control channels, PRACH control parameters, power control and cell reselection parameters, cell and mobile allocation lists, and BCCHs of neighbouring cells. The seventh message type is a system information message broadcast on the BCCH, which contains GPRS relevant information (e.g. the description of the channel carrying a PBCCH, if available, see Subsection 4.8.2). The content of this seventh message can also be sent as a PSI message type on the PACCH.

Finally, there are a substantial number of miscellaneous messages, namely:

- The *packet cell change order* message sent on the downlink PACCH or PCCCH to command a mobile station to leave the current cell and change to a new cell (in release 99, this may be a UMTS cell); and the *packet cell change failure* message sent on the uplink PACCH to indicate that a commanded cell change order has failed.

- The *packet downlink dummy control block* message, sent on the downlink PCCCH or PACCH as a fill message, when no other message is to be sent; and the *packet uplink dummy control block* message, sent on the uplink PACCH as a fill message, if the MS has no other block to transmit.

- The *packet measurement order* message, sent on the downlink PCCCH or PACCH, which specifies what measurements must be performed by the mobile station and provides information relating to network controlled cell reselection.

- The *packet measurement report* message sent on the uplink PACCH, which contains measurement reports on neighbouring cells and on the downlink channel quality in the serving cell.

- The *packet mobile TBF status* message sent on the uplink PACCH to indicate reception of erroneous messages (i.e. messages which passed the BCS, but are syntactically incorrect or not compatible with the current protocol state).

- The *packet PDCH release* message is sent on the downlink PACCH to all mobile stations listening to that PDCH, to indicate that one or more PDCHs will be released immediately and become unavailable for packet data traffic (in either link direction). This is the fast release mechanism required to share dynamically resources between GPRS and circuit-switched calls, as outlined in Subsection 4.7.3.

- The *packet polling request* message is sent to a mobile station on the downlink PCCCH or PACCH to solicit a packet control acknowledgement message from that mobile station.

- The *packet control acknowledgement* message, which is sent on the uplink PACCH to confirm reception of downlink control messages (or individual message blocks, if the downlink message spanned two RLC/MAC blocks). It is sent either in the shape of four identical access bursts (in the four time-slots pertaining to one radio block), or in the usual RLC/MAC control block format.

- The *packet power control/timing advance* message is sent on the downlink PACCH. It is used by the network to signal/update power control parameters. It can also be used to signal TA updates, if this cannot wait for the next occurrence of the relevant PTCCH/D block.

- The *packet PRACH parameters* message may be sent on the downlink PCCCH to update the PRACH parameters in-between packet system information messages containing these parameters.

- The *packet PSI status* message may be sent on the uplink PACCH to indicate which PSI messages the MS has received.

- Finally, the *packet time-slot reconfigure* message sent on the downlink PACCH to assign uplink and/or downlink resources to an MS, which is already in packet transfer mode.

4.10.7 Mobile Originated Packet Transfer

4.10.7.1 The Packet Access Procedure

To initiate a mobile originated packet transfer, the MS sends a (packet) channel request message either on its PRACH or, if none is available, on the RACH. Details on the packet access procedure are provided in the next section, so let us assume here that the network receives the request message somehow, whether this be in the first attempt, or after retransmissions. In the normal case, the network will respond with a packet uplink assignment message or immediate assignment message on the same (P)CCCH (i.e. either on the PAGCH or the AGCH) on which it received the request message. Alternatively, the network may also respond with a packet queuing notification message (and send a packet uplink assignment message later, possibly after a roundtrip of packet polling message and packet control acknowledgement message, to adjust an out-of-date TA). Finally, it could respond with a packet access reject message, which may tell the MS to reinitiate a packet access procedure after a certain amount of time.

Consider now the normal case again. There are two main types of packet access, namely *one-phase access* and *two-phase access*. In the one-phase access, the packet uplink assignment message specifies resources that can be used for data transfer. The two-phase access, as illustrated in Figure 4.31, requires the exchange of two additional control messages. The first uplink assignment message sent in response to the packet channel request message allocates only a single uplink PACCH block, on which the MS sends a packet resource request message, allowing a detailed specification of the resources required by the MS and of the MS capabilities. The network will then respond on the

```
                    MS                              Network

        Packet channel request (or channel request)
        ────────────────────────────────────────────▶   PRACH (or RACH)

          Packet uplink assignment (or immediate assignment)
        ◀────────────────────────────────────────────   PAGCH (or AGCH)

                    Packet resource request
        ─ ─ ─ ─ ─ ─ ─ ─ ─ ─ ─ ─ ─ ─ ─ ─ ─ ─ ─ ─ ─ ─▶   PACCH

                    Packet uplink assignment
        ◀─ ─ ─ ─ ─ ─ ─ ─ ─ ─ ─ ─ ─ ─ ─ ─ ─ ─ ─ ─ ─ ─   PACCH
```

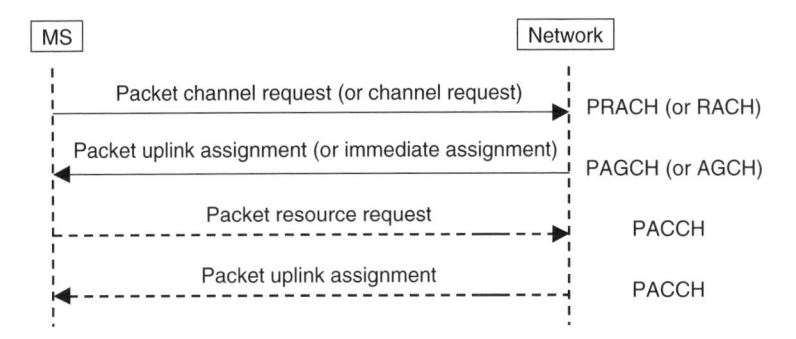

Figure 4.31 Message exchange for one-phase packet access (solid arrows) and two-phase packet access (solid and dashed arrows)

downlink PACCH (on the same physical channel as the uplink PACCH) with a further packet uplink assignment message, specifying the resources to be used for the uplink packet transfer, taking into account this additional information.

If the MS requests a one-phase access in the channel request message, the network may still decide to assign only one block and force the MS to perform the second phase. Otherwise, the network will have to assign the resources for the uplink transfer based on the limited amount of information which can be included in the packet channel request message. If the packet channel request message is sent on the RACH, one-phase access is limited to a single-slot assignment, and the network can only guess for how long the resources need to be assigned. If it is sent on the PRACH, at least the MS multi-slot class, and possibly also the radio priority can be signalled in the packet channel request message, as discussed in more detail in the next section. A one-phase access does not prevent the MS from sending a packet resource request message at a later stage (it can do so on any assigned uplink radio block, as the MS can decide whether to use it as a PACCH or a PDTCH block). From an MS QoS perspective, therefore, the one-phase access is the preferred option. The MS can start to transfer uplink packets earlier than it would be able to do with the two-phase access, and after having sent a resource request message, it can continue sending data packets on the initially assigned resources while the network is processing the request for further resources.

The detailed rules on when which type of packet request message shall be sent (there is also a 'short access request' on top of one-phase and two-phase request) can be found in GSM 04.60 [232].

4.10.7.2 Contention Resolution

As for 'plain' GSM, a contention resolution mechanism is required due to the fact that an unambiguous mobile identity cannot be included in the packet channel request message, such that two mobile terminals may perceive the same channel allocation as their own. The two-phase access procedure provides inherently for contention resolution, since the MS must include its unique TLLI in the packet resource request message, and will only perceive a subsequent uplink assignment message as its own, if the network included the same TLLI in this message.

If a one-phase access procedure is used, the MS must include its TLLI in every RLC/MAC block it sends on the specified allocation, until it has received a packet uplink

ACK/NACK message containing the same TLLI. Because of this, the network should respond with the first ACK/NACK message as soon as it received the first RLC/MAC block correctly, since inclusion by the MS of the 4-octet TLLI in multiple RLC/MAC blocks reduces the available payload and is thus wasteful. As we can see, the one-phase access has also its drawbacks.

4.10.7.3 (Extended) Dynamic Allocation

If the network chooses dynamic or extended dynamic allocation as the MAC mode, then the packet uplink assignment message signalling the resources to be used for the packet transfer must include the list of assigned PDCHs, and for each PDCH, the assigned USF value. A unique TFI is allocated and is thereafter included in each RLC/MAC block related to that TBF. With dynamic allocation, the MS must monitor downlink blocks on all allocated PDCHs to decode the USF values. Whenever an MS detects on a downlink block of an assigned PDCH its assigned USF value, it may either transmit a single RLC/MAC block or a sequence of four RLC/MAC blocks on the following uplink radio block(s) of the same PDCH. Whether a USF value is valid for a single block or four blocks is specified in the USF_GRANULARITY parameter, which is signalled in the packet uplink assignment message.

Since every PDCH assigned to an MS for an uplink TBF may be used by the network as a downlink PACCH for that MS, the MS must not only decode the USF values on these downlink blocks, but should also attempt to decode the full messages. If one turns out to be an RLC/MAC control message addressed to that MS, it shall act upon this message accordingly.

Extended dynamic allocation is only relevant for multi-slot operation. With conventional dynamic allocation, the MS would have to monitor all assigned downlink PDCHs for the occurrence of the right USF values and for PACCH blocks. Depending on the number of PDCHs assigned and the MS multi-slot class, the MS might not be able to do so. If extended dynamic allocation is applied instead, the monitoring rules are relaxed, and the MS may, under certain conditions, not have to monitor all downlink PDCHs. If the right USF value appears for a block on one PDCH, then the MS is, again under specific conditions, not only allowed to access the uplink block on that PDCH, but also the same uplink block on higher numbered PDCHs it was assigned to. Clearly, the network must all the same signal the USF values assigned to this specific MS on these other PDCHs, to avoid conflicts with other mobile stations assigned to the same PDCHs. For details, refer again to GSM 04.60.

USF-based resource allocation allows a population of mobile terminals to be dynamically multiplexed onto shared resources. It is possible to implement arbitrary USF scheduling mechanisms. For instance, it is possible to assign an uplink PDCH to the exclusive use of a single MS, e.g. as long as the MS has queued data to transmit (potentially comprising multiple LLC frames), by keeping the USF static. On the other hand, it is possible to multiplex several mobile stations onto an uplink PDCH on a time-sharing basis by switching the USF value regularly. USF-based multiplexing can also be used to pre-empt low priority packet transfers when high priority packets need resources, or when an urgent control message needs to be scheduled on the uplink, by simply switching the USF in the required instant. This flexibility in terms of resource allocation comes at a price in terms of downlink interference levels. This is firstly due to constraints when executing downlink power control, as explained in Subsection 4.9.8. Secondly, for very much the same reasons, it is difficult to apply other downlink interference reduction techniques such as smart antennas. These techniques rely on the transmission power being directed to a single mobile station for each radio block, which is inconsistent with the inclusion of broadcast-type information such as USF flags in every radio block [233].

A TBF may either be close-ended or open-ended. In the former case, the MS is only allowed to send a specific number of RLC data blocks, as signalled by the network (retransmission of data blocks and transmission of control blocks do not count towards this number). In the latter case, the MS may continue to use the TBF until its sending queue is empty. In both cases, the normal release procedure is initiated by the MS counting down the last few RLC data blocks (as signalled in the RLC header). Once it has sent the last data block, it is not allowed to make further use of its uplink assignment, unless the network responds with a packet uplink ACK/NACK message indicating that one or several blocks were received in error. Otherwise, the network will respond with an uplink ACK/NACK message with the so-called 'final ACK indicator bit' set, and a valid RRBP field reserving an uplink block period for the MS to acknowledge this final ACK/NACK message. If the MS fails to send an acknowledgement, the network must wait for the expiry of a guard timer, before it can safely reuse the same TFI for another assignment, otherwise, it may reuse the TFI as soon as an acknowledgement is received. In any case, as soon as the network has correctly received all the RLC data blocks belonging to the TBF in question, it may reallocate the same USF(s) to some other user. There are also means for the network to initiate a release of an uplink TBF, e.g. by using the packet TBF release message.

Figure 4.32 shows an example message sequence exchange for uplink data transfer. One resource reallocation is shown, which is acknowledged by a packet control acknowledgement message. This acknowledgement is only required if the network polls the MS by setting the S/P bit contained in the MAC header of the packet uplink assignment message. The reception of the final packet uplink ACK/NACK message is also acknowledged by the MS, enabling the network to reuse the TFI immediately.

4.10.7.4 Fixed Allocation

If the network chooses to allocate resources using the fixed allocation MAC mode, the packet uplink assignment message signalling the resources to be used for the packet transfer (i.e. the first or the second one depending on the type of access) must communicate the detailed uplink resource allocation to the MS. This consists of a start frame, slot assignments, and a block assignment bitmap specifying for each time-slot precisely which blocks can be used. The fixed allocation does not include the USF values; correspondingly, the MS does not need to monitor the USF to determine which uplink blocks it may use. However, if the network wants to mix fixed allocation for one set of mobiles and dynamic allocation for another set on the same PDCH, then an unused USF value must be used for blocks assigned using the fixed allocation scheme, to prevent mobiles holding dynamic allocations from transmitting during these blocks.

As in the case of dynamic allocation, a unique TFI is allocated and included in each RLC/MAC block related to the TBF in question.

The MS may request a close-ended TBF or an open-ended TBF. In the former case, the number of blocks requested should only account for the number of data and control blocks it intends to send, hence not making allowances for blocks it might have to retransmit due to transmission errors. By counting the blocks that pass the BCS, the

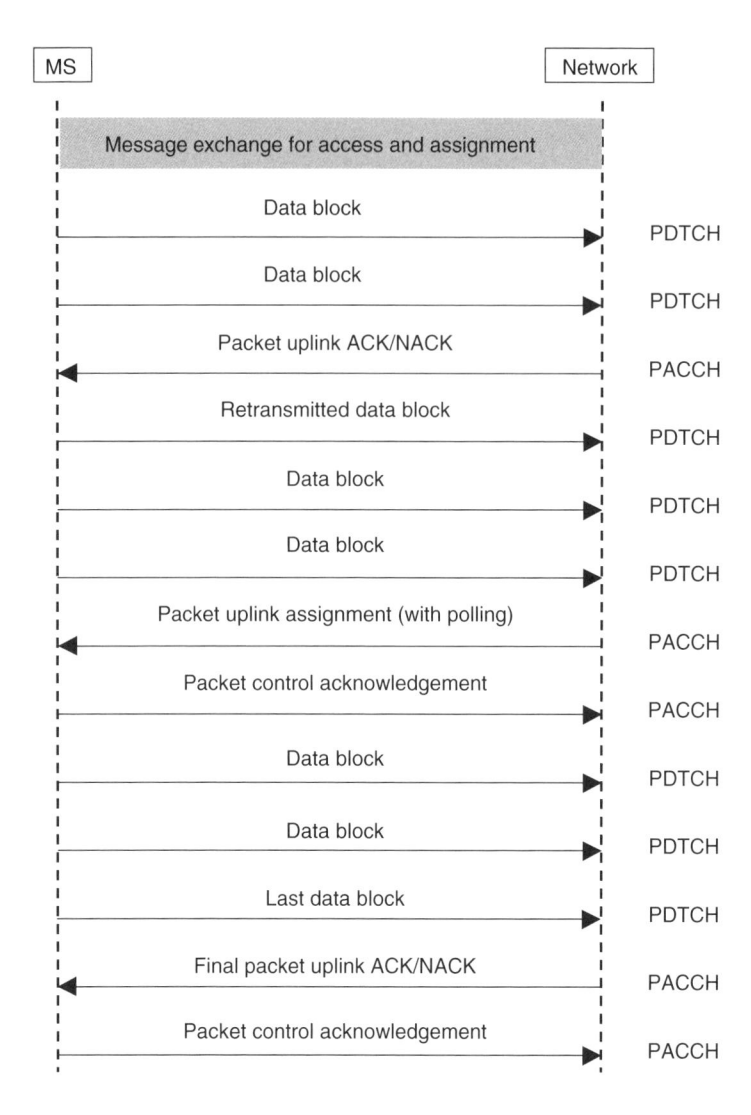

Figure 4.32 Message sequence example for uplink data transfer

network can keep track of whether additional resources must be assigned to the MS for block retransmissions. These additional resource assignments can be appended to packet uplink ACK/NACK messages, which the network must send anyway to acknowledge transferred blocks. Alternatively, the network can also allocate additional resources for the retransmissions using an unsolicited packet uplink assignment or packet time-slot reconfigure message. At any time during the uplink TBF, the network may send any of these three signalling messages on the downlink PACCH monitored by the MS to initiate a change of resources (e.g. to add a new PDCH to the current allocation or to remove a previously assigned PDCH from the current allocation).

In the case of an open-ended TBF, each time the MS receives a fixed allocation, and it wishes to continue the TBF beyond this allocation, it must send another packet resource request message to the network. Either type of TBF may be ended by the network before the number of requested octets has been transferred. In this case, the network indicates the end of the TBF by including a FINAL_ALLOCATION indication in a packet uplink assignment message. Upon receipt of such a message, the MS executes the countdown procedure in a manner that the countdown ends before the current allocation is exhausted.

As the MS does not need to monitor the USF values for its allocated PDCHs, it would not make sense to request that it monitored all downlink blocks on these PDCHs for possible occurrences of a PACCH block directed to this MS. In fact, a multi-slot class 'type 1' MS (one that cannot transmit and receive simultaneously) would often not be capable of doing so anyway. For fixed allocation, therefore, the assignment message specifies explicitly one PDCH to be used for PACCH blocks directed to the MS during this fixed allocation, so that a 'type 1' MS needs only monitor this PDCH. The network should schedule downlink PACCH blocks only on this time-slot and only during sufficiently large gaps in the uplink allocation (how large depends on the MS multi-slot class). However, a 'type 2' MS, which can transmit and receive simultaneously, should also monitor other PDCHs for possible occurrences of PACCH blocks.

In terms of message exchange, the fixed allocation is not much different than dynamic allocation, the sequence shown in Figure 4.32 is also a valid example for fixed allocation.

It is probably fair to say that fixed allocation is easier on the MS than dynamic allocation (since no USF flags need to be monitored and often only one PDCH for occurrences of dowlink PACCH blocks). On the other hand, the resource allocation entities in the network are kept a bit busier; they have to allocate block by block for every single MS and need to consider the exact MS capabilities whenever they do that. The same is not necessarily true for dynamic allocation, particularly with open-ended TBFs, where the USFs could be kept fairly static. Furthermore, although pre-emption of low priority users is also possible with fixed allocation, this is more robust and potentially quicker with dynamic allocation. Physical layer implications due to the use of USF have been discussed earlier. For a more detailed comparison of these two schemes, see also Reference [233].

4.10.7.5 Initiating a Downlink Packet Transfer while in Uplink Packet Transfer

During an ongoing uplink TBF, within certain constraints imposed by the MAC mode chosen and the multi-slot class of the MS in question, the MS monitors continuously downlink PDCHs on which the network may deliver RLC/MAC control messages directed to it. Put simply, the MS is reachable through its downlink PACCH. The PACCH may not only be used for packet uplink ACK/NACK and assignment messages, but also to send a packet downlink assignment message or a packet time-slot reconfigure message, which can be exploited to initiate downlink packet transfers while an uplink transfer is in progress. Provided that the number and relative position of the PDCHs assigned in uplink and downlink direction and the scheduling of downlink PACCH blocks respect the MS multi-slot capability, these transfers may take place in both link directions simultaneously.

For fixed allocation only, a *half-duplex mode* specific to GPRS is defined, which is characterised by the fact that downlink and uplink TBF are not active at the same time. This half-duplex mode is only applicable for MS multi-slot classes 19 to 29, as discussed in Subsection 4.7.6. If such an MS is assigned to operate in half-duplex mode, the network shall wait for the MS to finish its current uplink resource allocation, before sending RLC data blocks on the downlink. When operating in half-duplex mode and suspending an uplink TBF due to arrival of a downlink TBF, the RLC entities in the MS and the network save their state variables relating to the uplink TBF. They can then resume this TBF seamlessly once the downlink TBF is either completed or suspended.

4.10.8 Mobile Terminated Packet Transfer

For the network to be able to initiate a mobile terminated packet transfer by sending a packet downlink assignment message, the target MS must be in MM ready state. Trivially, this is the case if the MS is already engaged in a mobile originated packet transfer, but it may not be if it is in packet idle mode. If the MS is in MM standby state, then it needs to be moved into ready state through packet paging.

4.10.8.1 Packet Paging

The network initiates a packet transfer to an MS in MM standby state by sending one or more packet paging request messages on the downlink PPCH or PCH in all cells of the last stored routing area for this MS. The MS responds to such a message by sending a packet paging response message. To send this message, which must be formatted as a complete LLC frame (as opposed to an RLC/MAC control block), the MS needs to initiate a mobile originated packet transfer. This entails the usual message exchange already discussed in the previous subsection, as illustrated in Figure 4.33. Upon conclusion of the packet paging procedure, the MS is in MM ready state, as desired, allowing the network to assign the radio resources to the MS required to perform the downlink data transfer.

4.10.8.2 Downlink Packet Transfer

We have already explained in the previous subsection how a downlink packet transfer can be initiated by the network while the target MS is engaged in an uplink packet transfer by sending a packet downlink assignment message on this mobile's PACCH. If the MS is in MM ready state, but no packet transfer is in progress, then the network sends the packet downlink assignment message on the PCCCH, or more precisely on the

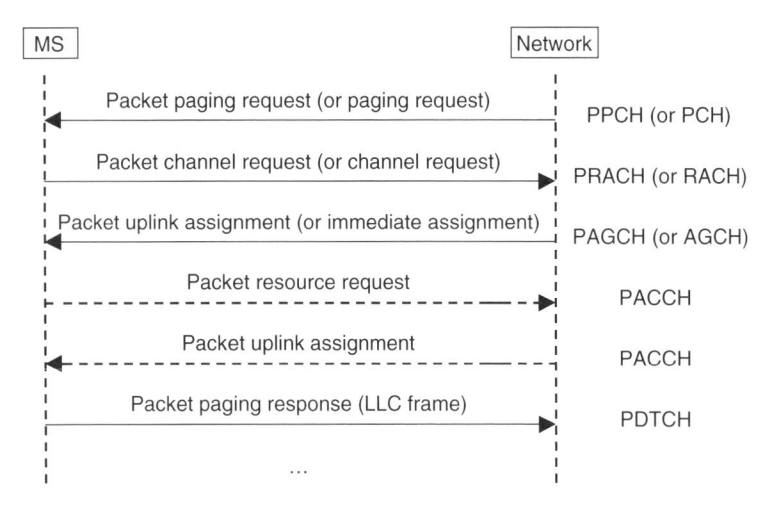

Figure 4.33 Message sequence required to perform a packet paging procedure (dashed arrows represent optional messages)

PAGCH (or alternatively, on the AGCH if there is no PCCCH in the cell). The packet downlink assignment message includes the list of PDCHs to be used for downlink transfer (obviously taking into account the MS multi-slot capability). Furthermore, if available, TA and power control information are also included. If the TA, which the MS needs to send uplink signalling messages, is not available, the MS may be requested to respond to either this downlink assignment message or to a subsequent packet polling request message with a packet control acknowledgement message. The acknowledgement message must be formatted as four access bursts, as discussed in Subsection 4.9.6.

Either immediately after reception of the downlink assignment message, or at a certain TBF starting time indicated in the assignment message, the MS starts to listen to the assigned PDCH(s). It will detect its RLC/MAC blocks by means of TFI. TFIs are included as identifier in every block to enable multiplexing of blocks destined to different mobile stations on the downlink part of a single PDCH.

If the acknowledged RLC mode is used, then the MS must send every so often packet downlink ACK/NACK messages required for selective ARQ operation. Unlike uplink transfer, where the receiving entity (i.e. the network) decides when to send uplink ACK/NACK messages, in this case, the transmitting side (again the network) must control the frequency and timing of acknowledgements, since there are no permanent uplink resources allocated to the MS for this purpose. This is achieved by means of polling, which can be performed by the network simply by setting the S/P bit in the MAC header of a downlink data block. The RRBP bits in the MAC header then specify which uplink radio block is reserved for the MS to send an ACK/NACK message. The USF relating to this uplink block must be set to a value not occupied by an uplink TBF transfer using dynamic allocation.

The MS includes channel quality reports in the packet downlink ACK/NACK message. It can also include a channel request description, either to initiate an uplink packet transfer, or to request additional uplink signalling resources (for PACCH blocks).

The normal release of resources is initiated by the network by setting the 'final block indicator' flag in the RLC header of the last block in a downlink TBF, and polling the MS for a final packet downlink ACK/NACK message. If the MS has received all blocks of this downlink TBF correctly, it will set the 'final ACK indicator' flag in this ACK/NACK message. If blocks are negatively acknowledged, then the 'final ACK indicator' is not set and the network must re-initiate the release of resources after having retransmitted these negatively acknowledged blocks. Timers are used which must expire, before the network is allowed to reuse the TFI for another TBF.

Alternatively, the network may initiate an immediate abnormal release of a downlink TBF by transmitting a packet TBF release message to the mobile station on the PACCH, in which case the MS immediately stops monitoring its assigned downlink PDCHs. If so requested in the release message (through S/P and RRBP bits in the MAC header), the MS will respond in the indicated radio block with a packet control acknowledgement message.

Figure 4.34 shows an example message sequence for downlink data transfer. One resource reallocation is taking place, which is signalled by sending a packet downlink assignment message during the transfer (alternatively, a packet time-slot reconfigure message could be sent). In this example, the network does not poll the MS in the first packet downlink assignment message, while it does in the second, causing the MS to respond with a packet control acknowledgement message. Polling for an acknowledgement

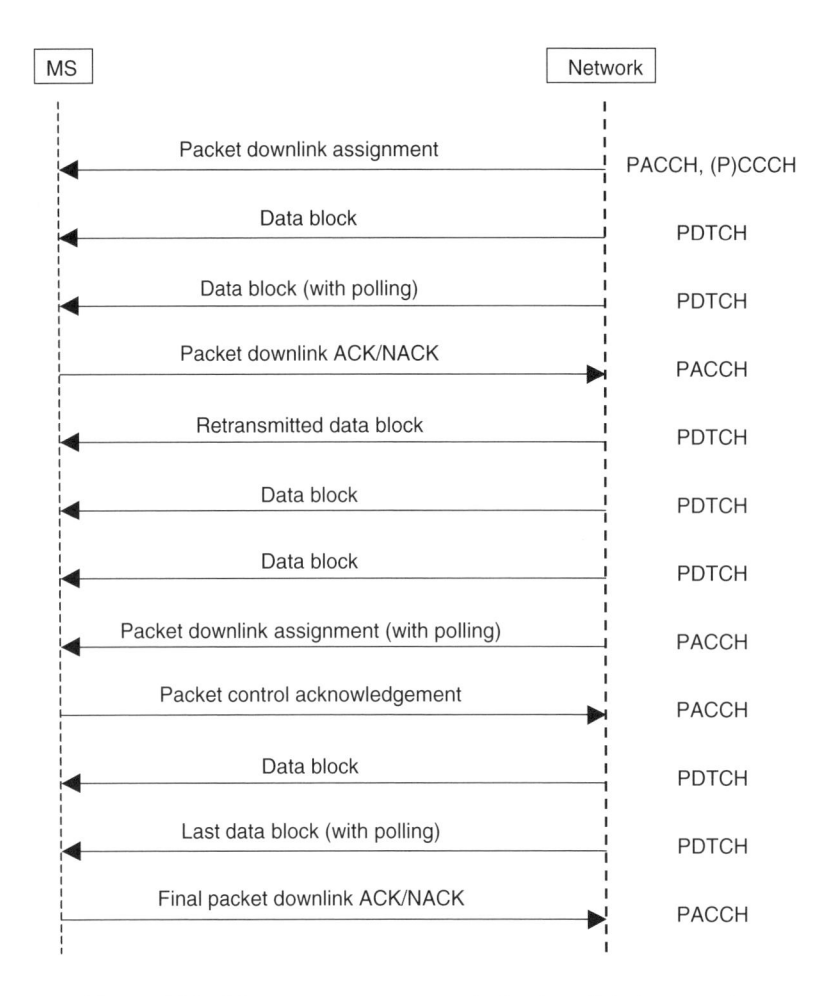

```
MS                                              Network

      ◄──── Packet downlink assignment ─────     PACCH, (P)CCCH

      ◄──── Data block ──────                    PDTCH

      ◄──── Data block (with polling) ──────     PDTCH

      ──── Packet downlink ACK/NACK ────►        PACCH

      ◄──── Retransmitted data block ──────      PDTCH

      ◄──── Data block ──────                    PDTCH

      ◄──── Data block ──────                    PDTCH

      ◄──── Packet downlink assignment (with polling) ──── PACCH

      ──── Packet control acknowledgement ────►  PACCH

      ◄──── Data block ──────                    PDTCH

      ◄──── Last data block (with polling) ──── PDTCH

      ──── Final packet downlink ACK/NACK ────► PACCH
```

Figure 4.34 Message sequence example for downlink data transfer

wastes an uplink block. On the other hand, it makes the protocol more robust. Without polling, the network would not know whether the MS either received a packet downlink assignment message incorrectly or missed out on it altogether and may start sending downlink data without the MS listening to the right PDCH(s), so wasting a lot of resources when such signalling errors occur.

4.10.8.3 Fixed Allocation MAC Mode with GPRS Half-Duplex Operation

Like for uplink TBFs, fixed and dynamic allocation MAC modes are also distinguished for downlink TBFs. The mode to be used is signalled in the downlink assignment message. However, the differences are rather limited in this case, and the distinction is mainly made to enable the GPRS-specific half-duplex operation described in subsection 4.10.7.5, for which fixed allocation is a prerequisite. The network may direct an MS operating in GPRS half-duplex mode to perform neighbour cell power measurements in predefined gaps in the downlink allocation, i.e. a subset of time-slots, which recur every several radio blocks. If this is the case, then no neighbour cell measurements need to be performed during other radio blocks, as a result of which mobile terminals of 'type 1' classes 24 to 29 may receive on up to eight time-slots simultaneously during these blocks (only seven, though, if SFH is applied). Obviously, this excludes blocks in which uplink signalling messages are sent, during which the network must leave a few additional gaps on the downlink. Without such tricks, reception on all eight time-slots would not be possible for an MS unable to receive and transmit simultaneously. Examples of fixed allocation time-slot assignments with and without half-duplex operation are provided in Appendix D of GSM 04.60.

4.10.8.4 Initiating an Uplink Packet Transfer while in Downlink Packet Transfer

If the MS wants to send packets to the network during an ongoing downlink TBF, it can indicate this by including a channel request description in one of the downlink ACK/NACK messages instead of going through the conventional packet access procedure discussed in the previous subsection. The network may then respond by transmitting a packet uplink assignment or packet timeslot reconfigure message on the PACCH, or may reject the request by sending a packet access reject message, again on the PACCH. The network has already knowledge of the PDCH(s) currently assigned to the MS for downlink transfer; it can therefore assign uplink resources in a manner that respects the multi-slot capability of the MS for concurrent TBFs in both link directions, if it chooses not to reject the request. Depending on the polling frequency for RLC acknowledgements, it may last longer to initiate the packet uplink transfer in this manner than through accessing the PRACH. On the plus side, there is no collision risk and no need for contention resolution.

As in the opposite case explained in the previous subsection, an MS operating in GPRS half-duplex mode must suspend the downlink TBF before starting (or resuming) the uplink TBF, and save the RLC state variables associated with the downlink TBF, such that it can be resumed later on.

4.11 The GPRS Random Access Algorithm

4.11.1 Why a New Random Access Scheme for GPRS?

We concluded in Section 4.4 that there should be spare capacity for GPRS on the existing GSM RACH. So, why do we need something different for GPRS?

On the GSM RACH, only few code-points are available for establishment causes due to the access burst format. Most of them were already occupied before GPRS was standardised, hence only two cause values could be assigned to GPRS. These are one-phase access requesting single-time-slot transmission, and single-block access, to send a packet resource request message (in other words, to request a two-phase access). The network does *not* know which MS is requesting resources and, in the case of the one-phase access, for how long resources are required. Also, the MS cannot request a one-phase access, if multi-slot transmission is desired. Often, therefore, a two-phase access will be required. A good packet data service should provide quick allocation of resources and, for efficiency reasons, the amount of signalling required to allocate resources should be small, particularly when traffic channel resources are only required for the transmission of a few hundred octets. Therefore, improvements to the GSM random access scheme are desirable.

Providing a separate PRACH specifically for GPRS has the obvious advantage of being able to redefine establishment cause values which relate to circuit-switched services on the RACH for GPRS purposes. On top of that, by redefining the channel coding on the access bursts, the number of information bits available for the channel request message can be increased. Indeed, GPRS provides an optional extended channel request message spanning 11 bits instead of eight. The price to be paid for these three additional bits is reduced robustness. Obviously, 11 bits are still not sufficient for inclusion of some kind of MS identity but, at least, more random bits aiding contention resolution can be included in the access request message.

The introduction of the PRACH provides also scope for improvements in a second area, namely the random access algorithm. For a start, the delay performance of the GSM algorithm is rather mediocre, which is mainly due to the S-parameter, indicating the desirability of a new or modified algorithm. Furthermore, for reasons outlined in Section 3.7, it was decided to introduce multiple access classes in GPRS, allowing for prioritisation at the random access, provided that a suitable algorithm is defined. The desirability of a stabilisation mechanism was also discussed in the SMG 2 GPRS ad hoc group, since due to the fact that resources are allocated on the basis of short packets rather than circuits, the RACH load will increase substantially. Again, this calls for a suitable algorithm. We submitted several contributions on stabilisation and prioritisation to this group, as discussed next.

4.11.2 Stabilisation of the Random Access Algorithm

We have already discussed potential stability problems with S-ALOHA before, both in this chapter and in the previous one. While we concluded in Section 4.4 that stability problems were likely not an issue in 'plain GSM', the situation is substantially different in GPRS. Due to the packet-based resource allocation, numerous channel request messages will have to be sent during a data session, such that the RACH load increases. Also, due to the dynamic sharing of physical resources between PRACH and PDTCH, every individual radio block not allocated to the PRACH can be used for the transfer of user traffic. To reduce the number of required PRACH slots, it would be desirable to operate the PRACH at as high a normalised load as possible. This in turn means that stability problems could arise and efforts should be invested in appropriate stabilisation methods.

For this purpose, we have compared the throughput and delay behaviour of various retransmission schemes for the S-ALOHA protocol, namely fixed retransmission probabilities, exponential backoff, and both centralised and decentralised implementations of Bayesian broadcast control in a submission to the SMG2 GPRS ad hoc group [132]. Subsequently, in Reference [234], we have investigated further aspects relating to our preferred scheme, namely Bayesian broadcast control, the only scheme among those considered with which stability problems can be avoided. Some of the results are summarised in the following. The results are of general validity: a channel with equally spaced slots of unit duration is considered, thus all results are normalised. The only GPRS-specific assumption is the consideration of the capture model derived for GPRS and incorporated in the GPRS radio channel evaluation criteria [133].

4.11.2.1 System Description and Performance Measures

We consider a system with a fixed number N of 100 mobile terminals. Terminals are either in *origination mode* or in *backlogged mode*. In any given slot, a terminal in origination mode creates a packet with a probability p_0 (for GPRS, this packet represents a packet channel request message). Unless otherwise mentioned, it is assumed that this packet is immediately transmitted. If it is received successfully by the BTS, the terminal stays in origination mode. If transmission in the first time-slot is not successful, the terminal transits to backlogged mode and attempts repeatedly to retransmit its packet until notified by the base station of its successful reception, in which case the terminal reverts back to origination mode. While backlogged, in each time-slot, a transmission attempt is made with probability p. This probability can be fixed or variable. In the latter case, control of p may either be centralised, in which case p will be the same for every backlogged terminal at any given time, or it could be decentralised, in which case p applied in a given time-slot by different terminals is not necessarily the same. Recall the discussion in Subsection 4.4.4 on slot-based transmission probability values considered here versus retransmission intervals defined by TX_INTEGER applied in GSM. The GSM approach does not require a Bernoulli experiment to be performed in each slot, but on the other hand, it does also not lend itself easily to mathematical analysis and is not very well suited for centralised retransmission schemes such as the Bayesian scheme introduced below.

The arrival rate per slot of newly generated packets λ_{ar} will depend on the number of terminals in backlogged mode (i.e. the *system state* n), it is $(N - n) \cdot p_0$. The total traffic offered to the channel G not only depends on n, but also on the transmission probability values p applied by the different backlogged terminals. For instance, if fixed values are applied for the retransmission probability p, $G = (N - n) \cdot p_0 + np$. Depicting the delay and throughput behaviour as a function of G or λ_{ar} would make it difficult to compare the performance of the different transmission schemes, since these will affect the backlog distribution. Therefore, delay and throughput behaviour are depicted as a function of the *input traffic* Np_0, as in References [235] and [236]. Note that Np_0 is a virtual quantity, unless the system state is zero.

The delay values reported are normalised to slots of unit duration. The definition in Reference [125] for the average delay is used, i.e.

$$\overline{D} = \lim_{T \to \infty} \frac{\left[\sum_{t=1}^{T} n(t)\right]/T}{\left[\sum_{t=1}^{T} \psi(t)\right]/T}, \qquad (4.4)$$

with t the time-slot, T the observation interval, and $\psi(t)$ the number of successfully transmitted packets in a slot (zero or one). This definition only accounts for delays due to retransmission of collided packets, but not for any other transmission delays. Hence under very low load, when the network state is almost always zero, the delay will be close to zero as well. It is assumed that the base station acknowledges successful reception immediately (e.g. through a packet uplink assignment message). To convert these delay values into 'real' access delay values relevant for GPRS, one would have to add first the average time from packet arrival to the first incidence of a PRACH time-slot. Next, $\overline{D}$ would have to be multiplied by the average interval between time-slot occurrences pertaining to the considered PRACH. Finally, one would have to add the time required by the BTS to respond to a received channel request message.

The throughput S as a function of the offered traffic G for S-ALOHA has already been shown in Figure 3.7 for an infinite number of terminals, i.e. according to Equation (3.8) provided in Section 3.5, which is written here again for convenience:

$$S = Ge^{-G} + \sum_{k=2}^{\infty} C_k \frac{G^k e^{-G}}{k!}. \qquad (4.5)$$

For a finite number of terminals, the throughput is obtained with the binomial formula

$$S(p_0) = \sum_{k=1}^{\infty} \binom{N}{k} C_k p_0^k (1-p_0)^{N-k} \qquad (4.6)$$

rather than the above Poisson formula. Note that this formula is only valid when either retransmissions are ignored, or when p is always set equal to p_0, such that the offered traffic $G = Np_0$ does not depend on the system state. At $N = 100$, results obtained through Equation (4.6) are virtually the same as those obtained through Equation (4.5). Without capture, for instance, the throughput behaves according to the well-known $G \cdot e^{-G}$ curve shown in Figure 3.7.

With capture, the capture probability C_i, i.e. the probability that the packet with the highest received signal level can be detected correctly when i packets are transmitted simultaneously, is assumed to be 1.0, 0.67, 0.48, 0.40, and 0.35 respectively, for $i = 1..5$ and 0 for $i > 5$ according to Reference [133]. Under these conditions, the maximum throughput S_0 is 0.595 at a channel load G_0 of 1.74 (where G_0 is the traffic level G at which S is maximised). This result corresponds closely to one case considered in Reference [131], namely one of the BCH-coding curves from Figure 3, for which the capture probabilities happen to be similar to those of the model used here.

The benchmark for the delay and throughput performance comparison of the different transmission schemes introduced below is the *optimum transmission scheme*. In this scheme, unlike the one with fixed permission probabilities, both terminals with newly arriving packets and backlogged terminals transmit with the same probability $p = G_0/n$ in each slot. In other words, the terminal is considered to be backlogged immediately upon arrival of a packet. In this optimum scheme, the system state n is assumed to be known. In practice, at best an estimate of n will be available. For the class of *global probabilistic control schemes* (where all terminals transmit with the same transmission probability values), setting $p = G_0/n$ will result in maximum throughput and therefore minimum backlog and delay over all p_0 and for arbitrary N.

4.11.2.2 Exponential Backoff and Fixed Retransmission Probabilities

A system in which newly arriving packets are immediately transmitted and a fixed retransmission probability p is applied to backlogged terminals can be designed such that it operates in a stable manner, provided that N and p_0 are given. However, in real systems, not the least in cells of mobile communication systems, where the cell population will fluctuate due to terminal mobility, N and p_0 will vary. If p_0 for instance falls below the value which determined the choice of p, backlogged packets will be delayed for an unnecessarily long time, since p is lower than necessary. The overall delay behaviour of such a system will therefore be poor. Conversely, if p_0 exceeds the initial design assumptions, the system can become unstable.

To alleviate stability problems, terminals could back off (i.e. lower the transmission probability) when they observe that the system is congested. One popular approach is to

4.11 THE GPRS RANDOM ACCESS ALGORITHM

let terminals determine the level of congestion according to the number of retransmissions required to successfully convey a message to the base station. The retransmission probability value p is reduced with each retransmission attempt. As a result, the probability value for the first retransmission, $p(1)$, can be higher than in a system applying fixed p while still ensuring stable behaviour. With the *backoff-rate b*, for the xth retransmission attempt, the retransmission probability is set to

$$p(x) = \begin{cases} p(1) \cdot b^{x-1} & \text{for } x \leq R, \\ p(R) & \text{for } x > R. \end{cases} \quad (4.7)$$

In this definition of exponential backoff, after the Rth retransmission attempt, the retransmission probability is not decreased any further. Often, exponential backoff is associated with $b = 0.5$ and $R = \infty$, although this choice of b and R might be too conservative in some cases. Exponential backoff could also be implemented in a GSM-type random access algorithm, by increasing TX_INTEGER with every retransmission attempt.

Exponential backoff allows the delay performance to be improved compared to fixed permission probabilities, but it does not solve all stability problems. As shown in Reference [126], such systems may suffer from the same bi-stable behaviour as non-adaptive ALOHA systems, which we discussed in Section 3.5. Hence, either optimal sets of parameters have to be chosen for given N and p_0, leaving the possibility for instability when either N or p_0 changes, or conservative parameter values have to be chosen which ensure stable operation for a range of different N and p_0 values. In the latter case, the delay performance will again be unnecessarily poor.

Figure 4.35 compares the throughput behaviour for various sets of values for the triplets $(p(1), b, R)$ with that for the optimum retransmission scheme. As could be expected, the optimum scheme reaches the maximum throughput S_0 at some given input traffic Np_0 (to be precise, at $G(Np_0) = G_0$), and maintains S_0 with increasing input traffic, whereas with exponential backoff, depending on the parameters, a more or less significant decrease in throughput can be observed when exceeding a certain Np_0 level. Figure 4.36 compares the delay behaviour for these $(p(1), b, R)$ triplets with the optimum retransmission scheme and with a system operating with the fixed retransmission probability values $p = 0.04$ and 0.1 respectively.

Clearly, p fixed at 0.04 results in considerable delay at low load, which is due to the average waiting time of 25 slots between retransmission attempts, hence even a small fraction of terminals suffering collisions will have a significant impact on the average delay performance. On the plus side, the throughput does not collapse even when all terminals are backlogged, since $G(n = N) = N \cdot p = 4$, yielding a throughput S of approximately 0.4 according to Figure 3.7. With $p = 0.1$, by contrast, lower delay is experienced at low load, but once n exceeds a certain value, the backlog will most likely continue to build up and tend towards N, where $G = 10$ and $S < 0.03$, that is, the system gets stuck in a low-throughput-high-delay equilibrium point. This explains the sudden jump of $\overline{D}$ between $Np_0 = 0.5$ and 0.6.

Similar observations can be made for the exponential backoff scheme. The good throughput behaviour with parameter-triplet (0.13, 0.5, 4) at high load has to be paid for by moderate delay performance at low load (worse than p fixed at 0.1). The delay performance of (0.15, 0.85, 4) is better, but saturation (i.e. transition to a low-throughput-high-delay equilibrium point) occurs at $Np_0 > 0.6$. In order to avoid early saturation with

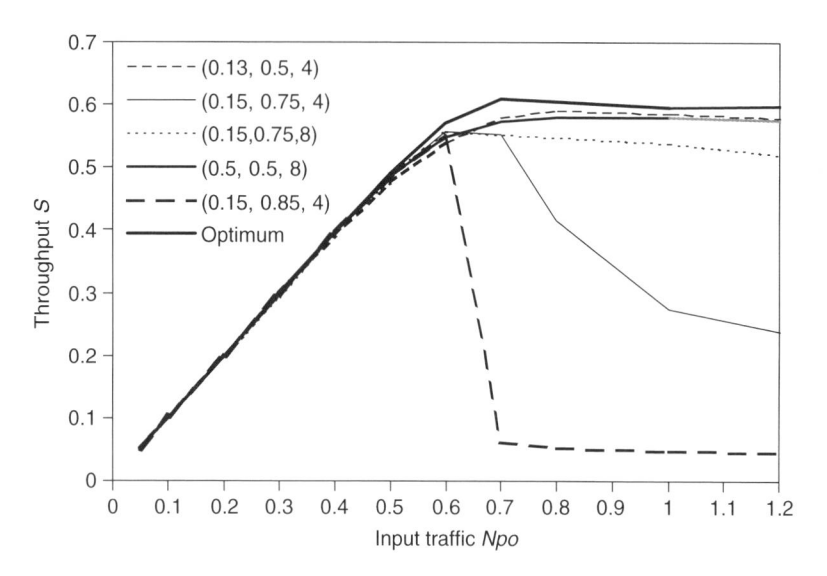

Figure 4.35 Throughput vs input traffic for exponential backoff and optimum scheme. The parameter values in parenthesis for exponential backoff represent $p(1)$, b and R respectively

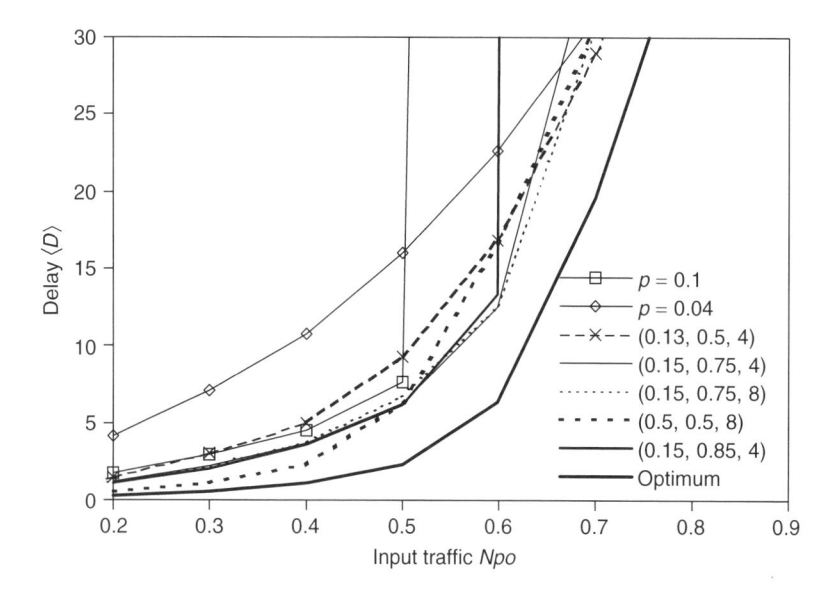

Figure 4.36 Delay comparison for fixed p, exponential backoff and optimum scheme

$p(1) = 0.15$ and $b = 0.75$, $R = 8$ has to be chosen instead of $R = 4$. This choice of R increases slightly the delay for an input traffic between 0.5 and 0.6. To obtain low delay values at low input traffic, a high $p(1)$ has to be chosen, which in turn requires a high R to avoid early saturation. (0.5, 0.5, 8) is such a choice. Again, due to the high R, the delay performance is sub-optimal in the input-traffic region around $Np_0 = 0.6$.

4.11.2.3 Bayesian Broadcast Control

The pseudo Bayesian broadcast (or retransmission) scheme was first proposed by Rivest in Reference [51] for conventional slotted ALOHA and adapted to capture by Robertson and Ha in Reference [131]. With Bayesian broadcast, the distribution of the system state or backlog n is estimated based on channel feedback observation (in terms of idle, success, collision time-slots) using Bayes' rule. With pseudo Bayesian broadcast, the distribution is approximated by a Poisson distribution with mean v. The retransmission probability is then set according to the most likely state v, that is to $p = G_0/v$, which is the same as in the optimum retransmission scheme, provided that n is correctly estimated, i.e. $v = n$. In the following, we consider pseudo Bayesian broadcast, but drop 'pseudo' for simplicity.

For the perfect collision channel (no capture), the relevant algorithm is discussed in detail in Section 6.5. Here, we simply write the algorithm adapted for the capture case, as derived in Reference [131].

Each station maintains a copy of v. At $t = 0$, all stations set v to G_0. Then, during each slot, each station:

(1) transmits with probability $p = G_0/v$, if it has a packet to transmit;

(2) decrements v by G_0, if the current slot is idle, decrements v by 1 if it is a success slot, or increments v by 1 if it is a collision slot;

(3) sets v to $\max(v + \hat{\lambda}_{ar}, G_0)$, and goes to step 1 for slot $t + 1$.

In the last step, $\hat{\lambda}_{ar}$ is an estimation of the arrival rate λ_{ar} of new packets per slot. Rather than estimating it, we will use the maximum arrival rate allowing for stable operation (i.e. the maximum throughput S_0) as suggested in Reference [51], that is, $\hat{\lambda}_{ar} = S_0$. Again, see the discussion in Section 6.5 for more details. Recall that $S_0 = 0.595$ and $G_0 = 1.74$ with the GPRS capture model. The beauty of this algorithm is that it does not depend on N and p_0 (p_0 obviously affects λ_{ar}, but the algorithm works fine with $\hat{\lambda}_{ar} = S_0$). What needs to be determined, however, is G_0 and S_0, hence the capture ratio appropriate for GPRS base stations needs to be established and statistics on the distribution of received power levels for access bursts need to be collected.

The question is now how such an algorithm could be implemented in a cellular communications system. Three options are identified here.

(1) In a completely decentralised implementation, each mobile terminal could establish individually the channel feedback, and independently estimate v and calculate p. This would save precious downlink resources.

(2) The channel feedback as perceived by the base station could be signalled to the mobile terminals, which again estimate v and calculate p independently, but on the basis of this signalled feedback rather than their own observation.

(3) The base station estimates v, calculates p and signals the value to be used to the mobile terminals in a fully centralised fashion.

If the signalling of the feedback in the second option or the transmission probability values in the third option occur error free, then all mobiles will use the same value for p, such that exactly the same throughput and delay behaviour should result. On the other hand, in a completely decentralised implementation according to the first option, we cannot realistically expect that the feedback perceived at different mobile terminals is the same as that at the BTS (due to range restrictions, shadowing, etc.). In Reference [132], we presented results for such a decentralised implementation, using a ternary symmetric channel to account for feedback errors. However, requesting the mobiles to be able to receive on their transmit frequencies (preferably even while transmitting, since channel feedback must also be assessed during own transmission attempts), only for the sake of saving a few downlink resources and improving the random access performance somewhat, would not go down well with handset manufacturers. We will therefore constrain ourselves to the other two options.

With option (2), the BTS would have to signal a ternary feedback value on every downlink time-slot paired with a PRACH slot. From the discussion so far, it should be clear that this would have serious implications on the existing GSM design (the only two downlink channels that convey information in a single time-slot are the FCCH and the SCH). What we can expect, at best, is a few bits set aside per downlink radio block paired with an uplink 'PRACH block'. From this perspective, feedback signalling is not easily feasible (which, by the way, is another argument on top of those listed in Section 3.5 against splitting algorithms). It would also mean that mobile terminals would have to listen constantly to the feedback channel to keep track of their estimate of v, which is not desirable from a battery saving point of view. What we can do, however, is signal p at limited resolution and with an update interval i_u of say once every four PRACH slots.

We consider first the signalling of p at 'full resolution' (i.e. at 'double' precision in a computer program), with various update intervals, including the ideal case $i_u = 1$. We assume that mobile terminals keep always track of the last signalled value of p, such that they can attempt to transmit immediately upon arrival of a packet with the appropriate parameter value. Again, this is not desirable in practice, so the terminal would either have to wait for the next occurrence of p after packet arrival before being allowed to access the PRACH, or use a low default value, both methods obviously affecting the delay behaviour. Note that with USF-based multiplexing, the mobile terminal would anyway have to listen to a 'free' USF flag before being able to schedule a packet channel request message, so an appropriate model of operation would be to signal p in every downlink block containing a free USF flag.

The delay performance of Bayesian broadcast with perfect signalling of p (that is, at 'full resolution' and updated prior to every PRACH slot), assuming immediate acknowledgement, comes close to that of the optimum scheme. As the update interval increases, so does the experienced delay. All the same, stability problems were not experienced for the update interval values and over the range of input traffic we considered. In Reference [237], it is reported that S-ALOHA without capture can be stabilised to 1/e, when the feedback is passed through a discrete memoryless channel, as long as its capacity is bigger than zero. If we assume error-free signalling, having $i_u > 1$ is kind of similar to having a feedback channel with reduced, but all the same non-zero capacity. With $i_u \leq 4$, the delay performance achieved with Bayesian broadcast is superior to that with exponential backoff with any of the $(p(1), b, R)$ triplets considered, let alone to that with fixed permission probabilities. At $i_u = 8$, on the other hand, the delay performance is similar to that of exponential backoff with carefully selected triplets, as illustrated in Figure 4.37.

Hence, from a random access perspective only, a centralised access control scheme such as Bayesian broadcast is the preferred solution if $i_u \leq 4$, i.e. if the necessary signalling

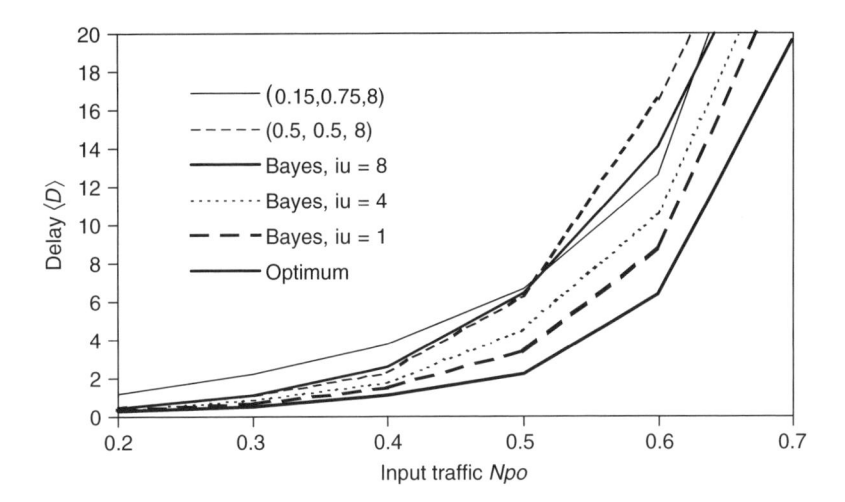

Figure 4.37 Comparison of the delay performance achieved with Bayesian broadcast to that with exponential backoff

resources can be expended on the downlink. Unfortunately, though, this is a major sticking point. In theory, $i_u = 4$ is feasible by signalling p in the same way as the USF. It should be clear, though, that at best only a few bits can be afforded, thus the resolution of p would have to be limited by applying quantisation.

4.11.2.4 Centralised Bayesian Broadcast with Quantisation of p

Consider $j = 2^k$ quantisation levels for p. These could be placed in an equidistant manner between 0 and 1 (that is, *linear quantisation* would be applied). However, as there will be $n = 0, 1, 2, 3, \ldots$ backlogged users, it would be advantageous to have a low resolution for high retransmission probability values and a high resolution for lower values. Hence, a non-linear quantisation scheme might be more efficient. One possibility is to quantise the estimated backlog v rather than p itself, resulting in j discrete levels $p_i = G_0/v_i$, $i = 0, 1, \ldots, j - 1$. For this purpose a geometric quantisation of the backlog is proposed with quantised backlog values $\{v_0, v_1, \ldots, v_j\} = \{1, q, q^2, \ldots, q^{j-1}\}$. The boundary between two adjacent levels v_{i-1}, v_i is $q^{i-1} + 0.5 \cdot (q^i - q^{i-1})$ for $i = 1, \ldots, j - 1$. This is henceforth referred to as *geometric quantisation*.

Consider again $N = 100$ transmitters, such that q^{j-1} is preferably set to 100 as well, thus

$$q = \sqrt[j-1]{100}. \qquad (4.8)$$

For $j = 8$ and 16, we have tested both linear and geometric quantisation against Bayesian broadcast with *full resolution*. As expected, we found that for the same amount of quantisation levels, the geometric scheme performed better than the linear scheme. This is particularly true at high input traffic levels, where the limited resolution for low retransmission probability values with linear quantisation causes particularly high delays. Furthermore, for $i_u = 1$, we found that the performance achieved with 16 geometric quantisation levels (that is, with four bits) almost matched that with full resolution, while only eight

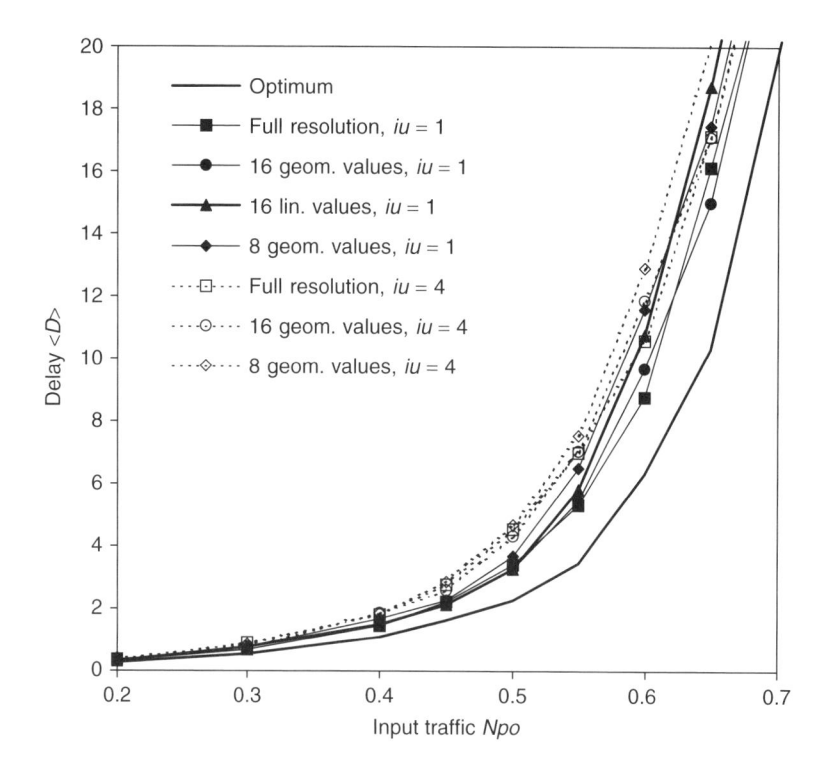

Figure 4.38 Impact of the quantisation of p on the performance of Bayesian broadcast

levels lead to slightly increased delays. The same is true for $i_u = 4$, but the performance difference between $j = 8$ and $j = 16$ is slightly smaller, as illustrated in Figure 4.38.

So far, we have considered the performance impact of quantisation and extended update intervals for the signalling of p, but we have always assumed immediate acknowledgement. Because of limited BTS processing power and because appropriate resources must be found on the downlink to signal acknowledgements, we cannot really achieve this in practice. Delayed acknowledgements will tend to increase the access delay values experienced with the various retransmission schemes (beyond the trivial component between reception of a request message and sending of an acknowledgement by the BTS). For instance, the accuracy of the backlog estimation will suffer in the case of Bayesian broadcast. Similarly, in the backoff scheme, a terminal adapts its behaviour based on a past channel load situation, which reflects the current situation the less accurately the larger the acknowledgement delay is. In both schemes, the precise impact will depend very much on implementation details, for instance whether a terminal must wait a minimum period before retransmission similar to the S-parameter in GSM. For a specific scenario with PRMA-based protocols, the impact of delayed acknowledgements on Bayesian broadcast is examined in Sections 6.5 and 8.3.

4.11.3 Prioritisation at the Random Access

In Section 3.7, we made a case for priority-class specific access control. Prioritisation can be introduced with all the retransmission schemes discussed so far. With fixed permission probabilities, different probability values can be chosen for different priority classes. However, the same stability problems are encountered with this scheme as in the case without prioritisation. With backoff schemes, prioritisation can be achieved by setting either $p(1)$ or the backoff rate b or both individually for each priority-class. Different approaches to prioritisation with Bayesian broadcast are discussed in detail in Chapter 9 and some of them are tested with MD PRMA. Finally, again as discussed in Chapter 9, stack-based schemes lend themselves also well to prioritisation.

To compare the 'prioritisation efficiency' of the different prioritised schemes, for which the class-specific behaviour is relevant rather than the average performance, we would really need to know the access delay requirements of the different classes. These are only specified indirectly in GSM 02.60 through specification of the required total transfer delay performance of packets with a certain length (as listed in Table 4.7), which includes other delay components on top of the access delay. We would therefore have to consider these other delay sources as well, for instance resource allocation delays, and account for the amount of PDTCH resources allocated during the transmission of these packets. Rather than presenting results for prioritised S-ALOHA schemes here, we refer to Reference [55], where we examined prioritised schemes based on exponential backoff and Bayesian broadcast as well as on a stack algorithm. In terms of average delay behaviour, we found (not surprisingly) that exponential backoff performed worst, and we also found the Bayesian algorithm to outperform the stack-based algorithm. Considering the problems associated with channel feedback signalling required for the stack-based scheme in GPRS, and given that various options exist to prioritise Bayesian broadcast (as discussed in Section 9.2) and thus to achieve whatever 'prioritisation efficiency' may be required, such a centralised control scheme is our preferred option also from a prioritisation point of view.

Regarding the signalling overhead for prioritised centralised access control schemes, the default assumption would be that if there are k priority classes, the overhead would also have to be k-fold compared to a scheme without prioritisation. However, we could argue that the range of probability values to be covered per class would be smaller, and we might get away with, for example, three or only two instead of four bits per class, but this would have to be tested in detail.

4.11.4 The GPRS Random Access Algorithm

Based on the above, centralised access control would appear to be the preferred option, both from a stabilisation and a prioritisation point of view. As far as stabilisation and delay performance is concerned, provided that the update interval of the relevant parameters is low (e.g. $i_u \leq 4$), centralised schemes outperform decentralised ones such as exponential backoff, which rely only on their own history of transmission failures. In terms of prioritisation, centralised schemes are the most flexible ones, as the degree of prioritisation can be controlled by the network with suitable ad hoc algorithms, which may be modified whenever required without affecting the population of already deployed mobile terminals.

For GPRS, a centralised solution was adopted, which distinguishes between four access classes or *radio priority classes*, class 1 having highest, class 4 lowest priority. This solution goes quite some way towards our preferred one.

According to GSM 03.64, a two-step approach is adopted by the network to calculate the optimal persistence on the PRACH for packets pertaining to different priority classes.

Based on the analysis of the incoming packet channel request messages, the network calculates both a short-term estimation and a long-term estimation of the 'persistence level' to be applied by the mobiles wanting to transmit packets of a given priority class. The actual 'persistence levels (or in our terminology, permission probabilities) should depend upon the priority class i of the packet to be transmitted, the amount of traffic within higher priority classes, and the amount of traffic within priority class i. The short-term levels, referred to as 'contention levels' in GSM 03.64, are signalled frequently to the mobiles, less frequently the long-term levels, the so-called 'persistence instruction set'. This is indeed a good approach, as it allows the downlink signalling load to be minimised. This is also illustrated in Section 9.2, where we discuss prioritisation algorithms in which certain parameters need to be signalled less frequently than others.

4.11.4.1 PRACH Control Parameters

The packet access procedure, the equivalent of the immediate assignment procedure in GSM, is specified in full detail in GSM 04.60. There, the following PRACH control parameters are identified, most of them already familiar from the RACH.

- A bitmap for the access control classes, indicating for each of 15 access classes plus emergency calls whether mobiles pertaining to that class are allowed to access the PRACH (the 15 classes are exactly the same classes as those discussed in Subsection 4.4.4 for the GSM RACH algorithm).

- The MAX_RETRANS value (1, 2, 4 or 7), which has the same evident meaning as on the RACH, however, now on a per-class basis, i.e. requiring two bits for each of the four radio priority classes.

- TX_INTEGER, again having the same meaning as on the RACH, i.e. specifying the distribution of the retransmission interval. Its coding, in four bits (not class-specific), is also similar, but rather than values from 3 to 50, it can assume 16 different values from 2 to 50.

- The S-parameter is also serving the same purpose as on the RACH, that is specifying the minimum retransmission interval, but here, it is explicitly coded using four bits (again not class-specific) rather than implicitly determined through TX_INTEGER and the CCCH type. Currently, only 10 values are defined, the minimum value being 12, the maximum 217 time-slots.

- The new and interesting aspect is the definition of optional *persistence levels* $P(i)$ for each radio priority class i. Four bits are used for each of the four classes to code the persistence level. The coding is linear from 0 to 14, but 15 is omitted, and the highest value is 16. Class 1 has the highest, class 4 the lowest radio priority. If the persistence levels are not signalled, a level of zero is assumed for each class.

The control parameters are broadcast on the PBCCH in the PSI type 1 message, which, although the most frequently broadcast PSI message type, is typically only signalled in integer multiples of the 52-multiframe, thus for our purposes clearly 'long term'. To realise the two-step approach with more frequent signalling of the 'short-term persistence levels' or 'contention levels' mentioned above, the persistence levels can optionally be included in various control messages sent on the PCCCH, thereby reducing the update interval for these levels. To be precise, they can be included on the packet uplink and downlink assignment messages, the packet paging request message, and the packet downlink dummy control message, while they cannot be included on the packet access reject message and the packet queuing notification message. As long as the system is not running out of radio resources, access reject and queuing notification messages should be rare, meaning that almost all downlink PCCCH blocks could signal persistence levels. Therefore, assuming that the number of PRACH blocks scheduled is not larger than that of downlink PCCCH blocks, it should be possible to update the persistence levels for almost every PRACH block. This is not exactly the same as $i_u = 4$, because the updated value does not reflect the backlog at the beginning of the PRACH block, but rather that at the beginning of the previous block due to the interleaving delay. However, within the constraints of the radio block structure applied on the downlink, where data is always interleaved over four time-slots, this is really as good as we can get. It would in theory be possible to meet $i_u \leq 4$ by never assigning two consecutive radio blocks to the PRACH, e.g. by alternating PRACH blocks on the uplink with PCCCH blocks on the downlink, thus being able to signal the persistence levels in between PRACH blocks.

4.11.4.2 Content of the Packet Channel Request Message

The content of the packet channel request message is either coded on 8 or 11 bits, depending on the access burst format to be used, which is specified in system information messages. Seven establishment causes are distinguished in both cases, namely one-phase access request, short access request, two-phase access request, page response, cell update (sent upon cell reselection when in MM ready state), MM procedure (e.g. RA update or GPRS attach), and single block transmission without TBF establishment.

If the request message signals one-phase access request, then 5 bits are used to code linearly the multi-slot class; if it signals short access request, 3 bits are used to code linearly the number of blocks required. The value-add of the 11-bit message is mainly in allowing the 2-bit radio priority to be indicated (for one-phase, two-phase and short access) and having more random bits, the purpose of which has been extensively discussed in Section 4.4. Additionally, there are more free code-points for the introduction of new establishment cause values.

4.11.4.3 The Packet Access Procedure

The mobile station initiates the packet access procedure by scheduling the sending of a packet channel request message on PRACH blocks pertaining to its PCCCH and simultaneously leaving the packet idle mode. The request message is scheduled based on the last PRACH control parameters received on the PBCCH. While waiting for a response, the MS monitors all downlink PCCCH messages sent on its PCCCH, and decodes any occurrence of the persistence-level parameter included in such messages. The latest persistence level values shall then be used for the scheduling of subsequent request messages (the maximum number of which is determined by MAX_RETRANS, exactly as on the RACH).

Unlike in GSM, the first attempt is scheduled on the very first PRACH slot occurring after initiation of the packet access procedure on the relevant MPDCH (as determined either by system information or by an USF set to 'free'). However, also unlike in GSM, for each attempt including the first one, the MS must determine through a random experiment whether it is allowed to send the access burst on the chosen slot. For this purpose, it draws a random value R according to a uniform probability distribution in the set $\{0, 1, \ldots, 15\}$. The MS is only allowed to transmit a request message, if $P(i) \leq R$, i.e. if the persistence level associated with the radio priority of the TBF to be established is less than or equal to this random number. Looking at the more or less linear coding of $P(i)$, as discussed when introducing the PRACH control parameters above, this corresponds to centralised control with linear quantisation rather than geometric quantisation, which we found to perform better (cf. Figure 4.38). By setting $P(i)$ to 16, it is possible to bar priority class i (e.g. the one with the lowest priority, class 4) from scheduling access attempts altogether (since $R \leq 15$), a useful feature during phases of temporary congestion. Using the control bitmap for the different access classes instead would be slower and, on top of that, it could also affect high-priority users, since the access classes are not related to radio priority classes.

Subsequent access attempts are scheduled as on the RACH, that is, the number of PRACH time-slots *between* two successive packet channel request messages (excluding the slots on which the messages are sent) is selected by drawing a random value for each new transmission according to a uniform probability distribution in the set $\{S, S+1, \ldots, S+\text{TX\_INTEGER} - 1\}$, where S is the value of the S-parameter. In essence, therefore, the algorithm applied on the PRACH is almost the same as that on the RACH. The main difference is that once an access attempt is scheduled according to S and TX_INTEGER, on the PRACH, the MS must also obtain permission to send an access burst in the scheduled slot using the random experiment described above. However, if the persistence levels are not signalled, and thus a default value of zero is assumed for each radio priority class, then permission is always obtained. In this case, the differences are reduced to the immediate scheduling of the first transmission attempt, class-specific MAX_RETRANS values, and (fortunately!) a different range of possible values for the S-parameter.

4.11.4.4 Discussion of the Procedure

This procedure comes fairly close to what we would have liked to see, i.e. centralised control with update interval $i_u \leq 4$ slots, with a few exceptions.

Firstly, while it is fair enough for the signalling of persistence levels to be optional, if these levels are not signalled, then prioritisation at the random access is not really possible (the class-specific coding of MAX_RETRANS does not help too much). It would therefore have been nice to see a class-specific coding of TX_INTEGER, but only as a fallback option with which to introduce some prioritisation when persistence levels are not signalled.

Secondly, the coding of the persistence levels is linear. With the same number of bits per class, the delay performance could have been improved by applying an optimised coding scheme, e.g. geometric coding. However, the performance improvement is only evident at high normalised load, as shown in Figure 4.38, at which the absolute delay performance may be insufficient anyway, so this is a minor issue.

Finally, while the coding of the S-parameter (and TX_INTEGER) has been changed compared to GSM, a minimum value of 12 slots is still comparatively high. Clearly, for the same reasons as already discussed in Section 4.4, it must be avoided that the MS retransmits a packet channel request message before the BTS had a chance to reply to a previous one it managed to decode. However, while there are limited occurrences of AGCH blocks in GSM, in GPRS, it is up to the BTS to schedule immediately a downlink

PCCH block containing a suitable response to the request message (be it an assignment, a queuing notification or a reject message), by pre-empting for example a PDTCH. In other words, we do not depend on the S-parameter to enable some kind of packet queuing. Admittedly, one has to allow for some processing time at the BTS. Furthermore, it might happen that four different mobile stations manage to send successfully a packet channel request message on the four time-slots pertaining to a PRACH block, whereas the network can only respond to one of them with an uplink assignment or queuing notification message on the subsequent downlink block. However, the network could respond with a packet access reject message, which cannot only address all four mobiles in one single block, but also indicate individual wait times during which the respective mobiles are not allowed to retransmit a packet channel request message. Alternatively, it could avoid scheduling new PRACH blocks until all incoming request messages could be served. In essence, therefore, a minimum value of four time-slots for the S-parameter would have been desirable to improve the access delay performance.

As an incidental remark, if the channel request message could carry enough information to allow for an unambiguous identification of the requesting MS, one would have to worry much less about the BTS response time. In fact, certain protocols, such as FRMA discussed in Chapter 6, are explicitly designed with a view to letting the MS retransmit channel requests before the BTS even has the possibility to respond to the first one.

4.12 EGPRS

So far, with the exception of the dual transfer mode and the associated exclusive allocation, which are release 1999 features, we have only dealt with GPRS features contained in release 1997. GPRS capabilities and features were considerably enhanced through release 1999, with most of the enhancements figuring under the 'EGPRS' heading. Additional extensions provide interoperability between GPRS and UMTS. This includes features such as cell reselection across GPRS and UMTS cells, for which additional packet system information messages containing non-GSM system information were introduced, and enhancements to the cell reselection algorithm were made. Similar enhancements were also made in the 'circuit-switched domain' of GSM, e.g. new system information messages were introduced and handover between GSM and UMTS was enabled.

The 'E' in EGPRS stands for 'enhanced'. The main enhancement is that data-rates can be increased on the PDTCHs owing to 8PSK modulation, a feature defined in the framework of the EDGE work item. The new coding schemes introduced for EGPRS and a new *Link Quality Control* (LQC) scheme, which combines link adaptation known from GPRS with incremental redundancy, are described in the following subsection. Other enhancements, which are briefly listed in Subsection 4.12.2, include modifications of the RLC protocol.

EDGE COMPACT is a special mode of operation, which allows the deployment of an EGPRS system with as little as 1 MHz of spectrum available to an operator. This is particularly important for those existing D-AMPS operators in the Americas, who have chosen EGPRS as part of their migration route to 3G, but who may, at least in a first step, have to deploy new access technologies in existing cellular or PCS spectrum used for D-AMPS. As a result, only gradual reuse of this spectrum for EGPRS purposes, for example, is possible, and it must be possible to launch EGPRS services with as little spectrum as possible. The situation is different for GSM operators, who do not have to put any dedicated (E)GPRS spectrum aside. They can share their existing spectrum flexibly between GSM, GPRS and EGPRS, owing to the capacity on demand principle and to the fact that GPRS and EGPRS handsets can share the same PDCHs. A selection of articles on capabilities of and migration to EGPRS is provided in Reference [238].

Considering how quickly the telecommunications industry changes direction, it is rather futile to make any forecasts on whether EGPRS and EDGE COMPACT will ever see large-scale deployment, but it is all the same worthwhile mentioning the following. While EGPRS is a natural evolution from GPRS, it implies hardware changes at the base stations. Almost all existing GSM/GPRS operators have opted for UMTS (i.e. UTRA FDD) as their chosen technology for 3G, involving a costly roll out of new radio infrastructure. At the time of writing, they appear to be rather reluctant to commit to additional significant investments in their existing GSM/GPRS infrastructure. It should also be noted that the main driving force behind EGPRS standardisation were D-AMPS operators. In fact, the acronym EDGE, initially standing for 'enhanced data-rates for GSM evolution', was later interpreted to mean 'enhanced data-rates for *global* evolution' for this very reason. However, a key D-AMPS operator decided in late 2000 to adopt a GSM/GPRS/UMTS evolution route rather than a D-AMPS/EGPRS route to 3G, and other operators are likely to adopt the same approach. All this taken together makes the future of EGPRS look somewhat uncertain.

4.12.1 EGPRS Coding Schemes and Link Quality Control

The main new feature in EGPRS is EDGE, i.e. 8PSK modulation leading to enhanced data-rates, where the signal quality permits it. 8PSK modulation may only be used on PDTCHs, on all other logical channels, GMSK with CS-1 is used, exactly as in GPRS R97. Unlike enhanced circuit-switched data, where the enhancement consists almost exclusively in the new modulation scheme, for EGPRS, 8PSK comes together with a number of other features, probably the most important one being a new link quality control scheme, which adds incremental redundancy to the link adaptation scheme already known in GPRS.

4.12.1.1 Link Adaptation and Incremental Redundancy

We pointed out previously that due to channel fluctuations, measurement delays and inaccuracies, we could not expect link adaptation to provide optimum performance. If the chosen code-rate is too low (that is, the error coding is too robust), capacity is wasted due to unnecessary redundancy. If it is too high, an excessive number of retransmissions may be required, which leads to delayed data transfer, and again to wasted capacity, since the type I ARQ scheme applied in GPRS R97 does not keep information from bad blocks, so RLC blocks received in error are simply lost.

If delay is an issue, then we will have to put up with some inefficiency and have to be willing to err on the robust side (although not too much, as this reduces the net data-rates available to the users and thus increases the transfer delay). However, where delay is not an issue, we can improve on the throughput performance achieved with link adaptation through a scheme, which does not depend on the speed and the accuracy of link measurements, namely a so-called *type II hybrid ARQ* scheme applying *Incremental Redundancy* (IR). In type II schemes, bad frames are saved by the receiver and combined in some way with subsequent retransmissions.

In pure IR schemes, blocks are initially sent with minimum redundancy. If a reception error occurs and the receiving side requests a retransmission, the transmitting side will add redundancy. Crucially, the coding and puncturing is such that there is value in the initially transmitted block; that is, the receiving side will decode the block based on a suitable combination of both copies of the block it has received rather than only the second copy. This means that no capacity is wasted. If reception fails again, the receiver will request the transmission of additional data that can again be combined with the data already received and, in theory, this process is repeated until it can correctly decode the data block. Pure IR schemes maximise data throughput especially with good link conditions and they do not require the link quality to be measured before transmission. However, the delay will increase when poor link conditions require numerous retransmissions to successfully decode each data block.

Through the combination of link adaptation and IR, it is possible to trade off efficiency against delay performance, by choosing the amount of redundancy contained in the initial block transmission according to the currently experienced link conditions rather than attempting transmission with minimum redundancy. The link quality control scheme chosen for EGPRS does exactly this. By exploiting the strengths of both link adaptation and IR, good data throughput can be achieved with minimum delay.

4.12.1.2 Modulation and Coding Schemes in EGPRS

Link adaptation on the PDTCH now entails the choice of the optimum combined *modulation and coding scheme* (MCS) rather than only the coding scheme: MCS1 to MCS4 use GMSK modulation, MCS5 to MCS9 use 8PSK modulation. Blind detection can be applied to determine the used modulation scheme, which means that the receiver can figure out which modulation was applied by looking at the waveform of the modulated signal. The MCS can be changed on a block-by-block basis.

The radio block structure known from GPRS has been retained, that is, a radio block is carried on four bursts, and there is on average one radio block every 20 ms. However, apart from that, there have been a lot of changes. An EGPRS RLC/MAC block used for data transfer is still carried in one radio block in the same manner as a conventional GPRS RLC/MAC block, but it is now composed of an RLC/MAC header and one *or two* RLC data blocks (Figure 4.39). The stealing bits are not signalling the coding scheme anymore, instead they signal the format of the RLC/MAC header. In this context, the header format not only specifies the length of the header payload and the *Header Check Sequence* (HCS), but also the type of coding that was applied. There are currently two header formats defined for 8PSK radio blocks, and one for GMSK blocks. Additional header types can be introduced, if required. The modulation and coding scheme employed is then signalled in the RLC/MAC header. In summary, the receiver detects the type of modulation, reads the stealing bits, decodes the content of the RLC/MAC header to understand which MCS was used and thus what type of error coding was applied, and can then read the rest of the block containing the RLC data.

RLC/MAC block		
RLC/MAC header	RLC data block 1	RLC data block 2 (conditional)

Figure 4.39 EGPRS RLC/MAC block format for data transfer (the second RLC data block is conditional on the MCS applied)

With the exception of the USF, which are encoded separately, all bits are always first encoded using a rate 1/3 convolutional code, irrespective of the modulation and coding scheme used. Additionally, and before convolutional coding, an 8-bit header check sequence is added to the RLC/MAC header. The USF bits are never punctured, while different puncturing schemes are applied to the remaining RLC/MAC header bits (including HCS bits) on one side, and RLC data bits on the other. The maximum code rate after puncturing is 0.53 for the header bits (in MCS-1 to 4) and 1.0 for data bits (in MCS-4 and MCS-9). So why go through the trouble of convolutional coding and then puncturing to achieve a code-rate of 1.0 rather than just use a 'no-coding scheme' such as CS-4 in GPRS R97? The reason is the *Two-Burst-Based Link Quality Control scheme* (2BB-LQC) applied in EGPRS, which combines link adaptation with IR, the latter requiring that convolutional coding be always performed. IR is also why the new GMSK coding schemes MCS-1 to 4 had to be defined. EGPRS mobiles can fall back to GPRS operation and use CS-1 to 4, however, in this case, only 'conventional' link adaptation is possible, while 2BB-LQC (and thus IR) cannot be applied.

The MCS are divided into the three families A, B and C. The MCS in each family use the same basic unit of payload, namely 37, 28, and 22 octets respectively, with one exception, namely only 34 octets for MCS-8 belonging to family A. The MCS in each family differ in how many payload units are transmitted in each radio block (in addition to header bits). For families A and B, one, two, or four units are possible, for family C only one or two units. When four units are transmitted (as is the case in MCS-7, 8 and 9), these are split into two separate RLC data blocks with separate sequence numbers and BCSs. In MCS-8 and MCS-9, the data bits and BCSs of the two blocks are interleaved separately over two bursts each, while in MCS-7, they are interleaved together over four bursts. RLC/MAC header bits (which include the sequence numbers of the data blocks) are in any case interleaved over four bursts. MCS-1 to 6 carry only one RLC data block per radio block, interleaved together with the RLC/MAC header over four bursts, hence following the approach known from GPRS R97.

The basic parameters of the different modulation and coding schemes are listed in Table 4.10. Readers wanting to hypothesise about data-rates 'seen' by the user (starting from a theoretical upper bound of $8 \times 59.2 = 473.6$ kbit/s) along similar lines as we did in Section 4.9 should note the following: unlike those reported in Table 4.9 for CS-1 to 4, the data-rates listed in the last column of this table account correctly for all MAC and the standard RLC header overhead.

The fact that data bits are interleaved over two rather than four bursts in MCS-8 and MCS-9 is why the LQC scheme is said to be 'two-burst-based'. The reason for this reduction in interleaving depth at a code-rate of close to one (MCS-8) or one (MCS-9) is that frequency hopping combined with interleaving has an adverse effect on error performance, if no (or little) error coding is applied. This is because unlike with error coding, a single 'bad burst' in a block (i.e. one containing errors) will for sure cause the block to be in error, so spreading bad bursts more or less equally over many blocks through frequency hopping will increase the average block error rate. By reducing the interleaving depth with MCS-8 or MCS-9 to two bursts for data bits only, this effect can be mitigated, while still letting the RLC/MAC header bits in such blocks (which are error protected), as well as blocks sent on the same PDCH at lower code-rates, benefit from frequency hopping.

The choice of the modulation and coding scheme for the initial block transmission is always a matter of link adaptation, but if retransmissions are required, they can either

Table 4.10 Modulation and coding schemes in EGPRS

Scheme	Code rate	Header Code rate	Modulation	RLC data blocks per radio block	Raw data in one radio block	Family	BCS	Tail bits	HCS	Data rate [kb/s]
MCS-9	1.0	0.36	8PSK	2	2 × 592	A	2 × 12	2 × 6	8	59.2
MCS-8	0.92	0.36	8PSK	2	2 × 544	A	2 × 12	2 × 6	8	54.4
MCS-7	0.76	0.36	8PSK	2	2 × 448	B	2 × 12	2 × 6	8	44.8
MCS-6	0.49	1/3	8PSK	1	592	A	12	6	8	29.6
MCS-5	0.37	1/3	8PSK	1	448	B	12	6	8	22.4
MCS-4	1.0	0.53	GMSK	1	352	C	12	6	8	17.6
MCS-3	0.80	0.53	GMSK	1	296	A	12	6	8	14.8
MCS-2	0.66	0.53	GMSK	1	224	B	12	6	8	11.2
MCS-1	0.53	0.53	GMSK	1	176	C	12	6	8	8.8

occur in *link adaptation mode* or in *incremental redundancy mode*. In link adaptation mode, the robustness may be increased by choosing another MCS in the same family with reduced code-rate, more robust GMSK modulation instead of 8PSK modulation, or both. This entails block splitting. Considering for instance family A, an RLC/MAC block sent initially with MCS-9 can be retransmitted using two MCS-6 blocks, an MCS-6 block in turn can be retransmitted using two MCS-3 blocks. If MCS-8 was chosen initially, because of the 34-octet instead of 37-octet payload units, blocks retransmitted using MCS-6 or MCS-3 will contain padding octets.

In incremental redundancy mode, blocks are retransmitted using the same MCS, but different *puncturing schemes* (PS) for the bits contained in the RLC data blocks, that is, different subsets of the coded data bits are transmitted. There are two to three different puncturing schemes per MCS, for instance three in MCS-9. If the receiver cannot decode a block based on a first transmission, e.g. with PS-1, it stores these bits and combines them with the retransmitted data bits sent using PS-2, for example. In the case of MCS-9, PS-1 and PS-2 bits are disjunctive. The effective code-rate after the first transmission is 1.0, hence the code-rate after the combination of the initially transmitted block with the retransmitted block reduces to 0.5. PS-2 and PS-3 bits are also disjunctive, and so are PS-1 and PS-3 bits, so if three copies of an MCS-9 block with the three different puncturing schemes are received, then the code-rate after combination amounts to 1/3. To complicate matters further, incremental redundancy is also possible across different MCS in one family, e.g. a block transmitted with MCS-7 followed by one transmitted with MCS-5. For further information, including justifications to some of the design choices, refer to Reference [239].

For the receiver to be able to combine the right blocks together, it must be capable of extracting information such as block sequence numbers from the RLC/MAC headers of all blocks to be combined, which will include by nature erroneously received blocks. This is why the HCS is required for IR and why higher code-rates are applied to RLC/MAC header bits than to data bits (by puncturing fewer bits). Also, for header bits, there is only one puncturing scheme per MCS, so that always the same bits are retransmitted. In other words, the information contained in the header bits enables the application of incremental redundancy, but is in itself not subject to the application of IR.

From a link adaptation perspective, some modulation and coding schemes are never optimal, that is, they would never be selected based on a graph equivalent to Figure 4.29. However, they are required for the IR mode.

4.12.2 Other EGPRS Additions and Issues

4.12.2.1 Modifications to the RLC

The RLC is the entity that has to perform the 2BB-LQC, so it is evident that a few modifications to the RLC were required. For instance, as already discussed above, a new EGPRS RLC/MAC block format for data transfer was adopted, containing a combined RLC/MAC header and one or two RLC data blocks (two for MCS-7 to -9). The distribution of RLC control information between the RLC/MAC header and the RLC data block(s) was governed by robustness requirements imposed by incremental redundancy, which may be applied to the data block part. For instance, the BSNs are contained in the header (one BSN per data block), while the BCSs are contained in the data blocks (again one per block).

Already in GPRS R97, there is a risk of RLC protocol stalling, particularly when a TBF is conveyed on multiple time-slots and when the PCU is remote from the BTS (resulting in delayed acknowledgements, see Reference [239] for details). Since an RLC/MAC block transmitted with MCS-7 to -9 contains two RLC data blocks, this risk is further increased. To avoid stalling, a variable window size was introduced. The maximum allowed window size depends on the number of time-slots allocated. For single-slot operation, it is equal to 192, for eight-slot allocation to 1024 RLC data blocks. Recall that the fixed window size in GPRS R97 is 64 RLC data blocks.

Increasing the window size has an impact on the bitmap format and in turn on the acknowledgement messages. For this and other reasons, a new EGPRS specific packet downlink ACK/NACK message was introduced. Since a control message sent in the uplink direction must fit into a single radio block, it might not always be possible to send the full bitmap in one message, even though bitmap compression is possible in EGPRS to make this more likely. Through an EGPRS S/P field included in the downlink EGPRS RLC/MAC header, which spans two bits rather than the single S/P bit known from GPRS R97, the network can poll the MS to send an EGPRS packet downlink ACK/NACK message containing partial bitmaps. Specifically, the MS may be polled to send either a first partial bitmap, but not the channel quality report usually included in a downlink ACK/NACK message, the next partial bitmap again without the channel quality report, or the next partial bitmap with a channel quality report. The channel quality report, when included, is enhanced compared to GPRS to facilitate link adaptation. In particular, it may contain bit error probability estimates both for GMSK and for 8PSK modulation.

The packet uplink ACK/NACK message, although sharing the message type with the respective GPRS R97 message, has a significantly modified message content, when addressed to an MS operating in EGPRS mode. Again, this is mostly due to the signalling of the acknowledgement bitmap. Unlike uplink messages, the packet uplink ACK/NACK message sent on the downlink can be segmented into two RLC/MAC blocks, increasing the likelihood of being able to signal the full bitmap, particularly since bitmap compression may also be applied on the downlink. Where this is not possible due to very large window sizes, partial bitmaps may be sent in packet uplink ACK/NACK messages in the same manner as in packet downlink ACK/NACK messages.

Because of the extended window size, also an extended BSN range for RLC data blocks (using 11 bits for coding instead of seven) is required. This is catered for by the new RLC/MAC header format, which had to be introduced anyway for other purposes, as already discussed above.

Regarding the window size, it should be noted that while a larger window size is convenient to avoid protocol stalling, there are also other considerations to be made such as MS memory requirements. When operating in IR mode, the MS may have to store several copies of a single RLC data block. Since memory is at a premium in small battery powered handsets, this may result in the MS running out of memory, particularly when the window size is large. The MS can indicate in packet downlink ACK/NACK messages when this is about to happen, at which point it should be reverted to pure link adaptation mode.

4.12.2.2 Additional RLC/MAC Control Messages

Several existing RLC/MAC control messages had to be extended for EGPRS operation, for instance assignment messages, which must signal (among other new parameters) the RLC window size to be used, or the packet uplink ACK/NACK message, as discussed above. A few new message types were also defined. These are additional packet system information message types, an *EGPRS packet channel request* message, the EGPRS specific packet downlink ACK/NACK message already discussed above, an additional *MS radio access capabilities* message, the *packet enhanced measurement report* message, and a *packet pause* message.

Six additional packet system information message types were defined for R99. Five of them can be sent either on the PBCCH or the PACCH, the sixth is only intended to be sent on the PACCH. None of these messages was specifically introduced for EGPRS, as EGPRS-related information is signalled in extended existing message types. The new message type sent on the PACCH only was introduced to support DTM operation. All other new message types were introduced to signal information relating to other systems, e.g. neighbouring UMTS cells, cdma2000 cells, or D-AMPS related information, hence they are not directly related to EGPRS.

The EGPRS packet channel request message, which should be used by EGPRS mobiles if so indicated in the relevant packet system information message, differs from the 11-bit GPRS packet channel request message in that one of two EGPRS specific training sequences is used. This allows the MS to signal whether it supports EGPRS capability, but without 8PSK capability in the uplink direction, or full EGPRS capability including 8PSK capability in the uplink. Everything else, e.g. the use of GMSK modulation, and the coding of establishment causes on the available 11 bits, remains the same.

The network may request the MS to send an additional MS radio access capabilities message, specifying all EGPRS-related capabilities such as the frequency bands which are supported. This message may be sent following the packet resource request message in a two-phase access, if the required information does not fit into the resource request message itself. As a consequence, the first packet uplink assignment message sent by the network during a two-phase access must allocate two uplink radio blocks rather than only one.

4.12.2.3 Multiplexing GPRS and EGPRS Users on One PDCH

Downlink blocks sent using 8PSK are 'invisible' for conventional GPRS mobiles, thus they cannot detect the USF. This has implications when wanting to multiplex GPRS and

EGPRS users on the same PDCH. One obvious solution is to avoid the USF in this case and use fixed allocation instead. Another solution is to use USF of granularity four for the GPRS user, requiring every fourth downlink block to be sent using GMSK. Note that in the latter case, EGPRS users could still operate at a granularity of one. Also, if these downlink GMSK blocks are directed to an EGPRS user, they can be sent using MCS-1 to 4 rather than one of the GPRS R97 coding schemes. The USF coding in MCS-1 to 4 has been done in a way that also GPRS R97 handsets can read the USF, although they will not understand the rest of the block due to the different block format and coding. For this to be possible, the GPRS MS must be able to read the stealing bits to determine the coding scheme, and then decode the USF. In EGPRS, where the stealing bits are intended to indicate the header type rather than the coding scheme, the following trick was applied. The number of stealing bits was extended from eight to 12 for MCS-1 to 4, so that the eight bits located in the usual position known from GSM and GPRS signal CS-4 to the GPRS MS, whereas the additional four stealing bits indicate the header type to the EGPRS MS.

The most radical solution is to send a downlink block using GMSK modulation only when directed to a GMSK-only mobile terminal or when wanting to assign the subsequent uplink block to such a terminal. This would require the probability to be negligible that a GMSK-only mobile terminal mistakenly interprets a USF sent with 8PSK modulation as its own USF.

4.12.2.4 EDGE on the BCCH-Carrier

Because of power limitations and amplitude variations resulting from the use of 8PSK modulation, cell reselection performance can suffer if EDGE (i.e. 8PSK) is allowed on the BCCH-carrier. On the other hand, forbidding it altogether would severely restrict the use of EDGE. While the standards allow for it, it is up to the operator to decide whether to use it or not.

4.12.3 EDGE Compact

EDGE COMPACT is a stand-alone packet data system that can be deployed in as little as 600 kHz of spectrum (excluding guard bands) by using three 200 kHz carriers in a 1/3 reuse pattern. The 1 MHz of required spectrum quoted earlier implies 200 kHz guard band on each side of the three carriers. Operating at a 1/3 reuse with only 600 kHz of spectrum clearly means that frequency hopping is not possible. From Section 4.6, it should be evident that normal GSM operation on traffic channels is only possible with such a reuse pattern, if frequency hopping over many carriers combined with fractional loading is applied. Worse, for proper system operation, control channels require much looser frequency planning, e.g. 4/12 reuse, implying 2.4 MHz of spectrum excluding guard bands.

Since EDGE COMPACT is intended to be a data-only solution, GSM traffic channels for real-time services need not be supported. On non-real-time channels such as the (E)GPRS PDTCHs and PACCHs, ARQ can be applied to improve error performance, hence 1/3 reuse is viable even without frequency hopping. The remaining problem is that of control channels, which is tackled through time synchronisation at the base stations and discontinuous transmission on control carriers (employing new logical control channel combinations). As a result of time synchronisation and through appropriate scheduling, co-channel carriers in neighbouring sites can transmit in an alternate fashion on separate so-called *time groups*, rather than simultaneously, resulting in up to 4/12 effective reuse for system-critical control information.

4.12.3.1 New Logical Channels for EDGE Compact

Packet traffic and related channels, i.e. PDTCH, PACCH, and PTCCH, are reused in COMPACT mode. Specifically for EDGE COMPACT, the following new logical channels are introduced:

- The *COMPACT Packet Broadcast Control CHannel* (CPBCCH), which is a downlink-only channel used to broadcast the same pieces of information as on the PBCCH, but it has a different physical structure.
- The *COMPACT Packet Paging CHannel* (CPPCH), *COMPACT Packet Random Access CHannel* (CPRACH), and *COMPACT Packet Access Grant CHannel* (CPAGCH), which are the equivalent of the PPCH, PRACH, and PAGCH respectively. Analogous to CCCH and PCCCH, the three channels taken together are referred to as *COMPACT Packet Common Control CHannel* (CPCCCH).
- The *COMPACT Frequency Correction CHannel* (CFCCH), with the same format as the FCCH, but different time-slot mapping.
- The *COMPACT Synchronisation CHannel* (CSCH), serving the same purpose as the SCH in GSM, but with different time-slot mapping, and using a different coding for the frame number, which indicates also the time group on which the carrier in question is operating.

These new control channels are mapped together onto a 52-multiframe. PDTCH, PACCH, and PTCCH may be mapped onto the same multiframe as well.

4.12.3.2 Inter-Base-Station Synchronisation and the Time Group Concept

Each BS site is allocated at least three frequencies (exactly three with only 600 kHz of spectrum), one per sector, using a 1/3 frequency reuse pattern. The key enabling factor to achieve higher effective reuse patterns is the time synchronisation of base stations, which makes it possible to allocate common control channels in a way that simultaneous transmission or reception at neighbouring sites can be avoided. This is realised by introducing time groups, which separate in time the usage of co-channel carriers in a manner that only sectors of one time group may transmit or receive during a specific radio block, while sectors belonging to other time groups remain idle (i.e. silent both on uplink and downlink). This 'time reuse' applied on top of frequency reuse increases the effective reuse factor from 1/3 to 4/12 for example, if all four time groups which are defined for EDGE COMPACT are used, or alternatively to 3/9, if only three groups are used. The 4/12 reuse example is depicted in Figure 4.40.

Only CPBCCH, CPCCCH, CFCCH and CSCH are subject to the time group concept, which is only applied on odd time-slot numbers. To facilitate neighbour cell measurements for mobiles engaged in packet transfer, these control channels are not mapped in a static fashion onto time-slot numbers, but rather switch their time-slot once per 52-multiframe, such that a time-group rotation over odd time-slot numbers results. For PDTCH, PACCH, and PTCCH, 1/3 reuse is applied as discussed above.

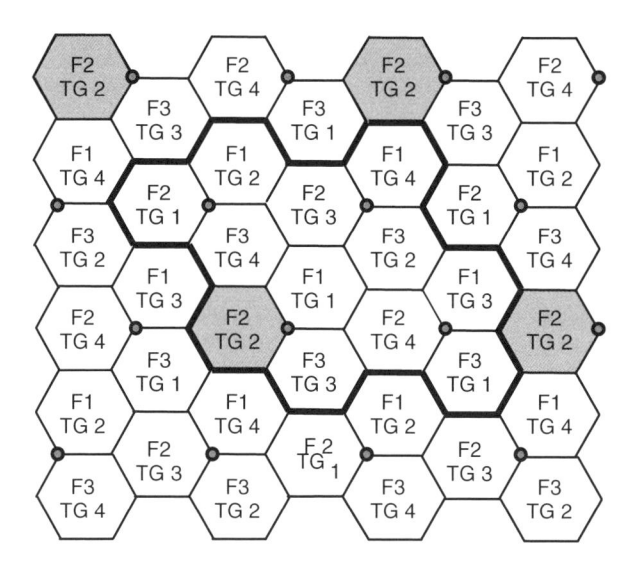

Figure 4.40 Example for a 4/12 cell pattern based on time and frequency reuse

The first three carriers in an EDGE COMPACT system are referred to as *primary COMPACT carriers*. The CPBCCH is only mapped onto such primary COMPACT carriers, and so are the CFCCH and the CSCH. If more than 600 kHz of spectrum is available and thus additional carriers can be deployed, these will act as *secondary COMPACT carriers*. Permitted channel combinations on secondary carrier time-slots are either PDTCH, PACCH and PTCCH, or CPCCCH, PDTCH, PACCH and PTCCH. In the latter case, the CPCCCH is assigned to the same time group as the CPBCCH on the primary carrier in the respective sector.

For more details, refer to GSM 03.64, GSM 05.02 or, where possible, to the EDGE COMPACT concept proposal submitted to SMG2 [240], which has most of the relevant information concentrated in one place. Publicly available resources providing a system overview and a performance assessment include References [82] and [241].

4.12.4 Further Evolution of GPRS

The evolution of the GSM system, and in particular of its GPRS components, does not stop with release 1999. The main efforts in enhancing GPRS capabilities further are directed towards the support of IP-based multimedia traffic, which requires proper real-time support both in the GPRS infrastructure and on the air-interface channels defined for GPRS. From a protocol stack perspective, the solutions chosen for this purpose are in alignment with similar solutions adopted for UMTS. It makes therefore sense to postpone the relevant discussion until after UMTS is introduced in Chapter 10, namely to Chapter 11, where both the further evolution of GSM/GPRS and that of UMTS beyond release 1999 are discussed.

5

MODELS FOR THE PHYSICAL LAYER AND FOR USER TRAFFIC GENERATION

In general, sophisticated communication systems are designed according to the concept of layering, often adhering to the OSI layering approach. The designer of a certain layer can then consider other layers as black boxes, which provide certain *services* defined in terms of functional relations between their respective inputs and outputs. She does not have to worry about the details of implementation of the next lower layer, but must only be aware of the services it provides. Conversely, she has to be sure that the layer being designed will cater for the services required by the next higher layer.

As we want to investigate the performance of multiple access protocols, we are interested in the MAC sub-layer, which will make use of the services provided by the physical layer. We must therefore assess the performance of the latter, or indeed, establish the relevant functional relations. Several options on how to model physical layer performance will be discussed and the models chosen for the performance assessment of a few multiple access protocols presented in Chapters 7 to 9 outlined in the following.

Regarding the relationship between the MAC sub-layer and higher (sub-)layers, of major concern here is the traffic coming from the latter, which has to be handled by the MAC making the best possible use of the available physical link(s). Where exactly (in terms of layers) this traffic is generated depends very much on the service considered, but is not of interest here. What is relevant is only the quantity and the temporal characteristics of this traffic as seen by the MAC sub-layer (and hence as delivered by the RLC sub-layer). For this purpose, traffic models are defined, which will then be used for the performance assessment of the MAC solutions investigated in later chapters. These include a model for packet-voice as an example of real-time packet data traffic, and models for Web browsing and email transfer as examples for non-real-time traffic. Furthermore, some aspects relating to video traffic are discussed.

5.1 How to Account for the Physical Layer?

5.1.1 What to Account For and How?

To carry traffic across the air interface, the MAC layer will use the services provided by the physical layer. The fundamental question that arises is: under what conditions can the physical layer be expected to deliver this traffic successfully to the peer MAC entity? This may not only depend on the input from the MAC layer (e.g. the number of bursts or packets to be carried at any one time), but also on conditions not directly under the influence of the MAC layer, e.g. the current state of the radio channel. As far as the latter is concerned, one would expect the physical layer designer to include means that allow provision of the required degree of reliability, such as appropriate FEC protection in combination with interleaving to combat the effects of fast fading.

Since increased reliability comes at the cost of reduced capacity, certain reliability problems will almost always prevail at the physical layer of a mobile communications system due to the typically adverse propagation conditions experienced on radio channels. It is possible to include these effects in the functional relations to be established by appropriate statistical modelling. However, for MAC layer performance optimisation of primary concern are the direct interdependencies between MAC and physical layer, and the main focus will be on modelling these. All the same, one has to be aware that the physical layer may fail to deliver information over the air irrespective of the behaviour of the MAC layer.

The physical layer model established in the following will predominantly be used for investigating MD PRMA performance on a hybrid CDMA/TDMA air interface. The fundamental functional relation of interest in this case is the error performance of the physical layer as a function of the number of bursts or packets carried in a time-slot. The error performance may also depend on particular spreading codes selected by the terminals (in particular, code collisions will affect error performance) and on the distribution of the power levels, at which the different bursts or packets are received by the base station. The latter is actually something that can only partially be controlled by the system, and is significantly affected by propagation characteristics.

There are two fundamental approaches to establish these functional relations:

- through use of appropriate mathematical approximations of the error performance; or
- through detailed assessment of physical layer performance, often via simulation.

5.1.2 Using Approximations for Error Performance Assessment

It would be nice if the physical layer performance could be approximated with reasonable accuracy by a set of formulae readily available from literature, preferably parameterised in a manner that permits different operating conditions to be investigated easily. MAC layer investigations could then be carried out without having to undertake detailed physical layer investigations first. Such approximations of the error performance for CDMA systems, such as the well known *standard Gaussian approximation* (SGA), have indeed been discussed widely in the literature and will be considered in the next section.

The simplest way to detect direct-sequence CDMA signals is to use a simple correlation receiver or matched filter, which detects a single path of the wanted signal. MAI is treated as noise. Multipath propagation does normally not result in undesired signal distortion, since the correlation receiver can lock onto and resolve paths individually[1]. Unfortunately,

---

[1] A path can be resolved by a Rake receiver, if its temporal separation from other paths is at least equal to the chip duration T_c.

replicas of the signal to be detected arriving through other paths will manifest themselves as self-interference, which will affect the error performance. To improve performance, Rake receivers are commonly included in CDMA system design concepts (e.g. Reference [12]). In a Rake configuration, several correlation receivers lock each onto a different path, and the individual signals are combined, which allows a path diversity gain to be achieved. Whether simple matched filters or Rake receivers are implemented, MAI is the main limiting factor to the error performance. Thus, to reduce errors further, the level of MAI has to be reduced, for instance through methods such as interference cancellation or joint detection (see below), or through the use of antenna arrays at the base station.

It is possible to invest arbitrary effort in error performance approximation, to account for the effect of multipath propagation, use of Rake receivers, and even antenna arrays [242,243]. However, the potential benefits of using such approximations in the context considered here do not justify the added complexity involved. Instead, when assessing the impact of MAI on MD PRMA performance, a standard Gaussian approximation for simple correlation receivers will be used, multipath propagation will be ignored, and it will be assumed that power fluctuations are compensated by power control. However, the impact of power control errors on error performance will be studied. This 'standard Gaussian model', which is described in Section 5.2, can be used on its own to establish the error performance of the physical layer, while ignoring code-assignment matters. Alternatively, as outlined below, it can be combined with a 'code-time-slot model' to account for code assignment and potential code collisions.

5.1.3 Modelling the UTRA TD/CDMA Physical Layer

A very important issue in CDMA systems is power control. In order to avoid capacity degradation due to the near-far effect, the power radiated by the different users on the uplink should be controlled tightly, such that each user's signal is received by the base station at a power level which correspond as closely as possible to a certain reference power level. Unfortunately, due to the fast power fluctuations caused by fast fading, it is impossible to control the power perfectly. Tight power control is particularly difficult to achieve in a hybrid CDMA/TDMA system, since closed-loop power control cannot be fast enough. In fact, Baier argues that the introduction of a TDMA component in a CDMA system with single-user detection (whether this be a single correlation receiver or a Rake receiver) will be virtually impossible for exactly this reason [109] and that multi-user detection should be used instead. In his research group, physical layer solutions were developed for hybrid CDMA/TDMA systems which incorporate *joint detection* (JD) of all signals transmitted in a time-slot, such that the near-far problem can be resolved without requiring tight power control (e.g. Reference [13]). Such an approach was also adopted for the UTRA TD/CDMA mode.

One could argue that with the fast and accurate open-loop power control possible in TDD configurations with alternating up- and downlink slots (see Section 6.3), joint or multi-user detection would not be required. However, TDD with alternating slots is limited to small cells, and multi-user detection schemes would still be required in all other cell types. Furthermore, according to Reference [86], the accuracy of open-loop power control is in general not very good due to terminal hardware limitations[2].

---
[2] For completeness, it is reported that a Japanese company proposed a wideband CDMA system with a TDMA element and without mandatory multi-user detection in the early phases of the UTRA standardisation, but also

Unfortunately, convenient approximations of the error performance of TD/CDMA with JD do not exist yet, since the detection algorithms used are rather complex. An alternative would be to establish the physical layer performance through simulation. The snapshots produced with such simulations, however, are only valid for very specific scenarios, thus not allowing for easy generalisation. It is possible to overcome this limitation, but at the expense of complex interfacing between physical layer simulations and higher layer simulations. These interfacing issues have actually resulted in a string of dedicated publications (e.g. references [227–229]). To adopt such approaches, it is necessary to process the results obtained during physical layer simulations in a particular manner.

In TD/CDMA, to perform joint detection, the receiver must be able to estimate reliably the channels of all users transmitting in the same time-slot. This requires inclusion of training sequences in the burst format, and limits the number of users that can simultaneously access a time-slot and the number of spreading codes available in this slot (note that a single user may transmit on more than one code in a particular time-slot). As a general rule, the fewer the number of users, the more codes are available, but the relationship is not straightforward [90]. Here it is assumed that every user is allocated only one code. This results in a fixed number of codes per time-slot, and thus in a rectangular grid of code-time-slots representing a TDMA frame, as for instance shown in Figure 3.13, each being able to carry a burst or a packet.

With the above considerations on the problems of assessing physical layer performance in mind, the simplest possible model is adopted for TD/CDMA in Section 5.3, namely that of the perfect-collision channel. This is probably also the most commonly used approach for MAC investigations. In this model, if only one user accesses a particular code-time-slot, its burst is assumed to be transmitted successfully, but if more than one user accesses that slot, a collision occurs and all bursts involved in this collision are assumed to be corrupted. This model is very basic; in particular it does not account for MAI. However, since JD will at least partially eliminate the dominant source of MAI, namely intracell interference, it is a reasonable approximation, provided that:

- strong FEC coding is used; and
- the number of code-slots provided per time-slot and the reuse factor are chosen such that the intercell interference level is tolerable even in case of fully loaded cells (i.e. the system is blocking limited).

The major drawback of this 'code-time-slot model' is that individual code-slots in a time-slot are considered to be mutually orthogonal, and it is assumed that even excessive intracell interference created by contending users in a particular time-slot will not affect users holding a reservation in the same slot. Unfortunately, JD cannot completely remove intracell interference, particularly not that of contending users. To model at least qualitatively the impact of non-orthogonality on the protocol operation, the models presented in Sections 5.2 and 5.3 will be combined in Section 5.4. Collisions on individual code-time-slots will again result in the erasure of the bursts involved in the collision, but on top of that, the error performance of all bursts in a particular time-slot will depend on the total level of interference in that time-slot.

---

pointed at the power control problem. Presumably, also the hybrid CDMA/TDMA candidate systems submitted in early phases of the GSM standardisation process (see Reference [3]) did not mandate multi-user detection.

5.1.4 On Capture and Required Accuracy of Physical Layer Modelling

The possibility of the receiver capturing one of several colliding bursts is ignored in the following. The qualitative behaviour of the protocols to be investigated would not be affected by capture. Quantitatively, protocol performance would improve without any particular precautions being required other than modifying update values in the case of backlog-based access control as outlined in Reference [131] (see also Sections 3.5 and 4.11).

On a more general note, it must be repeated that it would be beyond the scope of this book, which is concerned with multiple access protocols, to assess the physical layer performance in various environments in great detail. As long as it is made certain that the suggested protocol can cope conceptually with all possible effects affecting physical layer performance (whether for the better such as capture, or for the worse such as some residual errors), it is justifiable to focus on those effects that have a fundamental impact on protocol operation. These effects are collisions of bursts or packets and, where a CDMA component is considered, the impact of multiple access interference. Correspondingly, while protocol multiplexing efficiency and access delay performance will be investigated, with respect to the limitations of the physical layer models used, it would be unwise to quote any spectral efficiency figures.

5.2 Accounting for MAI Generated by Random Codes

5.2.1 On Gaussian Approximations for Error Performance Assessment

In CDMA systems, MAI is usually generated by a large number of users. Applying the *Central Limit Theorem* (CLT), one would therefore expect its distribution to be Gaussian. If this were the case, and if the variance were known, the approximate *bit error rate* P_e could be calculated using the error function. Pursley proposed to do exactly this in 1977. Furthermore, expanding on a paper from 1976 [244], he provided expressions for the variance in direct-sequence CDMA (DS/CDMA) systems, with random coding and BPSK modulation, as a function of the spreading factor or processing gain X, the number of simultaneous users K and additive white Gaussian noise [245]. This approximation is now commonly referred to as the *standard Gaussian approximation* (SGA) [246].

The CLT in its strictest form states that the sum of a sequence of n zero-mean independent and identically distributed (i.i.d.) random variables with finite variance σ^2 will converge to a Gaussian random variable as n grows large. Using random spreading sequences and assuming perfect power control, such that the power level received from each mobile user at the base station is the same, the CLT in its strictest form can indeed be applied, since each user looks statistically the same to the base station. We would therefore expect the SGA to deliver accurate results when the number of simultaneous users K is large, but might have to expect accuracy problems for small K. Indeed, Morrow and Lehnert found that for small K, when the phases and delays of the interfering signals are random, the MAI cannot be accurately modelled as a Gaussian random variable. Thus, SGA delivers only reliable P_e values when K is large [246]. Particularly inaccurate (overly optimistic) results are obtained for large values of the spreading factor X combined with small values of K.

The shortcomings of SGA can be overcome by an *improved Gaussian approximation* (IGA) proposed by Morrow and Lehnert in Reference [246]. In Reference [137], the same authors demonstrate how to reduce the computational complexity of IGA. However, even this simplified approach entails complex calculations to determine P_e. If a fixed X and perfect power control are considered, and only the MAI to the wanted user created by $K - 1$ other users needs to be accounted for to assess P_e, such $P_e(K)$ values can be calculated once for the desired range of K and then simply looked up when required. Courtesy of the authors of Reference [247], Perle and Rechberger, such results were available to us for the investigations reported in References [136] and [31]. In Reference [247], Perle and Rechberger propose an approach to extend IGA for unequal power levels, which requires establishing first the power level distribution, and then involves similar calculations as in Reference [137]. Again, this approach is quite complex, and would require separate calculations for every scenario that results in a different power level distribution. Furthermore, the approach would need to be further extended to account for interference created by users dwelling outside the test cell considered.

The interested reader is referred to a fairly recent letter by Morrow, which provides a good summary on the issues discussed above and the relevant results reported in Reference [248]. As a matter of fact, we would have welcomed this letter to appear some years earlier. The letter reports successful endeavours by Holtzmann to simplify IGA greatly for equal power reception without compromising too much on accuracy, and provides further simplifications to this approach. According to the letter, this *simplified IGA* (SIGA) can also be extended easily to the unequal power case. Such an extended SIGA could have been attractive for our investigations. However, it is perfectly justifiable to use SGA for our purposes for the following reasons.

- We are mostly interested in small X, in which case the difference between SGA and IGA is not so large (see Reference [246]).

- To maximise the normalised throughput, quite strong FEC coding is applied. In this case, the values of K for which SGA underestimates P_e will result in a packet success rate of 1 anyway, whether SGA or IGA is used [31].

- SGA can easily be extended to unequal power reception. While this violates the i.i.d. requirement of the strongest form of the CLT, a weaker form can be invoked, which requires that at least the variances of the individual contributors should be of the same order[3]. Unequal power reception of intracell interferers (which should dominate intercell interferers) will be due to power control errors. If these errors are small, the variances should indeed be of the same order. If they become too large, problems with limited accuracy of SGA will be overshadowed anyway by severe degradation of system performance to the point where the system being considered becomes useless.

- Finally, physical layer models in investigations dedicated to the MAC layer will always be subject to some simplifications (e.g. we are ignoring multipath fading).

---

[3] Actually, according to Reference [249], the weakest form of the CLT requires that every single contributor shall only make an insignificant contribution to the sum of contributions.

5.2 ACCOUNTING FOR MAI GENERATED BY RANDOM CODES

Small accuracy problems within the framework of the simplified model are negligible compared to the inaccuracies caused by these simplifications.

In the following, the physical layer performance is assessed in terms of the bit error rate or BER and of the packet success probabilities. SGA is used to evaluate the BER.

5.2.2 The Standard Gaussian Approximation

The standard Gaussian approximation for DS/CDMA systems is derived in detail in Reference [245] and makes use of results from Reference [244] to provide a value for the MAI variance for the case of *random coding* and *equal power reception*. Extending SGA to unequal power reception is fairly straightforward. Here, the system model considered is briefly introduced and the SGA expression used for performance assessment is reported. A detailed derivation of SGA in the unequal power case was provided in Reference [31].

The transmitted signal of user k, using BPSK modulation, is written as

$$s_k(t) = \sqrt{2P_k} a_k(t) b_k(t) \cos(\omega_c t + \theta_k), \tag{5.1}$$

with spreading sequence (also called direct or signature sequence) $a_k(t)$, data sequence $b_k(t)$, carrier frequency ω_c, transmitted power level P_k and carrier phase θ_k. The data sequence $b_k(t)$ is made up of positive or negative rectangular pulses of unit amplitude and bit duration T_b. The sequence $a_k(t)$, on the other hand, is also made up of rectangular pulses of unit amplitude, but now with duration T_c, the so-called *chip duration*. Random direct sequences are used, i.e. $\Pr\{a_j^{(\cdot)} = +1\} = \Pr\{a_j^{(\cdot)} = -1\} = 0.5$, where $a_j^{(\cdot)}$ is an arbitrary chip j of the direct sequence. This is why the section title refers to *random codes*. The length of the sequence is equal to the spreading factor or processing gain[4] $X = T_b/T_c$, with $X \in \mathbb{N}$. In other words, the sequence is randomly chosen to spread the first bit, but repeated for subsequent bits.

In a system in which K simultaneous mobile users transmit according to Equation (5.1), the total signal received at the base station can be written as

$$r(t) = n(t) + \sum_{k=1}^{K} \sqrt{2\frac{P_k}{\alpha_k}} a_k(t - \tau_k) b_k(t - \tau_k) \cos(\omega_c t + \varphi_k) \tag{5.2}$$

where τ_k is the propagation delay, α_k is the propagation attenuation experienced, such that the received power level amounts to P_k/α_k, and $n(t)$ is the additive white Gaussian noise (AWGN). Here, $\varphi_k = \theta_k - \omega_c \tau_k + \psi_k$ with $0 \le \theta_k < 2\pi$ and ψ_k the phase-shift due to fading. If user i is to be detected, we can assume $\varphi_i = \tau_i = 0$, as only relative delays and phase angles need to be considered. On the other hand, on the uplink of a mobile communication system it is rather difficult to achieve $\varphi_k = \tau_k = 0$ for $k = 1 \ldots K, k \ne i$. Instead, carrier phases φ_k are assumed to be uniformly distributed in the interval $[0, 2\pi)$, and chip delays τ_k in the interval $[0, T_b)$ for $k \ne i$.

---

[4] If FEC coding is used, the redundancy introduced through coding may be considered as part of the processing gain, in which case spreading factor X and processing gain are not equal.

In the following, it is assumed that the dominant interference contribution is MAI, and AWGN is ignored. The average BER or probability of bit error P_e can then be calculated using

$$P_e \approx Q(\overline{SNR}), \tag{5.3}$$

$$\text{with } Q(x) = \frac{1}{\sqrt{2\pi}} \int_x^\infty e^{\frac{-u^2}{2}} du, \tag{5.4}$$

and the short-term average signal-to-noise-ratio

$$\overline{SNR} = \sqrt{\frac{3X \cdot \frac{P_i}{\alpha_i}}{\sum_{\substack{k=1 \\ k \neq i}}^{K} \frac{P_k}{\alpha_k}}} \tag{5.5}$$

For equal power reception, that is if $P_k/\alpha_k = P$ for $k = 1, 2, \ldots, K$, Equation (5.5) reduces to the well known

$$\overline{SNR} = \sqrt{\frac{3X}{K-1}}. \tag{5.6}$$

Similar expressions reported in Reference [246] and summarised in Reference [248] can be obtained, if either phases, or chips, or both together are aligned. These cases are not relevant for the uplink considered here, but may be of interest for the downlink of a mobile communications system and, with limitations, for the uplink of synchronous CDMA systems such as the synchronous UTRA TDD mode envisaged to be introduced as part of further UMTS developments.

5.2.3 Deriving Packet Success Probabilities

Transmitting a packet of length L bits over a *memoryless* binary symmetric communication channel with average probability of data bit success $Q_e = 1 - P_e$ yields a *probability of packet success* Q_{pe} of

$$Q_{pe} = \sum_{i=0}^{e} \binom{L}{i} (1 - Q_e)^i (Q_e)^{L-i}, \tag{5.7}$$

when a block code is employed which can correct up to e errors. At this point, a problem ignored until now pertaining to both SGA and IGA needs to be addressed. Normally, physical layer design parameters will be chosen such that, at least for mobiles at moderate speed, the channel is quasi static during the transmission of a burst or packet[5]. Because of this, the assumption underlying these approximations, that delays and phases of the interfering users are random, is violated. While they may be randomly selected at the start of a packet, they will essentially remain constant over its duration. This in turn will introduce dependencies between bits in errors or, put differently, the channel will have memory. If one bit is in error, there is an increased likelihood that the next bit

---

[5] In the physical layer context, a packet is equivalent to a burst, i.e. a data unit transmitted in a single (code-)time-slot.

will also be in error, which has also implications on the calculation of the success rate of packets that are protected by error coding. Use of SGA or IGA to establish P_e and subsequent use of Equation (5.7) to establish Q_{pe} could therefore cause inaccurate results.

Such issues are addressed in detail in Reference [246], where a method for calculating packet success probabilities using IGA, which correctly accounts for bit-to-bit error dependence, is introduced. It is shown that, in systems appling error correction coding, the techniques that ignore error dependencies are optimistic for a lightly loaded channel and pessimistic for a heavily loaded channel. In other words, if error dependencies were correctly accounted for, the slope of $Q_{pe}[K]$ depicted in Figure 5.2 (see next subsection) would be flatter. For two reasons, these issues are ignored in the following and Equation (5.7) is resorted to for the calculation of Q_{pe}: application of interleaving and coding over several bursts should at least partially eliminate error dependencies[6]. Furthermore, the impact of flatter slopes on system performance will be investigated anyway in the context of power control errors.

5.2.4 Importance of FEC Coding in CDMA

According to Lee, coding is always beneficial and sometimes crucial in CDMA applications [6]. Results reported in References [137,247,250,251] confirm this statement[7].

In digital cellular communication systems currently operational, a combination of convolutional coding and block coding is often used. Typically, a Viterbi decoder carries the main burden of error correction at the receiving end, thus convolutional coding is applied for error protection. Block coding is then applied in the shape of cyclic redundancy checks, i.e. some parity bits are added, which allow in GSM for instance detection of whether a voice frame is bad or good.

For mathematical convenience, the focus here is on block FEC coding only. As in References [137] and [247], the *Gilbert–Varshamov-Bound* is used to account for the redundancy required to correct a certain number of errors and assess the code-rate r_c which maximises the normalised throughput S. Once r_c is determined, a BCH code [252] with appropriate parameters is selected.

The bandwidth-normalised throughput S is defined as

$$S = \frac{r_c \cdot K_{pe\,\max}}{X}, \qquad (5.8)$$

with

$$K_{pe\,\max} = \max_{K=1,2,\ldots} \left(K \,|\, P_{pe}[K] \leq \left(P_{pe}\right)_{\max} \right), \qquad (5.9)$$

where $P_{pe}[K] = 1 - Q_{pe}[K]$ is the *packet error probability*, and $K_{pe\,\max}$ is the number of users supported at a certain tolerated maximum packet error probability $(P_{pe})_{\max}$. Equal power reception is assumed to determine $P_{pe}[K]$ using Equation (5.7).

---

[6] For a detailed discussion on coding, channel memory, and interleaving, see e.g. Chapter 4 in Reference [3].
[7] Prasad identified certain scenarios in which it makes more sense to increase the processing gain through increase of the spreading factor X rather than to introduce FEC redundancy [26, p. 128]. However, in most scenarios considered, his investigations also underlined the benefits of FEC coding.

230   5   MODELS FOR THE PHYSICAL LAYER AND FOR USER TRAFFIC GENERATION

The Gilbert–Varshamov bound is employed to account for redundancy. According to Reference [253], for any integer d and L with $1 \leq d \leq L/2$, there is a binary (L, B) linear code with a minimum Hamming distance $d_{\min} \geq d$, such that

$$r_c \geq 1 - h\left(\frac{d-1}{L}\right), \tag{5.10}$$

where $h(p) = -p \log(p) - (1-p)\log(1-p)$ is the *binary entropy function*, L is the size of the packet and r_c the code-rate.

Such a code will correct at least

$$e = \frac{d_{\min} - 1}{2} \tag{5.11}$$

errors and have $B = r_c \cdot L$ message bits.

Figure 5.1 shows S as a function of r_c for $X = 7$ and $X = 63$, and values of one per cent and one per thousand for $(P_{pe})_{\max}$, respectively. For reasons outlined in Chapter 7, the number of message bits B is kept at 224. The steplike behaviour of the effective throughput can be explained by the fact that $K_{pe\,\max}$ can only be increased by steps of one user at a time, hence decreasing the code-rate will reduce S in spite of increasing e, as long as no additional user can be supported. The reason why the throughput is higher for $X = 7$ is because self-interference is ignored. Subtracting the desired user, e.g. in the denominator of Equation (5.6), increases the SNR the more, the lower X.

Irrespective of X, the optimal code-rate is in the range of 0.4 to 0.6. BCH codes are efficient block codes and therefore often used. A possible BCH code with $r_c = 0.45$ which supports around 224 message bits is one with $L = 511$ bits, $B = 229$ bits

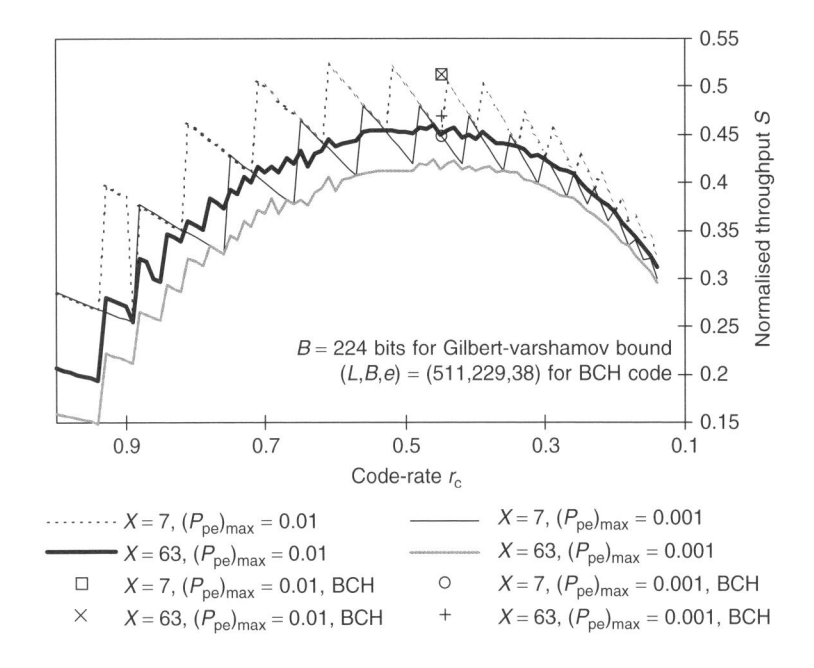

Figure 5.1   Impact of the code-rate on the normalised throughput

5.2 ACCOUNTING FOR MAI GENERATED BY RANDOM CODES

and the capability of correcting up to $e = 38$ errors [252], henceforth referred to as (511, 229, 38) BCH code. With this code and $X = 7$, the packet error rate is zero for $K \leq 7$, while $P_{pe}[8] = 2.6 \times 10^{-4}$, and $P_{pe}[9] = 1.40\%$. With $K \geq 10$, the performance degrades rapidly, and success rates are smaller than 1% for $K \geq 15$, as illustrated in Figure 5.2.

These calculations are intended to establish the best set of design parameters for MD PRMA as investigated in Chapters 7 and 8. With Φ_K a random variable describing the number of terminals per time-slot, S should be maximised taking the distribution of Φ_K experienced under MD PRMA operation into account, rather than trying to maximise $r_c \cdot K_{pe\,max}$ in Equation (5.8) as done here. However, given that access control in MD PRMA and therefore the distribution of Φ_K may depend on $Q_{pe}[K]$, while $Q_{pe}[K]$ in turn depends on r_c, this will prove rather difficult. On the other hand, efficient access control should ensure that K is equal to or close to $K_{pe\,max}$ most of the time. Note also that in References [137] and [251], throughput was maximised over the code-rate for slotted ALOHA and packet CDMA, respectively, thus taking protocol operation into account. The optimum range of values for r_c was found to be between 0.4 to 0.6 and 0.4 to 0.7 respectively, which is in agreement with the above findings.

5.2.5 Accounting for Intercell Interference

Consider a cellular environment with R equally loaded cells and K simultaneously active transmitters in each cell, all cells sharing the same spectrum, thus operating at a frequency reuse factor of one. Assume that perfect power control is employed, such that all the packets transmitted by terminals within any given cell, say for instance cell i, can be received by their base station at equal power level P_0. On the other hand, terminals served by cells other than cell i are power-controlled by their respective base station. The

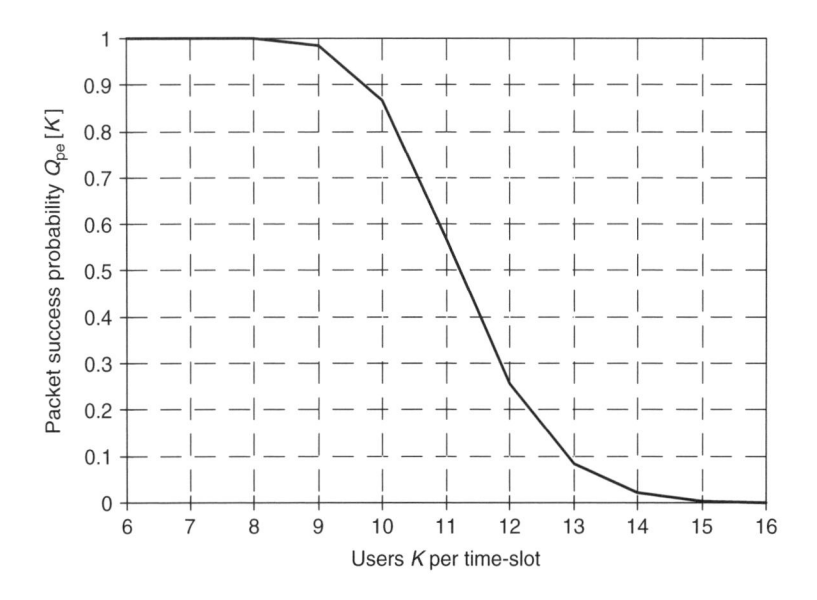

Figure 5.2 $Q_{pe}[K]$ for equal power reception, a (511, 229, 38) BCH code, and $X = 7$

power level $P_{(k,j)_i}$ received at base station i from a mobile k served by base station j will depend on the propagation attenuation from that mobile to both base stations i and j. Assuming perfect power control, the average SNR according to Equation (5.5) can be rewritten for test cell i as

$$\overline{SNR} = \sqrt{\frac{3P_0 X}{\underbrace{(K-1)P_0}_{\text{Intracell}} + \underbrace{\sum_{\substack{j=1 \\ j \neq i}}^{R} \sum_{k=1}^{K} P_{(k,j)_i}}_{\text{Intercell}}}}. \qquad (5.12)$$

If a propagation model with distance-independent pathloss coefficient γ_{pl} and log-normal shadowing is considered, the attenuation is [254]

$$\alpha(r, \zeta) = r^{\gamma_{pl}} 10^{\zeta/10}, \qquad (5.13)$$

where r is the distance between MS and BS, and ζ is the attenuation in dB due to shadowing. The spatial distribution of ζ is commonly modelled as a Gaussian random variable with zero mean and standard deviation σ_s. Typically, for mobile communications, $\gamma_{pl} = 4$ (lower for small cells) and σ_s is around 8 dB. With $r_{(k,j)_i}$ the distance from the mobile (k, j) being considered to the base station in cell i, and $r_{(k,j)_j}$ that to the base station in the serving cell j, as illustrated in Figure 5.3, $P_{(k,j)_i}$ amounts to

$$P_{(k,j)_i} = P_0 \frac{r_{(k,j)_j}^{\gamma_{pl}} \cdot 10^{\zeta_{(k,j)_j}/10}}{r_{(k,j)_i}^{\gamma_{pl}} \cdot 10^{\zeta_{(k,j)_i}/10}}. \qquad (5.14)$$

As mobile (k, j) moves, $P_{(k,j)_i}$ varies not only because of the changing distances, but also due to the values of the shadowing attenuation, which can change if the mobile moves far enough.

Consider a snapshot value of the level of intercell interference at base station i, normalised to the total received power of the K mobiles served by this base station,

$$i_{\text{intercell}} = \frac{1}{P_0 K} \sum_{\substack{j=1 \\ j \neq i}}^{R} \sum_{k=1}^{K} P_{(k,j)_i}. \qquad (5.15)$$

If each mobile is served by the base station with minimum attenuation, such that the intercell interference level is minimised, the normalised level of intercell interference $\overline{I}_{\text{intercell}}$ averaged over shadowing and the mobile locations will depend only on the spatial distribution of mobiles in the cell, the pathloss coefficient γ_{pl}, and the shadowing standard deviation σ_s. In particular, since γ_{pl} is assumed to be distance-independent, $\overline{I}_{\text{intercell}}$ does not depend on the cell radius r_0. It is therefore possible to account for the average intercell interference without having to consider more than one cell by evaluating $\overline{I}_{\text{intercell}}$ for the distribution of mobiles being considered and the values chosen for γ_{pl} and σ_s. Equation (5.12) then simplifies to

$$\overline{SNR} = \sqrt{\frac{3X}{(K-1) + K \cdot \overline{I}_{\text{intercell}}}}. \qquad (5.16)$$

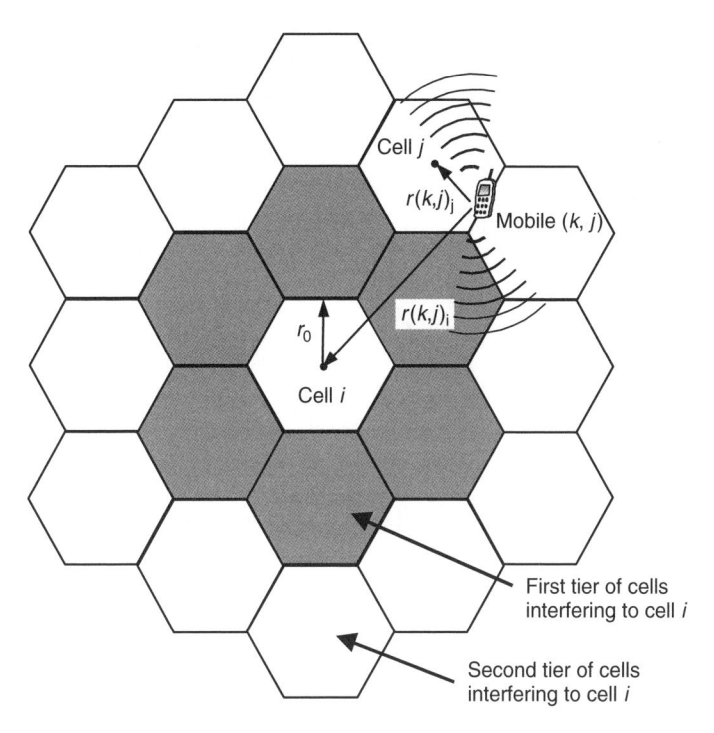

5.2 ACCOUNTING FOR MAI GENERATED BY RANDOM CODES 233

Figure 5.3 Centre test cell i with radius r_0 and two tiers of interfering cells. Mobile (k, j) is served by cell j and creates interference to cell i. Base stations are at cell centres

Due to perfect power control, the common reference power level P_0 disappears. $\overline{I}_{\text{intercell}}$ must be denormalised with $\overline{K}$, which is the average channel load or the average number of users per time-slot in each cell. The attentive reader will have observed a discontinuity: so far, K stood for the number of simultaneously active transmitters in each cell, whether test cell or interfering cells. Now, for Equation (5.16) to be useful for MAC performance investigations, a distinction is being made between the average channel load $\overline{K}$ (here to be assumed the same in the test cell and interfering cells) and the instantaneous number of users K in the test cell. Equation (5.16) can therefore be used to account for slot-to-slot load fluctuations caused by the MAC operation in the test cell, while ignoring such fluctuations in interfering cells.

In Reference [251], $\overline{I}_{\text{intercell}}$ was evaluated based on an approach outlined in Reference [9] for a cellular system with hexagonal cells, uniform distribution of mobiles and various values of the pathloss coefficient γ_{pl}. Shadowing was not considered, which means that the nearest base station is also the serving base station. Ganesh *et al.* report $\overline{I}_{\text{intercell}} = 0.37$ for $\gamma_{pl} = 4$, which is typical for large cells, and $\overline{I}_{\text{intercell}} = 0.75$ for $\gamma_{pl} = 3$, which is more representative for smaller cells, e.g. microcells. Similar values (slightly higher for $\gamma_{pl} = 4$) appear also in Reference [255], where Newson and Heath approximated the hexagonal cells by circular cells of equal area and integrated the interference numerically over two tiers of interfering cells. In Reference [254], on the other hand, $\overline{I}_{\text{intercell}} = 0.44$ and 0.77, respectively (i.e. considerably higher for $\gamma_{pl} = 4$), although the scenario considered by Viterbi *et al.* appears to be the same as that by Ganesh *et al.* at first glance.

A possible reason for these discrepancies could be the exact spatial distribution of the interference considered. Although in all references, uniform distribution is considered, it is not clear from References [9] and [251] whether this is on a per-cell-basis or over all cells. If Monte Carlo snapshot simulations are performed to assess the interference, during which mobiles are repeatedly redistributed over the test area, and the observed interference is averaged in the end, this matters. In Reference [254] for instance, the authors clearly specified that they considered a uniform distribution over the whole area of interfering cells, such that only the expected number of mobiles per cell is the same, while the actual value may fluctuate.

It is well known that CDMA capacity suffers under unequal cell load (see for instance Reference [112], where possible approaches to mitigate the problem through adaptive adjustment of the reference power level at each cell are discussed). Therefore, if interference is averaged over snapshots with unequally loaded cells, one will have to expect higher interference levels than when averaging over snapshots with an equal number of mobiles per cell, even if in both cases the cells are on average equally loaded. Indeed, we managed to reproduce the lower values reported in Reference [251] through Monte Carlo simulations considering the first two tiers of interfering cells[8] (see Figure 5.3) when for every snapshot the *same* number of mobiles were uniformly distributed in *each* cell.

Both in References [254] and [255] intercell interference levels are also evaluated when shadowing is considered. This case is more intricate, since the base station with the lowest attenuation is not necessarily the nearest base station, which complicates the evaluation of $\bar{I}_{\text{intercell}}$. In fact, while Newson and Heath carried out a numerical integration for the case without shadowing, they had to resort to Monte Carlo simulations when accounting for shadowing. In Reference [254], the selection of the serving base station is constrained to one of a limited set of N_c nearest base stations ($N_c = 1, 2, 3$ or 4), which may not include the base station with lowest attenuation. Shadowing from a given mobile to different base stations is assumed to be partially correlated, while no correlation is considered in Reference [255]. In both cases, $\bar{I}_{\text{intercell}} \approx 0.55$ for $\gamma_{pl} = 4$, $\sigma_s = 8$ dB, and $N_c = 4$ (the latter only relevant for Reference [254]). It appears that the constraint to $N_c = 4$ in Reference [254], which should increase $\bar{I}_{\text{intercell}}$ compared to the scenario considered by Newson and Heath, is offset by the partial correlation of the shadowing, which reduces $\bar{I}_{\text{intercell}}$.

For the results presented in Chapter 7, $\bar{I}_{\text{intercell}}$ values of 0.37 and 0.75 were used for $\gamma_{pl} = 4$ and 3 respectively, having the no-shadowing case in mind, where all cells are always equally loaded. In Reference [254], for $\gamma_{pl} = 4$, shadowing with $\sigma_s = 10$ dB, and the case where the serving base station must be among the three base stations closest to the mobile being considered (i.e. $N_c = 3$), $\bar{I}_{\text{intercell}}$ is also reported to be 0.75. Therefore, the results reported in Chapter 7 for $\gamma_{pl} = 3$ without shadowing could also be taken to stand for $\gamma_{pl} = 4$ with shadowing under the conditions just outlined.

Recall that $\bar{I}_{\text{intercell}}$ denotes the normalised averaged intercell interference level assuming the same constant number of simultaneously active mobiles $\bar{K}$ in every cell. Given the large number of interfering mobiles, we would expect $I_{\text{intercell}}$, the random variable describing instantaneous intercell interference levels, to be log-normally distributed. On top of movement of mobiles and fluctuating shadowing attenuation, the fact that the

---

[8] With $\gamma_{pl} \geq 3$, interference from outside the first two tiers of cells is negligible.

5.2 ACCOUNTING FOR MAI GENERATED BY RANDOM CODES 235

number of simultaneously active users in each cell varies from slot to slot will add further fluctuations to the intercell interference levels. These fluctuations will have a flattening effect on the slope of $Q_{pe}[K]$ depicted in Figure 5.2. As similar effects can be observed in the presence of power control errors, which results in fluctuating *intracell* interference even when K is fixed (see Figures 5.4 and 5.5), the assessment of these effects will be performed in the context of power-control errors, and only average intercell interference levels will be accounted for. To determine these levels, it is assumed that the test cell and interfering cells are equally loaded, hence $\overline{K}$ in Equation (5.16) reflects the mean number of users per time-slot in the test cell.

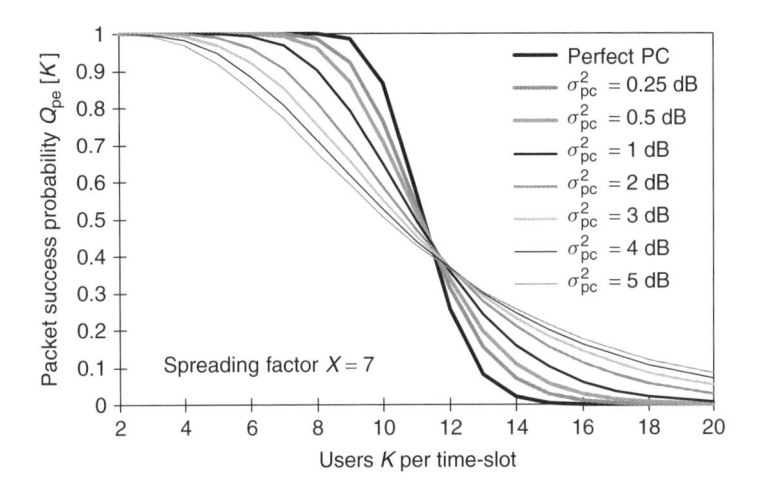

Figure 5.4 Impact of the power control error variance σ_{pc}^2 on $Q_{pe}[K]$ with $X = 7$

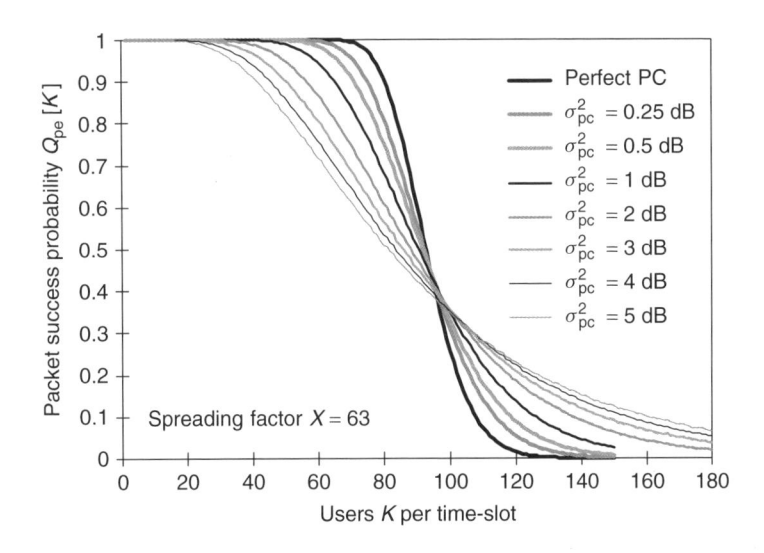

Figure 5.5 Impact of the power control error variance σ_{pc}^2 on $Q_{pe}[K]$ with $X = 63$

5.2.6 Impact of Power Control Errors

In CDMA, particularly with single-user detection, accurate fast and therefore normally closed-loop power control is required. Based on the difference between a target power level and the measured received power level, the base station orders the mobile user to increase or decrease the radiated power level in regularly sent power control commands. Until now, perfect power control has been assumed. In reality, however, power control errors will necessarily occur, since the power level measurement is affected by noise, the resolution and the precision of the power levels radiated by handsets are limited, and power level updates may not be fast enough to track the channel fluctuations. In a hybrid CDMA/TDMA system with closed-loop power control, the main source of power control errors is likely to be the power control delay of at least one TDMA frame[9], since the radiated power level in a particular time-slot will be determined by the measured power level in the same time-slot of the previous frame, or even earlier frames.

These errors are often jointly modelled in literature as a log-normal fluctuation around a reference power level with a certain power control error standard deviation σ_{pc} (e.g. Reference [255]), that is,

$$P_{(k,i)_i}(t) = P_0 \cdot 10^{\varepsilon(t)/10} \tag{5.17}$$

where $P_{(k,i)_i}$ is the power level received at its own base station from user k served by cell i, and $\varepsilon(t)$ is a Gaussian random variable with zero mean and standard deviation σ_{pc}. This model is also adopted here.

As far as the temporal behaviour of the power fluctuation is concerned, it is assumed that $\varepsilon(t)$ is roughly constant during the transmission of a packet. This is in line with the assumption that the channel is quasi-static during a time-slot and the error is mainly due to the power control delay. Define Z as a random variable describing the received power level according to Equation (5.17) and the intracell interference level W experienced as the sum of $K-1$ random variables Z, with z and w as particular realisations of Z and W respectively. If only *intracell* interference is considered, then

$$P_e(z, w) = Q\left(\sqrt{3Xz/w}\right). \tag{5.18}$$

With the quasi-static assumption, z and w can be treated as constant during a packet transmission, hence Equation (5.7) is applicable, which is rewritten here as

$$Q_{pe}(z, w) = \sum_{i=0}^{e} \binom{L}{i} P_e(z, w)^i (1 - P_e(z, w))^{L-i} \tag{5.19}$$

Assuming Z and W to be independent random variables[10], and with $f_Z(z)$ and $f_W(w)$ their respective probability density functions, the average packet success probability can be expressed as

$$\overline{Q_{pe}} = E[Q_{pe}(Z, W)] = \iint Q_{pe}(z, w) f_Z(z) f_W(w) dz dw. \tag{5.20}$$

---

[9] This assumes low-bit-rate users with a resource allocation of only one time-slot per frame.
[10] This is a reasonable assumption, since the processes governing fast channel fluctuations of different users are independent.

In Figures 5.4 and 5.5 average $Q_{pe}[K]$ values obtained through Monte Carlo simulations are depicted for spreading factors of $X = 7$ and 63 respectively, with perfect power control and with power control error variance values σ_{pc}^2 ranging from 0.25 to 5 dB.

It is often argued in the literature that for the purpose of performance assessment of multiple access protocols for CDMA, the packet success probability is modelled with sufficient accuracy by a step function (e.g. References [33,35,36]). Indeed, with perfect power control, and assuming bit-to-bit error independence, the slope of $Q_{pe}[K]$ of sufficiently large packets is relatively steep. However, as can be seen from the figures, the slope of $Q_{pe}[K]$ flattens considerably with increasing σ_{pc}^2, and a step function would be inappropriate in the presence of power control errors. The most important effect of power control errors is a severe degradation in the number of packets supported simultaneously at low error probabilities. For $X = 7$, for instance, with $(P_{pe})_{\max} = 1\%$, $K_{pe\,\max} = 8$ for $\sigma_{pc}^2 = 0$ dB, but only 3 for $\sigma_{pc}^2 = 5$ dB. Further results are listed in Table 5.1. Provided that the abscissa is appropriately normalised, the situation is very much the same with $X = 63$. In fact, Figures 5.4 and 5.5 are virtually indistinguishable.

While the discussion of the impact of power control errors on the multiple access protocols investigated will be limited in Chapter 7 to the single-cell case, in Reference [255] the impact of power control errors on *intercell* interference was also studied. The approximate increase in intercell interference levels, as compared to the perfect power control case read out from Figure 7(a) therein, are 10% and 15% for $\gamma_{pl} = 3$ and 4 respectively.

5.3 Perfect-collision Code-time-slot Model for TD/CDMA

The approach to modelling the UTRA TD/CDMA physical layer through perfect-collision code-time-slots has already been outlined in Section 5.1. In this section, more information on TD/CDMA is provided. For the assessment of multiplexing efficiency and delay performance of data, it is necessary to discuss the physical layer design parameters through which the payload available for user data transfer can be established. Furthermore, the suitability of UTRA TD/CDMA for in-slot protocols such as MD PRMA will be discussed.

5.3.1 TD/CDMA as a Mode for the UMTS Terrestrial Radio Access

As already mentioned in Chapter 2, TD/CDMA was adopted as the UTRA TDD mode. TD/CDMA will have to coexist with WCDMA, the UTRA FDD mode. Since currently the paired frequency allocation is significantly larger than the unpaired one, and since non-European parties mostly focused on wideband CDMA for 3G systems, one can expect WCDMA to be the dominant UTRA mode at least for the immediate future. It was therefore decided that the TD/CDMA parameters should be harmonised with the WCDMA ones, at the expense of losing commonality with GSM. For the research efforts discussed

Table 5.1 Impact of σ_{pc}^2 on $K_{pe\,\max}$

σ_{pc}^2 [dB]	0	0.25	0.5	1	2	3	4	5
$K_{pe\,\max}$ at $(P_{pe})_{\max} = 1\%$	8	7	6	5	5	4	3	3

in the following chapters, the focus is on TD/CDMA as a fully fledged 3G air-interface solution, including FDD and TDD mode, with the original parameters as defined in Reference [90]. This is due to our interest in both FDD and TDD modes of operation of the proposed protocol, and the fact that the relevant investigations were mostly carried out prior to any harmonisation efforts. The harmonised parameters are discussed in detail in Chapter 10. Note that the parameter harmonisation as such has no fundamental impact on MD PRMA performance, although the trunking efficiency will be increased slightly due to almost double the number of code-time-slots available per frame. However, there are a number of issues that arise due to the fact that TD/CDMA will only be used in TDD mode, which will be discussed below.

5.3.2 The TD/CDMA Physical Layer Design Parameters

TD/CDMA parameter values specified in Reference [90] are considered, i.e. prior to harmonisation with WCDMA parameters. $N = 8$ time-slots are grouped in a TDMA frame of duration $D_{tf} = 4.615$ ms, which is the same as in GSM. The time-slot duration D_{slot} of 577 µs corresponds to 1250 chip periods, resulting in a chip-rate R_c of 2.167 Mchips/s. The required carrier spacing is 1.6 MHz. Direct-sequence spread spectrum with a spreading factor $X = 16$ is used. Two burst types for data transfer are defined. Type one has a guard period of 58 chips and a training sequence of 296 chips, leaving 896 chips or 56 symbols as gross payload. Burst type two is only suitable for low delay-spread propagation environments, which allow shortening of the guard period to 55 chips and the training sequence to 107 chips, so increasing the payload to 68 symbols. These basic design parameters, most of them only indirectly relevant for protocol operation, are summarised in Table 5.2.

For the scenario that will be considered in Chapters 8 and 9, where every MS may make use of at most one code per time-slot, up to $E = 8$ spreading bursts or codes may be used in the same time-slot, subject to interference constraints. Focussing on one test cell only and on the understanding that frequency planning is carried out in such a way that the system is blocking limited, all eight codes can be used in a time-slot. This results in a resource matrix with $U = N \cdot E = 64$ resource units or code-time-slots, which are modelled as individual perfect-collision slots (implying that they are mutually orthogonal). It means that error-free transmission is guaranteed if only one user accesses a given code-time-slot, but that all information is lost if two or more users try to access this code-time-slot. The data-rate available for the user depends on the choice of data modulation, burst type, and the amount of forward error correction coding applied.

Table 5.2 Basic TD/CDMA design parameters

Description	Symbol	Parameter Value
TDMA Frame Duration	D_{tf}	4.615 ms
Time-slot Duration	D_{slot}	577 µs
Chip-rate	R_c	2.167 Mchips/s
Carrier Spacing		1.6 MHz
Spreading factor	X	16
Payload per Burst (in Symbols)		56 (type 1), 68 (type 2)

5.3.2 THE TD/CDMA PHYSICAL LAYER DESIGN PARAMETERS

For some of the results presented in Chapter 8, voice-only transmission with a basic MD PRMA scheme is considered, interleaving is ignored, and it is simply assumed that one code-time-slot per user and TDMA frame (i.e. up to 29.5 kbit/s gross bit-rate with QPSK modulation and burst type 2) is sufficient for all data to be transmitted including physical layer and MAC overheads. This scheme has two fundamental shortcomings, though. Firstly, the duration of packets or voice frames, D_{vf}, delivered by the voice coder is typically either 10 or 20 ms, and in both cases, $D_{vf} > D_{tf}$. Secondly, for FEC to be efficient in fast-fading environments, *interleaving* of data over several bursts needs to be applied, as discussed in detail in Chapter 4.

In extended schemes considered in Chapters 8 and 9, where these matters are accounted for, it is assumed that *rectangular* interleaving is applied, that is, adjacent data units are sent in separate and non-overlapping groups of bursts. The data unit is either referred to as an *RLC frame* or as an RLC protocol data unit (RLC-PDU). The group size is determined by the interleaving depth d_{il}, which in turn should be selected to provide the best trade-off between latency and error protection. In the case of voice, an RLC frame may consist of an integer multiple of voice frames, but to keep the delay low, interleaving is assumed to be carried out over the duration of a single voice frame only, i.e. $D_{vf} = d_{il} \cdot D_{tf}$. It is assumed that $D_{vf} = 18.462$ ms, hence $d_{il} = 4$, as suggested in [90][11]. Using QPSK for data modulation, burst type 1 with 56 symbols gross payload, and assuming an FEC code-rate r_c of 0.335 leaves a net payload of 150 bits for each voice frame, enough to support a net voice bit-rate of 8 kbit/s. Depending on the additional MAC overhead specifically required for packet-switched voice transfer, burst type 2 may have to be used, or else the FEC code-rate may have to be increased slightly.

For data services, the data transmitted in d_{il} bursts is referred to as a RLC-PDU. In Reference [90], for the so-called UDD 8 service (unconstrained delay data 8 kbit/s)[12], two PDU-types are suggested. The short PDU for the 'short UDD 8 data service', with $d_{il} = 4$, carries 150 bits subject to the same assumptions made in the case of voice. For services with relaxed delay requirements, d_{il} is increased to 96, resulting in a long PDU carrying 3600 bits for the 'long UDD 8 data service'. For an overview of the parameters relevant for MD PRMA operation, see Table 5.3.

Table 5.3 TD/CDMA Design parameters relevant for MD PRMA

Description	Symbol	Parameter Value
TDMA Frame Duration	D_{tf}	4.615 ms
Time-slots per Frame	N	8
Codes per Time-slot	E	8
Interleaving Depth (in bursts)	d_{il}	4 (voice, short), 96 (long)
Voice Frame Duration	D_{vf}	18.462 ms
Information Bits per Voice Frame		150
Information Bits per Data PDU		150 (short), 3600 (long)

---

[11] To accommodate a more typical D_{vf} of 20 ms with $D_{tf} = 4.615$ ms, an appropriate multi-frame structure similar to that in GSM would have to be adopted, see Chapter 4.
[12] UDD services with higher bit-rates are also defined in Reference [90], but these are not considered, since for performance results presented in later chapters, resource allocation is limited to one code-time-slot per TDMA frame per user.

5.3.3 In-Slot Protocols on TD/CDMA

A major feature of the UTRA TD/CDMA mode is joint detection. To carry out JD, knowledge of the radio channel of all users involved is required. For circuit-switched services, each user is assigned dedicated codes in specified time-slots and dedicated training sequences, which allow reliable channel estimation to be performed (subject to some delay-spread limits). Time alignment (known as timing advance in GSM) may be established upon initial access and maintained during transmission. In the original TD/CDMA concept as described in Reference [90], two approaches were proposed for the random access.

(1) *Random access without time alignment*: use of dedicated short access bursts with long guard periods similar to GSM on dedicated time-slots (the relevant logical channel is termed S-RACH), on which no JD is performed. The S-RACH would be used for initial access and by terminals that are handed-in from cell sites that are not synchronised with the target cell. In small cells, where long guard periods are not required, a shorter burst format occupying only half a time-slot could be used to double the RACH capacity.

(2) *Random access with time alignment*: Use of the normal bursts defined for data transfer. The so-called N-RACH can be mapped onto any resource unit (i.e. it can share a time-slot with traffic channels), and JD can be performed on the respective time-slot.

The main application for the N-RACH appears to be packet-data support, and its existence is in fact the enabling factor for in-slot protocols such as the MD PRMA protocol proposed. Packet-data users would access the S-RACH to establish a logical context and time alignment. Subsequent random accesses to request resources for individual packet transfers would be carried out on the N-RACH which, depending on the cell size, requires time alignment for users with a logical context to be maintained also during silence periods. The solution adopted for GPRS, which would also be suitable for TD/CDMA, is to let these silent users occasionally transmit a burst on what is called 'packet timing advance control channel' in GPRS (see Section 4.9).

There is one issue regarding the N-RACH, or more precisely the case where the N-RACH is mapped onto a code in a time-slot carrying also traffic channels, which is not addressed in Reference [90]. Two contending users may want to use this code at the same time, resulting in a collision. Apart from the increased MAI, which may negatively affect the traffic channels in the same time-slot, there is a problem specifically pertinent to JD in this case, which is a possible deterioration of the error rate of all users in this time-slot due to problems with channel estimation on the N-RACH. It may be argued that since all contending terminals selecting the same code will use the same training sequence, the base station should be able to estimate reliably the joint channel of these users. But to cancel the interference properly, also the user data of the contending terminals need to be known.

In release 1999 of the UTRA TDD specifications, no distinction between S-RACH and N-RACH is made. The burst format used on the RACH is similar to one of the normal burst formats, with the exception that the guard period is doubled at the expense of one of the data fields. The specifications neither explicitly rule out the sharing of a time-slot between RACH and traffic channels, as required for in-slot protocols, nor do they specifically mention such an option. For further information, refer to Section 10.4. For our purposes, it is sufficient to know that, although in-slot protocols can be used with TD/CDMA at least in principle, their performance may be affected negatively, if collisions occur often. Should this be the case, problems can be minimised by limiting the number of codes per time-slot on which a RACH may be mapped, and through appropriate access control, as discussed in Section 6.6.

5.4 Accounting for both Code-collisions and MAI

The model presented in Section 5.2 captures the MAI phenomenon, but since it is based on random coding, MD PRMA with code assignment cannot be studied. The perfect-collision code-time-slot model from Section 5.3, conversely, allows the study of problems related to code assignment, but MAI is not accounted for.

In his investigations on MAC for wideband CDMA, Cao [59] considers a packet to be successfully transmitted if the receiver first *acquires* the respective packet successfully, and then *retains* it. Acquisition is always successful if no other packet is transmitted using the same spreading sequence in the given time-slot. On top of that, if two or more packets use the same code, the receiver might still be able to capture one, if there is a certain minimum delay between these packets. This is referred to as *delay-capture*. The likelihood of the receiver being able to capture a packet is increased by deliberately randomising packet scheduling somewhat with respect to the slot boundaries. Retention depends on the packet success probability in the presence of MAI (both due to intracell and intercell inteference), which Cao assessed according to the approach outlined by us in Reference [30]. In other words, Cao combined our model from Section 5.2 with the notion of code-slots discussed in Section 5.3, and added the possibility of delay capture in the case of code collisions.

For some of the results presented in Chapter 8, we did the same to account for MAI with TD/CDMA, except for capture, which is ignored for reasons mentioned earlier. Perfect power control is assumed, and interleaving is not accounted for, such that BCH codes can be used on a burst level. With TD/CDMA burst type 2 and QPSK modulation, the payload available per burst amounts to 136 bits, which is enough to support BCH codes of length 127. A code suitable for provision of an 8 kbit/s service, which is listed in Reference [252], is the (127, 43, 14) code. In a deviation from the approach outlined in Section 5.2, the traffic load in interfering cells is considered to be constant irrespective of the load in the test cell. It is assumed that on average 80% of the E code-slots available per time-slot are loaded in interfering cells, i.e. $\overline{K} = 0.8 \cdot E$.

Evaluating the generic variance term of the MAI derived for QPSK in Reference [256] for asynchronous spreading sequences and rectangular spreading pulses (i.e. using the same assumptions as in Subsection 5.2.2)[13], an expression similar to Equation (5.16) can be obtained for the average SNR, namely

$$\overline{SNR} = \sqrt{\frac{3X}{2(K - 1 + 0.8 \cdot E \cdot \overline{I}_{\text{intercell}})}}. \tag{5.21}$$

---

[13] Note that the carrier phases disappear in the MAI variance term for QPSK derived in Reference [256].

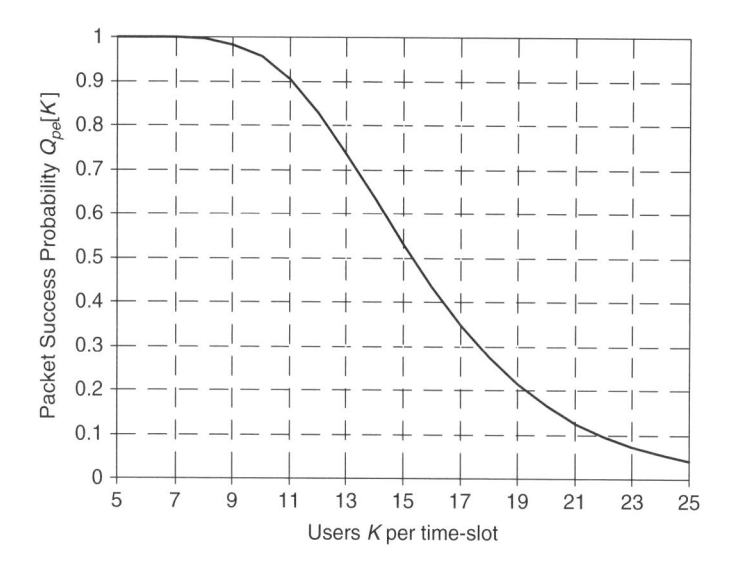

Figure 5.6 $Q_{pe}[K]$ for equal power reception, QPSK modulation, $X = 16$, a (127, 43, 14) BCH code, and constant intercell interference

A path loss coefficient $\gamma_{pl} = 4$ is considered and the reuse factor is assumed to be one, hence $\bar{I}_{intercell} = 0.37$, as reported earlier. *Pro memoria*, $X = 16$ and $E = 8$.

Note that K in Equation (5.21) is the total number of packets accessing a given time-slot (which may be more than the E code-slots available in this time-slot). A packet is successfully received by the base station if it does not collide with other packets on the same code-time-slot, and if it is not erased due to MAI as determined by the burst or packet success rate $Q_{pe}[K]$ obtained through Equations (5.21), (5.3) and (5.7). The corresponding erasure rate $P_{pe}[K]$ for bursts which did not suffer a code collision is below 10^{-5} for $K \leq 6$, 6.5×10^{-4} for $K = 7$, 4.1×10^{-3} for $K = 8$ and 1.6×10^{-2} for $K = 9$. As can be seen from Figure 5.6, the slope of this curve is considerably flatter than that of Figure 5.2, which is due to shorter packets, higher total processing gain (e.g. because of lower r_c) and the assumption of constant-level intercell interference (intercell interference was not considered for Figure 5.2).

This approach to determining physical layer performance accounts for neither the possible benefits of joint detection (i.e. suppression of intracell interference) nor the potential problems with JD in the case of collisions discussed in the previous section. It should therefore not be viewed as suitable for system performance assessment in terms of spectral efficiency. The main purpose is to investigate the impact of non-orthogonal code-slots on backlog-based access control, and the benefit of combining backlog-based access control with load-based access control in the presence of MAI (see Chapters 6 and 8).

5.5 The Voice Traffic Model

5.5.1 Choice of Model

In a conventional telephone conversation as much as in any other conversation between two people, normally only one of the speakers will talk at any one time while the other

5.5 THE VOICE TRAFFIC MODEL

listens. This means that most of the time either only the uplink or only the downlink is used to transfer speech, and the average activity cycle per link, the so-called *voice activity factor* α_v, is expected not to exceed 50%. Looking at activity patterns in more detail, even the participant in the conversation currently speaking will not be active permanently, instead there will be small gaps between individual sentences, individual words, possibly even individual syllables, such that α_v is even further reduced to, say, below 40%. To establish the possible statistical multiplexing gain on the link of interest, namely the uplink, the behaviour of individual speakers needs to be modelled. Brady's well known voice model, dating from the late 1960s, considers both speakers, and models the interdependencies between them by defining model states which depend on the joint state of both speakers, e.g. a state denoting the case where both speakers are talking. For a readily available reference, which expands on the Brady model, see Reference [257].

As we are investigating only the uplink of a mobile communications system in Chapters 7 to 9, interdependencies between the two speakers are not of interest, and a simpler voice model, which depends only on the state of the speaker at the mobile terminal, will suffice as source model for investigations on protocol performance. Furthermore, this model need not take into account the 'real' activity patterns of the speaker, but the patterns as seen by a voice activity detector which can actually be implemented without affecting voice quality significantly. Goodman and his co-authors used two models for this purpose:

- a two-state model, as depicted in Figure 5.8 (see next subsection), with a SILENCE state and a TALK state (the expressions 'talk gap' and 'talk spurt' may be used to indicate that a speaker is in SILENCE and TALK state respectively), as classified by a *slow* voice activity detector, taking into account talking, pausing and listening of the speaker; and

- a three-state model relevant for a *fast* voice activity detector, in which two SILENCE states are discerned, namely PRINCIPAL SILENCE and MINI SILENCE, the latter accounting for the punctuation of continuous speech.

For two reasons, we concentrate on the two-state model in the following.

- Currently operational voice activity detectors are slow, they cannot detect mini-silences or mini-gaps. This is partially due to the normally noisy environment in which communications with cellular phones take place, which makes a reliable detection of voice gaps and spurts difficult. According to Reference [257], better voice activity detectors can be expected for 3G systems, which can reliably detect mini-gaps even in noisy environments. However, we would expect that, from a physiological point of view, tightly gated speech might tend to create uneasiness, particularly when voice with background noise alternates with pure silence during voice gaps. In GSM, for instance, comfort noise parameters are transmitted during voice gaps in order to recreate the background noise at the receiving end [190]. These need to be extracted at the beginning of a gap. For these reasons, the GSM voice activity detector interrupts the generation of voice frames only after a so-called overhang period, which lasts roughly 100 ms.

- In PRMA, only limited gains in terms of number of simultaneous conversations supported can be achieved with a fast voice activity detector[14]. The fact that terminals

---
[14] The situation is slightly different with PRMA on hybrid CDMA/TDMA, as discussed in Chapter 7.

244 5 MODELS FOR THE PHYSICAL LAYER AND FOR USER TRAFFIC GENERATION

give up their reservations during a mini-gap and therefore have to contend for every mini-spurt, leading to increased dropping during a principal spurt, offsets the lower activity factor in certain cases [142].

For a complete description of the voice model, the distribution of the sojourn time in each state needs to be specified. Almost always, a (negative) exponential probability distribution function $f_T(t) = \lambda e^{-\lambda t}$ with mean $1/\lambda$ as shown in Figure 5.7 is used for investigations on multiple access protocol performance. The exponential distribution is indeed convenient. It is easy to handle both for simulations and analytical derivation of protocol performance, since it is memoryless.

Being convenient does not necessarily mean being accurate, though. For instance, we cannot realistically expect the mode[15] of a real talk spurt distribution to be at zero (as is the case with an exponential distribution). This was also confirmed by investigations carried out during the LINK ACS project on voice activity in GSM. On the other hand, a better model based on a mathematical description could not be identified in LINK ACS, and since we did not want to resort to sampled voice traces, we joined the large camp of researchers putting up with the exponential model. This still appears to be more realistic than the model used in Reference [154], where the sojourn time is assumed to be uniformly distributed with certain minimum and maximum values (although it may be worthwhile introducing a minimum value below which the duration of gaps and in particular spurts may not fall).

5.5.2 Description of the Chosen Source Model

The two-state voice model considered has two exponential sojourn times of mean duration D_{spurt} and D_{gap} for TALK and SILENCE state respectively, and can either be depicted as a continuous-time or a discrete-time Markov model. In Figure 5.8(a) a two-state

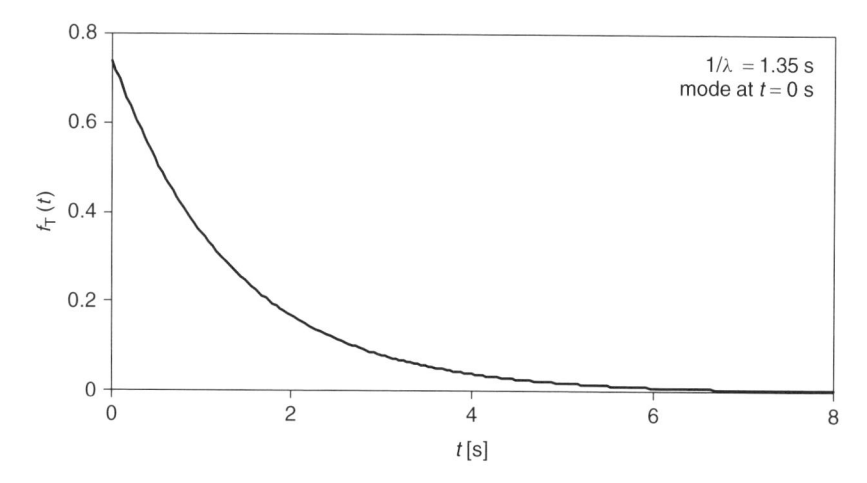

Figure 5.7 (Negative) exponential probability density function used to model voice state sojourn times

---

[15] The mode of a distribution is the abscissa value t at which $f_T(t)$ assumes its maximum.

5.5 THE VOICE TRAFFIC MODEL

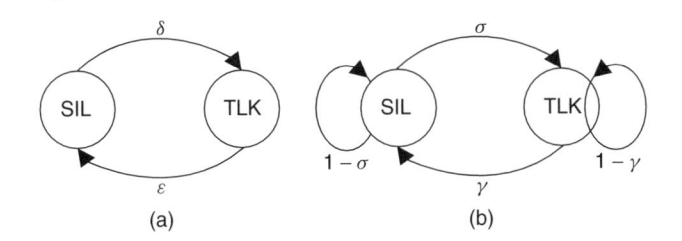

Figure 5.8 Two-state voice models. (a) Continuous-time Markov model. (b) Discrete-time Markov chain

continuous-time Markov model is depicted, with rate of change $\delta = 1/D_{\text{gap}}$ and $\varepsilon = 1/D_{\text{spurt}}$. Alternatively, in Figure 5.8(b) a discrete-time Markov chain representation of the voice source is depicted, with σ, γ, $1 - \sigma$, and $1 - \gamma$ being the probabilities of transition at the end of a time-slot of duration D_{slot}, where

$$\gamma = 1 - e^{(-D_{\text{slot}}/D_{\text{spurt}})}, \tag{5.22}$$

and

$$\sigma = 1 - e^{(-D_{\text{slot}}/D_{\text{gap}})}. \tag{5.23}$$

For simulation purposes, since an event-driven commercial simulation package is used, it is convenient to implement the continuous-time version depicted in Figure 5.8(a). Obviously, as a time-slotted system is considered, the time of arrival is eventually rounded up to the next slot-transition for access to the channel. However, for analysis purposes, the discrete-time version of the model is considered, as it is more convenient. Note, however, that as long as $D_{\text{slot}} \ll \min(D_{\text{spurt}}, D_{\text{gap}})$, the difference is marginal. For instance, with $D_{\text{spurt}} = 1$ s and $D_{\text{gap}} = 1.35$ s as in Reference [142], the resulting voice activity factor $\alpha_v = \delta/(\delta + \varepsilon) = 0.425532$ with the continuous time model is very well approximated by that of the discrete-time model with time-step $D_{\text{slot}} = 1$ ms, which is $\alpha_v = \sigma/(\sigma + \gamma) = 0.425564$.

Values found in the literature for D_{spurt} range typically from 1 to 3 s, for D_{gap} from 1.35 to 3 s for slow voice activity detectors. Goodman and his co-authors used 1 s and 1.35 s respectively, throughout their numerous publications on PRMA, resulting in the activity factor α_v of 0.426 derived above (note that the parameter chosen for the distribution shown in Figure 5.7 is $D_{\text{gap}} = 1/\lambda = 1.35$ s). The values used in the European RACE ATDMA project were 1.41 s and 1.74 s respectively [46,11], increasing α_v to 0.448. Finally, in Reference [56], 3 s is suggested for both spurt and gap duration. We used first the Goodman values, but during the course of our investigations, these other values were also considered, as specified in the relevant parts of this book. It is important to note that the packet dropping rate of PRMA as a function of traffic *normalised* to the activity factor does not depend significantly on the exact choice of these values. Exceptions are high load conditions, where the reduced number of spurts in the case of 3 s duration reduces dropping slightly.

5.5.3 Model of Aggregate Voice Traffic

A finite population of M voice terminals is considered, all involved in a conversation, as is common in investigations on PRMA multiplexing efficiency. This means that the number of ongoing conversations M per simulation run is fixed. Compared to an approach where call arrivals are modelled and the generated traffic is measured in Erlangs rather than in the number of ongoing conversations, this allows saving of simulation time. Furthermore, the results obtained can be converted to equivalent Erlangs, if required, as will be discussed in more detail in Section 7.5. According to Reference [258], the aggregate traffic of M such two-state or on–off sources can be represented through an $M+1$-state birth–death process depicted in Figure 5.9(a), where the state denotes the number of simultaneously active terminals T, and the rate of change between states is depicted. Given the properties of the Poisson process (e.g. Reference [104, p. 164]), this is immediately apparent.

Making use of results provided by Kleinrock in Reference [259], Maglaris et al. report in Reference [258] the steady-state distribution of this birth–death process given M terminals in total as

$$\Pr\{T = m\} = P_m = \binom{M}{m} \cdot \alpha_v^m \cdot (1 - \alpha_v)^{M-m}, \qquad (5.24)$$

where $\alpha_v = \delta/(\delta + \varepsilon)$. The mean $\overline{T}$ of this distribution is $M \cdot \alpha_v$. This result can easily be verified through application of the detailed balance equations valid for such a birth–death process [104, p. 261] to the discrete-time Markov chain shown in Figure 5.9(b). This is demonstrated in Reference [61, Appendix A].

The restriction of the discrete-time model depicted in Figure 5.9(b) is that it allows only changes between neighbouring states in a single time-slot. However, again provided that $D_{\text{slot}} \ll \min(D_{\text{spurt}}, D_{\text{gap}})$, the probability of a transition to a non-neighbouring state is negligible.

5.6 Traffic Models for NRT Data

5.6.1 Data Terminals

For data traffic, unlike for voice traffic, an infinite population of mobiles is assumed in the sense that data sessions are generated centrally according to a Poisson process and

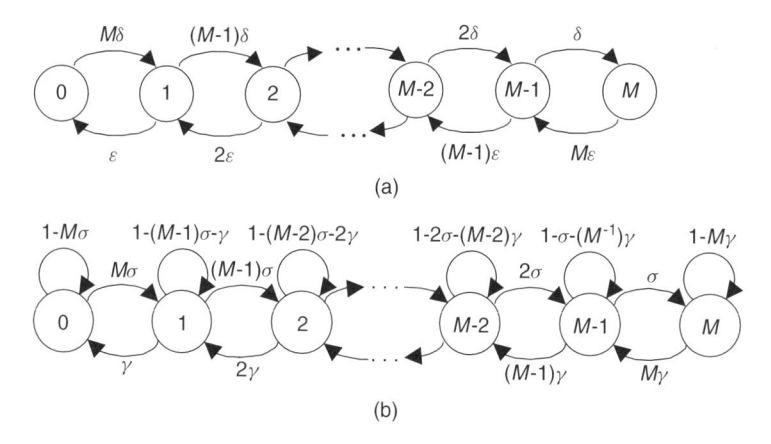

Figure 5.9 (a) Aggregate voice model as continuous-time $M+1$-state birth–death process. (b) Aggregate voice model as discrete-time $M+1$-state birth–death process

5.6 TRAFFIC MODELS FOR NRT DATA

each session is associated with a new mobile terminal. This is a necessity arising from the structure of the data model taken from Reference [56], and has implications on the simulation effort required, as will be discussed in more detail in Chapter 9.

Each new session is either a Web browsing session (specified in the next subsection), or corresponds to the transmission of a single email message, as specified in Subsection 5.6.3. This is determined by a random (Bernoulli) experiment carried out upon session arrival, with parameters according to the chosen fraction of traffic for each service type.

5.6.2 The UMTS Web Browsing Model

To select an air interface for UMTS, selection procedures were established as defined in Reference [56], which include the scenarios, channel, and traffic models to be considered for the overall performance assessment of the different UTRA candidates. Among the traffic models, a model for Web browsing was proposed in which a session is made up of several packet calls that in turn contain multiple packets or *datagrams* (Figure 5.10).

This model was suggested for both link directions, and will therefore be used for the uplink direction in our investigations. The number of packet call requests per session N_{pc} is a geometrically distributed random variable with a mean number of packet calls $\mu_{Npc} = 5$. The reading time, D_{pc}, between two consecutive packet call requests in a session, which starts when the last packet of a call is completely received by the receiving side, is distributed according to a geometrical distribution (in terms of simulation time steps, here D_{slot}) with mean $\mu_{Dpc} = 4$ s. Since $D_{slot} \ll \mu_{Dpc}$, an exponential distribution with mean μ_{Dpc} can be used instead. The number of packets in a packet call N_d is again geometrically distributed and has a mean μ_{Nd} of 25 packets. For the time interval D_d between the start instances of two consecutive packets inside a packet call, again an exponential distribution is used here with mean $\mu_{Dd} = 0.5$ s instead of the geometric distribution suggested in Reference [56].

Finally, the size S_d of packets or datagrams in bytes is modelled using a Pareto distribution with probability density function

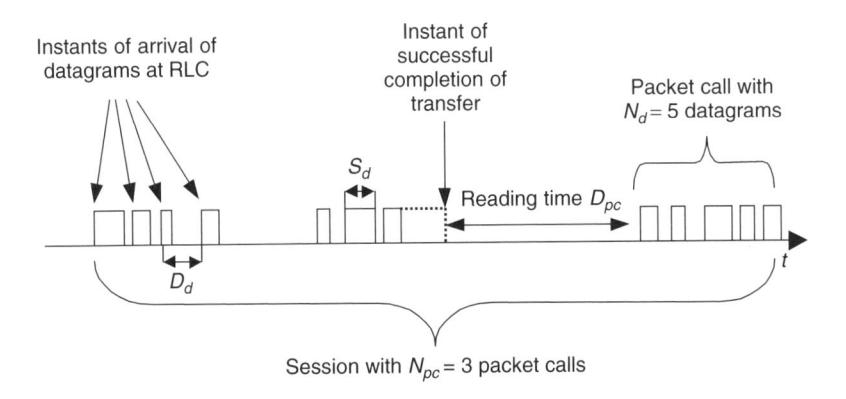

Figure 5.10 Session, packet calls, reading time between calls, and individual packets or datagrams with size S_d and interarrival time D_d

248 5 MODELS FOR THE PHYSICAL LAYER AND FOR USER TRAFFIC GENERATION

$$f_X(x) = \begin{cases} \dfrac{\lambda \cdot (e^\beta)^\lambda}{x^{\lambda+1}} & x \geq e^\beta, \\ 0 & x < e^\beta, \end{cases} \tag{5.25}$$

as shown in Figure 5.11 and cumulative distribution function

$$F_X(x) = \begin{cases} 1 - \left(\dfrac{e^\beta}{x}\right)^\lambda & x \geq e^\beta, \\ 0 & x < e^\beta. \end{cases} \tag{5.26}$$

Note that these packets or datagrams need to be segmented into smaller packets or RLC frames by the RLC sub-layer before they are handed down to the MAC sub-layer, and hence constitute a packet spurt for the multiple access protocols considered. To avoid confusion with the 'packets' known from PRMA, in the following, only the term *datagram* will be used in the context of Web browsing, although 'packet' is the newer term used with IP version 6 [169,260]. If X is Pareto distributed with parameters e^β and λ, then $\ln(X) - \beta$ is distributed according to an exponential distribution with parameter λ. The mean of the Pareto distribution is

$$\mu = \dfrac{\lambda(e^\beta)}{\lambda - 1}, \quad \lambda > 1, \tag{5.27}$$

and the variance is only finite for $\lambda > 2$.

In Reference [56], e^β is equal to 81.5 and λ is chosen to be 1.1, which would result in an infinite variance. In order to ensure finite variance, a *truncated* Pareto distribution is used, where all outcomes greater than a certain threshold value c are set equal to c, i.e.

$$f_X(x) = \begin{cases} 0 & x < e^\beta, x > c \\ \dfrac{\lambda \cdot (e^\beta)^\lambda}{x^{\lambda+1}} & e^\beta \leq x \leq c \\ P_c & x = c \end{cases} \tag{5.28}$$

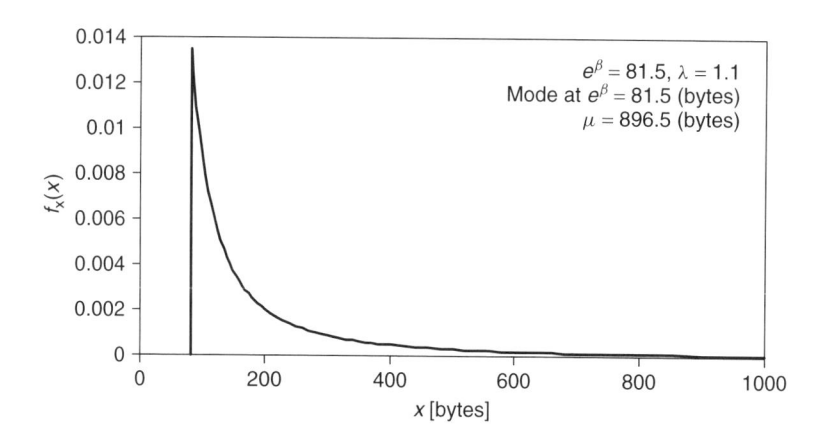

Figure 5.11 Probability density function of a Pareto distribution

where

$$P_c = \left(\frac{e^\beta}{c}\right)^\lambda. \qquad (5.29)$$

The mean of such a truncated Pareto distribution is

$$\mu = \frac{\lambda \cdot e^\beta - c(e^\beta/c)^\lambda}{\lambda - 1} \qquad (5.30)$$

and the variance (provided that $\lambda \neq 2$) is

$$\sigma^2 = \frac{\lambda \cdot e^{2\beta} - 2c^2(e^\beta/c)^\lambda}{\lambda - 2}. \qquad (5.31)$$

Accounting for the maximum IP packet size of 64 Kbytes reported in Reference [260] and presumably overhead which may be added by the layers between the IP-layer and the MAC, c is set to 66 666 bytes in Reference [56]. In this case, the mean before truncation (5.27) is $\mu = 896.5$ bytes and after truncation (5.30) $\mu_{Sd} = 481$ bytes, which shows the impact of a few very large packets on the mean of the Pareto distribution. In fact, the main feature of a Pareto distribution compared to a shifted exponential distribution is its elongated tail (note that the highest abscissa values shown in Figure 5.11 barely extend beyond the mean!). With these parameters, traffic is generated at an average bit-rate of roughly 8 kbit/s (481 bytes every 0.5 s) during a packet call. As our focus is restricted to the allocation of single physical channels, that is, in the case of TD/CDMA, a single code-time-slot, with parameters according to Table 5.3, this average bit-rate is equivalent to the allocated data-rate. A datagram with a maximum size of 67 Kbytes (we used a slightly higher value for c in our simulations than that listed in Reference [56]) will therefore require 68 s for transfer. Given the non-negligible probability ($>0.6 \times 10^{-4}$) that datagrams with this maximum size are generated, it is evident that queues of significant size have to be introduced to avoid the dropping of datagrams in a packet call.

A queue that can hold up to 50 datagrams (regardless of their size[16]) is included here with every data terminal. This does not completely avoid dropping of datagrams, but a dropping level of 1% is rarely exceeded. It is assumed that the RLC layer delivers its frames or PDUs to the MAC layer only immediately before the latter is ready to transmit them over the air interface (via physical layer, that is). The queue must therefore be situated above the MAC layer. Since the MAC performance is measured in the following chapters as a function of the traffic to be handled by the MAC, and not as a function of the session arrival rate, dropping of datagrams only affects the traffic model slightly, but does not distort the MAC performance assessment as such.

This UMTS Web browsing model was inspired by traffic models presented in References [261] and [262], both suggesting an activity burst to be composed of several datagrams. The model in Reference [261] can be represented as a semi-Markov chain and provides three levels of resolution in time, namely connection, action, and transmission. Here, the connection level represents session arrivals, the action level models the alternating behaviour between off-periods (i.e. reading) and on-periods (bursts of activities, here packet calls). Finally the transmission level describes the arrival process of

---

[16] This may appear strange, but for simulations, a datagram is simply an entity with a number of attributes associated, such as size and time of arrival. The memory required per datagram is independent of its size.

datagrams. Essentially, the UMTS model can be viewed as the infinite-terminal version of this model, since session arrivals are modelled globally as a single Poisson process rather than on a per-terminal basis as in Reference [261].

The main influence Reference [262] seems to have had on the definition of the UMTS Web browsing model is with regards to the reading time distribution. Anderlind and Zander suggest an exponential distribution with a mean value of 10 s. While values observed in practice may often be significantly longer, they state that 'the chosen value is sufficiently long to necessitate a release of unused resources and short enough to be practical for simulation studies'. In fact, the value first chosen for μ_{Dpc} reported in earlier versions of Reference [56] was only slightly larger, namely 12 s. However, in an attempt to limit simulation effort (which was well justified, judging from our experience), this value was subsequently reduced to the current value of 4 s.

Further modifications to the original set of traffic parameters were due to the fact that the truncation of the Pareto distribution, which results in a significant reduction of the mean size of a datagram, was not properly accounted for in early versions of Reference [56]. In order to generate the amount of data per session initially intended, the average number of datagrams per packet call had to be increased from 15 to 25, and to reach an average bit-rate of 8 kbit/s during packet calls, the mean time between arrivals of datagrams was reduced from 0.96 s to 0.5 s. While we started our investigations with the old set of parameters, to adhere to the latest version of the selection criteria, we subsequently adopted the new values. Given the effort required for simulations, it was not feasible to produce reliable results for both sets. However, based on the preliminary results obtained with the original set, it can be deduced that the qualitative behaviour of the multiple access protocol investigated is not affected by this alteration of parameters, and even the quantitative differences are small, if the results are normalised to the generated traffic.

Normally, the aggregate arrival process of traffic generated by a large number of sources is approximately Poisson, provided that the distribution of interarrival times at each user satisfies the condition $dF_X(0)/dx > 0$ and that interarrival times are independent [104, p. 165].

Here, interarrival times between the last datagram of a packet call and the first datagram of the following call depend on the service time, which in turn is Pareto distributed, since the service-rate used is constant. Therefore, $dF_X(0)/dx = 0$. Worse, since there is a significant probability that the transfer of a datagram is not completed before the next datagram in the same packet call arrives, there are even dependencies between interarrival times within a packet call. It will therefore be rather difficult to derive an appropriate aggregate traffic model. Even if it were not, since this traffic model is used to investigate protocol performance under mixed traffic scenarios, when prioritisation at the random access is employed, a protocol performance analysis would become prohibitively complex. The performance will therefore only be assessed through simulations.

5.6.3 Proposed Email Model

Several attempts were made in the LINK ACS project to model email traffic using email log files of the different partners involved. From these files, a minimum *email message size* of around 500 bytes can be deduced, which is mainly due to a minimum amount of header information required in an email. Furthermore, due to the increasing number of attachments in business traffic, histograms of these log files, such as the one depicted

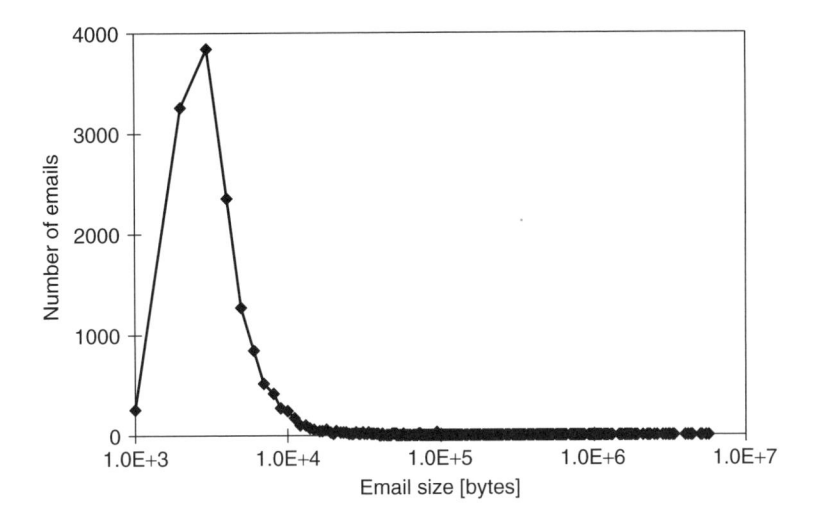

Figure 5.12 Example of a histogram for an email log file with 15 000 samples

in Figure 5.12 with S_d up to 5.6 Mbytes, normally feature a distinctly elongated tail. Correspondingly, mean email sizes are large. Therefore, a prime candidate distribution to model email size is the Pareto distribution. There are, however, some problems.

First, according to Equation (5.25) both minimum size and mode are equal to e^β, whereas from histograms with bin size 100 bytes one obtains values between 1600 and 1800 bytes for observed modes (note that the bin size in Figure 5.12 is 1000 bytes and, correspondingly, the mode shown is different). This is significantly larger than the minimum size. Next, a small minimum size e^β does not preclude an arbitrary large mean μ (choosing λ close to 1) according to Equation (5.27), which allows these two important features of observed email statistics to be captured. One would thereby, however, ignore the fact that emails are in practice subject to size limitation and have to live with an infinite variance, which is somewhat of an obstacle when trying to get meaningful performance results through simulations. With the Pareto distribution truncated at c, on the other hand, the choice of e^β imposes an upper limit on μ_{Sd}, which will make it impossible in practice to fit both minimum size and mean of observed distributions for any reasonable value of c, say 500 Kbytes. Choosing $e^\beta = 500$ bytes to fit the observed minimum size would result in $\mu_{Sd} < 4000$ bytes, which is much lower than for instance the mean size of 26 970 bytes over the 15 000 emails considered in Figure 5.12.

Summarising, it is not possible to fully fit minimum size, mode, and mean of the observed traffic concurrently with a truncated Pareto distribution. Correspondingly, the only successful attempt within LINK ACS to properly fit a Pareto distribution to the observed message size distribution was based on a small set of samples, and with a rough resolution of integer Kbytes for the message size, such that the mode and the minimum value happened to coincide.

At this point, it is worth mentioning that the data collected on email size is considerably affected by size limitations at the gateways of different companies. Furthermore, traffic statistics are likely to be affected by the change of medium from wire to air; the latter normally associated with rather low bit-rates and high costs. In particular, users will be more reluctant to transmit very large emails over the air than over the fixed infrastructure.

With these comments in mind, we might ignore emails larger than 500 Kbytes, which reduces the observed mean to 9450 bytes. With $\lambda = 1.01$, and $c = 500$ Kbytes, this mean can now be matched approximately by choosing $e^\beta = 1400$ bytes, which yields $\mu_{Sd} = 9423$ bytes according to Equation (5.30).

Obviously, this choice of e^β is a compromise somewhere between observed mode and observed minimum value, and what remains unsatisfactory is that no emails with $S_d < 1400$ bytes can be generated with this model. However, only long PDUs carrying 450 bytes each (see Table 5.3) will be used for email traffic, such that the smallest possible email (i.e. approximately 500 bytes in size) will have to be carried on two PDUs and effectively occupies 900 bytes (the last 400 consisting of padding bytes). In any case, compared with the fundamental limitations of a model for UTRA email traffic based on observations from computer networks, this compromise appears acceptable and allows adherence to the model structure for Web browsing. We assume that only a single email message is generated per session and do not account for the possible segmentation of this message into multiple packets at this stage, hence $\mu_{Npc} = \mu_{Nd} = 1$, and the values for μ_{Dpc} and μ_{Dd} are irrelevant. Table 5.4 lists the chosen parameters for Web browsing and email traffic.

5.6.4 A Word on Traffic Asymmetry

It is expected that unlike in 2G systems, which predominantly carry voice traffic, the distribution of the traffic load between uplink and downlink will be very unequal in 2.5G and 3G systems carrying a substantial fraction of data traffic. This expectation is based on the assumption that the dominant share of traffic to be carried will stem from information retrieval applications, for instance Web browsing. In such applications, the traffic generated by the retrieving entity consists mainly of short request and acknowledgement packets, while the dominant traffic flow is towards the retrieving entity. Since mobile terminals will in most cases be retrieving entities, traffic is expected to be downlink biased.

The traffic model in Reference [262] captures this asymmetry, while the UMTS traffic model described above does not. In fact, given the amount of generated data, it seems to be representative for the link to the retrieving entity. One could therefore object to the use of this model for the uplink. On the other hand, with increasing fixed-mobile convergence,

Table 5.4 Summary of model parameters for Web browsing and email traffic generation as relevant for TD/CDMA

Description	Symbol	Web-value	Email-value
Mean number of packet calls/session	μ_{Npc}	5	1
Mean reading time between packet calls	μ_{Dpc}	4 s	n/a
Mean number of packets/packet call	μ_{Nd}	25	1
Mean interarrival time of packets	μ_{Dd}	0.5 s	n/a
Minimum size of message/datagram	e^β	81.5 bytes	1400 bytes
'Scale factor' of Pareto distribution	λ	1.1	1.01
Maximum size of message/datagram	c	67 Kbytes	500 Kbytes
Mean size of message/datagram	μ_{Sd}	481 bytes	9423 bytes
Information bits per RLC-PDU		150	3600
Number of bursts per RLC-PDU		4	96

there is no reason to believe that the servers will never be at the mobile end. We would not be surprised to see relatively soon some of the very popular Web cameras installed on mountain peaks or other unwired exotic locations to entertain Web browsers. More importantly, future IP-based applications could become increasingly interactive, thereby increasing uplink traffic. This has already been observed recently with the advent of peer-to-peer computing and distributed information storage on many PCs, which lead to a significant increase of 'uplink' traffic in fixed access networks. The above does not suggest that the total traffic will necessarily be balanced, it merely justifies, at least to some extent, the use of the UMTS traffic model for uplink performance investigations.

5.6.5 Random Data Traffic

In PRMA, a random data traffic category is defined for traffic that is always transmitted in contention mode. In the traffic model considered for instance in Reference [144], every data terminal generates in each time-slot a packet with probability σ_d (independently from slot to slot). We used this model to assess the deterioration of voice quality with increasing fraction of random data traffic for load-based access control with MD PRMA, and published these results in Reference [30]. However, for two reasons they will not be reproduced here and random data will not be considered.

(1) The approach to access-control of data traffic was rather basic, more elaborate approaches were in the meantime proposed by Mori and Ogura in References [36] and [37], and by Wang *et al.* in Reference [40]. Compared to the earlier results in Reference [30], these approaches significantly improve voice packet-loss performance in the presence of random data traffic.

(2) Normally, data traffic is handed from higher layers to the MAC layer in units significantly exceeding the size of the 'packets' known from PRMA, which can be transmitted on a single time-slot. With TD/CDMA, for instance, with the values specified in Table 5.4, more than 100 bursts would be required on average for the transmission of an IP datagram, and more than 2000 for an average email message, hence reservations certainly make sense. The random data traffic model and correspondingly the random data category of PRMA appear only to be relevant for certain specific applications, such as transfer of some signalling messages (e.g. channel measurement parameters, or comfort noise parameters which need to be transmitted during a voice gap).

5.7 Some Considerations on Video Traffic Models

We used a very simple video traffic model for investigations of MD PRMA performance with mixed voice and video traffic in References [31] and [30]. This model was based on the random data traffic model above, with σ_d chosen in a manner such that the average bit-rate assumed a value of 144 kbit/s (i.e. the maximum theoretically possible bit-rate in narrowband ISDN). With the design parameters considered, σ_d had to be set to 0.9, and for access purposes, it was assumed that video terminals had the right to transmit a packet in every time-slot of every frame, hence effectively using circuit-switching with time-slot aggregation. The spare 10% of the time-slots were available for retransmission of packets corrupted by MAI. Possible implications on quality due to the additional delay inflicted by retransmissions were ignored, since an ideal scenario was considered, in which the terminal would instantaneously know whether its packet was transmitted successfully. If not, it would immediately schedule a retransmission for the next time-slot.

Based on the experience made in the wired world, in particular the observed general lack of customer acceptance for video telephony, one could question the success potential of real-time *interactive* video services in 3G systems[17]. However, already today, there are mobile phones with integrated video cameras on the market. Moreover, for the business market, video conferencing might be an application of potential interest. Certainly, a fundamental technical requirement for such services to become a success is a reasonable service quality at the bit-rates supported by 3G systems. With respect to the fact that a PCM-encoded video sequence would require bit-rates of hundreds of Mbits/s [258], it is obvious that compression algorithms will have to be applied to video signals which achieve considerable compression factors, without degrading quality too much. Even so, it will not be easily possible with 3G systems to match frame-rates and picture size (or picture resolution) of a standard TV signal.

Compression factors, which can be achieved within certain quality limits, depend very much on the complexity of the scenes and the speed of movements in the video sequence. In the case of real-time communication, if some fixed distortion compared to the uncompressed frames can be tolerated, the bit-rate of the coded sequence will be time-varying [258]. To carry such compressed variable-bit-rate (VBR) traffic streams efficiently on a communication network without having to reserve resources according to the peak bit-rate of each user, several video communications should ideally be multiplexed onto a common resource in packet-switched transmission mode. In order to keep delay and packet dropping at moderate levels, while still being able to achieve some statistical multiplexing gain, the mean bit-rate of video traffic sources must be considerably lower than the total bit-rate available on the common resource (e.g. a carrier) used. With carrier bandwidths of a few Megahertz, the mean bit-rates should rather not exceed the 144 kbit/s mentioned above, and preferably even be lower.

Statistical multiplexing of packet-video communications has received considerable attention in the last few years. Reference [258] is often referred to in more recent publications and provides a good starting point. However, most of the available literature is focused on compression schemes which achieve mean bit-rates of the order of a few Mbits/s, and are suitable for packet-switching on wired ATM networks. Only fairly recently, two suitable low- but variable-bit-rate video compression schemes were standardised, namely MPEG-4 and H.263. They are flexible in that the mean bit-rate can be traded off against quality. The output bit-rate of H.263, for instance, depends very much on rate-control, and can be constant, albeit at the cost of variable QoS. In the context of multimedia traffic streams supported by H.324, H.263 is used in CBR fashion, while with H.323 it is typically used in VBR fashion[18]. Mean bit-rates of 64 kbit/s are feasible, but allow decent frame-rates only for very low picture resolution, possibly suitable for hand-held devices with small screens.

---

[17] Video sequences downloaded from Websites as part of multimedia content may, but do not necessarily require real-time delivery, and are not considered to be relevant in this context.

[18] H.323 and H.324 are two ITU standards concerned with multimedia traffic over packet-switched and circuit-switched networks, respectively. They specify voice and video codecs to be used, as well as signalling protocols, such as those used for session establishment.

With these considerations in mind, the results reported in Reference [31] and published in Reference [30] for mixed voice and video traffic are of limited relevance and will not be reproduced here. For relevant investigations in this area, appropriate video traffic models would have to be derived and considerable effort would have to be invested in suitable resource allocation mechanisms. At the time when we contemplated investing some effort into this subject, the standardisation process of H.263 was only just completed and that of MPEG-4 not at all, so that suitable traffic models were not yet available in literature. As a result, instead of considering video traffic and investing effort into elaborate resource allocation schemes, we decided to focus on access control, and to consider only services that require data-rates that can be sustained on a single physical channel.

For completeness, it is reported that Uziel and Tummala used in Reference [263] a high-bit-rate traffic model available from the literature and scaled down the relevant parameters to obtain a model suitable for investigations on low-bit-rate video over wireless media. The question is whether the significantly larger compression-factors required would affect the statistics of the generated traffic in a manner causing significant aspects of the system behaviour not to be captured by this simple down-conversion of parameters. In fact, the authors state, that 'even though the original model parameters are based on 24 uncompressed frames/sec and we deal with highly compressed frames, we use this model due to its convenient mathematical tools'.

Finally, a resource allocation strategy for voice/data/video service integration on an advanced TDMA air interface was proposed and investigated in Reference [160]. As underlying multiple access protocol, PRMA++ is considered, but in an alteration of this protocol, conflict-free access for resource request messages of data and video services is provided, as already discussed in Section 3.6. To model the arrival process of a video terminal, the parameters of a discrete-state Markov chain were fitted according to traffic patterns from a real H.263-coded video sequence. This is the first such model we have seen in literature, but unfortunately, the parameter values are not included in the respective reference.

5.8 Summary and some Notes on Terminology

Two fundamentally different models for the assessment of the physical layer performance were discussed. The first model to be used in Chapter 7 is based on a *random coding assumption*, so that code assignment issues cannot be considered, but the error performance is assessed through a Gaussian approximation of the MAI. In the second model, which will be used in Chapters 8 and 9, resources are considered to be available in the shape of a rectangular grid of orthogonal code-time-slots. A collision of packets (or bursts) on the same code-time-slot will destroy the packets involved in the collision (capture is not considered), but mutual interference of packets transmitted simultaneously with different codes is ignored. Finally, a combination of these two models is also considered in Chapter 8, where individual code-time-slots are the resource units, and code collisions are possible, just as above, but at the same time, the impact of MAI is also accounted for through Gaussian approximations.

For traffic generation, three models will be considered: a two-state voice model with exponential sojourn time for voice gaps and voice spurts, a model for Web browsing composed of sessions, packet calls and individual datagrams, and a model for email message generation. Both datagram size and email message size are distributed according to a truncated Pareto distribution with appropriate parameters.

When considering voice traffic only and not accounting for interleaving, it is assumed that a 'packet' in MD PRMA is equivalent to the data transmitted in a burst, which in turn fits into a (code-)time-slot. In Chapter 9, on the other hand, where a mixed traffic scenario including Web browsing and email traffic is considered, and interleaving is accounted for, the following terminology (which is more or less in accordance with Reference [104]) is used: IP *datagrams* (for Web browsing) or email *messages* are delivered to the network layer (NWL), where they are optionally segmented into several *packets* before being handed down to the RLC layer[19]. An NWL packet is segmented by the RLC layer into *RLC frames* or *RLC-PDUs* (over which interleaving is applied). The output generated by the voice coder during a spurt is delivered to the RLC in the shape of *voice frames*, one voice frame fitting onto the payload of one RLC frame.

The term 'packets' is abandoned in this context at the MAC layer (i.e. for MD PRMA) to avoid confusion: The MAC layer delivers either the payload fitting into single *contention bursts* (or *access bursts*) to the physical layer, or integral RLC frames (including MAC overhead, where appropriate). The content of the latter is then error coded and interleaved over the appropriate number of *bursts*. Obviously, error coding is also applied to contention bursts.

---

[19] One could argue that email messages delivered over IP but exceeding IP datagram size limitations would have to be segmented already before being handed down to the NWL. This is ignored here.

6
MULTIDIMENSIONAL PRMA

The PRMA protocol extended for operation on a hybrid CDMA/TDMA air interface is defined in the following. This extended version of PRMA is referred to as multidimensional PRMA or MD PRMA. First, the basic protocol suitable for frequency-division duplexing is described. Then, the implications of different approaches to *time-division duplexing* on the protocol operation will be discussed. Finally, the two investigated approaches to access control, namely load-based access control and backlog-based access control, will be introduced. Before tackling the main issues of interest, a little digression is required to discuss the terminology used in conjunction with the research efforts presented here, or more precisely, the names used in previous publications when referring to this PRMA-based protocol.

6.1 A Word on Terminology

The following comments are provided to avoid potential confusion when looking at some of our earlier publications, since although the investigations documented in the next few chapters on MAC strategies are centred fundamentally on one protocol, this protocol has evolved over time, and so did the names we used when referring to it.

Initially, the protocol was referred to as the *Joint CDMA/PRMA* protocol in References [28–31], where random coding was considered, single time-slots could carry several packets, but individual code-slots were not discerned. In References [48] and [49] we suggested a protocol for operation on a rectangular grid of resource units, where the basic unit would normally be a code-time-slot, but it could also be a frequency-time-slot if the protocol were to be used with a hybrid FDMA/TDMA multiple access scheme. Apart from the different channel models considered, as discussed in detail in the previous chapter, and a different approach to channel access control, the protocol is essentially the same as Joint CDMA/PRMA, but since the focus was extended to hybrid FDMA/TDMA, a new 'umbrella name' was required.

Multidimensional PRMA (MD PRMA) was chosen as a name, with reference to the fact that resource units are defined in two dimensions rather than only one, and could in theory even be defined in three dimensions, when using FDMA, CDMA and TDMA all together. With a few exceptions, this is the only term which will be used in the following. While 'MD PRMA' does not specify which multiple access scheme is being used, the focus will be on CDMA/TDMA in the next few chapters. In the context of enhancements to EGPRS, the FDMA/TDMA version of the protocol could also be interesting, as discussed in Chapter 11.

6.2 Description of MD PRMA

6.2.1 Some Fundamental Considerations and Assumptions

In a cellular communications system, a certain amount of the downlink resources available in a cell will have to be reserved for signalling channels, which require resource units at regular intervals. These may be synchronisation or pilot channels, broadcast channels carrying system information, and common control channels, as known from 2G systems. Since traffic is normally symmetric or downlink biased, but rarely uplink biased, it is possible in FDD systems to reserve the corresponding resource units on the uplink as well, without wasting capacity. This resource could, for instance, be used to provide some guaranteed random access capacity for high-priority users or initial access purposes[1].

If both 'circuit-switched' and 'packet-switched' transmission modes are to be supported over the air interface, a common pool of physical resources should be shared, to enable efficient system operation. 'Circuit-switched traffic' (or rather: traffic carried on dedicated channels) can coexist without problems with 'packet-switched traffic' (traffic carried on shared or common channels) supported by MD PRMA. If a circuit is set up, one of the resource units will simply have to be reserved on a per-call basis rather than a per-packet-spurt basis. During the lifetime of the call, this resource unit will not be available for packet-switched traffic.

For the MD PRMA results reported in the next few chapters, the interest is exclusively in services supported on 'packet-switched' bearers. All considered terminals are already admitted to the system, such that initial access procedures need not be studied. Guaranteed random access capacity is not provided, and it is assumed that all the resources in a cell are available for MD PRMA operation.

6.2.2 The Channel Structure Considered

As in conventional PRMA [8], N time-slots of fixed length are grouped into frames (or TDMA frames, to distinguish them from voice frames). Depending on the context, a particular time-slot may either be specified using discrete time t (starting from $t = 0$, with unit increments for each time-slot), or by the time-slot number n_s (from 1 to N) together with the frame number n_f, where $n_s = (t \bmod N) + 1$. In the case of the physical layer model with code-time-slots described in Sections 5.3 and 5.4, each time-slot is subdivided into E code-slots, such that the basic resource unit is one of $U = N \cdot E$ code-time-slots or simply *slots* (see Figure 6.1)[2]. Since MD PRMA is an in-slot protocol, each such unit can either be a C-slot available for contention, or an I-slot used for information transfer. This implies that a particular time-slot can feature *both* C-slots and I-slots. If the 'pure' random coding model described in Section 5.2 is used and code-slots are not distinguished (but time-slots still are), then every time-slot may carry a number of packets irrespective of the codes selected, but subject to a packet error rate determined

---
[1] In GSM, for instance, the time-slot onto which synchronisation, broadcast, paging and access grant channels are mapped on the downlink, carries the random access channel on the uplink (see Sections 3.3 and 4.3).
[2] With $E = 1$, MD PRMA degenerates to conventional PRMA, and with $N = 1$, the protocol is essentially the same as a protocol proposed in Reference [35], which will be discussed further in Chapter 8.

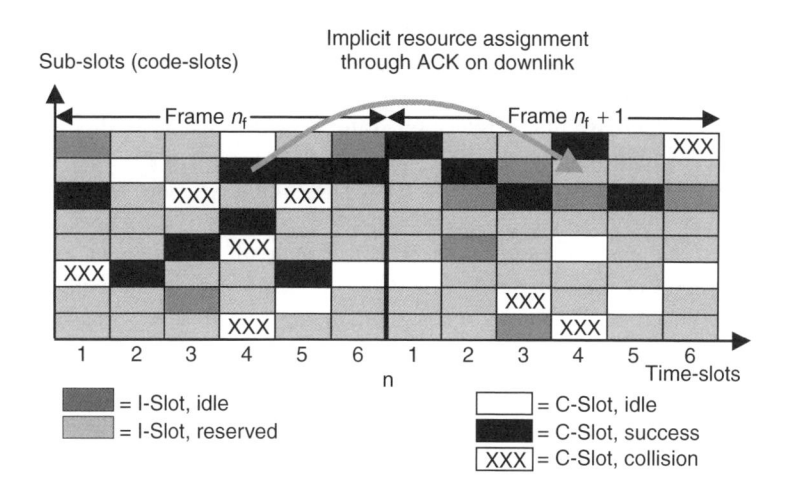

Figure 6.1 Code-time-slots and implicit resource assignment in MD PRMA

by the MAI experienced. Since there are no code-time-slots with this model, the notion of C-slots and I-slots does obviously not apply in this case.

In the case of time-division duplexing, the time-slots are shared between the two link directions, as discussed in more detail in Section 6.3. With frequency-division duplexing, the above description refers to the uplink channel only, while the exact structure of the downlink channel does not matter for MD PRMA operation, except for possible constraints regarding downlink signalling. However, as argued in Chapter 3, for complexity reasons it is considered desirable to use the same basic multiple access scheme and thus the same fundamental channel structure in both link directions.

The channel parameters are adapted to the bit-rate of the standard service (e.g. the rate of the full-rate voice codec) such that during a packet spurt with this service, one packet per frame is generated, which needs to be transmitted on one single slot. Due to this periodic resource requirement, such a source is termed a *periodic* information source[3].

6.2.3 Contention and Packet Dropping

On the uplink, resources are allocated on the basis of packet spurts. With the traffic models considered, packets to be transferred during a packet spurt will either carry data from a talk spurt, an IP datagram, or an email message. To obtain a resource reservation, terminals must go through a contention procedure. This procedure is first described for the code-time-slot case. Subtle differences in the random coding case are outlined subsequently.

6.2.3.1 The Code-Time-Slot Case

Terminals that are admitted to the system, but do not hold a reservation of resources, may only access C-Slots in contention mode with some time-slot and service or access-class specific *access permission probability* $p_x[t]$ signalled by the base station (for voice, $x = v$).

---
[3] Traffic generated by so-called *random* data sources defined in Reference [8] is not considered here for reasons outlined in Section 5.6.

A terminal with a new packet spurt will switch from idle mode to contention mode and wait for the next time-slot which carries at least one C-slot. It then determines whether it obtains permission to access this time-slot t by performing a Bernoulli experiment with parameter $p_x[t]$. In the case of a positive outcome, it will transmit the first packet of the spurt on a C-slot, which may have to be selected at random, if more than one such slot is available in the respective time-slot. In the CDMA context, selecting a C-slot means spreading the packet with the code-sequence which is assigned to the respective code-slot. If this packet is received correctly by the BS, it will send an acknowledgement, which implies a reservation of the same code-time-slot (now an I-slot) in subsequent frames for the remainder of the spurt. This way of assigning resources was already earlier referred to as *implicit resource assignment*, and is illustrated in Figure 6.1. The MS in turn switches to reservation mode and enjoys uncontested access to the channel to complete transmission of its packet spurt. In the case of a negative outcome of the random experiment, a collision on the channel with another contending terminal, or erasure of the packet due to excessive MAI, the contention procedure is repeated.

With delay-sensitive, but loss-insensitive services, packets are *dropped* when exceeding a delay threshold value $D_{\max}$, in which case contention will have to be repeated with the next packet in the spurt. As packet dropping will cause deterioration of the perceived quality of, for instance, voice or video, some maximum admissible *packet dropping ratio* P_{drop} will normally have to be specified.

The state diagram for the MAC entity of the mobile terminals is depicted in Figure 6.2. Note that the transition from CON to IDLE is only possible for a terminal that drops packets and may have to drop an entire packet spurt in exceptional cases. For loss-sensitive and delay-insensitive services (that is, NRT services such as email and Web browsing), packets are, at least in theory, never dropped at the MAC and therefore this transition is not possible.

6.2.3.2 Differences in the Random-Coding Case

There are subtle differences in the contention procedure for the 'pure' random-coding case. Since no code-slots are discriminated, the notion of C-slots and I-slots does not apply. The access permission probability to time-slots for contending users is controlled based on the number of users having a reservation on that time-slot, as outlined in Section 6.4. The equivalent of a time-slot without C-slot is a time-slot with access permission probability zero. If the probability is greater than zero, and the outcome of the Bernoulli experiment performed as a result is positive, contention may only fail due to the packet being erased by MAI, code-collisions are not possible.

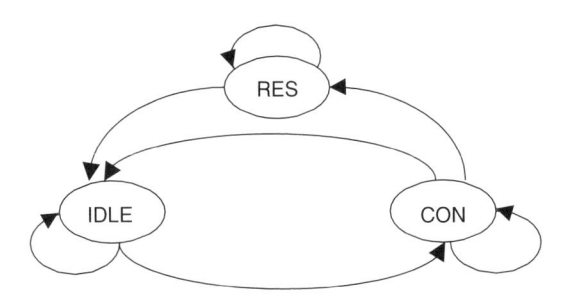

Figure 6.2 State diagram of mobile terminals (MAC entity)

6.2.4 Accounting for Coding and Interleaving

In conventional PRMA and basic MD PRMA introduced above, each packet, whether sent in contention or in reservation mode, carries an addressing header, some further signalling overhead and user data.

Once a logical context is established between a mobile terminal and the network and the latter knows for instance the destination of a mobile originated call, there is no need to transmit the full addressing information in every packet over the air interface[4]. The full header is therefore only required in the contention packet, if at all. In some cases, even only a temporary ID which identifies both the contending mobile and the relevant context unambiguously, will do. On the other hand, given the adverse propagation conditions in a mobile environment, data need to be error coded and interleaved over several time-slots to provide some protection against deep fades (see also Section 4.2).

These considerations lead to the following evolution of the basic protocol: when a packet spurt arrives, the MS generates a *dedicated request burst* for contention fitting into one slot and containing a temporary mobile ID, which is unambiguous in the considered context, and most of the signalling overhead required for the packet spurt, but no user data. Upon successful contention, the MS sends its user data *in groups of bursts* using rectangular interleaving, each burst again fitting into one slot (but for the standard data-rate, it sends again only one burst per TDMA frame, exactly as in the basic scheme). The group size is determined by the *interleaving depth* d_{il}. For the basic voice service, d_{il} is chosen here such that the transmission time of these bursts corresponds to the voice frame duration D_{vf}. The choice of air-interface parameters must then ensure that the data in one voice frame fits onto the payload of the bursts in one group. In Chapter 5, the term RLC frame was introduced for such a group of bursts. For data services, the data transmitted in d_{il} bursts is also referred to as an RLC protocol data unit or RLC-PDU.

In the case of the voice service, once a reservation is obtained, the voice frame most recently delivered by the RLC to the MAC is transmitted (no queuing is applied at the RLC or higher layers), while any older voice frame is dropped. This is equivalent to saying that D_{max} corresponds to D_{vf}. Dropping occurs frame-wise rather than packet-wise, such that P_{drop} denotes the frame dropping ratio.

In the case of NRT data services, the RLC delivers its PDUs either when the MAC is in IDLE state (in which case the delivery of a PDU triggers transition to the CON state) or, while in RES state, immediately after successful transmission of the previous PDU by the MAC. Dropping at the MAC does not occur.

At least as far as dedicated request bursts are concerned, this scheme bears some resemblance with burst reservation multiple access (BRMA) proposed in Reference [264].

6.2.5 Duration of a Reservation Phase

In PRMA as defined in Reference [8], an MS with periodic traffic may hold a reservation as long as needed to transmit successfully all packets in its spurt. If the MS leaves the allocated resource idle, the BS interprets this as the end of the spurt and

---
[4] In the case of IP traffic, address information may have to be transmitted with every single datagram. In this case, it is included in the datagram header (or IP header), which is considered to be part of the payload transmitted over the air interface. IP headers may be compressed, as discussed in Chapter 11.

terminates the reservation. While in practice, some protection against loss of reservation during deep fades will be required (see Section 3.6), this is not considered for our performance investigations and the PRMA approach is adopted. For MD PRMA on code-time-slots, the termination of a reservation involves changing the slot-status from I-slot to C-slot.

For NRT data, the reservation phase may be limited to an *allocation cycle*, as suggested in Reference [90] and discussed in Section 3.7. The allocation cycle length is indicated in terms of RLC-PDUs per cycle, and it is assumed that terminals need to re-contend for resources after expiration of a cycle. Upon successful transmission of the last PDU in a cycle, the MAC will therefore transit from RES to IDLE state. The alternative of piggybacking extension requests onto data transmitted on reserved slots is not considered. Within the constraints outlined in Subsection 6.2.7 regarding the resource allocation strategy used, the concept of allocation cycles would not make sense with piggybacking.

6.2.6 Downlink Signalling of Access Parameters and Acknowledgements

As discussed in Section 3.7, centralised access control is considered for MD PRMA. The BS will have to signal on the downlink the service or access-class specific access permission probability $p_x[t]$. For efficient access control, this probability value should be specific to each individual time-slot, thus it needs to be signalled on a per-time-slot basis. Furthermore, in the case of distinct code-time-slots, the base station will also have to indicate for each sub-slot individually, whether it is a C-slot or an I-Slot. It is assumed that all information relevant for access purposes is correctly available at every mobile terminal on a per-time-slot basis. To what extent this is required for proper protocol operation and what kind of overhead is involved quantitatively, depends also on the approach to access control considered. This will, as far as it has not yet been treated in Chapters 3 and 4 (in the context of the GPRS PRACH), be discussed in more detail in the relevant sections below.

The problem of acknowledgement delays was already discussed to some extent in Section 3.6. In particular, it was noted that, at least with FDD, when the same time-slot structure is used on the downlink as on the uplink, immediate acknowledgement is not possible. In the case of TDD, on the other hand, immediate acknowledgement may be possible, as outlined below. For MD PRMA performance assessment, in most cases, immediate acknowledgement of contention packets or request messages is considered, but the impact of the BS delaying acknowledgements is studied as well. For this purpose, it is assumed that a terminal that has sent a packet in contention mode in a particular time-slot will not be allowed to contend again in the next x time-slots (i.e. while waiting for an acknowledgement), regardless of whether there are resources for contention available in this time interval. The choice of the parameter x is influenced by processing delay, propagation delay and the structure of the downlink channel. It is assumed that successfully contending mobile terminals will receive their acknowledgement in time to make use of the first I-slot reserved for them, therefore $x \leq N$. The parameter x is very similar to the S-parameter in GSM and GPRS discussed in Sections 4.4 and 4.11 respectively.

6.2.7 Resource Allocation Strategies for Different Services

Some high-bit-rate services will require the allocation of multiple slots in a frame (be it an aggregation of time-slots, codes, or a combination thereof) to a single user. If an MS requests several slots, the BS will have to respond with a resource grant which specifies *explicitly* the resources reserved. A simple implicit assignment of resources through acknowledgements is insufficient. Using explicit resource assignment for all services would allow the BS to keep full control of if and when to allocate what kind of resources to which type of user. On the other hand, implicit resource assignment requires simpler acknowledgements (e.g. in the shape of a short, unambiguous terminal ID) and is particularly well suited for voice services, since their resource requests should always be satisfied to avoid a deterioration of the voice quality. Therefore, a hybrid approach may be preferred to cater for all the different needs while limiting complexity.

To keep it simple, we consider only implicit resource assignment for our MD PRMA performance investigations. As a consequence, multi-slot or multi-code allocation are out of scope. Furthermore, prioritisation of particular services in terms of resource allocation can only be achieved by controlling the access to C-slots as a function of the priority-class and choosing appropriate allocation cycle lengths. Pre-emption mechanisms are not considered either.

6.2.8 Performance Measures for MD PRMA

Assume that the quality impairment due to the packet (or frame) dropping probability P_{drop} and the probability P_{pe} of packets (or bursts) being erased due to MAI (as established in Sections 5.2 and 5.4) are perceived in a similar way[5]. Define the packet loss ratio P_{loss} as the sum of P_{pe} and P_{drop}. For real-time traffic, P_{loss} as a function of the traffic load can be used as the overall performance measure for MD PRMA. Since all terminals are assumed to experience the same propagation conditions or, to put it differently, since the location of terminals is assumed to have no impact on the physical layer performance, it is sufficient to assess average P_{loss} over all calls.

If only one type of real-time services is considered, and some admissible loss ratio $(P_{\mathrm{loss}})_{\mathrm{max}}$ is specified, the number of supported communications at this ratio can easily be established. For voice, a $(P_{\mathrm{loss}})_{\mathrm{max}}$ of 1% is typically considered to be admissible, but we will also consider a $(P_{\mathrm{loss}})_{\mathrm{max}}$ of 0.1%. Following the terminology used in the Goodman publications, $M_{0.01}$ stands for the number of communications supported at a $(P_{\mathrm{loss}})_{\mathrm{max}}$ of 1% averaged over all calls (accordingly, $M_{0.001}$ is used when $(P_{\mathrm{loss}})_{\mathrm{max}} = 0.1\%$). For voice with activity factor α_v, the multiplexing efficiency relative to perfect statistical multiplexing can then be calculated as

$$\eta_{\mathrm{mux}} = \frac{M_{0.01} \cdot \alpha_v}{N \cdot E}, \qquad (6.1)$$

---

[5] Whether 'front-end clipping' due for instance to PRMA operation is more disturbing than frame erasures spread over a conversation due to channel impairments is still a contentious issue. In Reference [8], Goodman *et al.* point to Reference [266] and state that front-end clipping is less harmful to subjective speech quality than other types of packet loss. However, in Reference [266], front-end clipping appears to be compared to 'mid-burst clipping' without considering sophisticated error concealment techniques other than 'gap closing'. Refer also to Chapter 9 in Reference [3] for detailed investigations on voice quality in PRMA-based systems.

where $U = N \cdot E$ is the number of resource units available for MD PRMA operation. If, on the other hand, calls experience different levels of quality depending on the location and movement of the respective terminals, one would have to establish the quality experienced by every call individually, and for instance require that no more than a given percentage of calls may suffer a packet-loss ratio exceeding the target ratio.

In the Goodman publications (for instance in Reference [142]), when the number of conversations per equivalent TDMA channel is established for PRMA, the packet header overhead is explicitly accounted for. In Equation (6.1) on the other hand, it is not, since the exact overhead specifically due to packet-switching may depend on the chosen implementation and is difficult to establish. However, when dedicated request bursts are generated, this overhead is implicitly accounted for, as these additional bursts may affect $M_{0.01}$.

For non-real-time traffic such as IP datagrams or email messages, packets need not be dropped, erased packets may be retransmitted, and adequate performance measures are access delay and total transmission delay. For two reasons, total delay performance is not evaluated in the following and only access delay performance is assessed. Firstly, the high variance of the Pareto distribution determining the size of email messages and IP datagrams makes it difficult to obtain reliable transmission delay results through simulations. Secondly, the focus is restricted to implicit assignment of a single resource unit per TDMA frame, and MAI is not accounted for in the mixed traffic scenario investigated in Chapter 9, which is the only scenario in which NRT traffic is considered. Therefore, retransmissions are never required in reservation mode and the average total transmission delay of an IP datagram or an email message is entirely determined by the average access delay and the average message length. For this to hold also for allocation cycles with limited duration, obviously, the access delay must not only include the delay experienced during the first access attempt, but also the time spent by a terminal in contention mode between individual cycles.

6.3 MD PRMA with Time-Division Duplexing

6.3.1 Approaches to Time-Division Duplexing

There are two fundamental approaches to providing time-division duplexing in a system with time-slots grouped into frames: either uplink and downlink time-slots alternate, as depicted in Figure 6.3, resulting in multiple switching-points per TDMA frame, or a train of successive uplink slots is followed by a train of successive downlink slots, such that there is only one switching-point per frame (Figure 6.4). For the TDD mode of the original TD/CDMA concept in Reference [90], only the latter approach was considered, while it is now envisaged to provide both alternatives for the UTRA TDD mode [84,265]. The following two issues need to be considered carefully when applying TDD.

- Overlapping between uplink and downlink bursts in one cell must be avoided, requiring an extra guard period at link switching-points, which is equivalent to at least the maximum one-way propagation delay in that cell. This is on top of guard periods required due to power ramping and timing advance inaccuracies. Therefore, it would appear that in medium and large cells, where propagation delay is not negligible, the single switching-point would be the preferred solution. However, to provide this guard period only at the switching-point, either the slots need to be spaced unequally, which

6.3 MD PRMA WITH TIME-DIVISION DUPLEXING

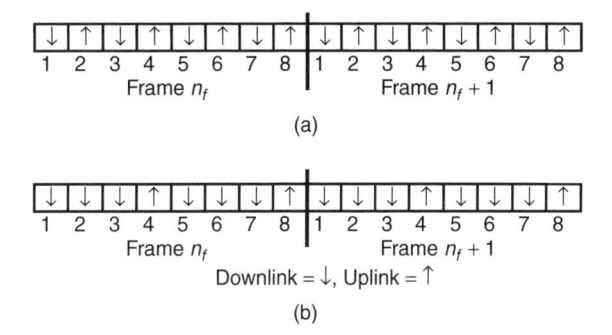

Figure 6.3 TDD with alternating uplink and downlink slots. (a) Symmetric resource allocation. (b) Asymmetric resource allocation

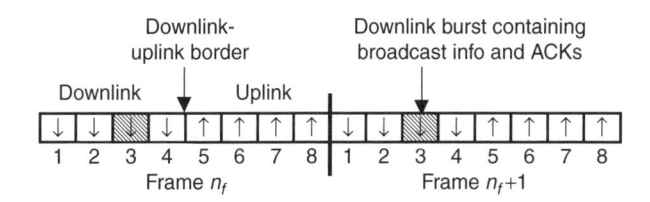

Figure 6.4 TDD with single switching-point (here shown with a symmetric resource split)

is inconvenient with respect to equipment clocks, or two burst formats would have to be defined, one with normal guard period, and one for slots adjacent to switching-points with reduced payload and extended guard period. The alternating slot option allows for relatively accurate open-loop power control owing to exploitation of channel reciprocity, if the duplex interval is smaller than the channel coherence time. This may offset (at least up to a certain cell size) any potential loss due to the additional guard periods required. Accuracy of power control may have significant implications on physical layer design, as already discussed in Section 5.1.

- To avoid interference between uplink and downlink (for instance between two terminals close to each other at cell fringes served by two different base stations), the same switching-points will likely have to be used in co-channel (and possibly even adjacent channel) cells of a contiguous coverage area. This may favour the single-switching-point approach, if guard periods not only have to cater for propagation delays within a cell, but also across cell boundaries. Alternatively, this problem could also be overcome by some clever slot scheduling (on the downlink) and access control (on the uplink) to avoid such 'collisions' causing significant interference. This will result in reduced capacity in cells affected by high interference levels (because certain slots cannot be used) and will require scheduling to be co-ordinated across multiple cells, which increases the system complexity. On the other hand, it could permit a more flexible scheduling of switching-points according to the traffic asymmetry ratio experienced in individual cells. On this topic, the reader is also referred to Section 2.3.

6.3.2 TDD with Alternating Uplink and Downlink Slots

While conventional PRMA was designed with an uplink channel structure exhibiting successive time-slots in mind, protocol operation is not significantly affected by the introduction of time-division duplexing with alternating uplink and downlink slots. In fact, only with such a channel structure can immediate acknowledgement (assuming zero processing delay) become conceptually possible, provided that every uplink slot is immediately followed by a downlink slot. As long as the number of time-slots in the uplink direction is the same in FDD and TDD mode, and these slots are equally spaced in the latter case, the behaviour of the ideal protocol with immediate acknowledgement is the same in both cases.

6.3.3 MD FRMA for TDD with a Single Switching-Point per Frame

A single switching-point between the two links limits downlink signalling conceptually to a frame-by-frame basis, which can have serious implications on the performance of both conventional PRMA and MD PRMA as defined earlier. In Reference [53], a scheme derived from PRMA called frame reservation multiple access (FRMA) was studied in which acknowledgements from the BS are only required at the end of a TDMA frame. The fundamental alteration to PRMA, which makes this protocol version suitable for such operating conditions, is that contending mobiles are allowed to contend repeatedly on C-slots in the same frame before receiving feedback. Should the BS receive several contention packets from the same MS during a single frame, it will acknowledge only one of them.

This scheme is particularly suitable for TDD with a single switching-point per frame. The BS can signal permission probabilities, slot status of the uplink slots, and acknowledgements in one of the downlink slots placed in a way that provides both MS and BS with suitable processing time, as illustrated in Figure 6.4. This strategy is adopted for MD PRMA with single-switching-point TDD, and is referred to as multidimensional FRMA (MD FRMA). Regarding broadcast information signalled in the downlink part of frame $n_f + 1$, which precedes the uplink part of this frame, it is assumed that:

- all received contention packets or resource requests sent in frame n_f are acknowledged (except for duplicate requests sent by one and the same MS, as outlined below); and

- all parameters relevant for access in the uplink part of frame $n_f + 1$ are signalled in a manner that they are available to all mobile terminals at or before the start of the uplink part.

In Reference [161], where a broadband PRMA-based TDD system with very short frame duration is considered, broadcast information signalled in frame $n_f + 1$ relates to frame $n_f + 2$ to allow further processing time.

With MD FRMA, an MS may send multiple request bursts on those uplink slots of a TDMA frame that are available for contention, if it obtains permission to do so, but at most one per time-slot. In the implementation chosen, if the BS receives multiple request bursts from a single MS, it will acknowledge only the first one.

6.4 Load-based Access Control

6.4.1 The Concept of Channel Access Functions

To protect reservation mode users from excessive MAI generated by contending users accessing the same time-slot, and to increase the probability of successful transmission of the latter, dynamic *load-based access control* can be used to restrict access rights for contending users in high load conditions. In this case, access is controlled through so-called channel access functions (CAF) proposed first in Reference [31] and further investigated in References [28–30].

With typical traffic scenarios in mind and given the burst lengths appropriate for cellular communications, most packet spurts are expected to be carried by a large number of individual packets or bursts, even when such a spurt carries only data of a relatively short (e.g. several 100 bytes) IP datagram. Hence, the probability that a spurt ends in any given frame is small. Therefore, $R'[n_f, n_s]$, defined as the sum of the number of users having transmitted in reservation mode R in a particular time-slot n_s in frame n_f, and those having contended successfully in that time-slot $C_s[n_f, n_s]$, is a good estimate for R in the same time-slot n_s in frame $n_f + 1$, i.e. $\hat{R}[n_f + 1, n_s] = R'[n_f, n_s] = R[n_f, n_s] + C_s[n_f, n_s]$. With channel access functions, this estimate of R for a certain time-slot n_s is related to the probability p with which contending users are allowed to access this slot[6]. For the case illustrated in Figure 6.5 for instance, $p(n_f + 1, n_s = 3) = f_{\text{CAF}}(R'[n_f, n_s = 3])$.

If the average number of packets per spurt is $\gg 1$, the choice of appropriate CAFs allows slot-to-slot load fluctuations (or in other words, the variance of the MAI) to be reduced. As a consequence, P_{loss} is reduced as well. Consider the packet success probability as a function of the number of users in a time-slot K, $Q_{pe}[K]$, as determined in Section 5.2 through SGA, for random coding, a spreading factor $X = 7$, perfect power control and a (511, 229, 38) BCH code. Since the packet error probability $P_{pe}[K]$ is larger than 10% for $K \geq 10$, it is obvious that access should be controlled so that always less than $K = 10$ users access simultaneously a time-slot. On the other hand, since $P_{pe}[K]$ is zero for $K \leq 7$, while $P_{pe}[8] = 2.6 \times 10^{-4}$, and $P_{pe}[9] = 1.40\%$, if a P_{loss} level of roughly

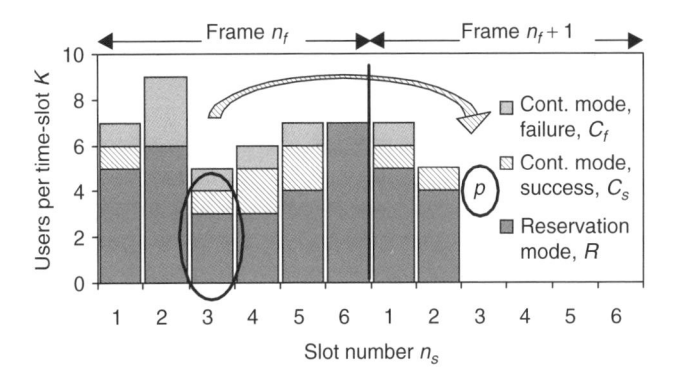

Figure 6.5 Illustration of load-based access control

---

[6] Here and in the remainder of this chapter, for simplicity and since only one access class is considered, p is used instead of $p_x[t]$.

1% is tolerated, K should preferably not fall below eight to make efficient use of the channel, while $K = 9$ can occasionally be tolerated.

To achieve this, it is obvious that if $\hat{R} \geq 9$, access should be denied for contending users. On the other hand, if $\hat{R}$ is low, since the probability that a large number of terminals will simultaneously switch from IDLE to CON is small, unrestricted access can be granted, i.e. $p = 1$, without risking stability problems. What remains to be established is the slope of the CAF. Considerations on how this can be done will be provided in Chapter 7. For illustration purposes, an efficient 'semi-empirical' channel access function for $Q_{pe}[K]$ as shown in Figure 6.6 is depicted in Figure 6.7.

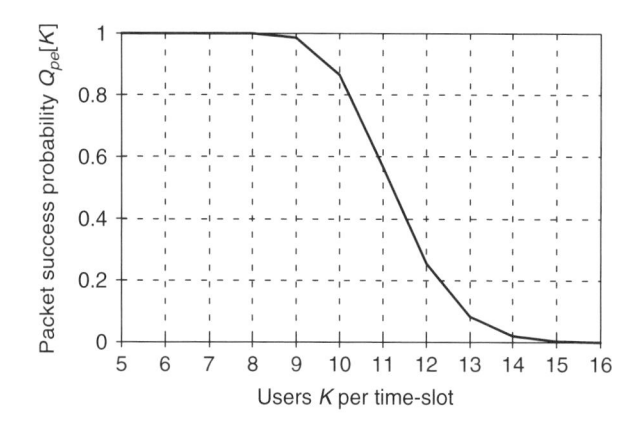

Figure 6.6 $Q_{pe}[K]$ for $X = 7$ and a (511, 229, 38) BCH code

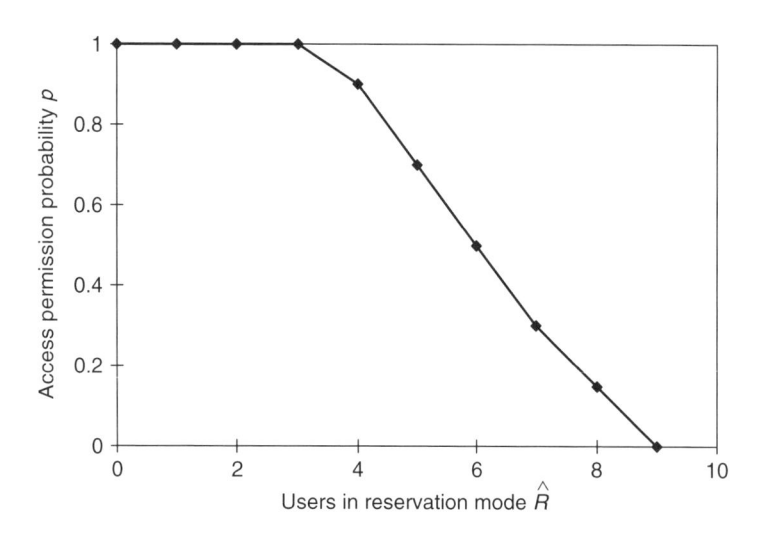

Figure 6.7 Example channel access function for $Q_{pe}[K]$ shown in Figure 6.6

6.4.2 Downlink Signalling with Load-based Access Control

Signalling of acknowledgements does not depend much on the access control strategy chosen and will not be discussed here. Note though, that the performance impact of delayed acknowledgements may depend on the particular strategy adopted. What is more interesting is the signalling of the permission probability values. Since here the value for a specific slot is determined by the load on the same slot in the previous frame, it can be signalled at any time between these two slots. In theory these permission probabilities could be signalled only once per TDMA frame. In any case, processing delay is not much of an issue and timely access-parameter-signalling is possible, if one is willing to expend the required signalling overhead[7]. Regarding the number of bits required per value, note that the number of users holding a reservation in a time-slot should not exceed $K_{pe\,\max}$ from Equation (5.9) in the case of random coding, and will not exceed E in the case of code-time-slots. Therefore, with non-prioritised access control, $\log_2(K_{pe\,\max})$ or $\log_2 E$ bits (rounded up to the next integer), respectively, are sufficient per time-slot. An appropriate signalling strategy could be to signal $\hat{R}$ on a per-time-slot-basis, while $p = f_{\text{CAF}}(\hat{R})$ is signalled less frequently together with other system information messages. Although not considered for our performance investigations, it can be advisable to change f_{CAF} adaptively, e.g. to counteract time-varying intercell interference levels, since this would cause time-varying $Q_{pe}[K]$.

6.4.3 Load-based Access Control in MD PRMA vs Channel Load Sensing Protocol for Spread Slotted ALOHA

In some respects, the approach to access control proposed here is similar to the channel load sensing protocol (CLSP) for spread slotted ALOHA in Reference [32], which was briefly introduced in Chapter 3. It is therefore appropriate to compare these two strategies here. In CLSP, the base station either allows or denies access to users centrally based on its observation of the uplink channel. If the air interface features equal-sized time-slots, then, in order to be able to benefit from channel sensing, the packets must span several slots.

Judging from the results presented in Reference [32], to gain in throughput compared to spread slotted ALOHA without sensing, the packet length should span at least 16 time-slots, preferably more, while the sensing delay should not exceed the time-slot duration. In the scenario considered here, the 'packet length' coincides with the slot size, but since reservations are used, the relevant entity for sensing is the packet spurt, which can be made up of several tens or hundreds of packets depending on traffic models and design parameters. Furthermore, owing to the TDMA channel structure, if the number of time-slots per frame N is large enough, sensing delay is not an issue, since the sensing information for a particular slot is not required to control access to the next slot, but only to the same slot in the next frame. Finally, note that we propose true adaptive probabilistic control, while in Reference [32], access is always granted to all new users with probability 1, if the current load is below a certain threshold value, otherwise, it is always denied.

---

[7] To allow for interleaving of downlink signalling messages over multiple frames, the whole protocol could be operated on a per-block basis as in GPRS (see Chapter 4, in particular Section 4.11) instead of the TDMA-frame basis considered here.

6.5 Backlog-based Access Control

6.5.1 Stabilisation of Slotted ALOHA with Ternary Feedback

In Reference [50], various stabilisation methods for slotted ALOHA based on ternary channel feedback were compared. It was found that the best delay-throughput performance was provided by methods that use deferred first transmission (DFT) and estimate the number of backlogged terminals (i.e. those having something to transmit) to calculate the access permission probability p. The *pseudo-Bayesian* algorithm [51] is one representative of this family of algorithms, along with others presented in the literature, which differ in how the update values for the backlog estimation are derived and how newly generated packets are accounted for. DFT means that a terminal with a newly arriving packet is considered to be backlogged immediately and is subject to the same permission probability p as a terminal that has to retransmit a packet following a collision. In the case of ternary feedback considered here, the feedback events are an idle slot (also referred to as a hole), a success slot, or a collision slot.

All these algorithms may be implemented in a centralised or a decentralised fashion. In the former case, the base station calculates and signals p, in the latter, the mobile terminals calculate p based on observed or signalled feedback (see also Section 4.11). In both cases (in the latter provided that terminals obtain correct feedback), every backlogged terminal will access a slot with the same probability p. Such schemes were referred to as global probabilistic control schemes in Chapter 3. On a perfect collision channel, their throughput cannot exceed 1/e.

6.5.2 Pseudo-Bayesian Broadcast for Slotted ALOHA

Let N_t denote the number of backlogged terminals at the start of time-slot t. In *Bayesian Broadcast*, Bayes' rule is used, after feedback observed in each time-slot, to update the estimated probability that $N_t = n$ stations are backlogged. Bayes' rule reads

$$P(H|E) = \frac{P(E|H)P(H)}{P(E)} \qquad (6.2)$$

where H stands for hypothesis and E for evidence. The hypothesis is that $N_t = n$, for each $n \geq 0$, and the evidence is the ternary feedback (hole, success, or collision). Therefore, for every n, $P(H)$ is the estimated probability that $N_t = n$ at the beginning of a slot, $P(E|H)$ is the probability of a hole, success, or collision, given n backlogged users, and $P(H|E)$ is the updated probability that $N_t = n$ at the end of the slot. Division by $P(E)$ can be viewed as the normalising operation required to ensure that the sum of $P(H|E)$ over all n adds up to one. To obtain the estimated probability that $N_{t+1} = n$ at the beginning of the *next* slot, the arrival rate of new packets λ_{ar} has to be accounted for. Furthermore, in the case of a success slot, the probability distribution of N_t at the end of slot t needs to be shifted by one position to the left, since one terminal leaves the population of backlogged users. The permission probability value p for slot $t+1$ is then chosen such as to maximise the expected success probability in that slot given the estimated backlog distribution.

In *pseudo*-Bayesian Broadcast, it is assumed that the probability distribution of N_t can reasonably well be approximated by a Poisson distribution, hence instead of individual probability values, only the mean v of the distribution needs to be estimated. Furthermore, the calculation of the optimum permission probability value p becomes very easy. It is simply the inverse of the mean of the distribution, as the expected probability of success

$$E[P_{\text{succ},t}] = \sum_n \frac{e^{-v}v^n}{n!} np(1-p)^{n-1} \qquad (6.3)$$

is maximised at $p = 1/v$. This is intuitively clear, since if $v = N_t$ (that is, if the estimated mean of the backlog distribution coincides with the real backlog), $p = 1/v$ will ensure that the expected offered traffic G assumes a value of one, which is the well-known optimum offered traffic for slotted ALOHA.

The pseudo-Bayesian algorithm for conventional slotted ALOHA reads [51]:

'Each station maintains a copy of v. At $t = 0$, all stations set v to 1. Then, during each slot, each station:

(1) transmits with probability $p = 1/v$, if it has a packet to transmit;

(2) decrements v by 1 if the current slot is a hole or success, or increments v by $(e-2)^{-1} = 1.3922$ if the current slot is a collision;

(3) sets v to $\max(v + \hat{\lambda}_{ar}, 1)$, and goes to step (1) for slot $t + 1$.'

Here, $\hat{\lambda}_{ar}$ is an estimation of the arrival rate λ_{ar} of new packets per slot, see Subsection 6.5.7. Observe that only binary collision/no-collision feedback is required, since the update value is the same for success and idle slots. If capture is accounted for, however, the full ternary feedback is required, as shown in Reference [131]. See also Section 4.11 for the modified algorithm taking capture into account.

For the detailed derivation of this algorithm, the reader is referred to Reference [51]. Alternatively, an analogous detailed derivation of the pseudo-Bayesian broadcast algorithm for two-carrier slotted ALOHA (see below) can be found in Reference [61], appendix B. In the following, for simplicity, pseudo-Bayesian broadcast is also referred to as Bayesian Broadcast (BB).

6.5.3 Bayesian Broadcast for Two-Carrier Slotted ALOHA

In MD PRMA, E code- or frequency-slots are available per time-slot. The *global* feedback or observation on the channel, that is, the observation for all sub-slots of a time-slot, will consist of a number of holes, successes, and collisions, the total number of which must add up to E. For $E = 2$, there are six different global observations (two holes, one hole and one success, etc.), while there are 10, 15, 21 for $E = 3, 4$, or 5 respectively, and 45 for $E = 8$. More generally, with E code-slots, $(E+1) + E + (E-1) + (E-2) \cdots + 1$ different global observations can be made. In Reference [61, Appendix B], we derived the update values for the pseudo-Bayesian broadcast algorithm based on such global observations for $E = 2$. This could be viewed as relevant for 'two-carrier slotted ALOHA' with aligned time-slots. The derivation is extremely tedious. Calculation of the update values along the same lines for $E = 8$, as required for MD PRMA with the parameters considered here, is not advisable. However, if the code-slots in a time-slot are orthogonal, there is a much easier way to find the appropriate update values, as shown next.

6.5.4 Bayesian Broadcast for MD PRMA with Orthogonal Code-Slots

Suppose for now that $A = E$ slots in a time-slot (i.e. all of them) are C-slots available for contention, as if we were considering a slotted ALOHA system with multiple carriers. The choice of a particular slot out of A in the case of a successful outcome of the Bernoulli experiment with parameter p is not conditioned on this Bernoulli experiment other than that it would normally not be made if the outcome was negative. Owing to this independence, we can invoke the 'splitting property' of the Poisson process [104, p. 165], i.e. if the total backlog is Poisson with mean v, then the 'backlog per slot' is Poisson with mean v/A. Next, recall that in the perfect-collision code-time-slot channel model described in Section 5.3, the slots are orthogonal. Finally, note that we are dealing with relative increments or decrements to the backlog in step (2) of the Bayesian broadcast algorithm, which do not depend on the value of v (except for the 'max' operation in step (3).

'Splitting property', orthogonality, and relative increments taken together, it is possible to update the backlog estimation for each sub-slot individually and add the resulting sum of update values in this time-slot to the previous estimation of v. Accordingly, to maximise the probability of success in every sub-slot individually, the 'permission probability per sub-slot' should be $1/v$, thus the total optimum permission probability is A/v. Therefore, the pseudo Bayesian algorithm for a slotted ALOHA system with A available sub-slots per time-slot can be written as follows:

(1) at $t = 0$, set v to A;

(2) each station that has a packet to transmit obtains permission to transmit with probability $p = A/v$ and selects a slot at random, if $A > 1$;

(3) observe the number of collision slots $C[t]$ in this *time*-slot and set v to $v + C[t]/(e - 2) - (A - C[t])$;

(4) set v to $\max(v + \hat{\lambda}_{ar}, A)$ and go to step (2) for *time*-slot $t + 1$.

The 'max'-operation in step (4) is required for arguments equivalent to those in Reference [51] for $A = 1$ and in Reference [61, Appendix B] for $A = 2$.

In MD PRMA, only a sub-set A of the E slots per time-slot may be available for contention, depending on the number of reserved slots in this time-slot $R[t]$. Therefore, the number of slots relevant for Bayesian broadcast is $A[t] = E - R[t]$, which varies with time. At desirable operating points of the protocol, packet dropping should be such that the probability of an entire packet spurt being dropped is negligible. This means that packet dropping has no impact on the backlog distribution, since a terminal will remain backlogged after dropping a packet. This leads to the following formulation of the pseudo-Bayesian broadcast algorithm for MD PRMA, if immediate acknowledgement is assumed (i.e. $x = 0$):

6.5 BACKLOG-BASED ACCESS CONTROL

Pseudo-Bayesian Broadcast Algorithm for MD PRMA

(1) At $t = 0$, set v to E;

(2) Each station which has a packet to transmit obtains permission to transmit with probability $p = \min(1, A[t]/v)$ and selects a slot at random, if $A[t] > 1$;

(3) Observe the number of collision slots $C[t]$ in this *time*-slot and set v to $v + C[t]/(e-2) - (A[t] - C[t])$;

(4) Set v to $\max(v + \hat{\lambda}_{ar}, A[t])$ and go to step (2) for *time*-slot $t+1$.

Considerations on the required update interval and the resolution of p were already provided in Section 4.11. In Chapters 8 and 9, when investigating the performance of MD PRMA with Bayesian broadcast, it is assumed that p is broadcast at the end of *each* time-slot in such a manner that it is available to all MS with full precision before the next time-slot starts.

6.5.5 Accounting for Acknowledgement Delays

The impact of delayed acknowledgements (i.e. $x > 0$) on the *performance* of the Bayesian broadcast algorithm is investigated in Chapter 8. This subsection considers how delayed acknowledgements will affect the *algorithm* as such. Since $x \le N$, acknowledgement delays have only an impact on terminals with colliding contention packets, but not on MS which contend successfully. The backlog estimation as such needs no modification, but when calculating p for a particular time-slot, one has to account for the fact that those backlogged terminals which suffered a collision during the last x time-slots will not attempt to obtain permission. If broadcast control is accurate, and there are enough contending terminals, the offered traffic per C-slot will be Poisson with rate $G = 1$, and consequently, the average number of terminals involved in a collision is

$$\mu_{\text{coll}} = \frac{1}{1 - 2/e} \sum_{k=2}^{\infty} k \frac{e^{-G} G^k}{k!} \bigg|_{G=1}$$

$$= \frac{1 - 1/e}{1 - 2/e} = 2.3922. \tag{6.4}$$

The number of terminals waiting can then be estimated using

$$w = C_x[t] \cdot \mu_{\text{coll}} \tag{6.5}$$

where $C_x[t]$ is the number of collision slots observed in the last x time-slots, that is

$$C_x[t] = \sum_{k=t-x}^{t-1} C[k]. \tag{6.6}$$

If $w > v - 1$, then w must be set to $v - 1$, and finally, the permission probability is set to

$$p = \min\left(\frac{A[t]}{v-w}, 1\right). \tag{6.7}$$

It is again assumed that this value is broadcast at the end of each time-slot in such a manner that it is available to all MS with full precision before the next time-slot starts. By doing so, we isolate the impact of delayed acknowledgements from that of non-ideal signalling of access parameters, and focus on the former.

6.5.6 Bayesian Broadcast for MD FRMA

To account for signalling and acknowledgements on a frame-by-frame basis instead of a slot-by-slot basis, and for the fact that a terminal may contend in several time-slots successfully before receiving acknowledgements, the algorithm is modified in the following manner for MD FRMA:

(1) at $t = 0$, set v and v' to E;

(2) each station which has a packet to transmit obtains permission to transmit with probability $p = \min(A[t]/v, 1)$ and selects a slot at random, if $A[t] > 1$;

(3) observe the number of collision slots $C[t]$ in this time-slot and set v' to $v' + C[t]/(e-2) - (A[t] - C[t])$;

(4) if time-slot $t+1$ is in the same frame as time-slot t, set v' to $v' + \hat{\lambda}_{ar}$ and go to step (2) for slot $t+1$, otherwise go to step (5);

(5) Set v and v' to $\max(v' + \hat{\lambda}_{ar} + s'', 1)$ and go to step (2) for time-slot $t+1$.

Since each successful MS can reduce the backlog only by 1, but is counted in step (3) for every successful contention, s'', which is the number of contention packets received other than the first one from each MS in this frame, is added in step (5) for compensation.

The backlog estimation v used to calculate p remains constant during the frame (unlike v'). Broadcasting v once per frame before its uplink section starts, together with status information of each code-time-slot, is therefore sufficient.

6.5.7 Estimation of the Arrival Rate

In all versions of the Bayesian broadcast algorithm listed above, the last step involves taking the estimated number of new packet arrivals during the respective time-slot into account. Rivest suggested in Reference [51] using either $\hat{\lambda}_{ar} = 1/e$ (which is the maximum arrival rate per slot allowing for stable operation of slotted ALOHA) or estimating the rate of new arrivals λ_{ar} according to

$$\hat{\lambda}_{ar}[t+1] = 0.995\hat{\lambda}_{ar}[t] + 0.005 S[t], \tag{6.8}$$

where $S[t] = 1$ if slot t is a success slot, otherwise $S[t] = 0$.

In Reference [50] it is reported that, for slotted ALOHA, the effort invested in estimating λ_{ar} using Equation (6.8) rather than setting $\hat{\lambda}_{ar}$ to 1/e is not rewarded by improved performance, which agrees with our observations reported in Reference [55]. The analogous approach in MD PRMA would be to set $\hat{\lambda}_{ar} = \overline{A}/e$ with $\overline{A}$ the number of C-slots per time-slot averaged over some time window. However, since C-slots and I-slots share the same resources in MD PRMA, such that random access and information transfer cannot be fully de-coupled, this approach yields unsatisfactory performance. Therefore, it is all the same necessary to estimate λ_{ar}, which is the rate per time-slot of new packet spurt arrivals. If the statistical parameters of the voice service are known to the BS and since the BS must have control over the admitted voice users M, the voice arrival rate λ_v can be estimated using

$$\hat{\lambda}_v = \frac{M \cdot D_{\text{slot}}}{D_{\text{spurt}} + D_{\text{gap}}}. \tag{6.9}$$

For data, $\hat{\lambda}_d$ is estimated using Equation (6.8), where $S[t]$ is the number of success slots in time-slot t (obviously counting only those success slots accessed by data terminals), and finally, $\hat{\lambda}_{ar} = \hat{\lambda}_v + \hat{\lambda}_d$.

That the backlog distribution can be approximated by the Poisson distribution and that $\hat{\lambda}_{ar}$ can simply be added to v for the backlog estimation, the arrival process of new packets will also need to be Poisson. While this is more or less the case with the voice traffic model considered, it is not with the traffic models used for IP datagram and email message generation. However, from the results reported in Chapter 9, it can be deduced that the algorithm as such is not, or at least not significantly, affected by this type of traffic (which is not to say that overall MD PRMA performance is not affected by the nature of the traffic generated).

6.5.8 Impact of MAI on Backlog Estimation

Bayesian Broadcast will be affected by MAI in the following two ways.

Firstly, while it was assumed in the derivation of the relevant algorithm above that code-slots were orthogonal, this is no longer true when MAI is present. It should in theory be possible to account properly for the resulting loss of orthogonality in the calculation of update values for the backlog estimation, but one may have to resort to global observations (i.e. on a per-time-slot-basis), which, as already outlined in Subsection 6.5.3, is bound to be cumbersome. To add to complexity, the level of MAI will also have to be included in the observation. Note, however, that backlog-based access control using ternary feedback seems not to be too sensitive to the exact update values chosen for backlog estimation. This can be deduced from a contour plot shown in Reference [50], where the mean number of backlogged terminals at a given input load is plotted as a function of the update values. This plot exhibits a rather flat minimum, that is, within a certain range of update values, the throughput-delay behaviour does not vary significantly, and remains close to the optimum value. For a moderate degree of non-orthogonality, it is therefore very likely that the performance achieved with update values derived for orthogonal slots is close to that achievable if these values were properly calculated taking MAI into account.

Secondly, as far as feedback is concerned, it is assumed that the BS will consider a code-slot carrying an erroneous packet as a collision slot, regardless of whether a code-collision has occurred or a single non-collided packet was erased due to MAI, although in the latter case v should not be increased. One would therefore expect increased dropping due to both reduced accuracy of the backlog estimation and packet erasure due to MAI. However, if the packet error probability $P_{pe}[K]$ is moderate for up to $K = E$ plus a few packets per time-slot, which is the case with parameters and conditions listed in Section 5.4, BB suffers only to a negligible extent from MAI, as will be shown in Section 8.4.

6.6 Combining Load- and Backlog-based Access Control

Backlog-based access control minimises packet dropping very efficiently. However, with single-user detection in realistic operating conditions, the P_{loss} performance of MD PRMA is dominated at low load by the packet error ratio P_{pe} rather than by the dropping rate P_{drop}. We proposed therefore in Reference [52] to combine backlog-based access control with load-based access control to allow the trading of P_{drop} against P_{pe} in a manner such that P_{loss} performance can be optimised.

This combination is achieved in the following manner: the basic permission probability p calculated with BB shall not exceed $p_{\max}$, that is

$$p = \min\left(p_{\max}, \frac{A[t]}{v}\right) \qquad (6.10)$$

where $p_{\max}$ is determined by a suitable channel access function. It will be shown in Section 8.4 that it is indeed possible, particularly at low load, to decrease P_{loss} with this hybrid approach compared to access control based on backlog only.

At first glance, combined access control does not make much sense in a scenario where MAI is (partially) eliminated through joint detection, and the P_{loss} performance is expected to be dominated by P_{drop}. However, backlog-based access control minimises the delay by attempting to control the offered normalised traffic per C-slot to unity. At unity traffic, if the traffic is Poisson, a collision will occur in 26% of the C-slots. It was outlined in Section 5.3, that such collisions might increase the packet error ratio, if they occur in time-slots in which joint detection is performed. Combining backlog-based access control with appropriate load-based access control allows the collision probability to be reduced. In other words, also in such a scenario, the P_{loss} performance may be optimised by the trading of P_{drop} against P_{pe} through a type of access control, which also exploits load information.

6.7 Summary

The basic mode of operation of the protocol under consideration in the next few chapters, here consistently called MD PRMA, was defined for frequency-division duplexing. Protocol operation remains the same in the case of TDD with alternating uplink- and downlink slots. On the other hand, if the TDD operation consists of a frame with a single switching-point, which separates a train of successive downlink slots from a train of successive uplink slots, signalling of access parameters and acknowledgements is conceptually only possible on a per-frame-basis. This requires modifications to the protocol. We adopted an approach along the lines of frame reservation multiple access for MD PRMA, thus referring to this protocol mode as MD FRMA.

6.7 SUMMARY

It was explained how load-based access control can be performed through use of channel access functions, and the similarities of this approach to a channel load sensing protocol proposed in Reference [32] were outlined. For backlog-based access control, the pseudo-Bayesian broadcast algorithm proposed by Rivest was adapted to MD PRMA and MD FRMA. In the case of MD PRMA, a variation of the algorithm, which caters for delayed acknowledgements, was also derived. It is expected (and will be shown in Chapter 8) that a small degree of loss of orthogonality between code-slots will only affect performance of these algorithms to a limited extent, although they were derived assuming orthogonal code-slots. Finally, it was outlined how backlog-based and load-based access control can be combined to reduce the total packet-loss ratio at a given traffic load, and to limit the number of code-collisions.

7
MD PRMA WITH LOAD-BASED ACCESS CONTROL

Chapter 5 provided descriptions of channel and traffic models used for investigations on the MD PRMA protocol, which was defined in detail in Chapter 6. Starting with this chapter, and continuing in Chapters 8 and 9, the outcomes of our research efforts on MD PRMA will be discussed.

In this chapter, the focus is on load-based access control (for MD PRMA), a technique adopted to protect reservation-mode users from multiple access interference generated by contending users. Only voice traffic is considered. We are investigating an interference-limited scenario, where code-slots are not distinguished, and random coding is assumed instead, such that 'classical' code-collisions cannot occur. However, users may still suffer 'collisions' due to excessive MAI, which will cause packet erasure. With access control, the overall packet-loss probability P_{loss} is composed of the packet-dropping ratio P_{drop} and the packet-erasure rate P_{pe}. Load-based access control is applied to trade off packet dropping against packet erasure in a manner which minimises P_{loss}.

To assess the benefits of load-based access control through so-called channel access functions (CAFs), several benchmarks are introduced. After a section defining the system considered, both analytical and simulation results for the first benchmark, a random access protocol, are presented. The analysis is expanded to underpin some of the comments made in the introductory chapter of this book on multiplexing efficiency. Other benchmarks include one used as a reference to assess multiplexing efficiency, and an ideal backlog-based access control scheme. Following considerations on channel access functions used for load-based access control, performances of the different schemes considered are compared for various scenarios. This includes a study of the impact of power control errors and the choice of the spreading factor.

7.1 System Definition and Choice of Design Parameters

7.1.1 System Definition and Simulation Approach

MD PRMA for frequency division duplexing as defined in Section 6.2 is considered, assuming immediate acknowledgements and using load-based access control as described in Section 6.4. Physical layer performance is accounted for assuming random coding and applying the standard Gaussian approximation to assess the error performance, as outlined in Section 5.2. When intercell interference is considered, all cells are assumed to be equally loaded, that is, $\overline{K}$ in Equation (5.16) is the average load per time-slot experienced in the test cell, as defined below, and $\overline{I}_{\text{intercell}}$ is taken to be 0.37 and 0.75, for values of the pathloss coefficient γ_{pl} of 4 and 3 respectively. Code-slots are not considered. Therefore, in Subsection 6.2.3, the specific considerations provided for the random-coding case apply. In other words, we are considering a purely interference-limited system, where, however, instantaneous interference levels are only considered for interference generated within the test cell, while intercell interference is assumed to be constant at its average level. To assess the benefit of load-based access control, MD PRMA performance is compared with that of various benchmarks, which are defined below.

The only traffic considered in this chapter is packet-voice traffic, using the two-state voice model specified in Section 5.5 with parameters $D_{\text{spurt}} = 1$ s and $D_{\text{gap}} = 1.35$ s, which results in a voice activity factor $\alpha_v = 0.426$. The number of conversations M supported simultaneously determines the system load. P_{loss} performance as a function of M is of interest here, and particularly, $M_{0.01}$ and $M_{0.001}$, the number of conversations which can be supported at tolerated maximum P_{loss} values, $(P_{\text{loss}})_{\text{max}}$, of 1% and 0.1%, respectively. A static scenario is considered, where P_{loss} is established as a function of M, and M remains fixed over the relevant period of observation. Therefore, the average number of users per time-slot $\overline{K}$ can be obtained through

$$\overline{K} = \frac{M \cdot \alpha_v}{N}. \qquad (7.1)$$

Simulations were performed using a commercial, event-driven and object-oriented tool for network simulations. Each simulation-run with fixed M covered 1000 s conversation time. Where required, several simulation-runs were performed for the same value of M, in which case the P_{loss} reported is the averaged result over these simulation-runs.

7.1.2 Choice of Design Parameters

The starting point for the choice of design parameters is to be found in Reference [146]. In this reference, a voice source rate R_s of 8 kb/s and a frame length D_{tf} of 20 ms are considered, yielding 160 information bits per packet[1], to which 64 header bits are added. With a PRMA channel rate R_p of 224 kbit/s, neglecting guard periods, a slot duration D_{slot} of 1 ms is required to accommodate a packet. A frame is therefore composed of $N = 20$ time-slots. The dropping delay threshold D_{max} is set to 20 ms, which is half the value considered in Reference [146]. This is to keep the total transfer delay low, to which also other sources of delay contribute, such as framing delay and processing delay.

It remains to specify the FEC code-rate r_c and the spreading factor X. In Section 5.2, the optimum value for r_c was established for packets with 224 message bits, applying the Gilbert–Varshamov bound. It was found that, irrespective of X, the bandwidth-normalised throughput was maximised when r_c was between 0.4 and 0.6. A suitable BCH code with a code-rate in this range of values is the (511, 229, 38) BCH code. It supports five more message bits than required (they will be attributed to the header), and has a code-rate r_c of 0.45. With this choice, R_p increases to 229 kbit/s before error coding, while the channel-rate after error-coding R_{ec} is 511 kbit/s. Interleaving is not applied, every packet

---

[1] In other Goodman publications, such as References [8] and [142], the voice source rate assumed was 32 kb/s, which is rather high for a basic voice service in cellular systems.

Table 7.1 Parameters relevant for the physical layer, protocol operation and traffic models

Description	Symbol	Parameter Value
TDMA Frame Duration	D_{tf}	20 ms
Time-Slots per Frame	N	20
Message bits per Packet	B	160 information bits + 69 header bits
Channel-Rate before Error-Coding	R_p	229 kbit/s
Channel-Rate after Error-Coding	R_{ec}	511 kbit/s
Chip-Rate	R_c	3.577 Mchip/s
Dropping Delay Threshold	$D_{\max}$	20 ms
Voice Terminal Source-Rate	R_s	8 kbit/s
Mean Talk Gap Duration	D_{gap}	1.35 s
Mean Talk Spurt Duration	D_{spurt}	1 s

is separately error-coded, and contention and reservation-mode packets have the same packet format. This also implies that contention packets contain the same amount of user data as those sent on reserved resources. No dedicated request bursts are generated.

In order to limit computer resource requirements for simulations, a rather low spreading factor of $X = 7$ was chosen when we started our investigations on PRMA-based protocols back in 1994. The resulting chip-rate R_c of 3.577 Mchip/s is surprisingly close to the one having been chosen for UTRA. Most of the results presented in the following are for $X = 7$. If larger spreading factors are considered, this is explicitly mentioned. The complete set of parameters used is listed in Table 7.1.

7.2 The Random Access Protocol as a Benchmark

7.2.1 Description of the Random Access Protocol

In References [28–31] we established the benefits of load-based access control through a performance comparison with what was referred to there as *random access CDMA*. Strictly speaking, the name 'random access *CDMA*' is somewhat misleading, since the same hybrid CDMA/*TDMA* channel structure as in MD PRMA is used. A more generic name will therefore be used for this protocol; it will be referred to here as random access protocol (RAP). In RAP, every user may access the channel at will. In other words, the access permission probability p is always set to one. For a voice user, this simply means that the next time-slot after the arrival of a talk spurt will be accessed. In Reference [31], it was assumed that a voice terminal needed to retransmit the first packet of a spurt until it was successfully received and acknowledged by the base station. This can be viewed as MD PRMA with $p = 1$. Here, in order to have completely unconstrained channel access, packets are never retransmitted, and the time-slot number used for all packets in a spurt depends only on the arrival instance of the first packet in that spurt. Therefore, P_{drop} is always zero, and the P_{loss} performance is entirely determined by P_{pe}.

7.2.2 Analysis of the Random Access Protocol

According to Section 5.5, the steady-state distribution for the number of simultaneously active terminals v given M voice sources is

$$\Pr\{V = v\} = P_V(v) = \binom{M}{v} \cdot \alpha_v^v \cdot (1 - \alpha_v)^{M-v} \qquad (7.2)$$

with mean $\overline{V} = M \cdot \alpha_v$. These v simultaneously active users will be distributed in some fashion over the N available time-slots. With completely unconstrained access as discussed above, there is no reason to expect that some time-slots are more likely to be chosen by any one of the users than others are. Furthermore, with exponentially distributed spurt and gap duration, any particular user will choose time-slots for successive spurts independently of each other. It can therefore be assumed that each slot in a TDMA frame is chosen with equal likelihood, i.e. $P_{\text{slot}} = 1/N$. The probability of k users accessing a slot conditioned on v active users is then

$$\Pr\{K = k | V = v\} = P_{K|V}(k|v) = \binom{v}{k} \cdot P_{\text{slot}}^k \cdot (1 - P_{\text{slot}})^{v-k}, \qquad (7.3)$$

and the unconditional probability can be calculated through

$$\Pr\{K = k\} = P_K(k) = \sum_{v=k}^{M} P_V(v) \cdot P_{K|V}(k|v). \qquad (7.4)$$

Note that the summation starts from k, since in order that k users access a certain time-slot, there must be at least k users active in total.

Finally, P_{loss} can easily be calculated according to

$$P_{\text{loss}} = \frac{1}{\overline{K}} \sum_{k=0}^{M} k \cdot P_K(k) \cdot P_{pe}[k], \qquad (7.5)$$

with $\overline{K}$ from Equation (7.1). To establish the packet erasure probability $P_{pe}[k]$, depending on the circumstances considered, Equations (5.7) and (5.3) together with either Equation (5.6) or Equation (5.16) are used[2]. Alternatively, Equation (5.20) may be used.

The steady-state distribution (Equation (7.2)) is a binomial distribution with parameters M and α_v. The Poisson distribution with mean $\overline{V} = M \cdot \alpha_v$ is a good approximation of the binomial distribution, provided that $\alpha_v \ll 1$ and M large (e.g. $\alpha_v < 0.05$ and $M > 10$). The first condition must hold for the variance of the binomial distribution, $M \cdot \alpha_v \cdot (1 - \alpha_v)$, to match roughly that of the Poisson distribution, $\overline{V}$. Here, α_v is significantly larger than 0.05, and thus, the variances of the two distributions cannot match, irrespective of M. Assume for now all the same, that Equation (7.2) can be approximated by a Poisson distribution with mean $\overline{V}$. In this case, since every slot is selected independently with probability P_{slot}, the probability distribution per slot is again Poisson with

---

[2] The attentive reader will have noticed that upper case 'K' was used for the number of users per time-slot in Chapter 5, while 'k' was used as an index for a particular user out of these K users. For consistency of notation in this chapter, 'k' is here the number of users per time-slot, and 'K' the respective random variable. Instead of Φ_K, we can write $P_K(k)$ for the probability distribution of this random variable.

mean $\overline{V}/N = \overline{K}$ owing to the 'splitting property' of the Poisson process discussed in Section 6.5. Therefore,

$$P_K(k) = \frac{\overline{K}^k e^{-\overline{K}}}{k!}. \tag{7.6}$$

This approximation is useful for the discussion on multiplexing efficiency provided in Subsection 7.2.4. Its accuracy is assessed below.

7.2.3 Analysis vs Simulation Results

In Figure 7.1, P_{loss} values resulting with the random access protocol are reported as a function of M for two cases, namely an isolated test cell and a test cell in a cellular environment, in both cases assuming perfect power control. In the single-cell case, there is no intercell interference, and Equation (5.6) is used for the average SNR which determines $P_{pe}[k]$. For the results shown for the cellular environment, a pathloss coefficient γ_{pl} of 4 is assumed, and average intercell interference is accounted for by using Equation (5.16), with $\overline{I}_{\text{intercell}} = 0.37$, instead of Equation (5.6). The curves with markers represent simulation results, whereas the solid and the dashed curves refer to analytical results with the binomial steady-state distribution according to Equation (7.2) and the Poisson approximation for the steady-state distribution respectively.

Two conclusions can be drawn from this figure. Firstly, judging from the P_{loss} values reported, the Poisson approximation models the P_{loss} performance obtained with the binomial distribution quite well for large values of M, regardless of the variance mismatch. With decreasing M, however, the gap between the P_{loss} values calculated widens. Secondly, in general a very good agreement between analysis and simulation results can be observed. Normally, even simulation results obtained from individual 1000 s simulation-runs closely match the analytical results, although most points shown in the

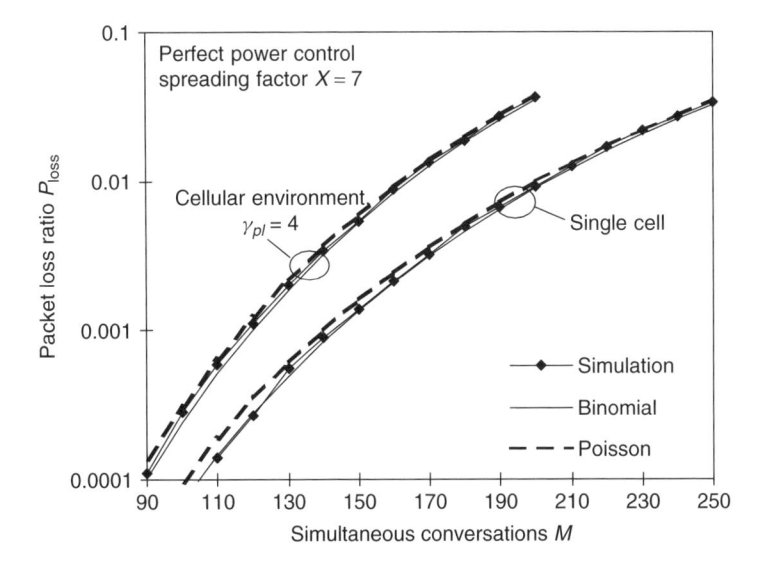

Figure 7.1 Performance of the random access protocol with perfect power control

simulated curves represent results averaged over several simulation-runs. As a result of this averaging, fairly smooth curves were obtained, but in rare occasions such as $M = 130$ in the single-cell case, even averaging over more than 50 simulation results did not allow the curve to smooth out perfectly.

Figure 7.2 shows equivalent results for the case when power control errors are accounted for through Equation (5.20). Here, due to the flatter $P_{pe}[k]$-slopes, the errors made with the Poisson approximation of the binomial steady-state distribution have a much smaller impact on the calculated P_{loss} values than in the case considered above. Correspondingly, the two analytical curves almost match even for small values of M. Again, a very good agreement between analytical and simulation results can be observed, which validates the simulation platform. Figure 7.3, showing only simulation results, summarises all cases considered for $X = 7$. Note that the impact of the intercell interference decreases with increasing error variance (which is partially due to the fact that fluctuations of the intercell interference level are not captured here). The two curves shown for $\sigma_{pc}^2 = 2$ dB (or $\sigma_{pc} = 1.41$ dB) even meet below a packet-loss ratio of 0.1%. A more detailed discussion of the performance degradation due to power control errors will be provided in Section 7.6.

For completeness, it is reported that simulations were also carried out for a spreading factor $X = 15$, obtaining similar agreement between simulation and analytical results as in the cases illustrated here.

7.2.4 On Multiplexing Efficiency with RAP

In Section 1.4, we claimed that the statistical multiplexing gain depended *essentially* on the standard deviation normalised to the mean of simultaneously active users (or almost equivalently: the standard deviation of the multiple access interference or MAI). In the following, this claim is first substantiated and, based on this, further observations on

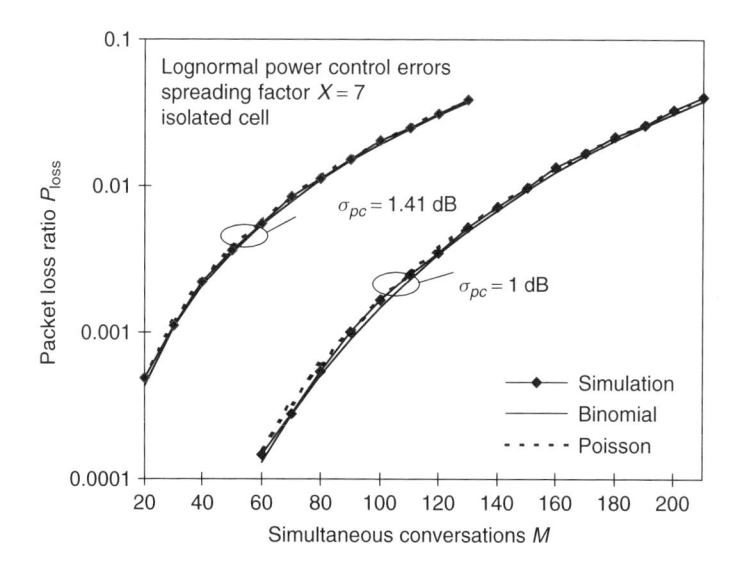

Figure 7.2 Performance of RAP when accounting for power control errors

7.2 THE RANDOM ACCESS PROTOCOL AS A BENCHMARK

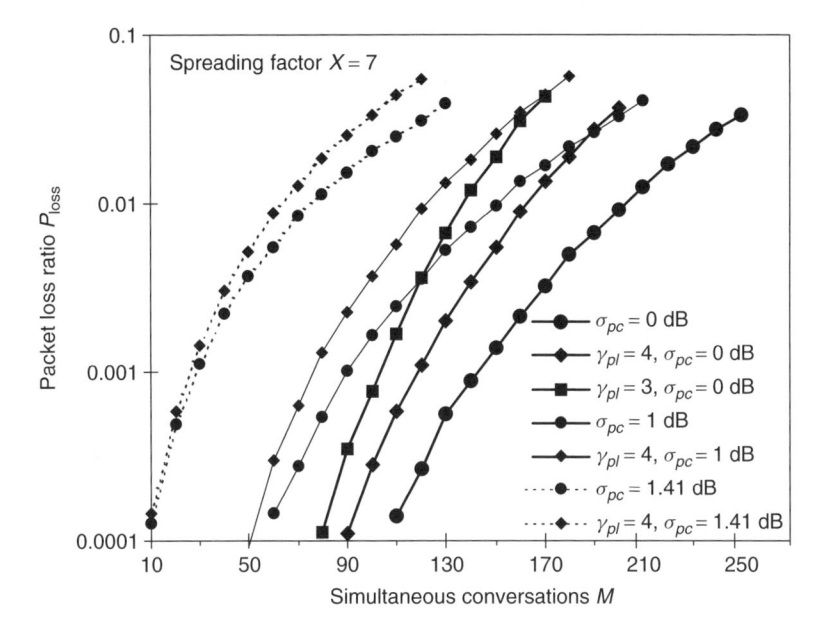

Figure 7.3 Simulation results for RAP obtained with different values for σ_{pc}, both for an isolated cell and two cellular scenarios with $\gamma_{pl} = 4$ and 3

multiplexing efficiency with pure CDMA and hybrid CDMA/TDMA air interfaces are provided.

For simplicity, assume a physical layer on which all packets are transmitted successfully, if no more than $K_{pe\,max}$ users access the channel simultaneously, and otherwise, all packets are erased. In other words, $P_{pe}[k]$ is approximated by a step-function,

$$P_{pe}[k] = \begin{cases} 0, & k \leq K_{pe\,max} \\ 1, & k > K_{pe\,max}, \end{cases} \quad (7.7)$$

and $K_{pe\,max}$ represents the number of resource units available.

With $P_K(k)$ describing the distribution of the random variable K, and with $P_{pe}[k]$ as above,

$$P_{\text{loss}} = \frac{1}{\overline{K}} \sum_{k=K_{pe\,max}+1}^{M} k \cdot P_K(k). \quad (7.8)$$

The appropriately weighted and normalised tail of the distribution of K, shaded in Figure 7.4 provided for illustration, determines the packet-loss ratio. Restricting the focus to bell-shaped distributions, at a given offset between $K_{pe\,max}$ and mean $\overline{K}$ of this distribution, it is obvious that the larger the standard deviation of K, σ_K, the larger P_{loss}. The more interesting question to ask is what kind of offset needs to be respected in order not to exceed a certain $(P_{\text{loss}})_{\max}$. Write this offset as a multiple of σ_K,

$$K_{pe\,max} - \overline{K} = c_{\text{offset}} \cdot \sigma_K, \quad (7.9)$$

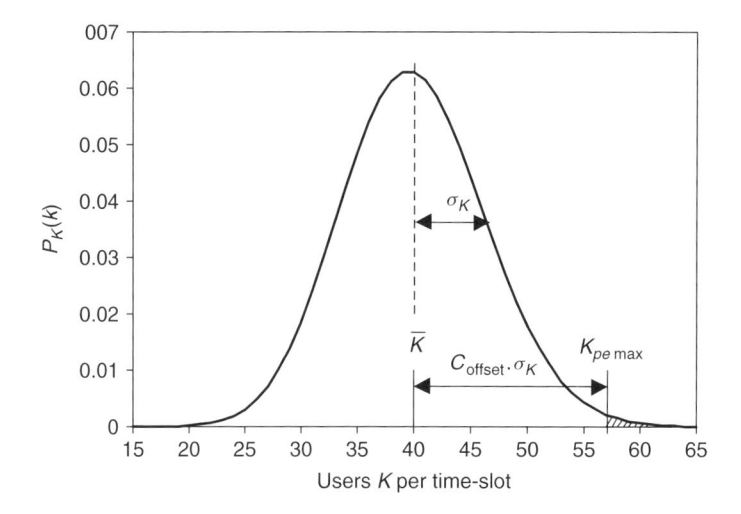

Figure 7.4 Illustration of $\overline{K}$ and $K_{pe\,max}$ when K is Poisson with mean $\overline{K} = 40$

as shown in Figure 7.4. Next, define multiplexing efficiency η_{mux} as the ratio of $\overline{K}$ over $K_{pe\,max}$, which is the normalised resource utilisation. Using Equation (7.9),

$$\eta_{mux} = \left(\frac{c_{offset} \cdot \sigma_K}{\overline{K}} + 1 \right)^{-1}. \tag{7.10}$$

If c_{offset} were a constant, we could indeed claim that the multiplexing efficiency only depends on the normalised standard deviation $\sigma_K/\overline{K}$. The smaller its value, the larger the multiplexing efficiency. But is c_{offset} constant? It would almost be, if the shape of the tail of the distribution of K were fully described by σ_K (as for instance with a normal distribution)[3]. Since it has just been found that P_{loss} performance of RAP is modelled accurately if K is assumed to be Poisson distributed, this distribution will be used in the following. If we ignore the fact that a continuous distribution is being compared with a discrete distribution, when $\overline{K} \gg 0$, the Poisson distribution resembles the normal distribution. In particular, it is nicely bell-shaped, as shown in Figure 7.4, although not completely symmetric with respect to $\overline{K}$. However, there is only one degree of freedom: the Poisson distribution is entirely specified by its mean $\overline{K}$, and the variance is equal to the mean, thus $\sigma_K = \sqrt{\overline{K}}$.

Figure 7.5 shows c_{offset} as a function of $\overline{K}$ for a Poisson distribution. Individual points in this graph were obtained by fixing $K_{pe\,max}$, imposing $(P_{loss})_{max} = 1\%$, and calculating the maximum value of $\overline{K}$ which is admissible to meet this P_{loss} requirement, as determined through Equation (7.8) together with Equation (7.6). As expected, c_{offset} is almost constant for $\overline{K}$ above 20, while there are somewhat stronger fluctuations for lower values of $\overline{K}$, where the Poisson distribution loses its bell shape. Since σ_K increases less than linearly with $\overline{K}$, η_{mux} increases with increasing $\overline{K}$ (the fact that

---
[3] Because of inevitable distortion effects due to the weighting and normalisation in Equation (7.8), c_{offset} will fluctuate slightly, even if a normal distribution is considered.

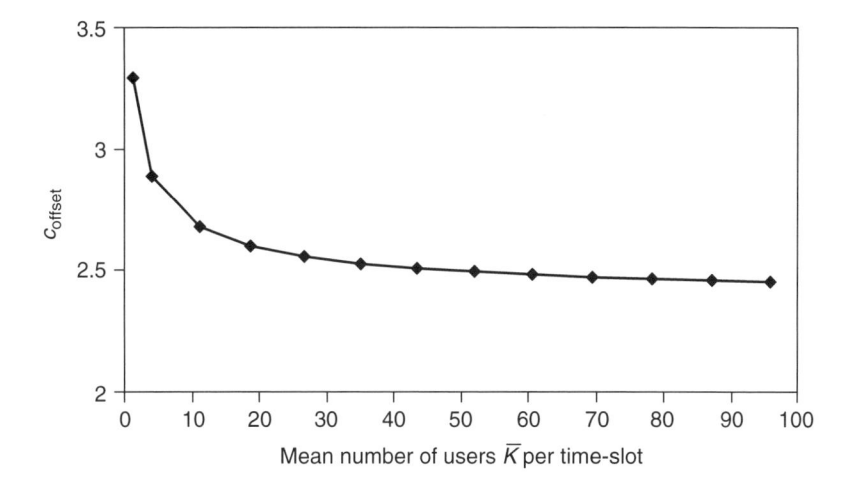

Figure 7.5 c_{offset} as a function of $\overline{K}$

c_{offset} decreases with increasing $\overline{K}$ further amplifies this effect). $\overline{K}$ can obviously only be increased, if $K_{pe\,\text{max}}$ is increased as well, that is, more bandwidth or resources must be provided.

Summarising, the multiplexing efficiency increases with increasing size of the population multiplexed onto a *common resource*. This is very similar to the so-called *trunking efficiency* in blocking-limited circuit-switched systems discussed in Section 4.6: the larger the number of channels provided, the higher the average channel utilisation at a given admissible blocking level.

Next, we need to ask how the relevant common resource is determined. In Section 1.4, we stated that the relevant resource for multiplexing was an entire carrier in wideband CDMA, but only a time-slot in hybrid CDMA/TDMA with unconstrained channel access. Consider a hybrid CDMA/TDMA system with N time-slots and $K_{pe\,\text{max}} = E$, and a wideband CDMA system with an equivalent amount of resources, but no time-slots (or rather, only one 'time-slot' per frame), i.e. $K_{pe\,\text{max}} = N \cdot E$. In the latter case, the steady-state distribution of the total number of users on the carrier being considered, which is approximately Poisson with mean $\overline{V}_C$, determines the packet-loss probability. Therefore, the trunking efficiency is determined by the total number of users. In the hybrid case, as just seen, the relevant distribution is the distribution per time-slot with mean $\overline{V}_{CT}/N$, thus the trunking efficiency is determined by the average number of users accessing a single time-slot. If the total amount of resources is the same as in the pure CDMA case, the trunking efficiency will be lower and, thus, the hybrid solution will support fewer users at a given $(P_{\text{loss}})_{\text{max}}$, i.e. $\overline{V}_{CT} < \overline{V}_C$.

For quantitative considerations on these matters, based on Gaussian approximations for physical layer modelling rather than the simple step function as per Equation (7.7), refer also to Section 7.6. These findings would suggest that a pure CDMA air interface is a better choice than a hybrid CDMA/TDMA air interface from a pure multiplexing point of view. However, if the load is balanced out between time-slots through access control, the relevant population becomes the total number of terminals admitted to this carrier in the hybrid CDMA/TDMA case as well. This is shown below.

7.3 Three More Benchmarks

7.3.1 The Minimum-Variance Benchmark

Suppose M simultaneous conversations are to be supported on N time-slots, such that the average load per slot amounts to $\overline{K} = M \cdot \alpha_v / N$ packets. Assume further that it is possible to schedule packets perfectly on the uplink, that is, the base station could have full control over how many users access any given time-slot, through whatever means may be necessary. Consider the case where the base station schedules either $k_1 = \lfloor \overline{K} \rfloor$ or $k_2 = \lceil \overline{K} \rceil$ packets per slot (in other words, k_1 and k_2 are the two consecutive integers embracing $\overline{K}$). Of all possible discrete distributions for K with mean $\overline{K}$, this is the one with minimum variance.

In Reference [30], to find the highest theoretically possible number of conversations which can be supported at a given $(P_{\text{loss}})_{\text{max}}$, the analysis focussed on such minimum-variance distributions, for which Equation (7.5) can be rewritten as

$$P_{\text{loss}} = \frac{P_K(k_1) \cdot k_1 \cdot P_{pe}[k_1] + P_K(k_2) \cdot k_2 \cdot P_{pe}[k_2]}{P_K(k_1) \cdot k_1 + P_K(k_2) \cdot k_2}. \quad (7.11)$$

In the above equation, the denominator is the mean $\overline{K}$ or expectation $E[K]$ of the distribution of K. Since $P_K(k_1) + P_K(k_2) = 1$, it can easily be shown that

$$P_K(k_1) = \frac{\overline{K} - k_2}{k_1 - k_2}. \quad (7.12)$$

$P_{\text{loss}}(M)$ can now be calculated using Equation (7.1) to establish $\overline{K}$, then Equations (7.12) and (7.11) with $P_K(k_2) = 1 - P_K(k_1)$. Note that these formulas imply perfect statistical multiplexing. In other words, it is assumed that arbitrary distributions of K can be shaped by scheduling and thus delaying packets as necessary, without having to drop packets.

We claimed in Reference [30] that M_x found at a given $(P_{\text{loss}})_{\text{max}}$ of x through the formulas above is a strict upper limit for M_x, and this benchmark was referred to as *perfect scheduling*. While we still believe that it will be rather difficult to exceed M_x with any other distribution than this minimum-variance distribution for the typical $P_{pe}[k]$ curves found in Chapter 5, which are the ones of interest here, we have invested no further efforts to prove this conjecture. On the other hand, if $P_{pe}[k]$ is approximated by the step-function (7.7), the distribution maximising M_x at a $(P_{\text{loss}})_{\text{max}}$ of x is *not* the distribution considered above with minimum variance, as shown in Reference [61, Appendix C]. Since 'perfect scheduling' may not only refer to the ability of the base station to schedule precisely the wanted number of packets in each time-slot, but could potentially also imply an optimum distribution for K, caution suggests that this benchmark should be referred to in the following as the minimum-variance benchmark (MVB) rather than the 'perfect-scheduling benchmark'.

Throughout the remainder of this chapter, the packet erasure rate $P_{pe}[k]$ will be modelled by Gaussian approximations and random coding is assumed, such that the number of code-slots E per time-slot is in theory unlimited. Equation (6.1) given in Subsection 6.2.8 can therefore not be used to assess multiplexing efficiency, since the number of available resource units is not clearly specified. However, $U = N \cdot \overline{K}$ can be used instead of $N \cdot E$

for the number of available resource units in Equation (6.1) with $\overline{K}$ obtained through MVB at the considered value for $(P_{\text{loss}})_{\text{max}}$. In other words, to assess the multiplexing efficiency achieved with load-based access control relative to (what we believe to be) perfect statistical multiplexing, the $M_{0.01}$ or $M_{0.001}$ values obtained thereby will be related simply to the respective numbers obtained with MVB.

7.3.1.1 Accounting for Packet Dropping

While MVB provides us with a tool to assess multiplexing efficiency, it is not suitable for assessing the quality of access control in MD PRMA. Even with the perfect multiple access protocol, that is, with perfect scheduling, if the service considered is delay constrained, packet dropping will necessarily occur, unless one is prepared to schedule more packets than the channel can carry reliably, thereby accepting increased P_{pe}. It would therefore be desirable to extend MVB in a manner that allows packet dropping to be accounted for. Such a benchmark, which will be referred to as MVB with dropping (MVBwd), is derived in the following. Assume that the scheduling is carried out frame-wise, and in every frame in which the number of active users v exceeds the number of resource units $U = \overline{K} \cdot N$, $v - U$ packets are dropped.

This is nearly, but not exactly equivalent to imposing a delay threshold D_{max} of one frame, as considered in this chapter for MD PRMA. The dropping probability can then be calculated according to

$$P_{\text{drop}} = \frac{1}{\overline{V}} \sum_{v=U+1}^{M} (v - U) P_V(v), \qquad (7.13)$$

with $P_V(v)$ from Equation (7.2) and $\overline{V} = M \cdot \alpha_v$. In this context, Equation (7.11) is used to calculate the packet erasure rate P_{pe} instead of the total loss rate P_{loss}, i.e.

$$P_{pe} = \frac{P_K(k_1) \cdot k_1 \cdot P_{pe}[k_1] + P_K(k_2) \cdot k_2 \cdot P_{pe}[k_2]}{P_K(k_1) \cdot k_1 + P_K(k_2) \cdot k_2}. \qquad (7.14)$$

The task is now to find an optimum trade-off between P_{pe} and P_{drop} using Equations (7.12) to (7.14), in order to maximise M at a given $(P_{\text{loss}})_{\text{max}}$. As a first step, over all U for which $P_{pe} < (P_{\text{loss}})_{\text{max}}$ according to Equation (7.14), find the largest M for which P_{drop} obtained through Equation (7.13) does not exceed $(P_{\text{drop}})_{\text{max}} = (P_{\text{loss}})_{\text{max}} - P_{pe}$. Since statistical multiplexing is not perfect, on average only $M \cdot \alpha_v < U$ resource units will carry packets, such that P_{pe} obtained with Equation (7.14) is conservative, or in other words, the $(P_{\text{drop}})_{\text{max}}$-limit used for Equation (7.13) is unnecessarily low. M maximised over U obtained in this manner is therefore referred to as *worst-case MVBwd* in the following.

To refine this benchmark, an iterative approach is adopted, where in an alternating fashion, P_{pe} is adapted based on the actual load M, and Equation (7.13) is used to update M based on the new $(P_{\text{drop}})_{\text{max}}$. To establish the correct load distribution needed for the calculation of P_{pe}, some further elaboration of the scheduling algorithm is required. Assume the base station knows the active number of users v at the beginning of a frame. If $v > U$, U packets are scheduled according to MVB, and the remaining packets are dropped, thus the mean number of packets per time-slot is in this frame $\overline{K}_f = U/N$, i.e. the same as $\overline{K}$ before. Additionally, if $v \leq U$, in which case no packet is dropped, the base station is assumed to carry out minimum-variance scheduling on the basis of v packets, with $\overline{K}_f = v/N$. Now, in a somewhat ugly fashion, P_{pe} can be written as

$$P_{pe} = \sum_{v=1}^{U-1} \frac{P_V(v)}{v/N} \left(\frac{v/N - k_2}{k_1 - k_2} \cdot k_1 \cdot P_{pe}[k_1] + \frac{k_1 - v/N}{k_1 - k_2} \cdot k_2 \cdot P_{pe}[k_2] \right)$$
$$+ \sum_{v=U}^{M} \frac{P_V(v)}{U/N} \left(\frac{U/N - k_2}{k_1 - k_2} \cdot k_1 \cdot P_{pe}[k_1] + \frac{k_1 - U/N}{k_1 - k_2} \cdot k_2 \cdot P_{pe}[k_2] \right) \quad (7.15)$$

with $k_1 = \lfloor \overline{K}_f \rfloor$ and $k_2 = \lceil \overline{K}_f \rceil$, where $\overline{K}_f = v/N$ for $v \leq U$, otherwise $\overline{K}_f = U/N$. $P_V(v)$ is obtained through Equation (7.2).

In summary, for a given U (and thus a given $\overline{K}$), first P_{pe} is calculated according to Equation (7.14), next (provided that $P_{pe} \leq (P_{loss})_{max}$) the largest M which satisfies $P_{drop} \leq (P_{drop})_{max} = (P_{loss})_{max} - P_{pe}$ using Equation (7.13), then P_{pe} is updated through Equation (7.15), and then Equations (7.13) and (7.15) are applied in an alternating fashion, until M settles. Maximising over U, the optimum trade-off between P_{pe} and P_{drop} can be found for any imposed $(P_{loss})_{max}$.

Figure 7.6 juxtaposes MVB, worst-case MVB with dropping, and MVB with dropping after several iterations. Calculations were carried out for an isolated cell, assuming perfect power control. Towards large values of M, the gap between the worst-case curve and that for plain MVB becomes smaller, which is due to increased multiplexing efficiency at high load. The MVBwd-curve obtained after several iterations lies in-between these two curves, as expected. A quick comparison with results for RAP (already shown in Figure 7.1) makes it immediately clear that uncontrolled channel access is not advisable in the scenario considered.

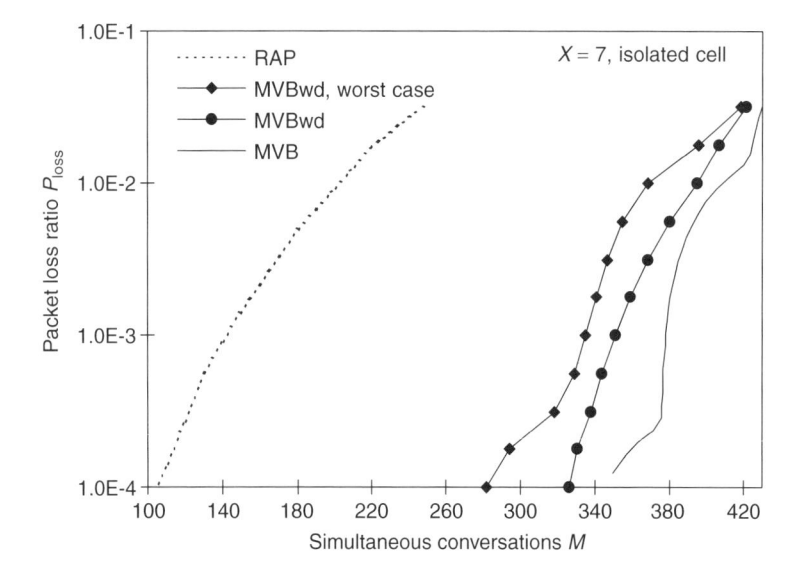

Figure 7.6 Comparison of MVB performance with and without dropping with that of the random access protocol

The 'Joint CDMA/NC-PRMA' scheme proposed by Wen et al. in Reference [45], which was briefly introduced in Subsection 3.6.5, attempts to provide a 'free' scheduling mechanism. Scheduling information is accommodated while all the same providing, within the same bandwidth, the same amount of resources for user traffic as in the scenario considered here. In their scheme, rather than scheduling at most U packets per frame and dropping excess packets, additional packets are accommodated on the least loaded slots, i.e. applying load balancing, which means that P_{drop} is kept at zero, while P_{pe} increases somewhat. Irrespective of whether we deem this scheme to be practicable, it is worth noting that the P_{loss}-results shown in Reference [45, Figure 9], for Joint CDMA/NC-PRMA with load balancing come close to our MVBwd results, as they should, given that the scheduling is very similar.

7.3.2 The 'Circuit-Switching' Benchmark

In the case of 'circuit switching', resources are allocated on a per-call-basis, i.e. dedicated channels are used. Packet dropping does not occur. Assume that every time-slot carries a fixed number of calls K', but note that, due to voice activity detection, only a subset k of these calls are in TALK state. Instead of Equation (7.5), P_{loss} is then

$$P_{\text{loss}}[K'] = \frac{1}{\overline{K}} \sum_{k=0}^{K'} k \cdot P_K(k) \cdot P_{pe}[k], \quad (7.16)$$

where $\overline{K} = K' \cdot \alpha_v$ and, from Equation (7.2),

$$P_K(k) = \binom{K'}{k} \cdot \alpha_v^k \cdot (1 - \alpha_v)^{K'-k}. \quad (7.17)$$

The maximum number of calls which can be served at a desired $(P_{\text{loss}})_{\text{max}}$ is $N \cdot K'_{\text{max}}$, with

$$K'_{\text{max}} = \max_{K'=1,2,\ldots} (K'|P_{\text{loss}}[K'] \leq (P_{\text{loss}})_{\text{max}}). \quad (7.18)$$

If the average P_{loss} over all calls is used as the relevant performance measure, it would normally be possible to assign $K'_{\text{max}} + 1$ calls to a certain fraction of time-slots. However, in a static scenario with a fixed number of calls, for calls assigned to such time-slots, P_{loss} averaged over an individual call would consistently exceed $(P_{\text{loss}})_{\text{max}}$, which is not acceptable. Therefore, all time-slots must be loaded with at most (or for the benchmark exactly) K'_{max} users. The same is true in a dynamic scenario with call arrivals and call departures, since the load fluctuations would be too slow to provide sufficient averaging in a manner which prevents calls from suffering insufficient quality over extended periods. In all other schemes considered, due to 'packet-switching', every talk spurt will be carried on a different time-slot and load fluctuations will be experienced on this time-slot during the talk spurt. This provides the necessary short-term quality averaging within and between calls which justifies considering average P_{loss} over all calls.

7.3.3 Access Control based on Known Backlog

With the last benchmark to be introduced here, the discussion of approaches to probabilistic access control for *reservation-based* protocols (where, unlike RAP, a terminal must contend successfully to obtain the right to access a time-slot repeatedly) finally starts. Assume that both the number of terminals in contention mode in slot t, $Y[t]$, and the number of terminals which will use that slot in reservation mode, $R[t]$, are known. Choose the access permission probability for this slot, $p_v[t]$ (recall that we are only considering voice traffic, thus index 'v'), as

$$p_v[t] = \min\left(\frac{K_{pe\,\max} - R[t]}{Y[t]}, 1\right). \qquad (7.19)$$

By doing so, in slots in which $K_{pe\,\max}\text{-}R[t] < Y[t]$, the expected number of terminals $E[K]$ will amount to $K_{pe\,\max}$. At the same time, the probability that exactly $K_{pe\,\max}$ users access that slot is also approximately maximised. $K_{pe\,\max}$ can therefore be considered as the target load for each time-slot.

As with the previous benchmarks, $P_{pe}[k]$ is modelled through Gaussian approximations which were discussed in Chapter 5. $K_{pe\,\max}$ is determined here by the admissible interference level, and needs to be calculated through Equation (5.9), namely

$$K_{pe\,\max} = \max_{k=1,2,\ldots}(k|P_{pe}[k] \leq (P_{pe})_{\max}). \qquad (7.20)$$

Even if the load could be controlled precisely to $K_{pe\,\max}$, $(P_{pe})_{\max}$ in Equation (7.20) should be smaller than the desired $(P_{\text{loss}})_{\max}$, since due to access control, P_{drop} will also contribute to the total packet-loss experienced.

In a somewhat cumbersome manner, this approach to access control was termed *optimized-a-posteriori-expectation-access-scheme* in Reference [30], because it is attempted to achieve a target value for E[K], based on knowledge (i.e. $R[t]$ and $Y[t]$) which is not *a priori* available. In the following, only slightly less cumbersome, it will be referred to as Known-Backlog-based Access Control (KBAC) instead. It should be noted that both $R[t]$ and $Y[t]$ can, at least under certain circumstances, be estimated reasonably well. As discussed in Section 6.4, depending on the traffic characteristics, the sum of the number of reservation mode users having accessed time-slot t-N (with the same time-slot number n_s as slot t) plus the number of successfully contending users C_s in that slot is a good estimate for $R[t]$, i.e. $\hat{R}[t] = R[t-N] + C_s[t-N]$. Furthermore, as shown in Chapter 8, the backlog estimation algorithm described in Chapter 6 allows an accurate estimation of the number of contending terminals $Y[t]$, if distinct code-slots are discriminated and feedback in terms of number of code-collisions in each time-slot is available.

$K_{pe\,\max}$ calculated through Equation (7.20) depends on $(P_{pe})_{\max}$, which in turn will be chosen with a certain $(P_{\text{loss}})_{\max}$, say 0.1%, in mind. Therefore, access control is tailored to specific values of M, in this case around $M_{0.001}$. If M is lower, access control might be too generous. On the other hand, at high load, when $\overline{K}$ approaches or even exceeds $K_{pe\,\max}$, packet dropping will be excessive, and the total packet loss may be reduced by increasing $K_{pe\,\max}$. Therefore, to achieve a good performance over a wide range of M, several values for $K_{pe\,\max}$ will have to be considered. Even so, the performance comparison below shows that KBAC, while providing good results, is not *the* optimum approach (in terms of minimising packet loss) to probabilistic access control. In particular, even with optimised $K_{pe\,\max}$, access control with KBAC can be too generous in certain conditions. With Equation (7.20), load is controlled in a 'symmetric manner around $K_{pe\,\max}$', that is, if $K_{pe\,\max} - R[t] < Y[t]$, the conditioned probability that $K_{pe\,\max} + 1$ users access the channel given $R[t]$ and $Y[t]$ is approximately equal to the conditioned probability that $K_{pe\,\text{max}} - 1$ users access the channel. If the packet success probability $Q_{pe}[k] = 1 - P_{pe}[k]$ decreases rapidly above $k = K_{pe\,\text{max}}$, choosing a value for $p_v[t]$ which is lower than that obtained through Equation (7.19) will in some cases reduce P_{pe} more than it will increase P_{drop}, thus improving the packet-loss performance.

Summarising, due to the complicated interdependencies between packet erasure and packet dropping performance, the optimum approach to probabilistic access control will not only depend on the shape of $Q_{pe}[k]$, but even at a given shape, it will be difficult to identify an approach that is optimum irrespective of M.

7.4 Choosing Channel Access Functions

Channel access functions relate the estimated number of reservation mode users in a given time-slot $\hat{R}$ to a suitable access permission probability value, i.e. $p_x = f_{\text{CAF}}(\hat{R})$.

The reader is reminded that only voice traffic is considered in this chapter. The purpose of the channel access functions discussed in the following is therefore to determine the access permission probability for contending voice users, p_v.

7.4.1 The Heuristic Approach

First results for load-based access control with channel access functions, which were published in References [31] and [30], were based on a purely heuristic approach to choosing these functions. To limit the degrees of freedom, the search for appropriate CAFs was limited to functions with two linear segments, such as the one shown in Figure 7.7. These functions are specified by an initial permission probability p_{vi}, the slope α_{ca} of the first linear segment, the position of the breakpoint, and the slope β_{ca} of the second linear segment. Note that since $p_v \geq 0$, $p_v(\hat{R}) = \max(f_{\text{CAF}}(\hat{R}), 0)$, for CAFs specified in this manner, as shown for $\hat{R} = 9$ in Figure 7.7.

Since the optimum permission probability for conventional PRMA with 20 timeslots per frame was found to be around 0.3 in Reference [142], p_{vi} was fixed to this value. Furthermore, α_{ca} was limited to small values, since otherwise, the definition of a breakpoint would not make sense. For reasons immediately apparent, but also explained in Section 6.4, the breakpoint and β_{ca} were chosen such that $p_v(K_{pe\,\text{max}}) = \max(f_{\text{CAF}}(\hat{R} = K_{pe\,\text{max}}), 0) = 0$, where $K_{pe\,\text{max}}$ is calculated through Equation (7.20). Within these constraints, the sets of values shown in table 7.2 were found to deliver good performances for the three cases with perfect power control considered, namely an isolated cell, and a cellular environment with pathloss coefficients γ_{pl} of 4 and 3:

Note that Figure 7.7 illustrates the first case, namely a single, isolated cell. All three functions together (the first, by the way, already used in Reference [31]) are juxtaposed in Figure 7.9, and the respective packet success probabilities are shown in Figure 7.11.

7.4.2 Semi-empirical Channel Access Functions

For the results reported in Reference [29], an improved set of channel access functions was used. While running simulations in which $p_v[t]$ was controlled based on known

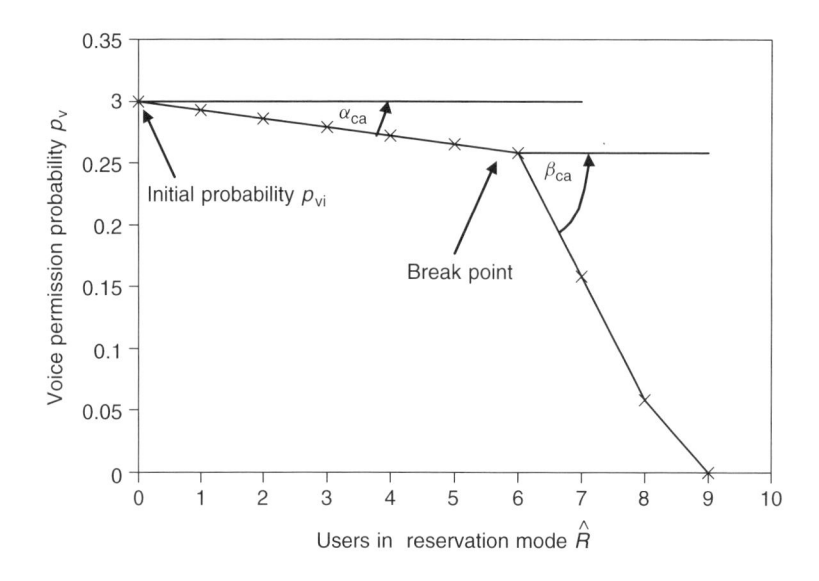

Figure 7.7 Channel access function composed of two linear segments

backlog according to Equation (7.19), statistics were collected to determine $\overline{p}_v[R]$, that is, $p_v[t]$ values were first classified according to the observed $R[t]$ values, and these classified values were subsequently averaged. Like this, backlog-based $p_v[t]$ values were mapped to load-based $p_v[t]$ values.

One could be tempted to use these $\overline{p}_v[R]$ values directly for access control. However, looking for instance at Figure 7.8 provided for the single-cell case, such access functions are extremely generous in comparison to the heuristic function used previously. In particular, $\overline{p}_v[K_{pe\,max} - 1] > 0.6$ for $K_{pe\,max} = 9$, and >0.7 for $K_{pe\,max} = 8$. These rather high values will cause time-slots with R close to $K_{pe\,max}$, for which the number of contending terminals happens to be high, to be heavily overloaded. In such slots, both the reserved packets will be erased and no new reservations can be granted. This will cause a temporary accumulation of contending terminals, which increases the probability for subsequent time-slots to suffer from overload, too. Eventually, the number of contending terminals will have grown to such an extent that all slots will be affected, new reservations cannot be granted and the system throughput breaks down. To avoid such system instability, only appropriately adjusted $\overline{p}_v[R]$ values should be used for channel access functions. The chosen adjustment process was again heuristic, as it simply consisted in reducing $f_{CAF}(\hat{R})$ values, particularly those for $\hat{R}$ close to $K_{pe\,max}$, until the system operated in a stable manner over a wide range of values for M. It is for this heuristic adjustment of empirical data that these access functions are referred to as *semi*-empirical.

It should be noted that the collected $\overline{p}_v[R]$ values are specific to the chosen value for M. To find channel access functions, M was chosen to be the number of conversations that can be supported at a tolerable level of P_{loss}, which in turn depends on the chosen functions. Determining suitable access functions may therefore involve several iterations. Furthermore, if intercell interference is accounted for and all cells are equally loaded, as assumed here, the intercell interference level increases with M, such that $K_{pe\,max}$ depends also on M. With intercell interference, therefore, even in the purely heuristic

7.4 CHOOSING CHANNEL ACCESS FUNCTIONS

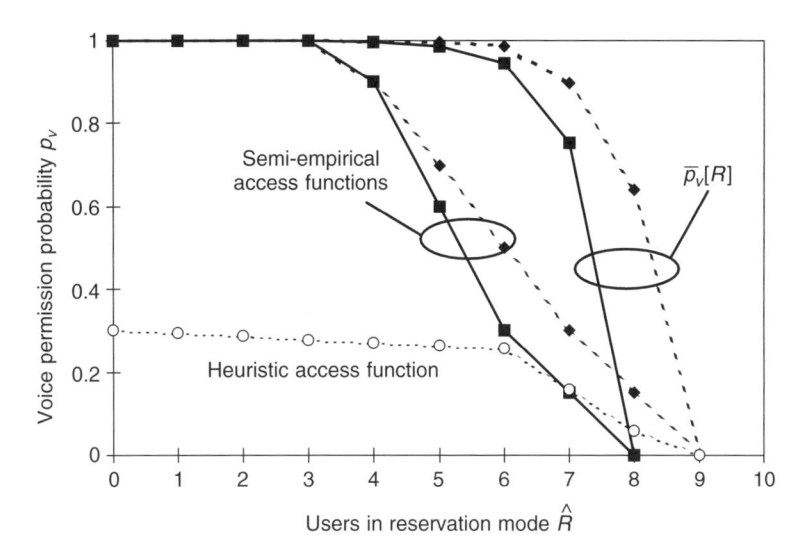

Figure 7.8 Heuristic and semi-empirical access functions used for an isolated cell, and the underlying $\overline{p}_v[R]$ values. Solid lines are for $K_{pe\,max} = 8$, dashed lines for $K_{pe\,max} = 9$

Table 7.2 Heuristic channel access functions used for MD PRMA with perfect power control

Environment	Single Cell	Cellular, $\gamma_{pl} = 4$	Cellular, $\gamma_{pl} = 3$
p_{vi}	0.3	0.3	0.3
α_{ca}	0.007	0.008	0.009
Breakpoint	6	4	3
β_{ca}	0.1	0.1	0.12

case discussed in the previous subsection, several iterations were required to determine the target $K_{pe\,max}$.

For the case of an isolated cell, Figure 7.8 compares the $\overline{p}_v[R]$ values obtained for $M = 400$ and 350 (for $K_{pe\,max} = 9$ and 8, respectively) with the adjusted values used for the semi-empirical access function, and with the heuristic access function specified in Table 7.2 and already depicted in Figure 7.7. The semi-empirical function established for $K_{pe\,max} = 8$ will be used for low values of M, that for $K_{pe\,max} = 9$ for high values.

Figures 7.9 and 7.10 show the heuristic and the semi-empirical channel access functions, respectively, for the three cases considered, and Figure 7.11 illustrates the packet success probabilities experienced in these three cases.

Finally, Figure 7.12 compares the packet-loss performance achieved with these two types of access functions. The performance improvement with Semi-Empirical CAFs (SECAF) is particularly evident for low values of M, in the single-cell case and the cellular case with $\gamma_{pl} = 4$, mainly owing to the separate access functions used for these low values. On the other hand, $M_{0.01}$ is only increased to a very limited extent (cf. Table 7.3 in the next section). The heuristic functions (HCAF) were in fact derived in an effort to maximise $M_{0.01}$, and not to minimise P_{loss} at $M < M_{0.01}$. From this point of view, these early results based on heuristic access functions were surprisingly good.

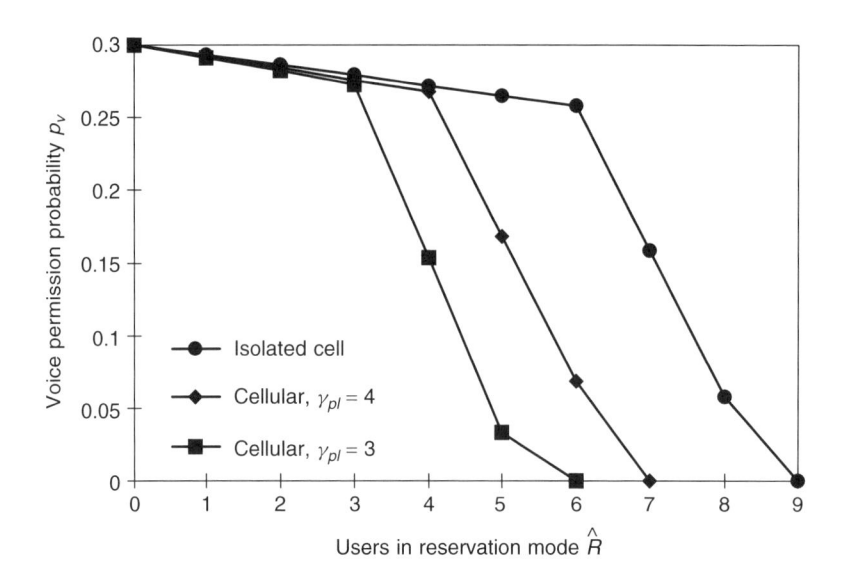

Figure 7.9 Heuristic channel access functions used for the three cases considered

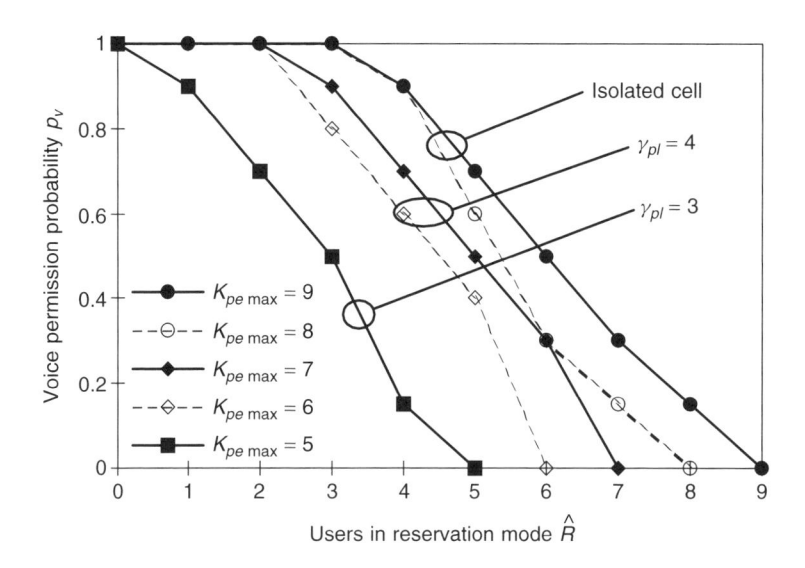

Figure 7.10 Semi-empirical CAFs with $K_{pe\,max} = 8$ and 9 in the case of an isolated cell, $K_{pe\,max} = 6$ and 7 for $\gamma_{pl} = 4$, and $K_{pe\,max} = 5$ for $\gamma_{pl} = 3$

In the cellular case with $\gamma_{pl} = 3$, only a single semi-empirical function was used for all values of M considered. In this case, the significantly reduced P_{loss} at low M has to be paid for by reduced performance at high load, such that the respective results cross with those for the heuristic CAF at $M = 210$. A slightly different semi-empirical function used for the results published in Reference [29] would avoid this crossover, but at the expense of instability at $M = 230$.

7.4 CHOOSING CHANNEL ACCESS FUNCTIONS 297

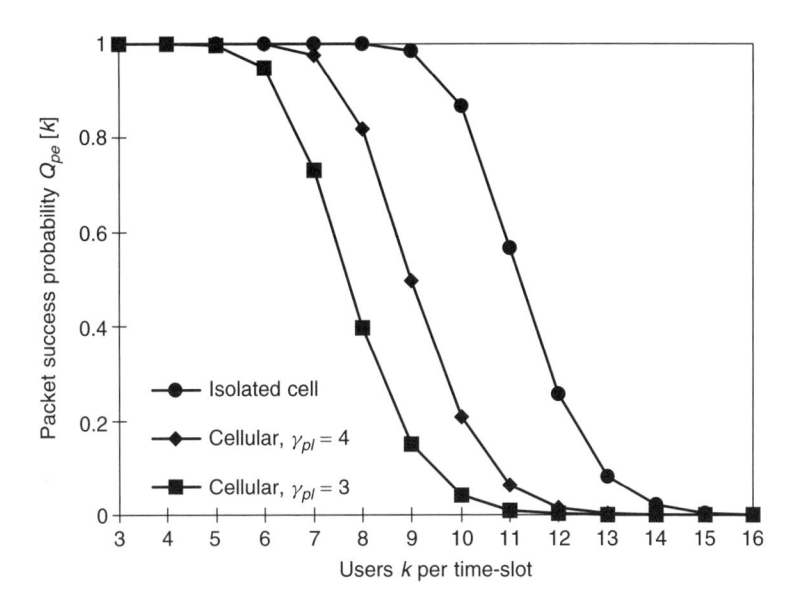

Figure 7.11 Packet success probabilities for the cases considered, in the cellular cases at a load of $M = 280$ for $\gamma_{pl} = 4$, and $M = 220$ for $\gamma_{pl} = 3$

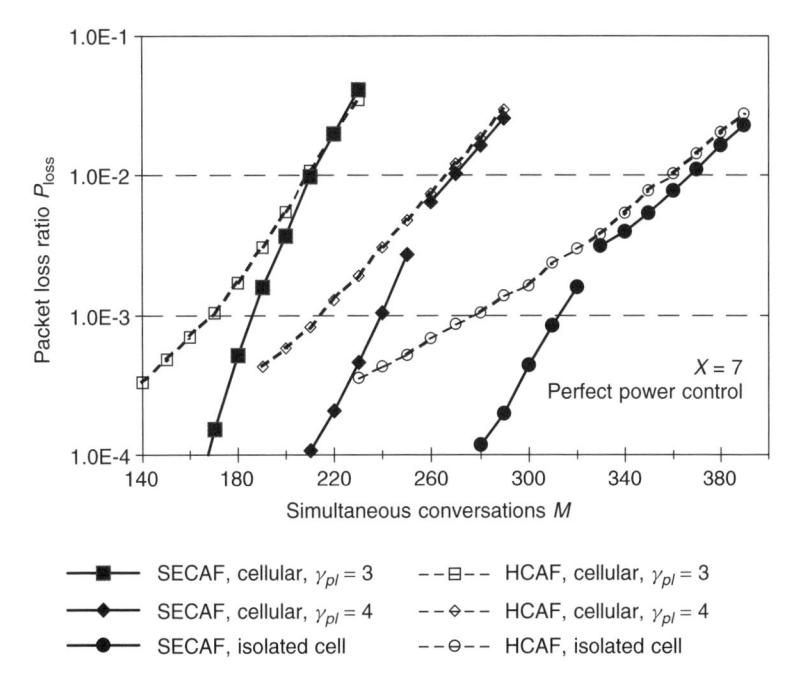

Figure 7.12 Comparison of the packet-loss performance achieved with heuristic and semi-empirical channel access functions (HCAF and SECAF respectively)

Table 7.3 Multiplexing performance summary

Environment Quality of Service	Single Cell $M_{0.001}$	$M_{0.01}$	Cellular, $\gamma_{pl} = 4$ $M_{0.001}$	$M_{0.01}$	Cellular, $\gamma_{pl} = 3$ $M_{0.001}$	$M_{0.01}$
RAP	143	203	119	164	103	137
CSB	220	260	180	200	140	160
HCAF	277	358	214	266	168	209
SECAF	312	367	239	269	185	210
MVB	378	408	282	293	213	236
Δ RAP-SECAF	118%	81%	101%	64%	80%	53%
η_{mux} = SECAF/MVB	0.83	0.90	0.85	0.92	0.87	0.89

To avoid using CAFs which depend on the intercell interference level, p_v could be expressed as a function of the total power level received by the base station from reservation-mode users both in the test cell and in the interfering cells. If this power level were normalised to the reference power level P_0 in the test cell, the CAFs used here for the single-cell case could be applied directly. This would also allow intercell interference fluctuations to be taken into account, to react for instance to congestion in neighbouring cells. Strictly speaking, the BS would have to be able to distinguish between interference stemming from reservation-mode users and that from colliding contention-mode users not only in the test cell, but also in interfering cells, and measure only the former. In practise, however, it may be sufficient to measure the total received power and deduct the contribution of colliding contention-mode users in the test cell (assuming that this contribution can be identified somehow), since the interference contribution from colliding contention-mode users in interfering cells should be small.

7.5 On the Benefit of Channel Access Control

7.5.1 Simulation Results vs Benchmarks

After these extensive considerations on channel access functions and benchmarks, the benefit of channel access control will finally be established quantitatively. Figure 7.13 compares the simulated performance of the random access protocol with that of MD PRMA using the semi-empirical channel access functions depicted in Figure 7.10. Table 7.3 summarises the performances of the different schemes considered in terms of $M_{0.01}$ and $M_{0.001}$. The substantial performance improvement, in certain cases exceeding 100%, is immediately apparent and clearly justifies channel access control *in the scenario considered*, that is, with the given set of design parameters and assuming perfect power control. In all fairness, with reference to Table 7.3, it should be added that the circuit-switching benchmark (denoted CSB in the table), for which no channel access control is required either, outperforms RAP, which reduces the potential gain to be achieved through access control. CSB performs better than RAP because the maximum number of users accessing a single time-slot is constrained to M/N in CSB, while it can in theory assume values up to M with RAP. The impact of altering crucial design parameters and introducing power control errors remains to be studied, but this is postponed until the next section.

For the case of an isolated cell only, Figure 7.14 shows results for access control based on known backlog (KBAC) and for the 'minimum-variance benchmark' (MVB)

7.5 ON THE BENEFIT OF CHANNEL ACCESS CONTROL

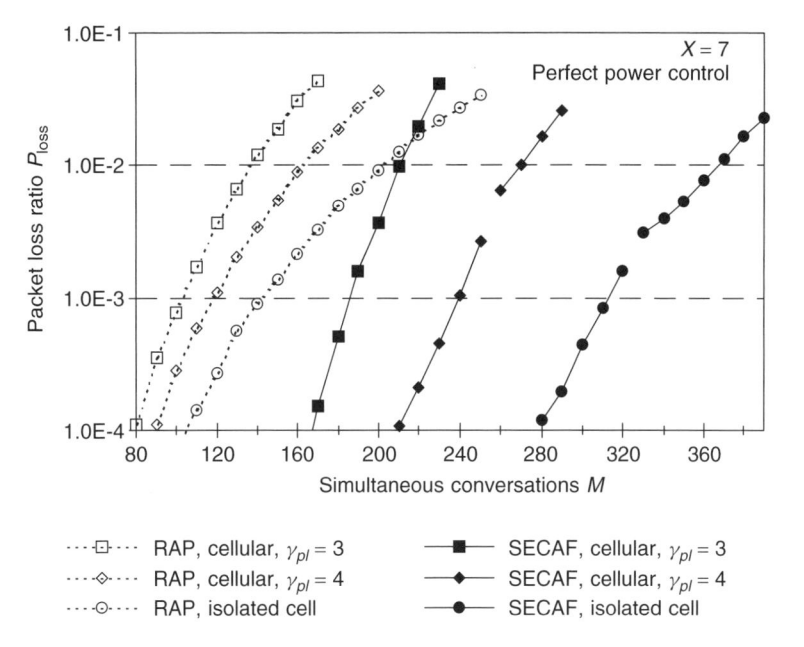

Figure 7.13 Comparison of the packet-loss performance experienced with the random access protocol (RAP) and semi-empirical channel access functions (SECAF)

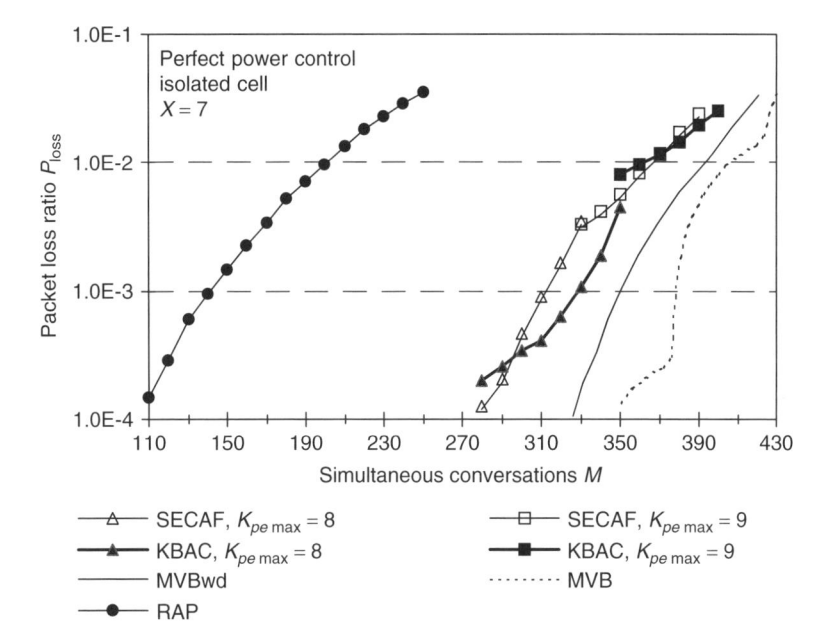

Figure 7.14 Comparison of the performances of RAP, MVB, and KBAC benchmarks with SECAF-results, only for an isolated cell

with and without dropping, on top of SECAF and RAP results. With the exception of MVB, all results are from simulations. Comparing first KBAC and SECAF-results for $M \geq 350$, where the target $K_{pe\,max}$ is in both cases equal to nine, it can be noted that KBAC outperforms SECAF for $M \geq 370$, but the opposite is true for $M < 370$. This is consistent with the discussion in Subsection 7.3.3, where it was stated that access control with KBAC could be too generous in some cases. In fact, as shown in Figure 7.8, the average permission probability values are lower with semi-empirical channel access functions than with KBAC. Similar observations can be made for lower values of M, where decreasing the target $K_{pe\,max}$ to eight provides better performance. Note also that the 'crossover-points' between the two segments with $K_{pe\,max} = 9$ and 8, respectively, are not the same for these two approaches to access control.

Next, looking at the results for MVB, for which the reader is also referred to Figure 7.15, there seems to be a certain headroom for further improvement. The same impression can be gained when consulting the entries for η_{mux} in Table 7.3, where η_{mux} is the multiplexing efficiency relative to perfect statistical multiplexing (i.e. relative to MVB performance). At first glance this cannot be a surprise, since the CAFs used were based on considerable guesswork. On the other hand, although there is a gap between SECAF and MVB results, it is much smaller than that between the SECAF and RAP results (this can be seen particularly well in Figure 7.15). Furthermore, at least a part of the headroom between SECAF and MVB is of a purely theoretical nature, since MVB assumes that scheduling

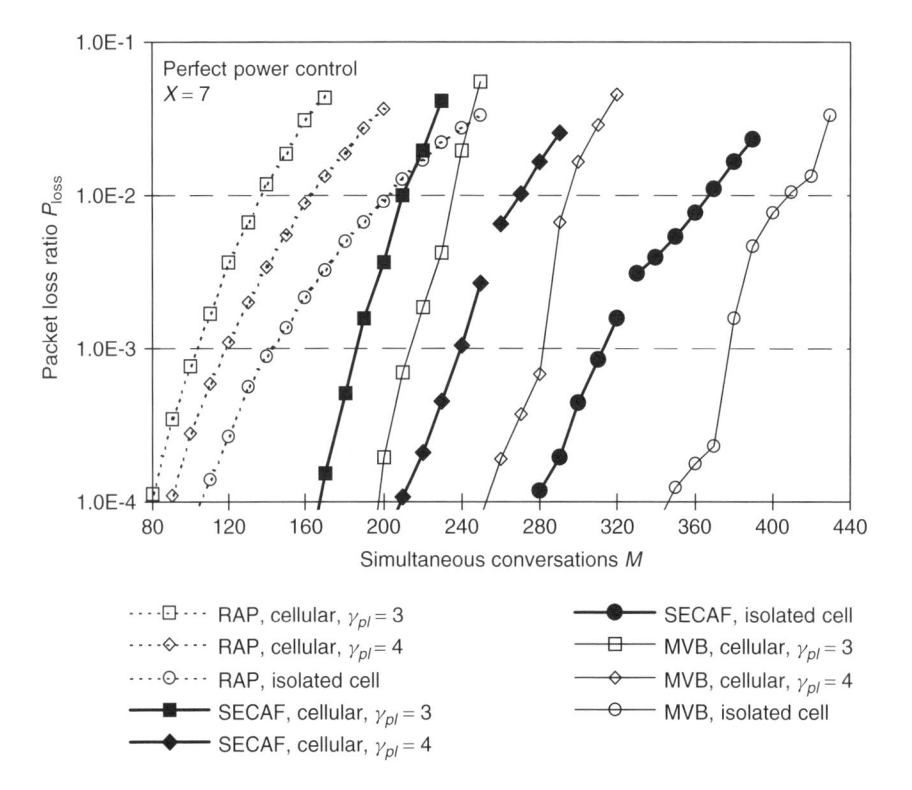

Figure 7.15 Performance of MVB in comparison with results for RAP and SECAF

and statistical multiplexing are both perfect. To assess the 'access control efficiency' or 'multiple access protocol efficiency' η_{map} achieved with SECAF, the performance of MVB *with dropping* should be considered as a reference instead, where imperfect statistical multiplexing is properly accounted for.

From the MVBwd-curve included in Figure 7.14, which is the same as that shown in Figure 7.6, it can be seen that $M_{0.01}$ is equal to 395 and $M_{0.001}$ to 351, resulting in impressive efficiency values η_{map} for SECAF of 0.93 and 0.89 respectively. In conclusion, at least as far as any kind of practicable access control is concerned, there appears to be only limited potential for further improvement.

For completeness, it is pointed out that the curves shown in Figure 7.15 for the two cellular cases are steeper than those for the isolated test cell, because as M increases, both the level of intracell *and* intercell interference increases, the latter due to the assumption of equally loaded cells. Finally, note that we provide only simulation results for MD PRMA with CAFs. Apart from observing that our SECAF and HCAF curves lie consistently between the RAP curves verified by analytical results and the theoretical MVB curves, as it should be, we also report that most of the HCAF and SECAF results were reproduced by Hoefel and de Almeida in Reference [38]. Hoefel and de Almeida have recently provided an equilibrium point analysis (EPA) for MD PRMA with HCAFs in Reference [267], which they claim matches well with their simulation results. However, for reasons outlined in Reference [61], we are reluctant to use EPA for MD PRMA with load-based access control and have not attempted to provide analytical results for MD PRMA with HCAFs and SECAFs.

7.5.2 Benefits of Fast Voice Activity Detection

For reasons listed in Section 5.5, only slow voice activity detection is considered. In Reference [38], Hoefel and de Almeida have also investigated the performance of MD PRMA with SECAF using a fast detector, with the small modification compared to PRMA that users in the MINI SILENCE state keep their reservation, but otherwise under exactly the same conditions as considered here. Such an approach would not make sense in conventional PRMA, since P_{drop} can only be decreased if users relinquish their resource during mini-gaps. In a CDMA environment, on the other hand, at least when random coding is assumed, it does make sense, since not transmitting during mini-gaps reduces the MAI in the respective slot, such that P_{pe} and consequently also P_{loss} decrease.

Hoefel and de Almeida report a capacity increase (in terms of $M_{0.01}$) of 8% achieved in this manner, which compares to a voice activity reduction of 12% through detection of mini-gaps (i.e. from 0.426 to 0.375)[4]. There is, however, an issue to consider, which will make it difficult to achieve this gain in practise. In the hybrid CDMA/TDMA environment considered here, with rather low spreading factors, it will normally not be possible to assign a code to every user admitted to the system. Instead, users will have to dynamically share the limited code resources available, for instance in the manner considered in the next two chapters, where every time-slot carries a limited number of code-slots.

In such a scenario, for the protocol to work efficiently, the terminals have to release their code-slot immediately after termination of a spurt, i.e. when a principal gap starts.

---

[4] From the text in Reference [38], it appears that they compare HCAF-results for the slow detector with SECAF-results for the fast detector, which is somewhat unfair. Using our SECAF-results for the slow detector, the potential gain reduces to 6%.

Unfortunately, when a gap starts, the voice activity detector may not know whether this is a mini-gap or a principal gap, and therefore the terminal does not know whether to keep the reservation or not. It may be that elaborate signal processing methods will allow a reliable estimation of the nature of a gap. Another solution would be to delay the release of a code-slot by roughly the average mini-gap-duration. In this case, however, increased P_{drop} is likely to offset any reduction in P_{pe}, since resources are occupied longer than necessary and terminals have to contend again for resources following longer than average mini-gaps, such that it may not be possible to increase the capacity.

7.5.3 Interpretation of the Results and the 'Soft Capacity' Issue

On cellular communication networks, voice calls last typically only a few minutes, and the number of ongoing calls varies with time. One could therefore question the wisdom of performing simulations with a fixed number of calls per simulation-run, and with simulation times exceeding by far the average call duration, and argue that a result such as $M_{0.01}$ is not particularly meaningful. Some elaboration is here in order.

As discussed in Section 4.6, in blocking-limited circuit-switched cellular communication systems, the quantity of interest is the number of Erlangs, which can be supported at a certain blocking level. This quantity depends on the number of calls that can be served simultaneously, which is equivalent to the number of traffic channels N_{tc} available. In a single-service scenario (e.g. full-rate voice only), it can either be calculated using Equation (4.1), namely the Erlang B formula, or looked up in so-called Erlang tables (see e.g. Reference [2]).

In packet-switched systems, the number of calls that can be supported simultaneously is not equivalent to the number of channels (or here, resource units) U available. This is exactly where $M_{0.01}$ (or $M_{0.001}$, depending on the quality required) comes into play. Since this result tells us how many calls can be supported at any one instant at a given quality level, it could in theory be used instead of N_{tc} in Equation (4.1), to determine the number of Erlangs supported by the system. Note though that with increasing M, the quality decreases gradually rather than suddenly (as long as the system remains stable), a feature which is often referred to as *soft capacity*. Hence, it is possible to exceed temporarily $M_{0.01}$ instead of blocking a call, if this is advantageous in terms of customer perception. In other words, the system is not strictly blocking-limited; it exhibits *soft blocking* rather than hard blocking. In fact, in this particular case, where we are considering an interference-limited CDMA system, already 'U is soft', that is, we would get soft capacity even if we were to carry voice on dedicated traffic channels. On the other hand, even with conventional PRMA on an otherwise blocking-limited system, where 'U is hard', 'M is soft' in that exceeding 'for example' $M_{0.01}$ slightly results in only a slight increase in P_{drop} (provided that the system remains stable). The peculiarity of the scenario considered in this chapter is that the soft capacity feature is due to both the CDMA and the PRMA elements of the multiple access protocol considered.

In any case, $M_{0.01}$ has a role to play for admission control, whether it is used as a hard upper limit or as a reference value, which may be exceeded temporarily. In the former case, the *average* number of supported calls will be smaller than $M_{0.01}$, which in turn means that the average P_{loss} will also be below 1%. One could therefore be tempted to increase the threshold level for admission control to a point at which the average P_{loss} just meets the 1% level. At this point, however, it is important to recall that with a dynamic number of calls, average P_{loss} over all calls is not a satisfactory performance measure anymore. The perceived quality will fluctuate with the currently active number of calls, such that not only the quality of calls would have to be assessed individually, but even the quality of individual call segments, to account for the fluctuations during the lifetime of each call.

Summarising, imposing a hard limit of $M_{0.01}$ for call admission purposes may be rather conservative, but this limit should not be increased without refined QoS measures. Furthermore, although the system is not strictly blocking-limited, using $M_{0.01}$ as an input value for the Erlang B formula or Erlang tables would give a good indication on the number of Erlangs it can support.

7.6 Impact of Power Control Errors and the Spreading Factor on Multiplexing Efficiency

7.6.1 Impact of Power Control Errors on Access Control

Owing to the perfect power control assumed so far in conjunction with access control, the slopes of the $Q_{pe}[k]$ curves shown in Figure 7.11 are rather steep. Above $K_{pe\,max}$, the probability of success decreases rapidly. If power control errors, bound to occur in a real system, are introduced, these slopes become flatter, and it is to be tested whether this has an adverse impact on the benefit that can be attained through access control.

To do this, simulations were performed with a power control error standard deviation σ_{pc} of 1 dB according to the model outlined in Subsection 5.2.6 and with CAFs (both heuristic and semi-empirical) optimised for this specific value of σ_{pc}. Figure 7.16 shows these results, together with the respective results for MVB and the ones for RAP, the latter corresponding to those in Figure 7.2. Note that in case of SECAF and $\sigma_{pc} = 1$ dB, three different access functions were used for $K_{pe\,max} = 5$, 6, and 7, as listed in Table 7.4, which shows the values for $f_{CAF}[\hat{R}]$ in the respective cells. Correspondingly, the curve shown in Figure 7.16 for this case is composed of three segments, which are connected with fine dashed lines. The parameters used for the single heuristic CAF considered were $p_{vi} = 0.3$, $\alpha_{ca} = 0.0095$, breakpoint $= 4$ and $\beta_{ca} = 0.13$.

Table 7.5 summarises the $M_{0.01}$ and $M_{0.001}$ values obtained. With all schemes considered, the capacity loss due to power control errors is larger for $M_{0.001}$ than for $M_{0.01}$. The relative loss is smallest for RAP (36% and 25% respectively), slightly larger for MVB, and 5% larger for SECAF than for RAP. Not surprisingly, given the optimisation criterion of $M_{0.01}$ applied to find a suitable HCAF, it is particularly large, namely 46%, with HCAF at $M_{0.001}$. In summary, the capacity gain that can be achieved through access control is slightly reduced, but still exceeds 100% for $M_{0.001}$, and the η_{mux} values achieved are still more than satisfactory.

Hoefel and de Almeida carried out a similar study, the results of which can be found again in Reference [38]. Instead of RAP and MVB, they compared their findings for load-based access control with results reported in the literature for circuit-switched CDMA systems. Their findings are similar to those reported here, namely that the performance loss is only slightly larger with MD PRMA than with schemes not relying on probabilistic access control. While their results for perfect power control correspond almost exactly to the results reported in Reference [29], since they used the access function parameters published therein, results shown here are slightly better in the case of power control

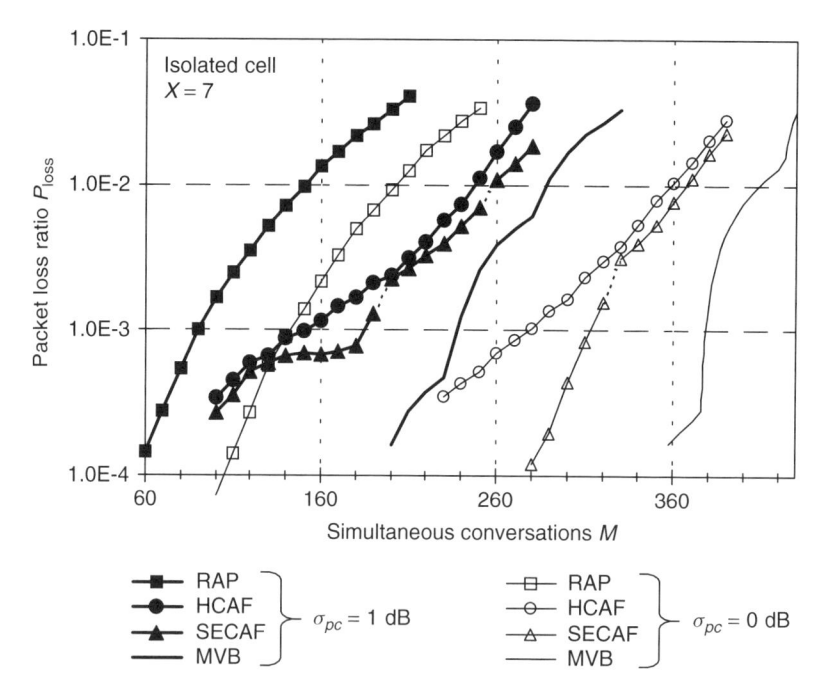

Figure 7.16 Load-based access control vs. benchmarks with perfect power control ($\sigma_{pc} = 0$ dB) and with power control errors ($\sigma_{pc} = 1$ dB)

Table 7.4 Semi-empirical channel access functions, isolated test cell, power control errors with $\sigma_{pc} = 1$ dB

	$\hat{R}=1$	$\hat{R}=2$	$\hat{R}=3$	$\hat{R}=4$	$\hat{R}=5$	$\hat{R}=6$	$\hat{R}=7$
$K_{pe\,max} = 7$	1.0	1.0	0.9	0.6	0.3	0.1	0
$K_{pe\,max} = 6$	1.0	1.0	0.8	0.6	0.4	0	0
$K_{pe\,max} = 5$	1.0	0.8	0.55	0.2	0	0	0

Table 7.5 Impact of power control errors

Quality of Service	$\sigma_{pc}=0$ dB $M_{0.001}$	$M_{0.01}$	$\sigma_{pc}=1$ dB $M_{0.001}$	$M_{0.01}$	Capacity Loss $M_{0.001}$	$M_{0.01}$
RAP	143	203	92	152	36%	25%
HCAF	277	358	150	247	46%	31%
SECAF	312	367	185	257	41%	30%
MVB	378	408	238	288	37%	29%
Δ RAP-SECAF	118%	81%	101%	69%	—	—
η_{mux} = SECAF/MVB	0.83	0.90	0.78	0.89	—	—

7.6 IMPACT OF POWER CONTROL ERRORS

errors. This is most likely due to better access functions, but cannot be verified, since the parameters of the access functions used are not listed in Reference [38].

7.6.2 A Theoretical Study on the Impact of Power Control Errors and the Spreading Factor

Finding efficient channel access functions is rather a time-consuming process, and since it is to a large degree heuristic, it is not clear when to stop, that is, when no further significant improvements are possible. For a more generic study on the combined impact of power control errors and the choice of spreading factors, only RAP and MVB will be considered in the following. In doing so, since MVB is a calculated benchmark, and analytical results for RAP agree very well with simulation results, not only can the process of finding channel access functions for every case considered be avoided, but simulations are avoided altogether. Still keeping the spreading factor at seven, Figure 7.17 shows the performance in terms of $M_{0.01}$ and $M_{0.001}$, normalised to the spreading factor, for several values of the power control error *variance*, σ_{pc}^2. In the case of a 5-dB variance, at $M_{0.01}$, the capacity loss compared to perfect power control amounts to 67% for both RAP and MVB, and exceeds 75% at $M_{0.001}$. This illustrates the need for tight power control in a CDMA system with single-user detection, which is well known from the literature. It also shows once again that the potential benefit that can be obtained through access control is not severely affected by power control errors.

In Section 7.2 it was outlined that the multiplexing efficiency with RAP depends on the average number of users per time-slot, which is low at the rather low spreading factor of seven considered so far. This is why the performance can be improved significantly

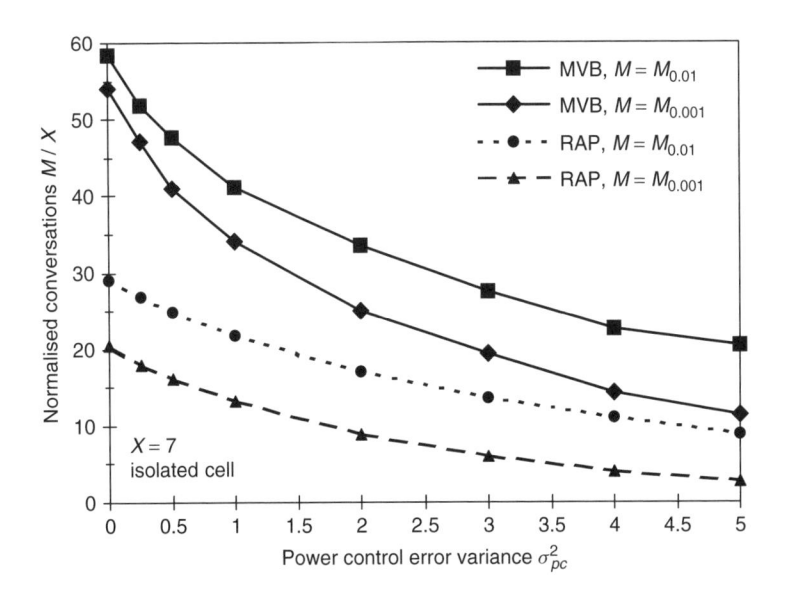

Figure 7.17 $M_{0.01}$ and $M_{0.001}$ normalised to the spreading factor with MVB and RAP, as a function of the power control error variance

through access control. If the spreading factor is increased, whether this be achieved by increasing the total bandwidth, or by reducing the number of time-slots per frame (potentially to a single slot, thus relinquishing the TDMA feature), multiplexing efficiency with RAP increases as well. This in turn must reduce the gain that is achievable through access control, which is demonstrated by Figure 7.18. For MVB, $M_{0.001}$ normalised to X is roughly the same for $X = 7$ and 63, while RAP-performance is significantly better at $X = 63$. Correspondingly, the performance difference between MVB and RAP, while much more substantial at $X = 7$, is only around 20% at $X = 63$, reducing further to slightly more than 10% for $M_{0.01}$ (not shown in this figure, but cf. Figure 7.20 for perfect power control).

For perfect power control, Figure 7.19 compares results for MVB and RAP as a function of X, which again illustrates the very limited sensitivity of MVB to X, while the RAP performance improves with increasing X. It also shows what is termed zero-variance benchmark (ZVB), where the base station schedules a constant number of packets each slot, namely $K_{pe\,max}$ from Equation (7.20). In the case of ZVB, X has to be increased by more than one in order to increase the admissible $K_{pe\,max}$ by one, which explains the steps in the respective curves.

Finally, again for perfect power control, Figure 7.20 illustrates how the performance difference between MVB and RAP decreases. As already discussed earlier, for $M_{0.001}$, this difference amounts to a mere 20% at $X = 63$. More representative of the capacity gain realistically obtainable through access control, the difference between worst-case MVBwd and RAP (not shown in the figure) is only 14%. The Figure also shows how η_{mux} of RAP increases with increasing X: at $X = 63$, η_{mux} achieved with RAP amounts to an impressive 0.88 for $M_{0.01}$, and still to 0.84 for $M_{0.001}$.

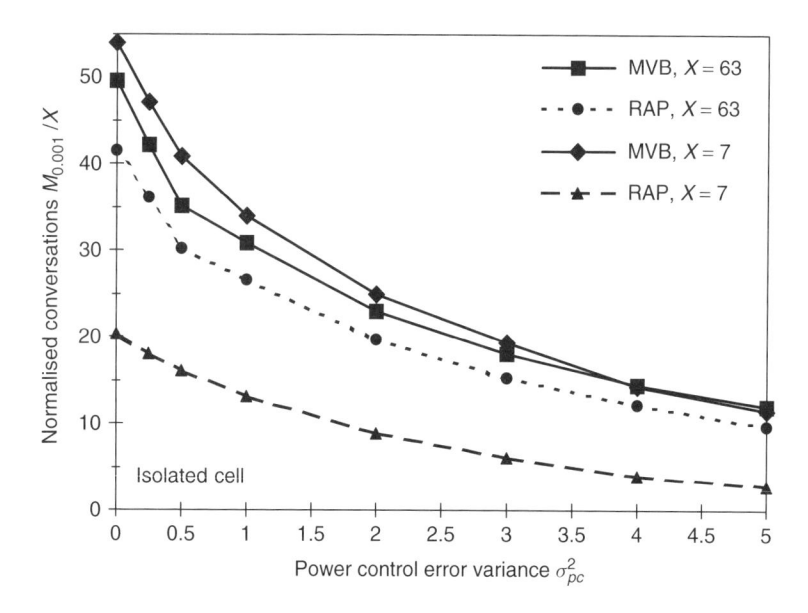

Figure 7.18 $M_{0.001}$ normalised to X with MVB and RAP, as a function of the power control error variance, for $X = 7$ and 63

7.6 IMPACT OF POWER CONTROL ERRORS

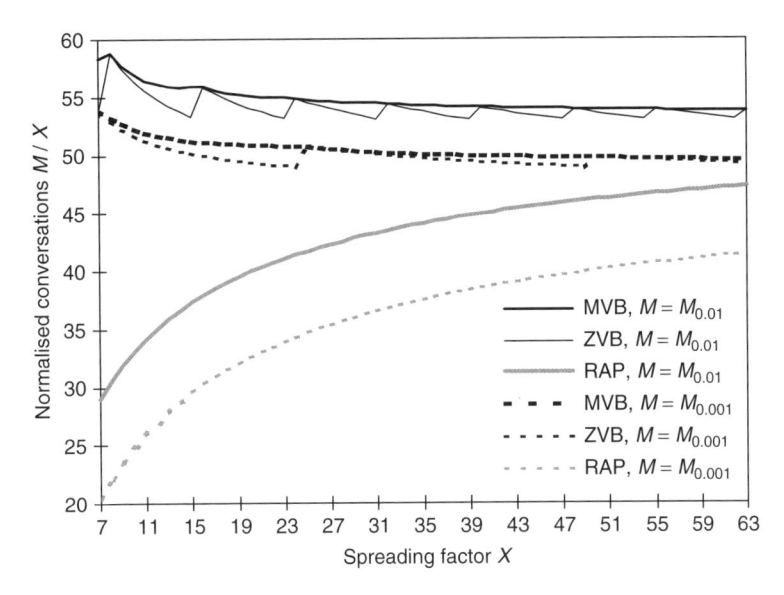

Figure 7.19 $M_{0.001}$ and $M_{0.001}$ normalised to X with MVB, ZVB and RAP, as a function of the spreading factor X

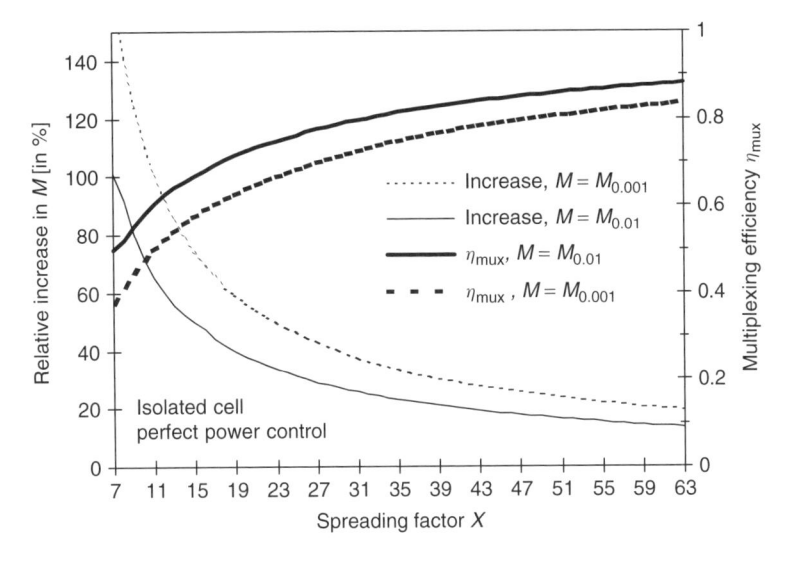

Figure 7.20 Potential capacity increase through access control (left scale) and η_{mux} of RAP (right scale) as a function of the spreading factor

For CDMA systems without a TDMA feature, spreading factors of 63 and higher are possible (see for instance the description of the UTRA FDD mode provided in Chapter 10). In fact, with respect to a system with 20 time-slots per frame and $X = 7$ considered here, the 'equivalent spreading factor' of a system 'without time-slots' would be $X = 140$. This explains why it is sometimes mentioned that CDMA systems provide an inherent near-perfect statistical multiplexing capability.

7.6.3 'Power Grouping': Another Way to Combat Power Control Errors?

For the 'Joint CDMA/NC-PRMA' scheme proposed in Reference [45], which was briefly introduced in Subsection 3.6.5 and mentioned in the context of the minimum-variance benchmark in Subsection 7.3.1, Wen *et al.* suggested 'power grouping' to combat power control errors. The 20 ms TDMA frame starts with a 2 ms control-slot split into mini-slots, followed by twenty 0.9 ms information slots. Each admitted terminal indicates at the beginning of every frame through the sending of a tone in its request mini-slot whether it has a packet to transmit or not, allowing the base station to schedule every packet individually according to the most appropriate policy. The idea of power grouping is that the base station measures the power levels received by all requesting terminals in the mini-slots, and then schedules in each time-slot terminals with as similar power levels as possible. From a MAC perspective, this is an elegant idea to eliminate the negative impact of power control errors, as it minimises the power variance on each time-slot. In practice, however, there are a few problems, mostly to do with the physical layer, not all of them with an obvious solution.

Firstly, immediate slot assignment is assumed, the first batch of terminals are assumed to know their assignment in time to transmit in the information-slot immediately following the control slot. Given that the base station must receive all requests to perform power grouping before it can send assignments in downlink mini-slots, this is not possible even when ignoring propagation delays. This problem could be overcome by regrouping control and information slots and allowing for some scheduling delay.

Secondly, and quite fundamentally, we have ignored in our investigations the temporal behaviour of the power fluctuations, and so have Wen *et al.* in Reference [45]. However, when power level measurements at the beginning of a 20 ms frame are taken as the basis for the scheduling of terminals towards the end of the frame, then temporal fluctuations occurring in the meantime cannot be ignored. Particularly for fast moving mobiles, the beneficial effects of power grouping are expected to be reduced significantly, hence the results shown in Reference [45, Figures 14 and 15], are very optimistic.

Thirdly, there are two problems associated with measuring power levels in the request mini-slots. The first is the limited statistical significance of the measurement in the presence of noise, given that terminals transmit a single tone in a short request mini-slot. The second and more severe one is that the measurements are carried out in control slots using a narrowband physical layer, while a wideband DS-CDMA scheme is used for the information slots. It is unwise to measure the power level in a small segment of the total spectrum used for information transmission and then to expect this measurement to be indicative for the received power level on the wideband carrier.

7.7 Summary

In this chapter, P_{loss} performances of several different schemes were compared for voice traffic only in an environment where both P_{pe} (due to MAI) and P_{drop} (due to access control, where applied) contributed to P_{loss}. As a first reference point, the performance of RAP was investigated, a protocol based purely on random access (hence $P_{\text{drop}} = 0$). It was illustrated that multiplexing efficiency in pure CDMA systems depends essentially on the standard deviation normalised to the mean of simultaneously active users. With the

7.7 SUMMARY

analysis provided for this purpose, it could also be shown that for a hybrid CDMA/TDMA air interface, the multiplexing efficiency depends on the average number of users multiplexed onto a *single time-slot*, if neither access control is applied nor is a reservation-based multiple access protocol used. This explains why the efficiency of RAP is rather low, in fact even lower than that of circuit-switching.

To improve multiplexing efficiency with hybrid CDMA/PRMA, two approaches to dynamic probabilistic access control for the contention phase in reservation ALOHA-based protocols were considered. Appropriate access control leads to load balancing between time-slots, which is where the improvement stems from. At the centre of interest, the benefit of load-based access control through heuristic and semi-empirical channel access functions was assessed. It is this PRMA-based protocol with load-based access control, which we introduced initially as the 'Joint CDMA/PRMA' protocol back in 1995, and which we here refer to as MD PRMA. As a benchmark, access control based on known backlog was also looked at. These two access control schemes perform similarly, with, depending on the load, slight advantages for the backlog-based scheme. It needs to be pointed out, though, that the backlog is assumed known in this scheme, that is, backlog estimation errors are not accounted for. Importantly, compared to RAP, the capacity can be increased significantly with both schemes. At the low spreading factor of seven considered, the capacity (in terms of $M_{0.001}$) is more than doubled in certain conditions.

To assess the multiplexing efficiency achieved with load-based access control, the capacity was compared with that of a so-called 'minimum-variance benchmark', which is based on the assumption of perfect scheduling and perfect statistical multiplexing. With perfect power control, for $M_{0.001}$ relative statistical multiplexing efficiency values η_{mux} from 0.83 to 0.87 were found, and for $M_{0.01}$, from 0.89 to 0.92. Note though that with delay-constrained services, e.g. real-time services, $\eta_{mux} = 1$ cannot be achieved. Therefore, in addition to *multiplexing efficiency*, we also established what we termed *protocol efficiency*, namely the efficiency relative to an extended minimum-variance benchmark, which takes delay constraints and, correspondingly, the resulting packet dropping into account. For an isolated cell, protocol efficiency values η_{map} of 0.89 and 0.93 were found for $M_{0.001}$ and $M_{0.01}$, respectively. This compares to η_{mux} values of 0.83 and 0.90. At least for voice-only traffic, one can therefore conclude that the performance with load-based access control comes close to the optimum performance achievable with any kind of probabilistic access control scheme. This is further confirmed by comparing the results obtained with those for known backlog-based access control as just discussed, and with other efforts to access control reported in the literature, such as those in Reference [36].

The performances of all schemes considered (particularly in terms of $M_{0.001}$) suffer from power control errors. From the results presented here it appears that MD PRMA with load-based access control suffers slightly more than for instance RAP, but, nevertheless, the capacity gain which can be achieved through load-based access control remains substantial. This is consistent with the findings reported in Reference [38]. For the power control error standard deviation σ_{pc} of 1 dB considered, η_{mux} of MD PRMA decreases from 0.83 to 0.78 for $M_{0.001}$, but only from 0.9 to 0.89 for $M_{0.01}$.

Having found that the multiplexing efficiency of RAP depends on the average number of users multiplexed onto a single time-slot, it is obvious that increasing the spreading factor will increase the efficiency as well, since the population per time-slot becomes larger. With perfect power control, at a spreading factor $X = 63$, η_{mux} for RAP amounts to an impressive 0.88 for $M_{0.01}$, and still to as much as 0.84 for $M_{0.001}$.

Since in a pure CDMA system (with only one 'time-slot' per frame) typically spreading factors even larger than 63 are used for low-bit-rate users, probabilistic access control cannot improve the capacity significantly. Besides, irrespective of the spreading factor, the beneficial effect of access control, namely load balancing to reduce the variance of the MAI, would now have to occur between entire frames rather than time-slots, which would only be possible with services exhibiting more relaxed delay constraints than voice. In fact, in CDMA systems, for voice traffic, near-perfect statistical multiplexing can be achieved using 'circuit-switching' or dedicated traffic channels, if voice activity detection or variable-bit-rate voice codecs are applied to reduce interference. A hybrid CDMA/TDMA system using RAP-like protocols or based on circuit-switching is therefore clearly inferior in terms of multiplexing efficiency to a pure CDMA system operating in the same manner[5] and using the same carrier bandwidth. On the other hand, the η_{mux} values found for MD PRMA with load-based access control, $N = 20$ and $X = 7$ are almost the same as those for RAP with $X = 63$ (the latter do not depend on the number of time-slots N)[6]. This means that access control efficiently compensates for the poor multiplexing efficiency experienced in a hybrid CDMA/TDMA system without access control.

Finally, note that 3G systems will also support high-bit-rate users. Using CAFs with two linear segments similar to those used here for HCAF, it was shown in Reference [60] for UTRA FDD that probabilistic access control increases the capacity and decreases the transfer delay of high-bit-rate (i.e. 144 kbit/s) packet-data users with relaxed delay requirements, as discussed in more detail in Section 10.3. This is due to the fact that only a few high-bit-rate users can be accommodated on a carrier, such that the size of the population multiplexed onto this carrier is too small to provide efficient statistical multiplexing without access control.

---

[5] Note that with only one time-slot per frame and on–off activity detection, RAP and circuit-switching are the same.
[6] A fair comparison would require $X = 140$ for pure CDMA. Interpolating the RAP results shown in Figure 7.20 to $X = 140$, we expect the respective η_{mux} values to be slightly higher than those for MD PRMA.

8
MD PRMA ON CODE-TIME-SLOTS

This chapter is concerned with MD PRMA on perfect-collision code-time-slot channels. The simple and abstract channel model used, representative for a blocking-limited system, allows one to consider an arbitrary number of code-slots E per time-slot, without having to worry about the spreading factor required to meet a certain packet erasure performance. In this framework, the scope of investigations can conveniently be extended to two extreme cases, namely only one code-slot per time-slot, but numerous time-slots N per TDMA frame, and only one time-slot per frame carrying numerous code-slots. In the first case, the CDMA feature is relinquished, and MD PRMA degenerates to pure PRMA. In the second case, the TDMA feature is relinquished. While this configuration (and in fact also PRMA itself) can simply be viewed as a special case of MD PRMA, it actually corresponds to the Reservation-Code Multiple Access (RCMA) protocol proposed in Reference [35].

As in Chapter 7, only voice-traffic will be considered. However, the focus shifts from load-based access control to fixed permission probabilities and backlog-based access control (the latter in the shape of Bayesian broadcast). The performances of pure PRMA, MD PRMA and RCMA will be compared, all with the same number of resource units $U = N \cdot E$. For MD PRMA with $N = 8$ and $E = 8$ (i.e. the original UTRA TD/CDMA parameters), the impact of acknowledgement delays and TDD operation on voice dropping performance is also studied. Furthermore, the code-time-slot channel model is enhanced to account for multiple access interference (MAI). In this scenario, unlike the perfect-collision case, load-based access control can make sense. Therefore, on top of 'conventional' Bayesian broadcast, a scheme combining Bayesian broadcast with a channel access function is considered.

8.1 System Definition and Simulation Approach

8.1.1 System Definition and Choice of Design Parameters

The common thread in this chapter is the consideration of code-time-slots based on the TDMA frame duration specified in Section 5.3, namely the 4.615 ms used in GSM and originally proposed for TD/CDMA. However, the focus is not limited to the TD/CDMA scenario with $N = 8$ time-slots and $E = 8$ codes per time-slot. Instead, on top of this balanced case, two extreme cases are also considered, namely one with $N = 64$ time-slots, but only one 'code-slot' per time-slot, and one with $E = 64$ codes on a single 'time-slot'. Effectively, the first case represents pure TDMA, where MD PRMA degenerates to conventional PRMA, and the second case is pure CDMA, for which MD PRMA corresponds to RCMA proposed in Reference [35]. Choosing the same frame duration and the same number of resource units U (namely 64) for all three schemes allows for a fair comparison of their respective performance.

For these three cases, MD PRMA for frequency division duplexing as defined in Section 6.2 is investigated, assuming immediate acknowledgement and using either fixed permission probabilities for voice (again the only traffic considered), or backlog-based access control. In the latter case, the voice permission probability p_v (or simply p) is calculated according to the Bayesian algorithm adapted for MD PRMA, as outlined in Subsection 6.5.4. Equation (6.9) is used to carry out the estimation of the arrival rate required for this algorithm. Considering an ideal case, the value of p_v is broadcast at the end of each time-slot in such a manner that it is available to all mobile stations with full precision before the next time-slot starts.

For the scenario with $N = 8$ time-slots and $E = 8$ codes per time-slot, the impact of acknowledgement delays is also studied by varying the parameter x introduced in Subsection 6.2.6. This parameter determines how many time-slots a terminal must wait for an acknowledgement following the time-slot in which it sent a packet in contention mode. While waiting, it is not allowed to contend again. In the case of Bayesian broadcast, if $x > 0$ (i.e. acknowledgement is not immediate), the Bayesian algorithm needs to be modified, that is, p_v needs to be calculated through Equation (6.7). Unlike the acknowledgements, p_v is assumed to be broadcast immediately at the end of each time-slot. For the same configuration of resource units, the performance of MD FRMA for TDD with a single switching-point per frame, as specified in Subsection 6.3.3, is assessed. From one to eight time-slots per TDMA frame are assumed to be assigned to the uplink direction, where the last case is obviously only of academic interest, since no resources would be available for the downlink in this case.

In the following two sections, when more than one code-slot is considered, these slots are assumed to be mutually orthogonal, which means that MAI is ignored. If dedicated channels were used, the system would exhibit hard-blocking, but owing to the PRMA element, it features soft-blocking or soft-capacity. In Section 8.4, on the other hand, MAI is accounted for in the manner specified therein, in order to assess the impact of the loss of orthogonality on access control. In this case, depending on the quality of service requirements, we are dealing with an interference-limited system; that is, excessive packet erasure may prevent all U resource units from being used. In the terminology used in Subsection 7.5.3, 'U is soft up to an upper limit of $N \cdot E$'. When interleaving is applied, it is rectangular interleaving over the length of a voice frame, which in turn is carried on four bursts (see Subsection 6.2.4). In this case, request bursts sent in contention mode are dedicated signalling bursts, transmitted on a single code-time-slot. By contrast, when interleaving is not applied, they carry not only signalling, but also user data, namely the same amount as carried by information bursts.

For the basic scheme without interleaving, the delay threshold $D_{\max}$ is normally set to a small value of 4.615 ms, which is equal to the length of a single TDMA frame. In the case of interleaving, $D_{\max}$ is set to the length of a voice frame, i.e. 18.462 ms. To isolate the impact of interleaving and dedicated request bursts, the basic scheme is also operated

8.1 SYSTEM DEFINITION AND SIMULATION APPROACH

Table 8.1 Parameters relevant for the physical layer, protocol operation and traffic models

Description	Symbol	Parameter Value
TDMA Frame Duration	D_{tf}	4.615 ms
Time-slots per Frame	N	8 (or 64, or 1)
Code-slots per Time-slot	E	8 (or 1, or 64)
Dropping Delay Threshold	D_{max}	4.615 ms (no interleaving)
		18.462 ms (with interleaving)
Mean Talk Gap Duration	D_{gap}	1.74 s (or 3 s)
Mean Talk Spurt Duration	D_{spurt}	1.41 s (or 3 s)

with a D_{max} of 18.462 ms in one case. Together with the traffic parameters discussed in the next subsection, all parameters mentioned so far are summarised in Table 8.1.

8.1.2 Simulation Approach, Traffic Parameters and Performance Measures

As in the previous chapter, the only traffic considered in the following is packet-voice traffic, using the two-state voice model specified in Section 5.5. Two different parameter sets are considered. The first set, namely $D_{spurt} = 1.4$ s and $D_{gap} = 1.74$ s, is from the RACE ATDMA project [46], and results in a voice activity factor α_v of 0.448, which is slightly higher than that in Chapter 7. As a second set, $D_{spurt} = D_{gap} = 3$ s taken from Reference [56] is used. This is to establish a link with Chapter 9, where mixed voice and data traffic is considered, and parameters from Reference [56] are used for both voice and Web browsing traffic.

The system load is determined by the number of conversations M simultaneously supported, and we are interested in P_{drop} performance as a function of M. Analogous to Chapter 7, $M_{0.01}$ and $M_{0.001}$ stand for the number of conversations which can be supported at a tolerated P_{drop}, $(P_{drop})_{max}$, of 1% and 0.1% respectively. A static scenario is considered, where P_{drop} is established as a function of M, and M remains fixed over the relevant period of observation. Multiplexing efficiency η_{mux} relative to perfect statistical multiplexing can easily be calculated using Equation (6.1). In Section 8.4, where MAI is accounted for, the relevant figure of merit is P_{loss} instead of P_{drop}, exactly as in Chapter 7.

Each simulation-run with fixed M covers 1000 s conversation time. Where required, several simulation-runs were performed for the same value of M, in which case P_{drop} and P_{loss} reported are the averaged result over these simulation-runs.

8.1.3 Analysis of MD PRMA

Pure and modified PRMA systems were analysed for instance in References [135,143,144, 149,150,268,269]. Most of these articles provide a full Markov analysis, some an equilibrium point analysis (EPA). Due to the dimension of the state space with the here considered design parameters, a full Markov analysis is rather challenging. In Reference [61], we provided an EPA for MD PRMA, which expanded on the EPA for PRMA provided in Reference [143] and adopted a few elements of Reference [149]. In certain scenarios, we found EPA to be satisfactory, in others not. In the following, we focus on protocol performance assessment through simulation studies.

8.2 Comparison of PRMA, MD PRMA and RCMA Performances

8.2.1 Simulation Results, No Interleaving

Figures 8.1 to 8.3 show P_{drop} performance of MD PRMA, PRMA, and RCMA respectively, with different fixed p_v values (in the figures simply referred to as p) on one hand, and p_v calculated through the Bayesian algorithm on the other. In all cases, the basic scheme without interleaving and a very short packet dropping delay threshold D_{max} equal to D_{tf}, namely 4.615 ms, was considered.

With MD PRMA (Figure 8.1) and Bayesian Broadcast (BB), $M_{0.01} = 131$ and $\eta_{\text{mux}} = 0.92$, while $M_{0.001} = 119$ (in which case $\eta_{\text{mux}} = 0.83$). With fixed p_v, $M_{0.01}$ lies between 121 (for $p_v = 0.1$) and 131 ($p_v = 0.3$), and $M_{0.001}$ peaks at 118 (with $p_v = 0.5$). This seems to indicate that if $M_{0.01}$ (or $M_{0.001}$) were the only performance measure of interest, there would not be much benefit in implementing adaptive access control. However, while it is possible to achieve high capacity with a fixed p_v, it is not possible to achieve high capacity with the same p_v value which gives low packet dropping at lower load. Furthermore, if p_v is too large, MD PRMA can become unstable. With the values considered here for M, this was experienced for $p_v \geq 0.6$ and $M = 140$.

In cases in which instability is experienced, P_{drop} results established through simulations are heavily affected by the instance in time in which the system first experienced congestion. Once caught in a congested equilibrium point, it is almost certain that the system remains in this state for the remainder of the simulation run and, from then on,

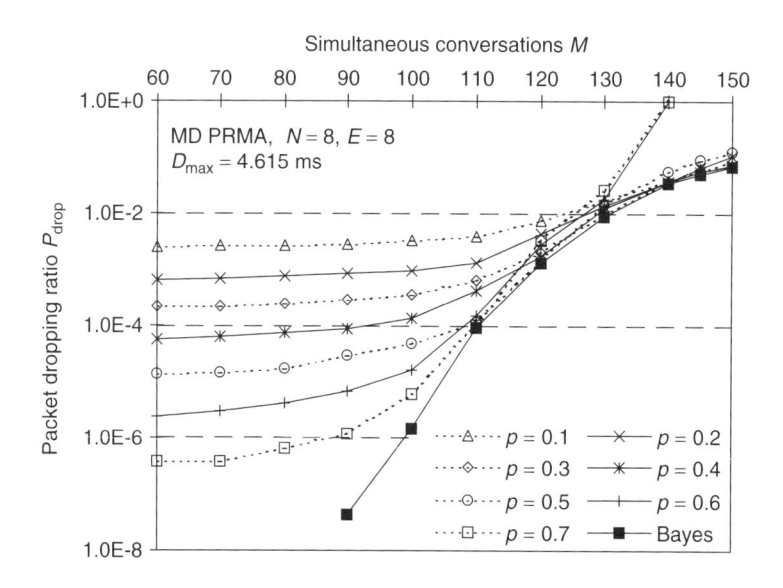

Figure 8.1 Simulated MD PRMA performance, overview

8.2 COMPARISON OF PRMA, MD PRMA AND RCMA PERFORMANCES

the dropping probability is close to one. For values of M for which stability problems were experienced, rather than reporting the average P_{drop} measured over a few simulation runs, which would not deliver statistically relevant results, P_{drop} was simply set to one in Figures 8.1 and 8.2. A better performance measure in such cases would be the so-called First Exit Time (FET) proposed in Reference [194] for slotted ALOHA and applied to PRMA in Reference [149]. The FET is the average first exit time into the unsafe region (i.e. a system state beyond the unstable equilibrium point, see Figure 3.6) starting from an initially empty channel or system.

Choosing $p_v = 0.5$ offers the best compromise between capacity ($M_{0.01} = 128$) and low dropping at low load, while appearing to allow for stable operation up to $M = 140$ (that is, the FET is much larger than the duration of an individual simulation-run). BB on the other hand allows for stable operation at high load while ensuring low packet dropping at low load and performs at least as well as the fixed p_v approach over the entire range of M considered.

One could argue that the performance of BB could be met by choosing a semi-adaptive approach, i.e. selecting p_v depending on M. However, such an approach cannot easily be extended to a mixed traffic scenario, possibly with unknown traffic statistics, whereas BB adapts automatically to different traffic mixes. Furthermore, it would also require regular signalling of p_v, leaving reduced computational complexity as the only potential argument in its favour. In view of the very small complexity of BB, this advantage is of no relevance in practice, though.

Similar considerations apply in the case of pure PRMA. In fact, looking at Figure 8.2, to avoid stability problems, p_v has to be selected even more carefully. Here, with BB, $M_{0.01} = 129$ ($\eta_{mux} = 0.9$), and $M_{0.001} = 119$. With fixed p_v, $M_{0.01}$ lies between 123 (for $p_v = 0.05$) and 129 ($p_v = 0.2$). $M_{0.001}$, on the other hand, although assuming 118 for $p_v = 0.4$, is limited to 114 ($p_v = 0.2$), if the only values of p_v considered are those for which the system remains stable up to $M = 140$.

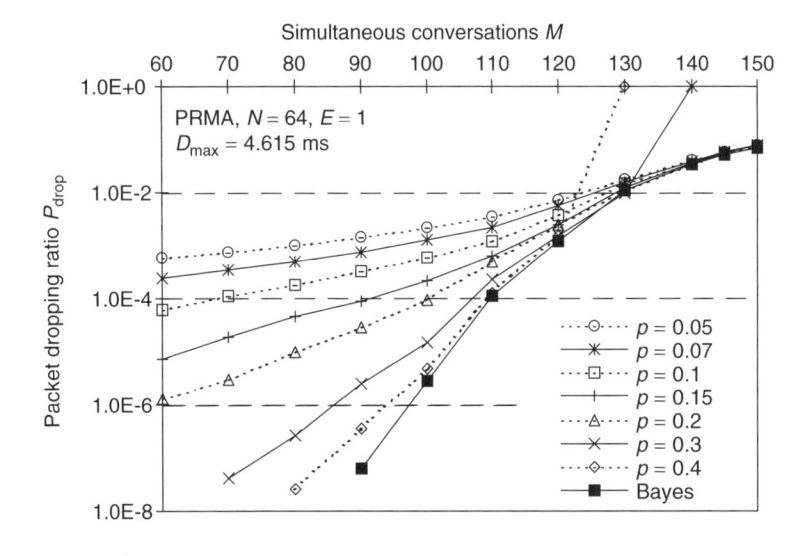

Figure 8.2 Simulated PRMA performance, overview

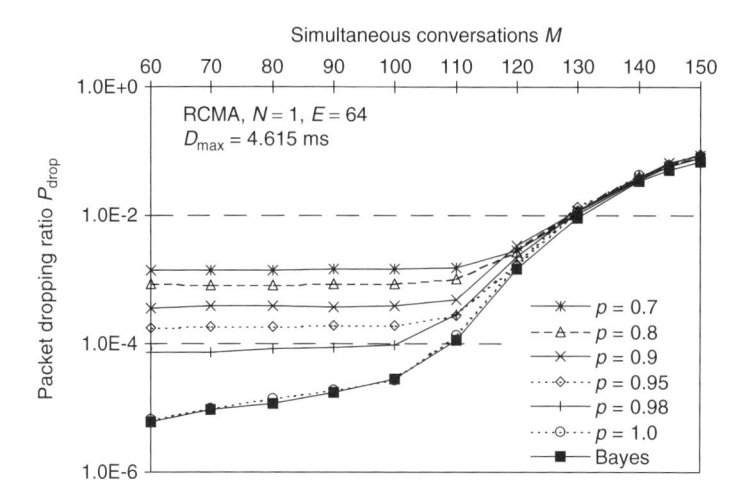

Figure 8.3 Simulated RCMA performance, overview

Finally, with RCMA, the situation is slightly different, as illustrated in Figure 8.3. Note first that, since $D_{\max} = D_{tf}$, there is only one contention opportunity for a terminal until the first packet in a spurt is dropped, such that there will be significant dropping irrespective of the load, as soon as $p_v < 1$. In fact, for $p_v < 0.98$ and $M \leq 100$, P_{drop} is almost uniquely determined by the waiting probability $1 - p_v$, which explains the flat segment of the respective curves. On the other hand, even if there is a temporary accumulation of contending terminals, they will normally be able to choose between numerous code-slots available for contention, such that the collision risk is small. Therefore, stability is not an issue even for $p_v = 1$. This in turn means that there is limited benefit in controlling access dynamically, e.g. through Bayesian broadcast, which is also shown in the figure. This is very much in contrast to pure PRMA (and to a lesser extent to MD PRMA), where the accumulation of a few contending terminals C, such that $C > 1/p_v$, can result in a number of successive collisions. During these collision slots, C will grow further, and eventually, $C \gg 1/p_v$ (or in the case of MD PRMA, $C \gg E/p_v$), such that the system is bound to become unstable. To complete the discussion of the results for RCMA, with the p_v values considered, $M_{0.01}$ is between 128 and 129, while with BB, $M_{0.01} = 130$. $M_{0.001}$ assumes a value of 118 for both BB and $p_v = 1$.

8.2.2 Performance Comparison and Impact of Interleaving

In Reference [35], it is claimed that 'RCMA is superior to PRMA in terms of system capacity even when a median size of code set is used'. To come to this conclusion, the authors of Reference [35] applied a frequency reuse factor of seven to PRMA, which may be considered conservative, but is probably not completely unrealistic. At the same time however, and curiously enough, the authors spent not a single word on where the 'median number of codes' should come from and what kind of bandwidth or spreading factor would be required to support the corresponding number of simultaneous users.

For reasons outlined in detail in Sections 3.2 and 5.1, we have no intention of stepping onto a field full of mines by trying to assess the spectral efficiency of TDMA, hybrid

8.3 DETAILED ASSESSMENT OF MD PRMA AND MD FRMA PERFORMANCES

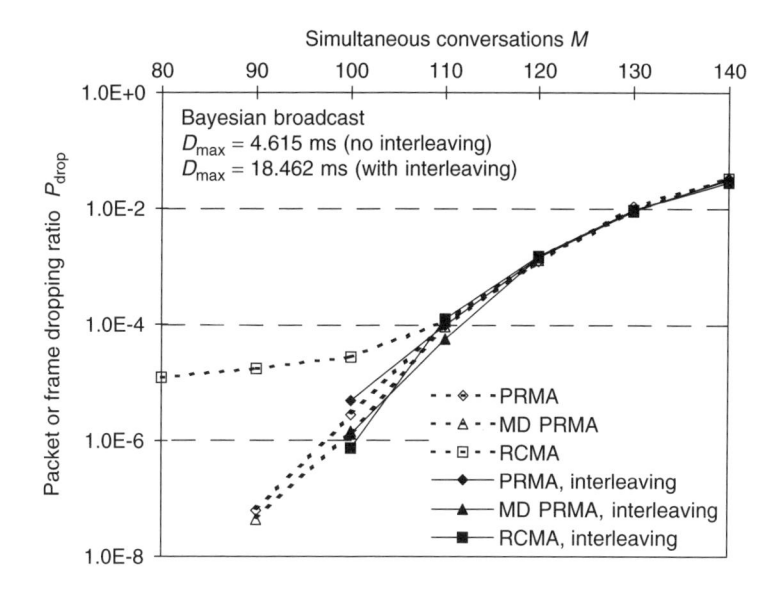

Figure 8.4 PRMA, MD PRMA, and RCMA with Bayesian broadcast

CDMA/TDMA, and CDMA systems operating with PRMA, MD PRMA, and RCMA respectively. Here, the focus is exclusively on the efficiency of the multiple access protocols as such. From this point of view, the only fair comparison appears to be one based on an equal number of resource units, equal frame duration, and assuming a perfect collision channel for individual units (whether these be code, time, or code-time-slots).

This is exactly how Figures 8.1 to 8.3 were obtained, and while the exact P_{drop} behaviour depends on the scheme and the p_v value considered, capacity in terms of $M_{0.01}$ is virtually identical for these three schemes, namely 130 ± 1. With Bayesian broadcast, even P_{drop} is the same for $M > 110$. The only notable difference is the somewhat higher dropping ratio at low load in the case of RCMA, which is due to the single contention opportunity available before packets are dropped, since $D_{max} = D_{tf}$. If interleaving over four bursts is applied, and D_{max} is chosen to be D_{vf} (i.e. the length of a voice frame), the excellent agreement between the performance of these three schemes can even be extended to $M \leq 110$. This is shown in Figure 8.4.

For completeness, as we did already in Section 1.4, we point again at Reference [17], where CDMA, TDMA and hybrid systems are compared from a packet queuing perspective.

8.3 Detailed Assessment of MD PRMA and MD FRMA Performances

8.3.1 Impact of Acknowledgement Delays on MD PRMA Performance

It was explained in Section 3.6 why immediate acknowledgements are conceptually impossible with PRMA protocols using frequency division for duplexing. In Subsection 6.2.6 a parameter x was introduced to model acknowledgement delays (see also Subsection 8.1.1). With increasing x, one would expect increased P_{drop}, since unsuccessfully contending terminals spend extra time to get a reservation. On top of that, a value of x greater than zero can also have a negative impact on the accuracy of the backlog estimation through the Bayesian algorithm, even though appropriately enhanced to cope with this situation. This is illustrated in Figures 8.5 and 8.6 for the basic MD PRMA scheme without interleaving, setting D_{max} to 4.615 ms.

Figure 8.5 shows the performance achieved with Bayesian broadcast for $x = 0$ to 6, and Figure 8.6 compares the performance of Bayesian broadcast with that of perfect backlog estimation (in Chapter 7 referred to as known-backlog-based access control, KBAC) for selected values of x. In both cases, as expected, voice dropping increases with increasing x, particularly at low load. However, while the performance, as far as statistically relevant, is exactly the same for $x = 0$ (i.e. immediate acknowledgement), the Bayesian algorithm suffers much more with increasing x.

The reason is as follows: at low load, the backlog is close to zero in most time-slots, and consequently, p_v is set to one. A sudden increase in backlog during a single time-slot period will normally result in a few successive collisions, if the number of available C-slots $A[t]$ is low. This will cause the algorithm to adapt to the situation and lower p_v. With $x = 0$, this can be achieved quickly and packets are rarely dropped. With $x > 0$, collisions will occur with period $x + 1$ time-slots, so spreading them in time, and the algorithm will need longer to adapt.

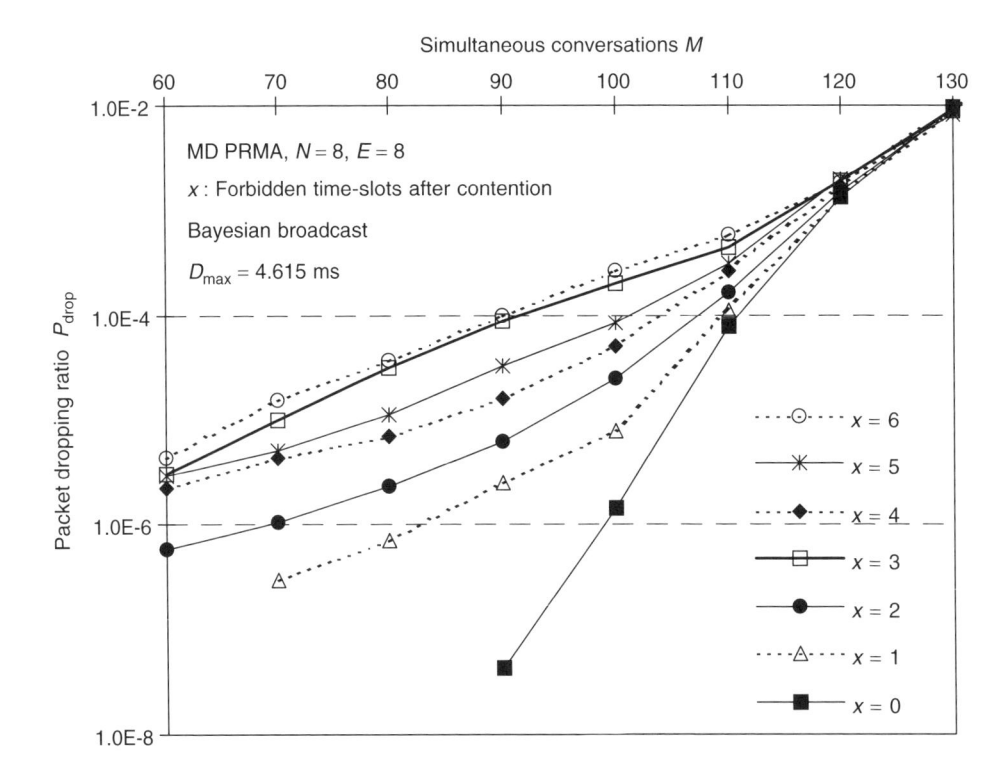

Figure 8.5 Impact of acknowledgement delays on Bayesian broadcast

8.3 DETAILED ASSESSMENT OF MD PRMA AND MD FRMA PERFORMANCES

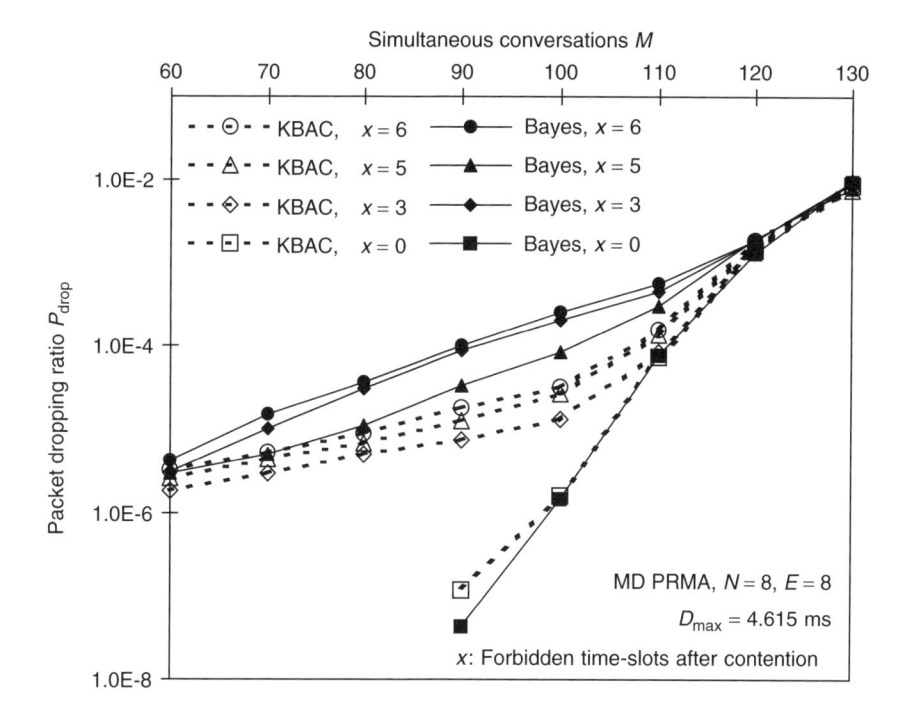

Figure 8.6 Backlog estimation (Bayes) vs known backlog (KBAC)

Even worse, it may be deceived by time-slots with numerous idle and success C-slots lying in-between the 'collision-time-slots' (due to other terminals accessing the system in time-slots not affected by collisions). This will further delay the tracking of the real backlog, resulting in packet dropping for those terminals caught in the 'collision slots'.

Interestingly, Bayesian broadcast produces particularly bad results with $x = 3$, which are even worse than those for $x = 4$ and 5. Here, a further factor comes into play: not only are collisions repeated every four time-slots, but also the $A[t]$ patterns will exhibit some repetitive behaviour with double this period (i.e. $N = 8$ time-slots), thus a 'bad slot' in terms of backlog may coincide regularly with 'bad slots' in terms of low $A[t]$ values. This could probably be described as 'resonant behaviour' or 'local catastrophes'. The comparatively large average P_{drop} in such circumstances is due to a few MS suffering considerable dropping, while those MS never caught in a 'bad slot' experience very moderate dropping.

Figure 8.7 compares the impact of acknowledgement delays on the performance when using fixed permission probabilities with that when using Bayesian broadcast. The dropping performance with $p_v = 0.2$ and 0.3 does not depend on x, since dropping is almost uniquely due to mobile stations not getting permission to send contention packets in this case (because of a large waiting probability $1 - p_v$), while C-slot collisions occur rarely. With $p_v = 0.5$, the dropping performance depends to some extent on x, although to a far lesser extent than observed for Bayesian broadcast. With $x = 5$, any performance advantage of the Bayesian approach dwindles away and $x = 3$ must obviously be avoided for BB, for the reasons just discussed.

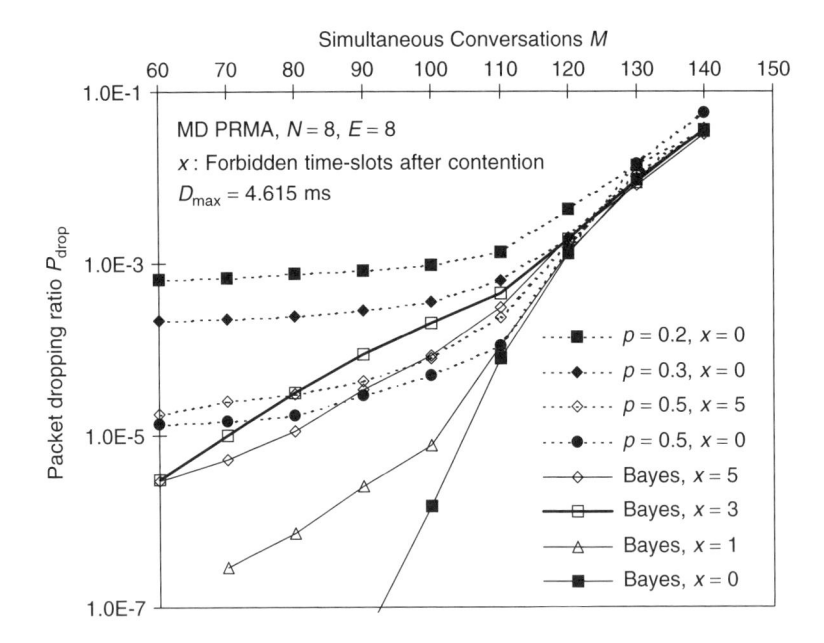

Figure 8.7 Bayesian broadcast vs fixed permission probabilities

To mitigate the problem experienced with $x = 3$, p_v calculated with BB could be limited to a maximum value $p_{v\,\max}$ below one, to disrupt the regular coincidence of the two types of 'bad slots'. However, this would defeat the purpose of broadcast control, which is to stabilise the protocol and ensure efficient operation for all possible traffic scenarios, without having to choose values for traffic dependent parameters, such as $p_{v\,\max}$.

8.3.2 MD FRMA vs MD PRMA

Results shown in Figure 8.8 serve several purposes. They allow the assessment of the impact of:

- increasing $D_{\max}$ from 4.615 ms to 18.462 ms in the basic MD PRMA scheme (assuming immediate acknowledgement);
- the added traffic in the case of interleaving (I/L) due to dedicated request bursts and on average two additional bursts per voice spurt due to rounding up the spurt duration to an integer number of voice frames (again assuming immediate acknowledgement); and
- inherent delay of acknowledgements (by an average of slightly less than $N/2$ time-slots) in the case of MD FRMA (again with interleaving).

Note that MD FRMA is considered here with eight uplink time-slots per frame, to compare all schemes with an equal number of uplink time-slots. This effectively means using MD FRMA in FDD mode, which would in practice not leave any time to signal acknowledgements for the entire frame before the subsequent frame starts.

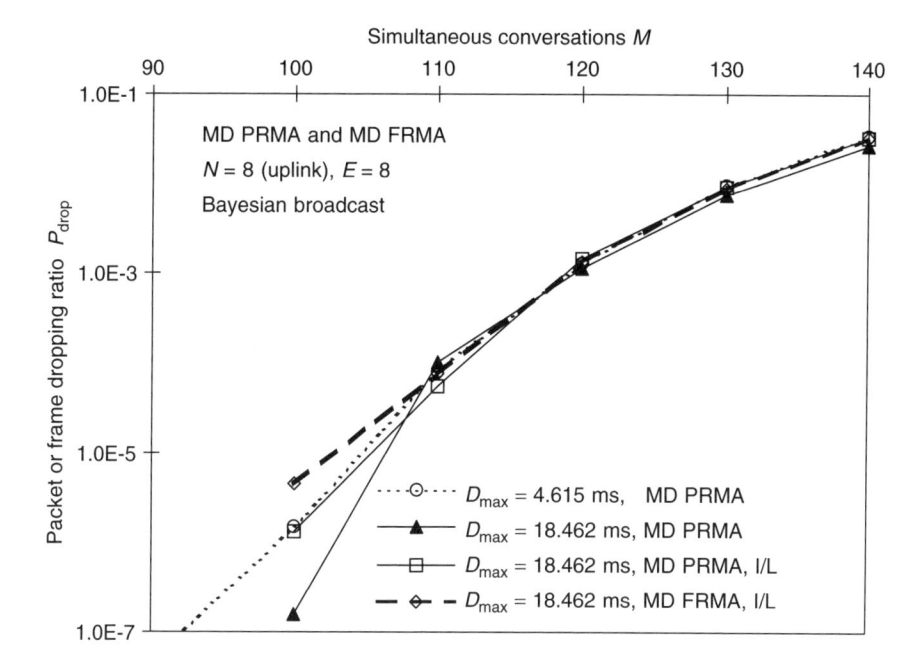

8.3 DETAILED ASSESSMENT OF MD PRMA AND MD FRMA PERFORMANCES

Figure 8.8 MD PRMA with and without interleaving vs MD FRMA

Judging from Figure 8.8, choosing D_{max} anywhere in-between D_{tf} and D_{vf} has very limited impact on P_{drop}. The same can be said about the additional load created by the dedicated request bursts and the impact of interleaving and voice-frame-wise traffic generation and dropping. In fact, with $D_{spurt} = 1.41$ s, there are on average 305 bursts per talk spurt in the basic scheme, such that the additional traffic load due to one request burst and on average two more information bursts per spurt amounts to less than 1%. The small gain achieved by increasing D_{max} is almost exactly compensated by this additional traffic.

Most interestingly, unlike MD PRMA, MD FRMA does not seem to suffer from acknowledgement delays. Recall the explanation for the bad performance of MD PRMA with delayed acknowledgements. It is clear that allowing terminals to contend repeatedly before receiving feedback in MD FRMA is advantageous, as it allows BB to track the backlog more quickly.

As an intermediate conclusion, we can state that statistical multiplexing can be exploited to a considerable extent while keeping packet dropping at very moderate levels owing to Bayesian broadcast. With $M = 110$ conversations sharing 64 slots, P_{drop} is below 10^{-4}. With $M = 90$, in most schemes considered, even MD FRMA, where acknowledgements do not occur immediately, not a single packet or frame was dropped during the simulation period. The additional delay incurred by using such packet-based reservation protocols rather than circuit reservations is moderate, it never exceeds D_{vf}.

8.3.3 Performance of MD FRMA in TDD Mode

Figure 8.9 depicts the voice dropping performance of MD FRMA in TDD mode with 1, 2, 4, 6, and 8 time-slots in the uplink direction (out of eight time-slots in total for

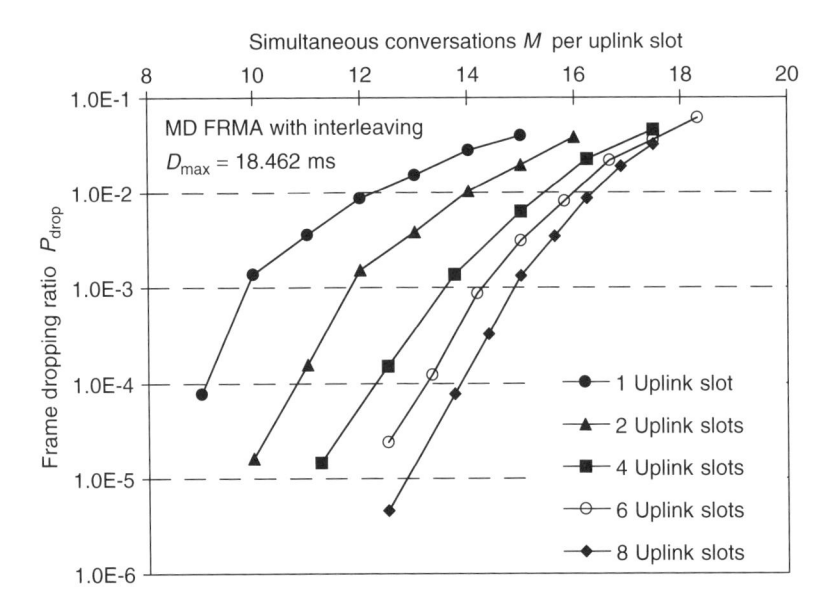

Figure 8.9 MD FRMA with Bayesian broadcast and a variable number of uplink slots constellation

both links). Interleaving is applied, thus $D_{\max} = D_{vf} = 18.462$ ms. P_{drop} is reported as a function of M normalised to the number of uplink time-slots. By doing so, the impact of trunking efficiency on dropping performance becomes immediately apparent. Looking for instance at $M_{0.01}$, 12 conversations can be supported with only one time-slot, whereas more than 16 can be supported per time-slot in the case of eight uplink time-slots per frame. The findings are very similar to those for the complementary case reported in Reference [49], where $E = 1, 2, 4, 6,$ and 8 (frequency-)slots per time-slot were considered for GSM, but the number of time-slots was kept fixed at eight (see also Section 11.3).

8.3.4 Impact of Voice Model Parameters on MD PRMA Performance

As indicated earlier, when considering in Chapter 9 mixed traffic scenarios, the parameters used for the voice model are $D_{\text{spurt}} = D_{\text{gap}} = 3$ s. Compared with $D_{\text{spurt}} = 1.41$ s and $D_{\text{gap}} = 1.74$ s used so far, this increases the voice activity factor α_v from 0.448 to 0.5. It is therefore no surprise that at a given value for M, P_{drop} is higher with these new parameters (Figure 8.10). More insight is provided if the performance is normalised to the voice activity factor or, as in Figure 8.11, if P_{drop} is reported as a function of the normalised throughput S (defined in detail in Section 9.3). In this case, P_{drop} performances are very similar. The slightly lower dropping with the new parameters is due to the lower fraction of contention traffic (since the average spurt duration is longer).

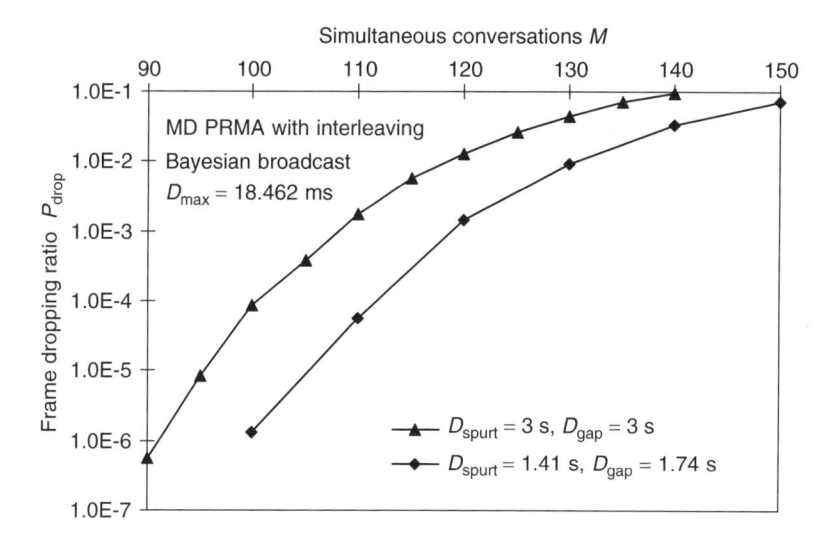

Figure 8.10 Performance of MD PRMA with two different parameter sets for the voice model

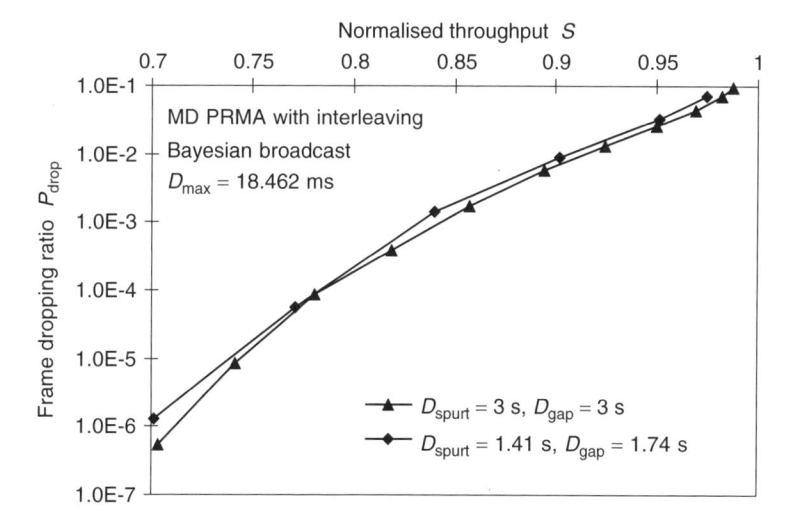

Figure 8.11 As Figure 8.10, but results reported as a function of the normalised throughput

8.4 Combining Backlog-based and Load-based Access Control

8.4.1 Accounting for Multiple Access Interference

In the following, multiple access interference will be accounted for to study its impact on Bayesian broadcast (cf. the respective discussion in Subsection 6.5.8), and to assess the benefits of combined backlog-based and load-based access control. Regarding the former, recall that increased dropping is expected due to both reduced accuracy of the backlog estimation and packet erasure due to MAI.

The model outlined in Section 5.4 is used, that is, the physical layer performance is assessed through the standard Gaussian approximation, assuming QPSK modulation. If on average 80% of the eight code-slots available per time-slot are loaded in interfering cells, the average SNR can be calculated with Equation (5.21). As seen earlier, this load level allows a P_{drop} level below 0.1% to be achieved. No interleaving is assumed (that is, the basic MD PRMA scheme is considered), and each burst is individually error coded with a (127, 43, 14) BCH code. The resulting burst or packet erasure rate $P_{pe}[K]$ is shown in Figure 5.6. Pro memoria, $P_{pe}[K]$ is below 10^{-5} for $K \leq 6$, 6.5×10^{-4} for $K = 7$, 4.1×10^{-3} for $K = 8$, and 1.6×10^{-2} for $K = 9$ packets per time-slot. The value for $K = 9$ is indicated because, due to contention, more than eight packets might access a time-slot, despite there being only $E = 8$ codes available. With this $P_{pe}[K]$ performance, it is quite clear that packet erasure will be an issue at full resource utilisation. In fact, depending on the QoS requirements in terms of $(P_{\text{loss}})_{\text{max}}$, the system will reach its interference limit below a resource utilisation of 100% (note though, that due to the interdependencies between P_{drop} and P_{pe}, it is difficult to establish a precise interference limit for this system). It is assumed that terminals never lose their reservation before completing a packet spurt transfer, although with $P_{pe}[K] > 0$, this could in theory happen[1,2]. To avoid

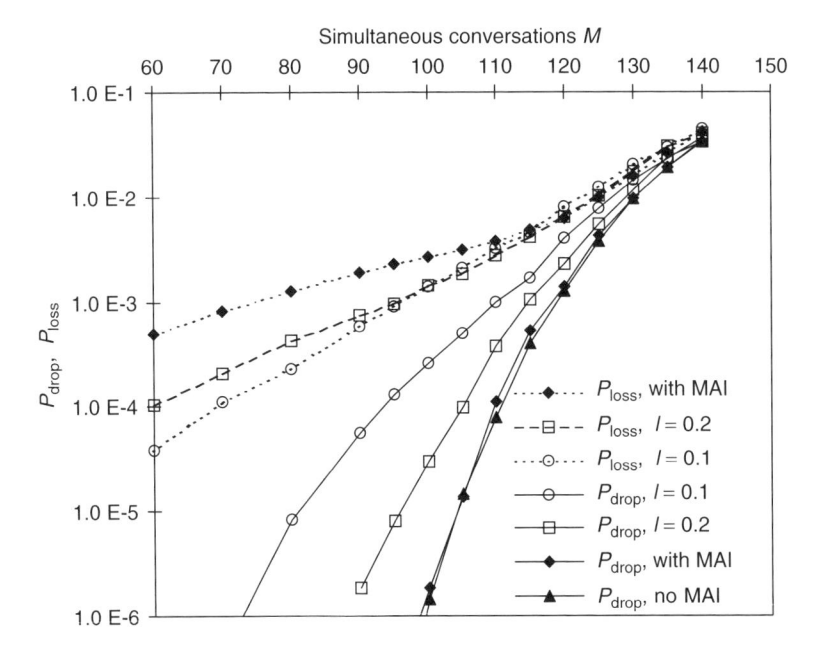

Figure 8.12 Impact of MAI on BB and performance of BB enhanced with load-based access control with two values for the parameter l

---

[1] The effects of loss of reservations on the performance of this protocol are discussed in Reference [37].
[2] This problem did not arise in Chapter 7, since a terminal, while holding a reservation in a certain time-slot, did not reserve a specific code-time-slot it could lose due to transmission failure. A corrupted reservation-mode

8.4 COMBINING BACKLOG-BASED AND LOAD-BASED ACCESS CONTROL

loss of reservations in practice, appropriate means such as end-of-reservation flags and reservation timers would be required.

Consider for now only those curves in Figure 8.12, for which no value of the parameter l is specified. According to the figure, BB suffers only to a negligible extent from MAI, that is, P_{drop} is only marginally increased if MAI is accounted for (the curve without MAI shown for comparison is from Figure 8.1). This is due to the fact that $P_{pe}[K]$ is moderate for up to $K = E$ plus a few packets per time-slot. Figure 8.12 also lists the total packet-loss ratio P_{loss}, which is the sum of P_{drop} and P_{pe} (where P_{pe} is measured both for packets transmitted in C-slots *and* I-slots[3]). Using P_{loss} as the relevant quality measure determining $M_{0.01}$ and $M_{0.001}$, these reduce from 131 to 124 and from 119 to 74 respectively, due to MAI. Observe that, with the parameters chosen, P_{pe} is dominant for $M < 130$, and it is indeed because of packet erasure that capacity in terms of $M_{0.001}$ is particularly seriously affected.

8.4.2 Performance of Combined Load- and Backlog-based Access Control

It was shown in Chapter 7 that load-based access control outperformed the KBAC benchmark, a backlog-based scheme, in certain load conditions. Adding to that the above observation on P_{pe} being dominant, one could wonder whether a combination of backlog-based and load-based access control would not provide a better trade-off between erasure and dropping performance at low load than pure backlog-based access control. This has already been discussed in Section 6.6, where we proposed to combine these two schemes through

$$p = \min\left(p_{\max}, \frac{A[t]}{v}\right), \quad (8.1)$$

where $p_{\max}$ is determined by a suitable channel access function.

To avoid having to play with multiple parameters, the channel access functions considered in the following obey

$$p_{\max} = \min(1, 2^{A[t]-1} \cdot l), \quad (8.2)$$

which leaves only l to be optimised. Although the structure of Equation (8.2) was chosen based on experience gained through the investigations described in Chapter 7, the prime concern here was simply to demonstrate the concept rather than maximising the performance. In fact, having only l to play with means that the point at which $p_{\max} = 0$ is fixed at $A[t] = 0$, i.e. access is only blocked if none of the eight code-slots is available, while $p_{\max} > 0$ for $A[t] > 0$, since $l > 0$ must hold for Equation (8.2) to make sense. With respect to the fact that $P_{pe}[7]$ is close to 10^{-3}, it might be possible to improve on $M_{0.001}$ with an additional parameter x allowing access to be blocked at $A[t] \leq x$ (here e.g. $x = 1$) rather than only at $A[t] = 0$.

In Figure 8.12, both P_{drop} and P_{loss} are drawn for $l = 0.1$ and 0.2, respectively. While P_{drop} increases considerably with $l = 0.1$ compared to pure BB, total P_{loss} is significantly reduced at low load. As a result, $M_{0.001}$ increases from 74 to 95. On the other hand,

---

packet simply meant that the base station would underestimate the number of users with a reservation and thus use the wrong input value for the channel access function following such transmission failures.
[3] Recall that $P_{pe}[K]$ is the packet erasure rate if K users access the channel (per time-slot), while P_{pe} is the ratio of erased packets resulting with MD PRMA at a given M.

P_{loss} is lower with pure BB for $M \geq 115$: $M_{0.01}$ is reduced from 124 to 122 due to the introduction of load-based access control. The better trade-off between P_{drop} and P_{pe} seems to be offered by $l = 0.2$. While P_{loss} is somewhat higher at low load than with $l = 0.1$, $M_{0.001}$ remains the same at 95, and $l = 0.1$ is outperformed for $M > 100$. Furthermore, with $l = 0.2$, performance of pure BB is almost met at high load, with $M_{0.01}$ the same at 124 in both cases.

Note that $p = p_{\max}$ in every single slot with $l = 0.1$ for $M < 120$, and with $l = 0.2$ for $M < 110$. In other words, access control is essentially load-based, and the purpose of BB is to contain occasional backlog excursions at high load, avoiding the stability problems experienced with some of the semi-empirical channel access functions (see Section 7.4).

8.5 Summary

Assuming immediate acknowledgement and not accounting for interleaving, performances of PRMA, MD PRMA, and RCMA (using TDMA, hybrid CDMA/TDMA, and CDMA as basic multiple access schemes respectively) were compared. Given an equal number of resource units, namely 64, and not accounting for 'side-effects' of the basic multiple access schemes (e.g. MAI if there is a CDMA component), $M_{0.01}$ was in all cases found to be almost the same, namely 130 ± 1. More impressive still, if the protocols are stabilised through Bayesian broadcast, interleaving is applied over a voice frame, and the delay threshold $D_{\max}$ is equal to the voice frame duration D_{vf}, P_{drop} over the whole range of values considered for M is virtually the same. Therefore, from a pure multiplexing point of view, the choice of the *basic* multiple access scheme does not matter. Clearly, such a comparison based on a perfect-collision code-time-slot channel model is quite simplistic, and the peculiarities of the multiple access schemes will affect operation and performance of the MAC layer (see below).

Interestingly, an observation could be repeated here which was already made in Chapter 7, albeit under considerably different circumstances, namely that access control is not important in the case of pure CDMA (i.e. RCMA) with voice traffic only. By contrast, in the case of PRMA and MD PRMA, dynamic control of access to C-slots through Bayesian broadcast is clearly beneficial. Firstly, stability problems experienced when using fixed permission probability values p_v call for stabilisation, which is provided by Bayesian broadcast. In the case of RCMA, on the other hand, no stability problems were experienced even with unconstrained access to C-slots, i.e. $p_v = 1$. Secondly, with immediate acknowledgement, P_{drop} performance of (MD) PRMA with Bayesian broadcast meets or exceeds that achieved with fixed permission probabilities, even if the values chosen for p_v are individually optimised for each value of M in the latter case. This contrasts again with RCMA, where the performance achieved with a single value of one for p_v is as good as that achieved with Bayesian broadcast, irrespective of M.

For the CDMA/TDMA case only, the impact of delayed acknowledgement was investigated. It was noted that MD PRMA with Bayesian broadcast suffered considerably from delayed acknowledgements, and could exhibit 'resonant behaviour' in certain circumstances. Therefore, depending on the conditions considered, the performance advantage over fixed probabilities can dwindle away. These problems are avoided in the MD FRMA scheme, where downlink signalling occurs only once per frame, but mobile stations are allowed to contend repeatedly without having to wait for acknowledgements. This protocol has the further advantage of being suitable for TDD with only one link switching-point per

TDMA frame. With MD FRMA (as well as with MD PRMA with immediate acknowledgements), when serving $M = 110$ conversations on 64 code-time-slots, P_{drop} is below 10^{-4}. If a P_{drop} of 10^{-2} is acceptable, more than 130 conversations can be sustained. At the activity factor α_v of 0.448 considered, this corresponds to a multiplexing efficiency η_{mux} in excess of 0.9.

Finally, again only for hybrid CDMA/TDMA, the impact of MAI on P_{drop} obtained with Bayesian broadcast was studied (maintaining, however, the code-time-slot channel structure not considered in Chapter 7). It was found that the impact is very limited, as long as the packet erasure rate $P_{pe}[K]$ remains small for up to $K = E$ plus a few packets per time-slot. However, in this case, P_{loss} is the relevant performance measure instead of P_{drop}, and at low load, the dominant contribution to P_{loss} is P_{pe}. In such a scenario, access control should be used to achieve an optimum trade-off between P_{drop} and P_{pe}, rather than to minimise P_{drop}. It has been shown that a combination of load-based and backlog-based access control provides better trade-offs between P_{pe} and P_{drop} than pure backlog-based access control. An increase in P_{drop} resulting from more restrictive access control can be more than compensated by reduced P_{pe}, such that the total packet-loss ratio experienced decreases as well.

9
MD PRMA WITH PRIORITISED BAYESIAN BROADCAST

This last of the three chapters dealing exclusively with MD PRMA-related research results finally treats scenarios that are not limited to voice traffic only. To assess the performance of the prioritised Bayesian broadcast algorithm proposed in the following, a simulation study for different mixtures of voice and data traffic is conducted, where the latter consists either of Web traffic, email traffic, or a combination of the two. Preceding the discussion of the simulation results, possible approaches to prioritisation at the random access stage are evaluated, and a number of different algorithms which combine Bayesian broadcast with prioritisation are presented.

9.1 Prioritisation at the Random Access Stage

The reader questioning the motivation for introducing prioritisation at the random access stage is referred to Section 3.7, where this topic was treated in considerable detail, and to the priority-class specific access control feature of GPRS described in Section 4.11. UMTS provides also random access prioritisation, see Chapter 10.

For a mobile terminal to successfully send a channel request message in a given time-slot, three events must occur simultaneously:

(1) there must be a random access slot (or C-slot, in the terminology used so far) in this time-slot;
(2) the terminal must gain permission to access this slot; and
(3) the request packet transmitted in this slot must be successfully received by the base station.

If the capture effect and transmission failures due to bad channel conditions are ignored, the last event essentially means that the packet must not collide with another packet transmitted in the same slot. Since the collision probability depends on how access to C-slots is controlled, events (2) and (3) cannot be de-coupled. On the other hand, the first event can be considered separately from the other two for the purposes considered here[1]. If we want to introduce prioritisation at the random access stage to discriminate the

---
[1] Strictly speaking, there is also a dependency between event one and the other two events. At a given arrival rate, with increasing interval between C-slots, the number of newly arriving packets between two such slots increases. Thus, the collision probability increases as well, if access to C-slots is not controlled adaptively.

access delay experienced by different priority classes, we therefore fundamentally have two options. We could define different classes of C-slots, each class only to be accessed by the respective priority class, and schedule these slots according to the QoS requirements of the respective services associated with the different classes. Alternatively, we could let all users access the same C-slots, but control the access to these slots according to the priority class. High-priority users will obtain permission to access a certain C-slot with higher probability than low-priority users. With the first approach, collisions do not occur between users of different priority classes, with the second, they do.

Recall that with in-slot protocols such as MD PRMA considered here, every slot that is currently not used for information transfer is a C-slot available for contention. Since C-slots are not explicitly scheduled, the first approach to random access prioritisation does not easily lend itself to MD PRMA. In the following, therefore, the focus is limited to the second approach, namely service-class specific access control to a single class of C-slots.

In Sections 3.5 and 4.11, different approaches to access control for S-ALOHA protocols and their derivatives were discussed. These are essentially:

- use of fixed permission probabilities;
- retransmission backoff schemes;
- stack-based schemes such as splitting or collision resolution algorithms; and
- (global) probabilistic access control such as Bayesian broadcast.

It is possible to introduce prioritisation with all these approaches. In the case of fixed permission probabilities, different values can be chosen for different priority classes. However, whether with or without prioritisation, stability problems are encountered with such a scheme. With backoff schemes, prioritisation can be achieved by choosing priority-class specific initial permission probability values (i.e. those relevant for the first transmission attempt), but also class specific backoff rates.

It was reported in Section 3.5 that it is possible to exceed the 1/e throughput-limit of S-ALOHA, for instance with tree-based collision resolution algorithms. This throughput advantage is also maintained in prioritised versions of such protocols, as shown for example in Reference [270] (for a selection of other relevant references in this field, the reader is advised to consult Reference [271]). However, these stack-based algorithms have the disadvantage of requiring immediate (although normally only binary) acknowledgements to work properly, which, as discussed at length earlier, is difficult to achieve in a real implementation.

Our interest in prioritisation was triggered by a submission to the ETSI SMG 2 GPRS ad-hoc group responsible for the standardisation of the GPRS air interface [272]. In this document, a prioritised pseudo-Bayesian broadcast algorithm with proportional priority distribution between two priority classes was proposed for the GPRS random access. This algorithm and subsequent enhancements are described in detail in the next section. In Reference [55], we compared the performance of a four-class pseudo-Bayesian broadcast algorithm with semi-proportional priority distribution with two other four-class algorithms. The first one is based on exponential backoff, where, on top of priority-class specific initial permission probabilities, as in Reference [273], the backoff rates chosen were also class specific. The second one, a stack-based algorithm, was first proposed in Reference [274]. It is an enhancement of a two-class algorithm due to Stavrakakis and Kazakos [275] and supports an arbitrary number of priority classes. The Stavrakakis–Kazakos algorithm has a maximum stable throughput from 0.32 to 0.357, depending on the traffic composition. We found that the exponential backoff algorithm, while not requiring any feedback other than acknowledgements, clearly performs worst, and exhibits stability problems. The Bayesian algorithm outperforms the stack-based algorithm, and is (unlike the latter) inherently fair, but may, depending on the chosen implementation, require slightly more signalling overhead, since individual permission probabilities need to be conveyed to the terminals. However, they need not necessarily be signalled for every slot (cf. Section 4.11). A further advantage is that the proposed algorithm can easily be adapted for operation with frame-based protocols.

With these considerations in mind, and given the successful adaptation of Bayesian broadcast to MD PRMA discussed earlier, it cannot be a surprise that the prioritised algorithm chosen for the following investigations is indeed based on Bayesian broadcast.

9.2 Prioritised Bayesian Broadcast

So far, the focus has been restricted to homogeneous voice traffic, thus for access control only a single (access) permission probability value p_v needed to be calculated. Consider now the generic (single-class) permission probability value p calculated according to the (pseudo-)Bayesian broadcast algorithm outlined in Subsection 6.5.4. In the following, starting with the initial proposal in Reference [272] for two-class proportional priority distribution, several algorithms will be introduced which, based on p, calculate individual access probability values p_i for each priority class i. Class 1 has highest priority, and voice traffic is always assigned to class 1, thus $p_v = p_1$. Most of these algorithms were initially proposed and investigated for S-ALOHA, but adaptation to MD PRMA is straightforward.

9.2.1 Bayesian Scheme with Two Priority Classes and Proportional Priority Distribution

On a perfect collision channel as considered here, the optimum traffic level G_0, at which the throughput curve of S-ALOHA peaks, is $G_0 = 1$. Therefore, p should be chosen such that the expected traffic assumes a value of one. If the estimated mean backlog v is equal to the real backlog n, this is achieved by choosing

$$p = \min(1, G_0/v), \qquad (9.1)$$

as already pointed out in Section 6.5. For MD PRMA with $A[t]$ C-slots in time-slot t, the only modification required is that G_0 is now $A[t]$.

In Reference [272] it was suggested to extend the Bayesian algorithm to support two priority classes by assigning different transmission probabilities to the users of the high and the low priority class based on p from the single-class scheme as follows:

$$p_1 = \min(1, (1+\alpha) \cdot p), \qquad (9.2a)$$

$$p_2 = \alpha \cdot p, \qquad (9.2b)$$

with the proportion of successfully received access bursts of class-two users to all users

$$\alpha = S_2/S, \qquad (9.3)$$

averaged over a suitable time-window. The degree of prioritisation cannot be chosen, it is determined by α, which is why this approach was termed 'proportional priority distribution' in Reference [272].

Provided that the backlog estimation is accurate, i.e. $v \approx n$, and that α reflects the backlog proportion of low priority users to all users, which is n_2/n (where $n = n_1 + n_2$), the offered traffic $G = p_1 \cdot n_1 + p_2 \cdot n_2$ assumes the value of the optimum offered traffic G_0, as desired. However, since $p_1 > p_2$, the throughput proportion α is expected to underestimate the backlog proportion n_2/n. Interestingly, together with David Sanchez, an M.Sc. student at King's College London in 1995/1996, we found that while this is indeed the case, at the same time also the backlog is underestimated. Taken together, these two effects compensate in a manner which causes the offered traffic to assume its optimum value all the same.

9.2.2 Bayesian Scheme with Two Priority Classes and Non-proportional Priority Distribution

In an internal LINK ACS document, Jason Brown pointed out that the algorithm proposed in Reference [272] could easily be extended to allow for non-proportional priority distribution. This is achieved by choosing

$$p_1 = \min(1, m \cdot p), \tag{9.4a}$$

$$p_2 = k \cdot p, \tag{9.4b}$$

with

$$m = \frac{1 - \alpha \cdot k}{1 - \alpha}, \tag{9.5}$$

and α as above. Either m or k can be chosen arbitrarily and thus used to control the degree of prioritisation of high-priority users. In the following, k will be used as the main prioritisation parameter.

9.2.3 Bayesian Scheme with Four Priority Classes and Semi-proportional Priority Distribution

In 1996/1997, Celia Fresco Diez, an exchange student at King's College London, continued our earlier investigations on the GPRS random access. Some of her findings are summarised in Reference [55]. The four-class semi-proportional algorithm described here is due to her. In this algorithm, the permission probability values of the four classes are set as follows:

$$p_1 = \min(1, m \cdot p), \tag{9.6a}$$

$$p_2 = \min\left(1, \frac{2 \cdot m + k}{3} p\right), \tag{9.6b}$$

$$p_3 = \min\left(1, \frac{m + 2 \cdot k}{3} p\right), \tag{9.6c}$$

$$p_4 = k \cdot p, \tag{9.6d}$$

with p and m as before and

$$\alpha = \frac{(S_2/3) + 2 \cdot (S_3/3) + S_4}{S}. \tag{9.7}$$

The parameter k allows the delay spread between priority classes to be chosen, whereas the degree of prioritisation between each priority class cannot be chosen individually, which is why this approach was termed *semi-proportional* rather than 'non-proportional' priority distribution. The lower the value of k, the larger the difference between the transmission probabilities and thus the stronger the degree of prioritisation. For $k = 1$ the algorithm degenerates to single-class Bayesian broadcast control.

9.2.4 Bayesian Scheme with Four Priority Classes and Non-proportional Priority Distribution

In some cases, it may be desirable to control the degree of prioritisation for each class individually. For Reference [52], we extended the above algorithm to a full non-proportional algorithm according to the following:

$$p_1 = \min(1, m \cdot p), \tag{9.8a}$$

$$p_2 = \min\left(1, \frac{z_1 m + z_2 k}{z} p\right), \tag{9.8b}$$

$$p_3 = \min\left(1, \frac{z_2 m + z_1 k}{z} p\right), \tag{9.8c}$$

$$p_4 = k \cdot p, \tag{9.8d}$$

with p and m as before. In this algorithm, k determines the delay spread between classes 1 and 4, while the parameters z_1 and z_2, with $z_1 > z_2$ and

$$z = z_1 + z_2, \tag{9.9}$$

determine the relative degree of prioritisation of classes 2 and 3. The relevant throughput proportion must now be calculated as

$$\alpha = \left(\frac{z_2}{z} S_2 + \frac{z_1}{z} S_3 + S_4\right) \frac{1}{S}. \tag{9.10}$$

Note that the semi-proportional algorithm is simply a special case of the non-proportional algorithm with $z_1 = 2$ and $z_2 = 1$. Under assumptions equivalent to those made for the two-class proportional algorithm above, the offered traffic G is controlled to the optimum traffic $G_0 = 1$, as desired. This can be verified through relatively simple arithmetic, as shown in Reference [61, Appendix E].

In Subsection 3.7.4, a case was made for a centralised implementation of such access control schemes, and the resources for downlink signalling required with such an approach were discussed in Section 4.11. It was pointed out that the permission probability p would have to be transmitted regularly (although not in every slot), but that a 4-bit resolution was sufficient. On the other hand, with the structure of the prioritisation algorithm considered here, m (which fluctuates with α, which in turn depends on the chosen time-window for averaging) can be broadcast less frequently. Finally, only infrequent signalling of values for k, z_1, and z_2 is required. Obviously, exploiting the structure of a specific algorithm to reduce signalling load means limited flexibility once the system is deployed. By contrast, if p_1 to p_4 were signalled individually (e.g. by three to four bits each), a network operator could change the algorithm arbitrarily through a software update at the base station, without affecting mobile terminals already in use.

9.2.5 Priority-class-specific Backlog Estimation

The algorithms proposed above have the two following things in common.

(1) No attempt is made to estimate the backlog of individual priority classes separately, instead, the backlog proportion is estimated based on the total estimated backlog and the throughput proportion.

(2) The individual permission probability values are expressed as a function of the single-class permission probability value G_0/v.

While the second feature is deliberate to limit the signalling overhead required on the downlink, as just discussed, the first was initially identified as a shortcoming of these algorithms, which one might wish to overcome. However, since we found that, for S-ALOHA, the average access-delay performance over all classes with these prioritised algorithms exactly matched the performance of the single-class algorithm, we did not invest any further effort in attempting to estimate the backlog of each class individually.

In the mean time, Frigon and Leung proposed in Reference [271] a prioritised version of Bayesian broadcast for x priority classes, which relies on such individual backlog estimation. In this algorithm, a prioritisation parameter γ_i can be selected for each priority class individually, which in turn determines the access permission probability to be used for each class through $p_i = \min(1, \gamma_i/v_i)$. Since γ_i can be viewed as the expected offered traffic for each class, the sum over all x γ_i-values must amount to $G_0 = 1$. If for any class $v_i < \gamma_i$ in any given slot, and p_i were simply set to $\min(1, \gamma_i/v_i)$, the expected total offered traffic in that slot would be less than one, which is not optimum. Frigon and Leung account for this fact, and assign such leftover capacity to other priority classes (in order of priority) by temporarily increasing the traffic fraction assigned to these classes. Doing so ensures that low priority users send their packet immediately, if currently no high priority users are backlogged and, conversely, that no capacity is reserved for low priority users if there are none. Taking it to the extreme, γ_x can be set to zero, such that the lowest priority class obtains only 'leftover capacity'. This improves the average delay performance compared to the single-class algorithm slightly, and thus also compared to the algorithms we proposed above. To exploit this performance advantage, the backlog needs to be estimated for each class individually, an extra effort which could well be justified. As a further attractive feature, this algorithm could also be extended by setting γ_1 adaptively given v_1, to help meeting a specific desired access delay performance for class 1 (as a result, $\gamma_2..\gamma_x$ would obviously also need to be readjusted). However, it would not be possible to link p_i to p any more as in Equation (9.8). Instead, individual

p_i values would have to be signalled for every time-slot carrying C-slots, adding overhead on the downlink.

Rivest's derivation of the pseudo-Bayesian broadcast algorithm in Reference [51] is based on the assumption that the arrival process of newly generated packets is Poisson, and the same also holds true for all prioritised versions of this algorithm. Frigon and Leung therefore looked at the impact of non-Poisson arrivals, by assessing the performance of their algorithm with self-similar traffic in Reference [271]. While self-similar traffic resulted, not surprisingly, in increased total average access delay, they found that their algorithm still reduced effectively the average access delay of the high-priority users. This is consistent with our findings reported later for email and Web traffic, which is similar in nature to the traffic Frigon and Leung considered.

9.2.6 Algorithms for Frame-based Protocols

In Reference [271], Frigon and Leung adapted their prioritised algorithm also for frame-based protocols, that is, TDD protocols with a single switching-point between link directions per TDMA frame. In this modified algorithm, every mobile terminal may pick only one C-slot per TDMA frame. This seems also to be the case in the PRMA/TDD protocol proposed by Delli Priscoli in Reference [161], which also features adaptive and priority-class dependent parameter calculation for access control (but the parameter calculation is not based on Bayesian reasoning, instead it relies heavily on the knowledge of traffic statistics). Compared to the frame-based FRMA protocol discussed in Section 6.3, where a terminal is allowed to contend repeatedly in a frame before receiving acknowledgements, this approach increases access delay unnecessarily. We believe therefore that terminals should be allowed to contend more than once per frame. With the semi-proportional and non-proportional prioritisation algorithms discussed above and the adaptation of the single-class Bayesian broadcast algorithm to FRMA described in Subsection 6.5.6, all tools required for prioritisation with FRMA are available.

Since every terminal is allowed to contend repeatedly in a frame, it is expected that rather low values will need to be chosen for the prioritisation parameter k in Equation (9.6) or (9.8) to achieve worthwhile access delay discrimination. This is because with FRMA, as with every frame-based protocol, access delay discrimination through access control alone can only be achieved in terms of multiples of the TDMA frame duration. To influence the delay behaviour beyond the resolution of an entire frame, appropriate priority-class dependent resource allocation algorithms could be added (for instance, the time-slots at the start of a frame could be assigned to high-priority users with preference).

Note that both the slot-based and frame-based protocols considered in Reference [271] are purely contention-based. By contrast, both the slot-based MD PRMA protocol and the frame-based MD FRMA protocol (the latter not being investigated any further here) are reservation-based, as they belong to the family of in-slot R-ALOHA protocols (cf. Subsection 3.6.4). Finally, the PRMA/TDD protocol in Reference [161] is an out-slot R-ALOHA protocol, which not only features adaptive parameter calculation for the C-slot permission probability, but also adaptive calculation of the required number of C-slots in each frame to guarantee a certain success probability. This in turn affects the position of the switching-point between link directions.

9.3 System Definition and Simulation Approach

9.3.1 System Definition

The system considered here is based on the original TD/CDMA design parameters, as summarised in Tables 5.2 and 5.3. *Pro memoria*, a TDMA frame lasts 4.615 ms and carries $N = 8$ time-slots, as in GSM. $E = 8$ codes are available on each time-slot and, as in most parts of Chapter 8, these code-slots are assumed to be mutually orthogonal, which means that MAI is ignored. Accordingly, considerations provided in Chapter 8 on the system being blocking-limited apply also here. At most one code-time-slot per TDMA frame is allocated to each user, thus limiting the available net data-rate to 8.125 kbit/s. Rectangular interleaving over the length of one RLC-PDU is applied for all traffic types considered. For the transmission of voice frames and the short RLC-PDUs used for Web traffic, which carry 150 user bits each, four bursts are required. The long RLC-PDU used for email traffic carries 3600 user bits, which fit into the payload of 96 bursts. Padding is used to fill the last of a sequence of RLC-PDUs carrying a datagram or message. Hence, on average, 1800 bits per email message are padding bits.

As far as voice and Web traffic are concerned, the traffic model parameters considered are those suggested in Reference [56]. For voice traffic, this means $D_{spurt} = D_{gap} = 3$ s. The Web traffic parameters are listed in Table 5.4 together with the email traffic parameters we derived.

For access control, prioritised Bayesian broadcast control with four priority classes and either semi-proportional or non-proportional priority distribution according to Equations (9.5) to (9.10) is used. The basic permission probability p is calculated according to the Bayesian algorithm adapted for MD PRMA, as outlined in Subsection 6.5.4. In the implementation chosen, to calculate the throughput proportion with Equation (9.7) or (9.10), the running average from $t = 0$ is taken for every simulation-run[2]. To carry out the required estimation of the arrival rate, Equations (6.8) and (6.9) are used for data and voice traffic respectively[3]. Immediate acknowledgement and full signalling of the relevant access parameters for every time-slot are assumed. For duplexing, FDD is considered, that is, the FRMA-based version of the protocol is not investigated further.

Finally, with regards to the duration of the reservation phase, voice and Web terminals hold their reservations until they have emptied their transmission buffer (i.e. upon completion of the transfer of a talk spurt or an IP datagram). In the case of email traffic, limited *allocation cycles* are considered as well, in which case the duration of the reservation phase is limited to the transfer of a few long RLC-PDUs, as indicated. In accordance with the discussion on terminology provided in Section 5.8, this could also be viewed in the following terms. In the case of Web traffic, the network layer (NWL) does not perform segmentation, thus every packet delivered to the RLC layer contains an entire datagram. In the case of emails however, if allocation cycles are limited, a message is segmented by the NWL into several packets equivalent to the allocation cycle length. In both cases, the duration of the reservation phase corresponds simply to the length of the NWL packet. In the implementation considered, it is assumed that terminals with non-empty transmission buffers at the end of an allocation cycle need to contend for further cycles, rather

---

[2] To cater for traffic mix fluctuations, it may be better to average over an appropriate time-window instead.
[3] Note that Equation (6.8) could also be used for voice traffic, if the traffic statistics required for Equation (6.9) were not known.

than send an extension request message in reservation mode. Again, this could be viewed as the lower layers first releasing the reservation and only then notifying the NWL of successful transfer of its packet, in which case the NWL immediately submits the next packet to the RLC.

9.3.2 Simulation Approach

As discussed in Section 5.6, due to the intricacies of the data traffic models considered, no attempt is made to analyse the delay and dropping performance for mixed voice and data traffic, thus only simulation results will be provided in this chapter.

Performances are investigated using the global mean data session interarrival time μ_{Dsess} as the main simulation parameter, while keeping the number of ongoing voice conversations fixed at a certain value M. The values considered for μ_{Dsess} are between 0.5 and 12 s. Given the relatively few data sessions per 1000 s simulation-run, and the large variance σ_{Sd}^2 of the size of Web datagrams S_d, but particularly of email messages, as determined by Equation (5.31), the amount of data generated per simulation-run, even with constant μ_{Dsess}, fluctuates considerably. Results presented as a function of μ_{Dsess} would therefore be meaningless. Instead, results are reported as a function of the *normalised throughput* S (that is, normalised to the total user channel rate of $N \cdot E \cdot 8.125$ kbit/s $=$ 520 kbit/s). This has two added benefits: it allows one to relate P_{drop} for voice in the mixed-traffic-scenarios considered to P_{drop} for voice-only traffic. Furthermore, datagrams dropped due to overflow of the queue of data terminals serving WWW traffic do not distort results, as discussed in Subsection 5.6.2[4]. Recall that such dropping due to memory constraints has nothing to do with dropping a voice frame at the MAC level due to delay constraints. From a MAC perspective, for non-real-time traffic, $P_{drop} = 0$. The throughput includes all bursts received error free by the BS including request bursts and fill-bursts (due to padding) in the last RLC-PDU.

For voice traffic, P_{drop} is reported as a function of S. For data traffic, for reasons outlined in Subsection 6.2.8, only access-delay performance is assessed, separately for each priority class, and again as a function of S. Access delay is defined as the time between arrival of a request at the MS MAC entity (hence excluding queuing delay while this entity is busy transmitting previous NWL packets) and the end of the successfully accessed C-slot. The time between successful contention and actual start of user data transmission, which could also be considered as part of the access delay, is ignored, since it is the same for all packets here, namely $N - 1$ time-slots.

Figure 9.1 shows that the results obtained, before appropriate processing, fluctuate considerably (the processed results for the example shown can be found in Subsection 9.6.2). Typically, P_{drop} fluctuations occur predominantly at low traffic, mainly due to limited statistical significance as a result of the low number of dropped frames. On the other hand, data access delay fluctuates most at high load. This is once more due to the intricacies of the data traffic models, in particular due to the high variance of the packet size distribution σ_{Sd}^2. To increase the reliability of the results and to obtain smooth

---

[4] As indicated there, with the queues considered (holding 50 datagrams irrespective of their size), rarely more than 1% of the generated datagrams are dropped. This may appear significant from a QoS perspective and would call for longer queues, or better still, for the allocation of multiple slots to reduce the transmission time and therefore the dropping probability. However, it is only relevant here in terms of the impact on the traffic model, which is moderate.

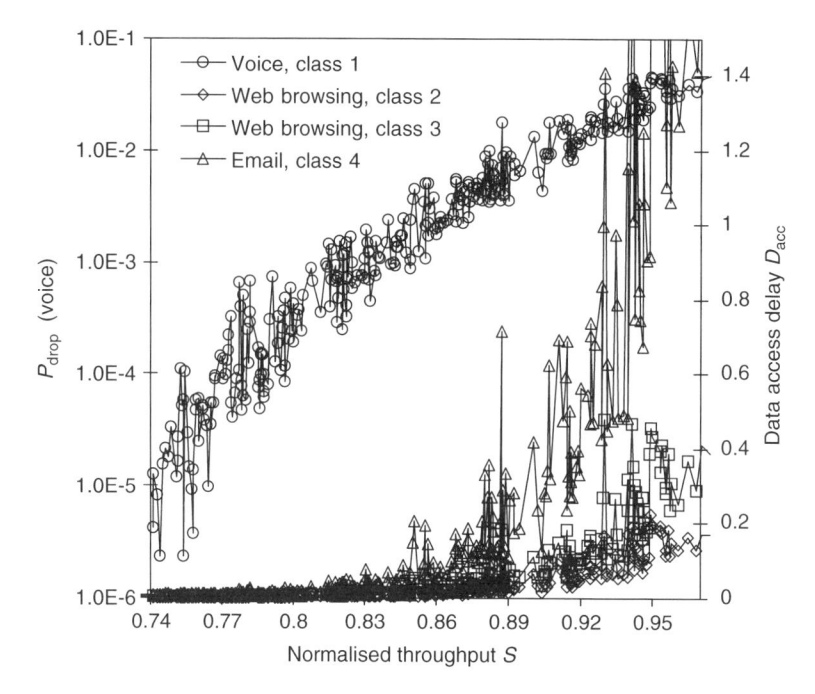

Figure 9.1 Illustration of the fluctuations in unprocessed voice dropping and data access delay results

graphs, a large number of simulations had to be performed in a very time-consuming exercise. Every point shown in the following figures is a result of averaging individual points obtained in several 1000 s simulation-runs. The grouping of points for averaging was performed manually. Care was taken that the range of individual throughput values spanned by the averaged points did not exceed 1.5% of the average throughput over these points. In retrospective, obviously, it would have been worthwhile to invest some effort in accelerating the simulation process, and in automating the processing of the results.

9.3.3 Traffic Scenarios Considered

To investigate the impact of data traffic on the voice dropping performance and of prioritised Bayesian broadcast on the access delay performance of data assigned to different priority classes, four traffic scenarios are considered. These are two different scenarios with a mixture of voice and Web browsing traffic, a scenario with voice and email traffic and, finally, voice and both data traffic types together. With voice and a single type of data traffic assigned to only one priority class, voice uses priority class 1 and data priority class 4. For the second scenario with mixed voice and Web traffic, half of the Web traffic is associated with priority class 2, and the other half with priority class 4. Every new Web session is assigned to either of these classes, according to the outcome of a Bernoulli experiment with parameter $p = 0.5$.

In scenario 4 with mixed data sources, voice is associated with priority class 1, Web browsing with priority classes 2 and 3, and emails with priority class 4. The fraction of sessions for each priority class is chosen such that on average a third of the data bursts generated carry email traffic, a third Web class 2 and the final third Web class 3 traffic. A further set of parameters is also considered with scenario 4, where the average email traffic generated amounts to half of the total data traffic, while the other 50% is shared equally between Web classes 2 and 3. While the performance is investigated with several values of k, z_1 and z_2 are normally set to 2 and 1 respectively (which corresponds to the semi-proportional algorithm), unless explicitly mentioned otherwise.

9.4 Simulation Results for Mixed Voice and Web Traffic

9.4.1 Voice and a Single Class of Web Traffic

In Figure 9.2, the voice dropping performance with $M = 90$ conversations and a variable amount of Web traffic is depicted for several values of the prioritisation parameter k and compared with the performance of homogeneous voice traffic.

Without prioritisation of voice over data (i.e. $k = 1.0$), P_{drop} experienced in the heterogeneous traffic case is higher than that at the same throughput levels in the homogeneous case. However, the prioritisation parameter k not only influences the access delay depicted in Figure 9.3, as desired, but also allows voice dropping to be traded off against data access delay. It is therefore possible to achieve significantly lower P_{drop} at a given throughput level compared to the voice-only case, if one is willing to pay the price of higher data access delay. Here, for $k \leq 0.3$, P_{drop} is below that experienced with voice-only traffic at the expense of high data access delay, and vice versa for $k \geq 0.7$. Note that $S = 0.76$ roughly corresponds to 8% data traffic (as a fraction of the total traffic) and $S = 0.95$ to 27%.

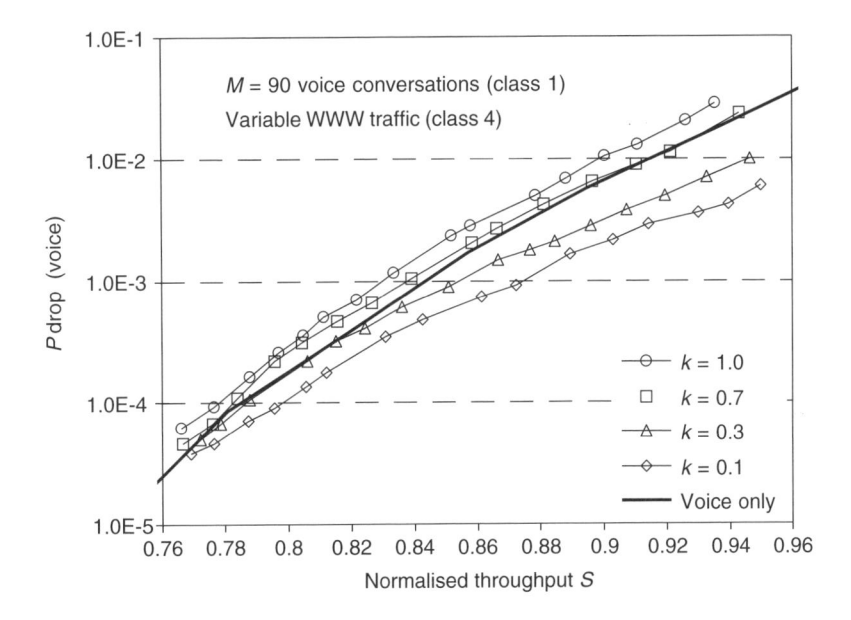

Figure 9.2 Voice dropping performance with mixed voice and single-class Web traffic

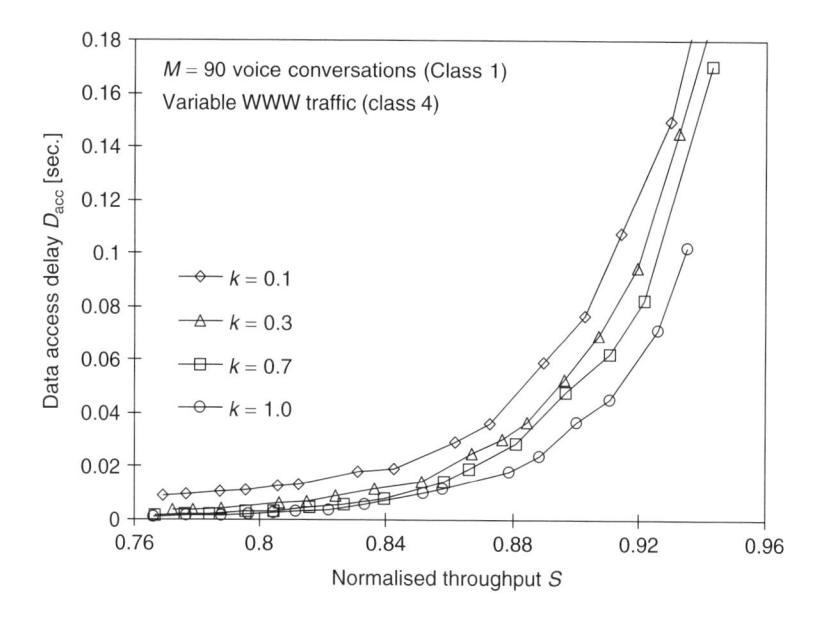

Figure 9.3 Delay performance of data with mixed voice and single-class Web traffic

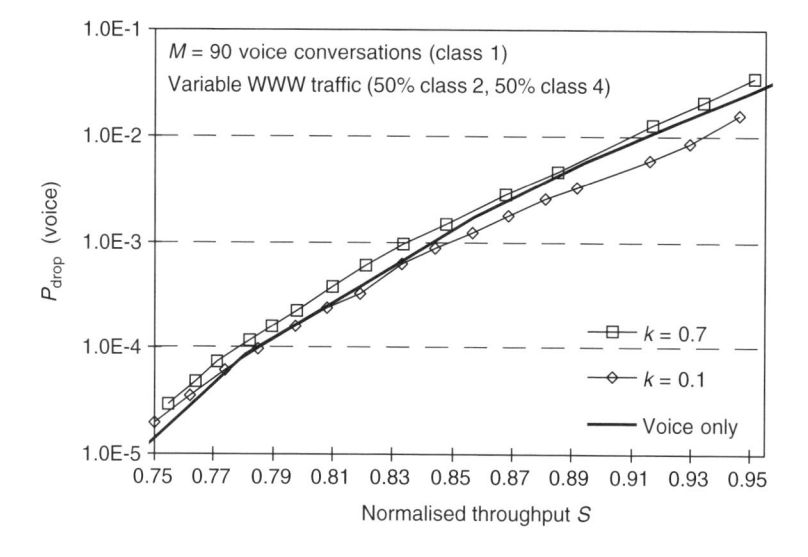

Figure 9.4 Voice dropping performance with mixed voice and two-class Web traffic

9.4.2 Voice and Two Classes of Web Traffic

Figures 9.4 and 9.5 show results equivalent to those in the previous two figures, for a scenario, where on average 50% of data sessions are assigned to priority class 2, while the remaining 50% are assigned to priority class 4. Similar observations as above can be

9.5 SIMULATION RESULTS FOR MIXED VOICE AND EMAIL TRAFFIC

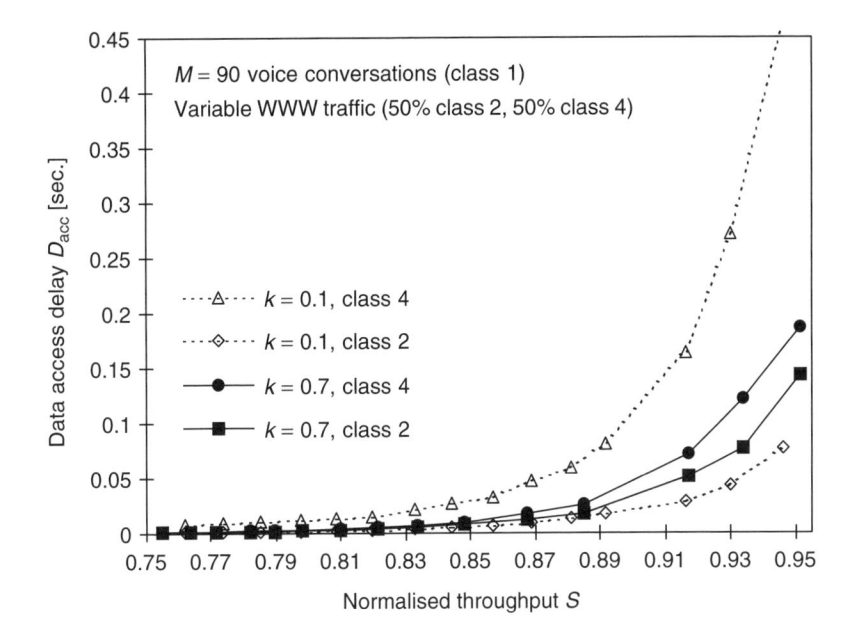

Figure 9.5 Delay performance of data with mixed voice and two-class Web traffic

made. With $k = 0.1$, P_{drop} in the mixed traffic case is below that experienced with voice-only traffic, while the reverse is true for $k = 0.7$. However, the spread between the two curves in Figure 9.4 is smaller than that between the relevant two curves in Figure 9.2, mainly because of increased dropping compared to the single-class case for $k = 0.1$. This is due to the fact that a low k will not only prioritise voice over class-four-data, but also class-two-data over class-four-data, such that the overall prioritisation of voice is smaller than in a single-data-class scenario.

Figure 9.5 illustrates how the spread between the access delay values experienced by the two data classes increases with a decreasing value of k. With $k = 0.7$, below $S = 0.86$, the access delay performance of the two data classes is very similar, while with $k = 0.1$, a delay discrimination can already be observed at much lower throughput levels than with $k = 0.7$.

9.5 Simulation Results for Mixed Voice and Email Traffic

9.5.1 Performance with Unlimited Allocation Cycle Length

Figure 9.6 reports voice-results in the case of a mixture of voice and email traffic, again with $M = 90$. At $S = 0.74$, the fraction of data traffic is approximately 5% and at $S = 0.95$, 27%. Without limiting allocation cycle length, mainly due to the high variance of the size of email messages generated, the voice dropping performance in the heterogeneous case is slightly worse than that in the homogeneous case for both $k = 0.1$ and $k = 0.7$. Also, due to the large amount of data contained in a single email message, only a relatively low message arrival rate can be supported (from 1200 to 7200 email messages per hour

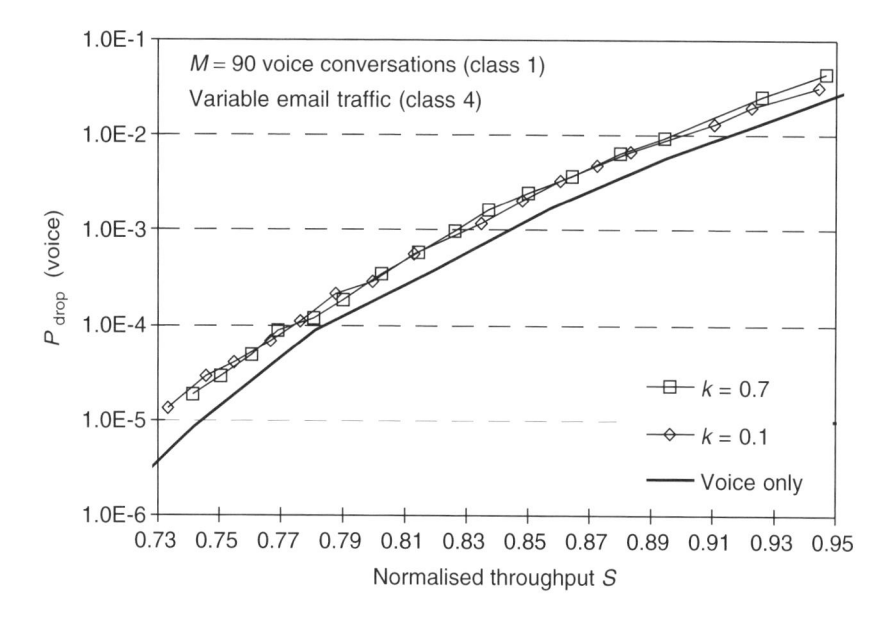

Figure 9.6 Voice dropping performance with mixed voice and email traffic

in the scenario considered in Figure 9.6). Furthermore, since only one request burst per email message is generated, prioritisation at the random access stage through varying k has no impact on P_{drop}. This can be corrected by limiting the allocation cycle length, which is discussed below.

Again due to the high message size variance σ_{Sd}^2, it was rather difficult to obtain smooth P_{drop} curves. In fact, we surrendered in our attempts to eliminate some unwanted crossovers between the two curves for $S < 0.8$, after having simulated more than 400 points in total.

Note that this case without allocation cycles is an extreme case with little practical relevance. Typical NWL implementations are expected to segment large email messages.

9.5.2 Impact of Limiting Allocation Cycle Lengths

With limited allocation cycle length, both P_{drop} can be reduced and a spread between the P_{drop} performances with $k = 0.1$ and $k = 0.7$ achieved, as depicted in Figure 9.7 for allocation cycle lengths of three PDUs and one PDU. Obviously, this has to be paid for by increased delay for the data service. Figure 9.8 shows the average access delay of data for all three cases considered. In the case of a limited allocation cycle length, this is the cumulative 'access' delay experienced until completion of the message transfer, which includes the time spent in the contention state between allocation cycles. These access-delay values relate to an average transfer delay, ignoring padding, of approximately 9.3 s (i.e. 9423 bytes mean message size transmitted at 8.125 kbit/s).

9.6 SIMULATION RESULTS FOR MIXED VOICE, WEB AND EMAIL TRAFFIC

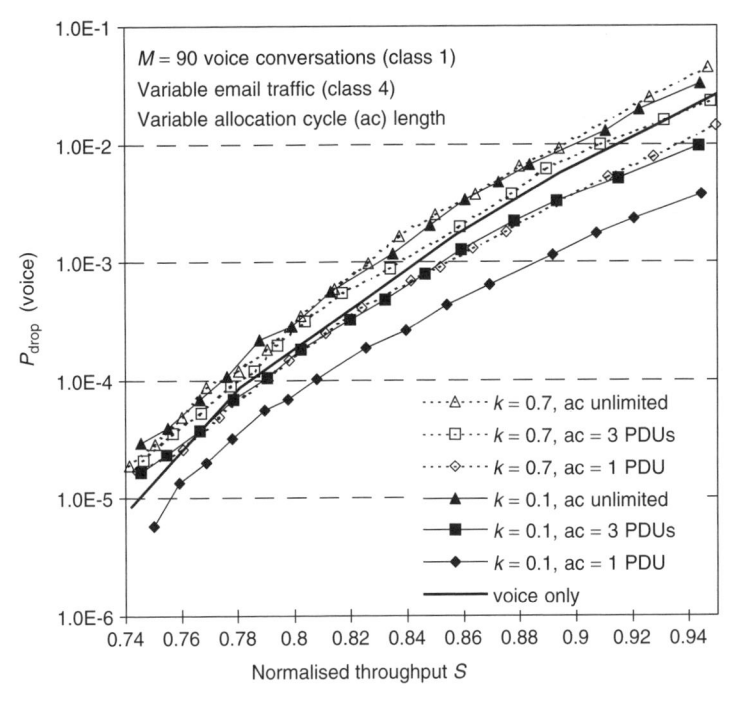

Figure 9.7 Voice dropping performance with mixed voice and email traffic, considering variable allocation cycle lengths (in terms of long PDUs per cycle)

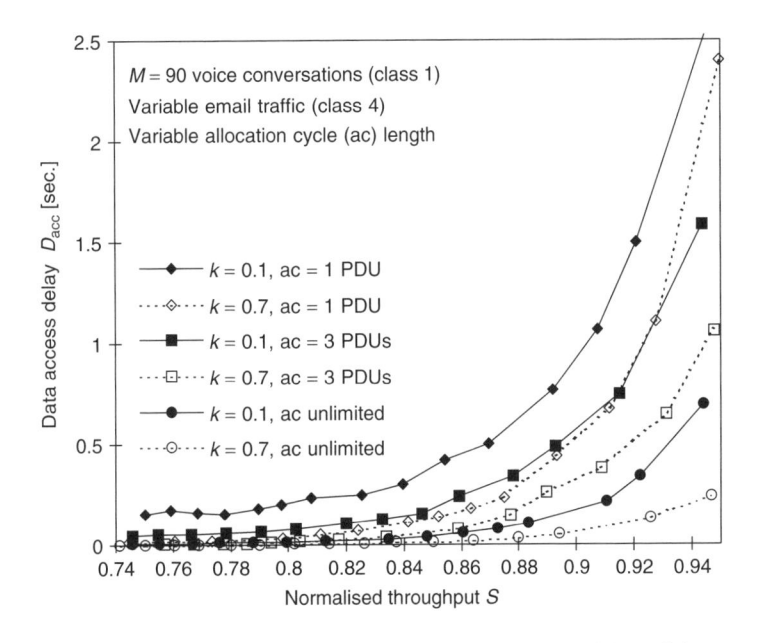

Figure 9.8 Mixed voice and email data traffic, access delay of data

9.6 Simulation Results for Mixed Voice, Web and Email Traffic

9.6.1 Equal Share of Data Traffic per Priority Class

Figures 9.9 and 9.10 report the results for the mixed scenario 4 with $M = 80$ voice conversations, and an equal share of data generated by each priority class. Here, $S = 0.75$ and 0.95 correspond to 17% and 36% data traffic respectively.

As far as voice dropping is concerned, the performance degradation due to heterogeneous traffic seems rather worse than that shown in Figures 9.4 and 9.6, let alone Figure 9.2. This is due to the combination of two effects already discussed above. Firstly, data is assigned to priority classes 2 to 4 rather than to 4 only, such that the 'average access penalty' for data at a given value of k is smaller. Secondly, since the traffic mixture contains email traffic, and unlimited allocation cycle lengths are considered, the respective considerations provided above apply in this case as well. These two effects also explain the small spread between the two curves shown in Figure 9.9.

In Figure 9.10, the effectiveness of the prioritised Bayesian broadcast algorithm in terms of discriminating the access delays experienced by the different classes is illustrated. With $k = 0.7$, a significant spread between all three data classes can only be observed for $S > 0.88$, while with $k = 0.1$ class 4 suffers already at low load from significantly larger access delay values than the other classes.

9.6.2 Unequal Share of Data Traffic per Priority class

Again for $M = 80$, but now for an unequal share of NRT data associated with each of the relevant priority classes, namely 50% email traffic (class 4), and 50% Web traffic shared

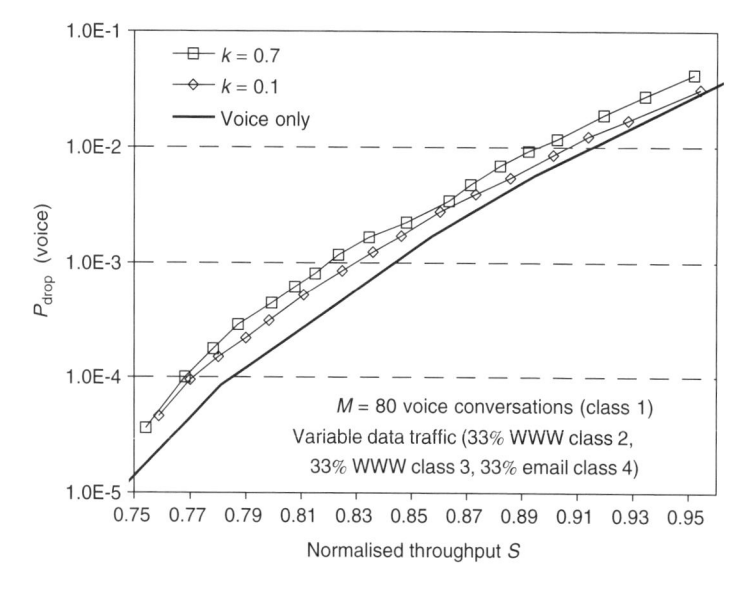

Figure 9.9 Voice dropping performance with mixed voice, Web, and email traffic, with an equal share of data traffic for each priority class

9.6 SIMULATION RESULTS FOR MIXED VOICE, WEB AND EMAIL TRAFFIC

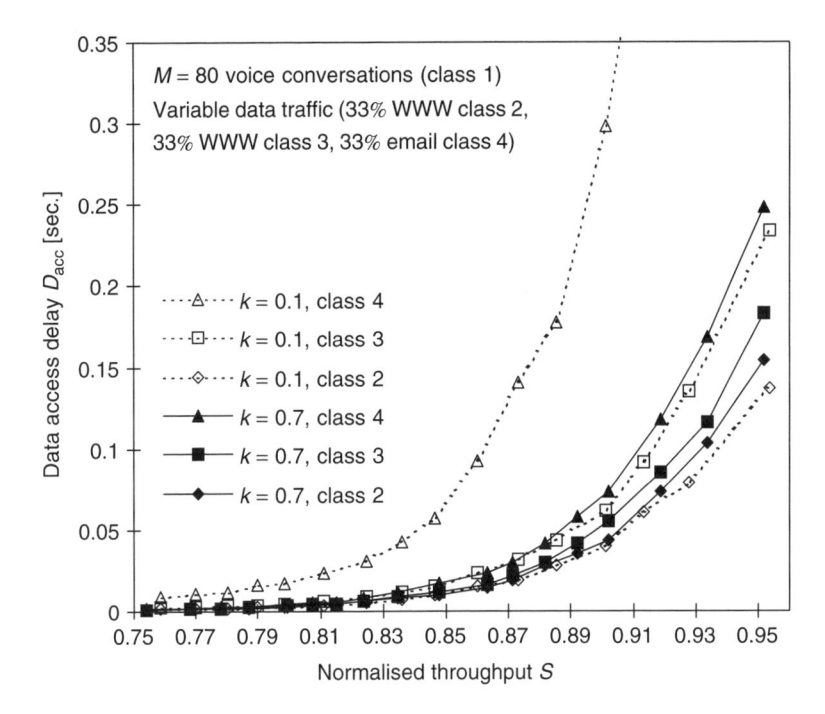

Figure 9.10 Delay performance of data with mixed voice, Web, and email traffic, with an equal share of data traffic for each priority class

equally between classes 2 and 3, P_{drop} performance of voice is shown in Figure 9.11. Consider for now the two curves equivalent to those shown in Figure 9.9, namely for $k = 0.1$ and $k = 0.7$, respectively. Due to the higher fraction of email traffic, for $k = 0.1$, P_{drop} increases, while it remains roughly the same for $k = 0.7$, such that the spread between these two curves is even smaller than that in Figure 9.9.

With the semi-proportional priority distribution algorithm used until now, the degree of prioritisation between individual access classes depends on the proportion of contention traffic (or access bursts) per class, which determines α in Equation (9.7), thus cannot be chosen. In the two traffic scenarios considered in this section, the proportion of access bursts is fairly similar. For instance, the fraction of email access bursts is below 2% in both cases. Correspondingly, unlike the voice dropping performance, the data access delay remains virtually unaffected by the changing traffic composition and therefore, the equivalent to Figure 9.10 is not shown.

Instead, for an unequal traffic-split between priority classes 2 to 4 and only for $k = 0.1$, Figure 9.12 shows the impact of non-proportional priority distribution, according to Equations (9.8) to (9.10), on the data access delay performance. Choosing $z_1 = 3$ and $z_2 = 1$ increases the relative degree of prioritisation of class 2 (i.e. increases p_2), but decreases that of class 3, while p_1 and p_4 remain the same as with the semi-proportional algorithm. As a consequence, the access delay of class 2 is slightly decreased at the expense of increased access delay of class 3, while both the access delay of class 4 and P_{drop} of voice remain unaltered (the latter shown in Figure 9.11). In short, the algorithm behaves as desired.

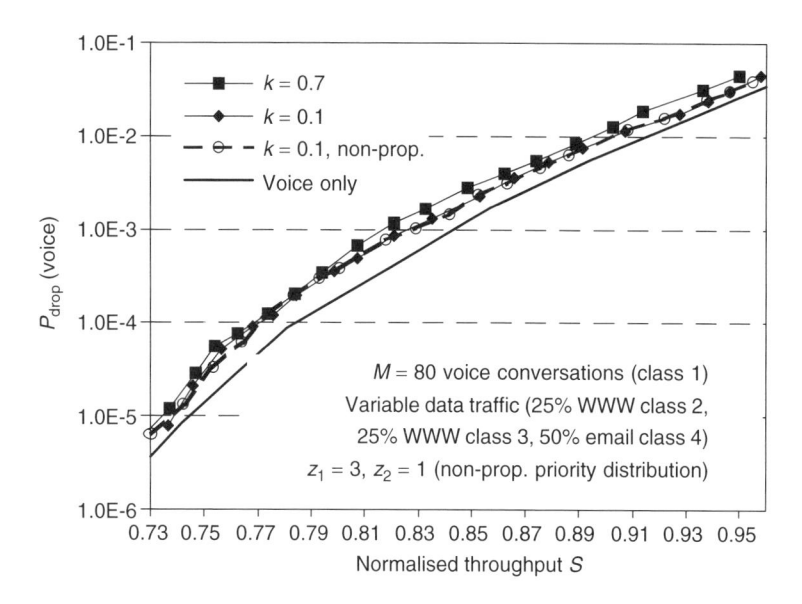

346 9 MD PRMA WITH PRIORITISED BAYESIAN BROADCAST

Figure 9.11 Voice dropping performance with mixed voice, Web, and email traffic, with an unequal traffic-split between priority classes two to four, using both semi- and non-proportional priority distribution

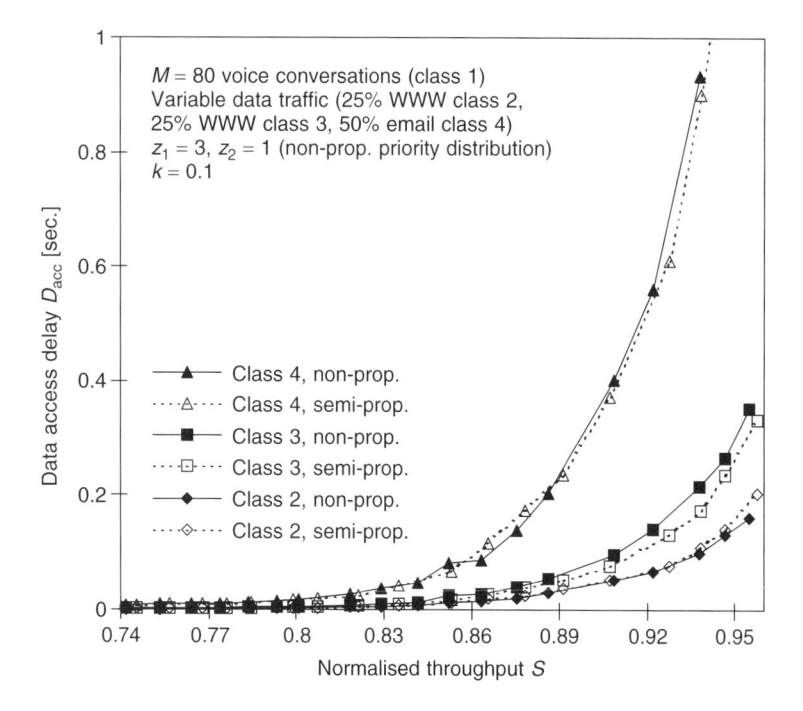

Figure 9.12 Delay performance of data with mixed voice, Web, and email traffic, with an unequal traffic-split between priority classes two to four, using both semi- and non-proportional priority distribution

9.7 Summary

Recognising the benefit of prioritisation at the random access stage, one is left with two fundamental options on how to achieve such prioritisation. Different classes of C-slots can be defined, which are scheduled according to the QoS requirements of the different services considered. Alternatively, a single class of C-slots is provided, but access to these slots is controlled individually for each priority class (hybrid approaches are obviously also possible). With in-slot protocols such as MD PRMA, where C-slots are not explicitly scheduled, the second approach is clearly more convenient. A case was made for the use of an appropriately enhanced Bayesian broadcast algorithm for such priority-class specific access control. This algorithm combines elegantly prioritisation with stabilisation. The latter, as a matter of fact, would also be required in the case where several classes of C-slots are provided.

The proposed four-class algorithm, with main prioritisation parameter k and optional parameters z_1 and z_2, works well with the traffic scenarios considered. On top of influencing the access delay experienced by the data services assigned to the different priority classes, it allows voice dropping performance to be traded off against data access delay. If one is prepared to put up with increased data access delay, at a given throughput level, it is even possible to achieve better voice dropping performance than in the case of homogeneous voice traffic. In certain traffic scenarios, for this to be possible, allocation cycle lengths for data services need to be limited to a few PDUs. In the case of mixed voice and email traffic for instance, due to the nature of the email traffic, P_{drop} is higher with heterogeneous traffic than with homogeneous traffic in the case of an unlimited allocation cycle length, irrespective of the choice of k. However, limiting the allocation cycle length to three long RLC-PDUs, or even only one PDU, allows both a significant reduction of P_{drop} to be achieved, and the respective trade-off to be controlled through appropriate selection of k.

In short, by choosing k (optionally also z_1 and z_2) appropriately and, depending on the service, limiting the length of allocation cycles, one can effectively control the amount of prioritisation, and also trade-off voice dropping probability against data access delay.

The proposed four-class algorithm features functional relations between the different access permission probability values to enable efficient downlink signalling. Where a predictive access delay performance needs to be achieved for a certain delay class, it might be necessary to abandon the functional relations and to control the access of each class individually.

10
PACKET ACCESS IN UTRA FDD AND UTRA TDD

In this chapter, first a brief introduction to UMTS Terrestrial Radio Access (UTRA) matters is provided, such as fundamental radio access network concepts, basics of, for example, the physical and the MAC layer, and the types of channels defined (namely logical, transport and physical channels). This is followed by a discussion of certain UTRA FDD features, such as soft handover, fast power control and compressed mode operation. The main focus is on the mechanisms that are available for packet access on UTRA FDD and UTRA TDD air interfaces, as provided by release 1999 of the 3GPP specifications. Improvements being considered for further releases, currently mostly dealt with in 3GPP under the heading of High Speed Downlink Packet Access (HSDPA), are also discussed.

For more general information on UMTS, the reader is referred to dedicated texts such as Reference [86].

10.1 UTRAN and Radio Interface Protocol Architecture

10.1.1 UTRAN Architecture

The UMTS terrestrial radio access network (UTRAN) consists of one or more Radio Network Subsystems (RNS), which in turn are composed of a Radio Network Controller (RNC) and multiple base stations. In UMTS terminology, base stations are referred to as *node B*; in the following, both terms will be used. A single node B may serve one or more cells (e.g. different sectors served from one site). A node B is connected to its RNC via the I_{ub} interface. The RNC is said to be the Controlling RNC (CRNC) of that node B. The RNC is connected to the Core Network (CN) via the I_u interface (see Figure 10.1). To be precise, two variants of the latter are discerned, namely I_u-CS, which provides the connection to the circuit-switched core network (i.e. to an MSC), and I_u-PS, providing the connection to the packet-switched core network (i.e. to an SGSN). Equivalent interfaces in GSM are the A_{bis} interface between a BTS and a BSC, the A interface between BSC and MSC, and the G_b interface between BSC and SGSN.

Compared to GSM, UTRA FDD supports two new handover types, namely *soft handover* and *softer handover*. In both cases, communication between a mobile terminal and the network takes place over two (or more) air interface channels concurrently. With softer handover, the two channels are associated with two different sectors served

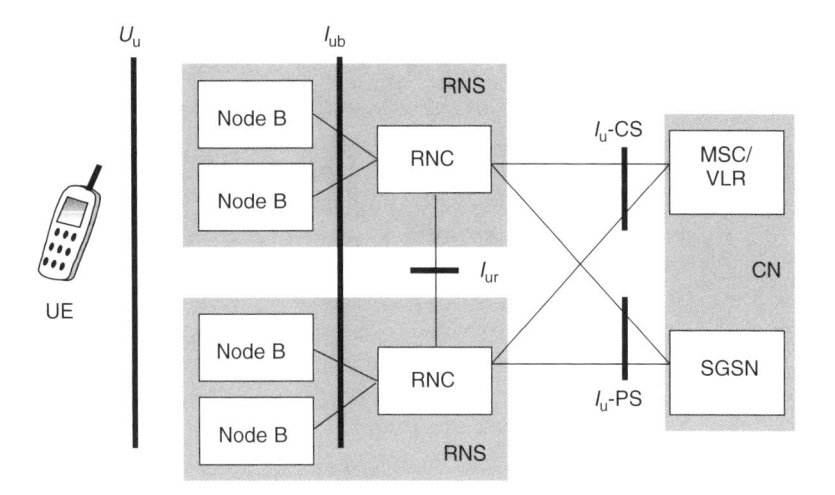

Figure 10.1 The UTRAN architecture

by the same node B, which has only 'local' implications not affecting the fundamental UTRAN architecture. During soft handover, instead, the mobile terminal is connected to the network via multiple node Bs, which may not all be controlled by the same RNC. In this case, a means for communication between RNCs is required, which is the main reason why a new interface is defined in UMTS to connect two RNCs, namely the I_{ur} interface. Being connected to multiple cells served by different antenna sites allows one to benefit from so-called *macro-diversity*, a technique which improves the transmission quality and helps, together with fast power control, to combat the near-far problem typical of CDMA systems. One RNC, the Serving RNC (SRNC), must ensure that the right signals are sent by the relevant node Bs on the downlink, and must combine the signals from multiple node Bs on the uplink in order to deliver only one signal stream onwards to the core network. If node Bs involved in the soft handover are controlled by other RNCs, then these are referred to as Drift RNC (DRNC). For further information on this subject, the reader is referred to 3GPP technical report 25.832 [276] on handover manifestations.

The UTRAN architecture is shown in Figure 10.1. This figure shows also the so-called User Equipment (UE), which is the combination of a mobile terminal or Mobile Equipment (ME) with a Universal Subscriber Identity Module (USIM), the UMTS version of the well know GSM SIM. The radio interface, that is the interface between UE and node B, is denoted U_u. In the following, we stick to the terminology known from GSM, i.e. we continue to refer to a UE as a mobile terminal or mobile station (MS).

10.1.2 Radio Interface Protocol Architecture

As in GSM, three layers are relevant for the radio interface, namely the *physical layer* (layer 1 or PHY), the *data link layer* (layer 2) and the *network layer* (layer 3), the last two featuring several sub-layers. However, in contrast to the rather confusing situation in GSM depicted in Figure 4.3, the radio interface protocol architecture has been rationalised in UMTS — at least as far as terminology is concerned. The lowest three (sub-)layers

10.1 UTRAN AND RADIO INTERFACE PROTOCOL ARCHITECTURE

are uniformly referred to as physical layer, MAC, and RLC, the last two being sub-layers of layer 2. In the so-called *control-plane* or *C-plane* dealing with signalling, the Radio Resource Control (RRC) sits on top of the RLC. The RRC is the lowest sub-layer of layer 3, and is the only sub-layer of layer 3 fully associated with and terminated in the UTRAN. In the *user-plane* or *U-plane*, additional sub-layers may be required at layer 2 depending on the services supported, namely the Packet Data Convergence Protocol (PDCP) in the 'packet domain', which replaces the LLC and the SNDCP known from GPRS, and the Broadcast/Multicast Control Protocol (BMC). The UMTS protocol architecture is illustrated in Figure 10.2. As shown, the PHY offers its services to the MAC in the shape of *transport channels*, and the MAC to the RLC in that of *logical channels*. A transport channel is characterised by *how* the information is transferred over the radio interface, while a logical channel by the *type of information* transferred. This distinction is not made in GSM, where the PHY offers logical channels to the upper layers.

Layer 2 provides *radio bearers* to higher layers. The C-plane radio bearers provided by the RLC to the RRC are *signalling radio bearers*. The RRC interfaces not only the RLC, but also all other layers below it for control purposes, quite like RR in GSM. For more information on the radio interface protocol architecture, the reader is referred to 3GPP technical specification 25.301 [277].

One reason why the GSM protocol architecture is somewhat confusing is that the system was designed initially for circuit-switched services, in particular voice, so MAC and RLC with associated header overheads were not really required at first and only added later for GPRS. In UMTS instead, for consistency, MAC and RLC are always defined, but they can both be operated in different modes, depending on what MAC and RLC features are

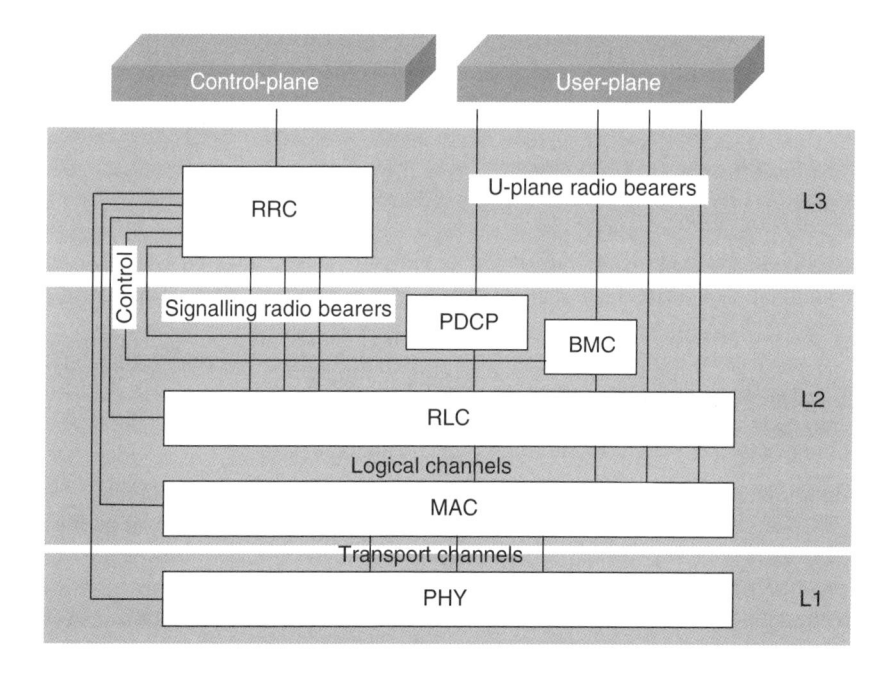

Figure 10.2 Protocol architecture on the radio interface

required for a specific service. For instance, when no MAC header is required, the MAC operates in transparent mode.

Before delving into some of the details pertaining to PHY, MAC, and RLC, let us reiterate a definition hidden in a footnote in Chapter 4 and add a new one, both listed in Reference [213]. A Protocol Data Unit (PDU) of protocol X is the unit of data specified at the X-protocol layer consisting of X-protocol control information and possibly X-protocol layer user data. A Service Data Unit (SDU) of protocol X is a certain amount of information whose identity is preserved when transferred between peer $(X + 1)$-layer entities and which is not interpreted by the supporting X-layer entities. In simple terms, taking as an example layer X to be the MAC and $X + 1$ the RLC, a MAC PDU is composed of a MAC header and an RLC PDU. From a MAC perspective, the RLC PDU represents the MAC SDU.

10.1.3 3GPP Document Structure for UTRAN

The 3GPP Technical Specifications (TS) relevant for UTRAN are the 25-series of specifications. Documents numbered 25.1xy deal with radio frequency matters, 25.2xy with the physical layer of the air interface, 25.3xy with radio layers 2 and 3 (i.e. MAC, RLC and RRC) and 25.4xy with the radio access network architecture. Additional information can be found in Technical *Reports* (TR) numbered 25.8xy and 25.9xy. The information presented in the following was mostly derived from 25.2xy and 25.3xy documents, in some cases complemented by 25.8xy and 25.9xy reports, as referenced in the text. For further information on the 3GPP document structure, refer also to the appendix.

10.1.4 Physical Layer Basics

10.1.4.1 Physical Layer Functions

The physical layer performs numerous functions as listed in TS 25.201 [278]. Among them are:

- macro-diversity distribution/combining and soft handover execution;
- FEC encoding/decoding of transport channels, error detection on transport channels and indication of errors to higher layers;
- multiplexing of transport channels onto so-called Coded Composite Transport CHannels (CCTrCH) at the transmit side, demultiplexing from CCTrCHs to transport channels on the receive side;
- mapping between CCTrCHs and physical channels;
- modulation/spreading and demodulation/despreading of physical channels;
- frequency and time synchronisation, the latter on the level of chips, bits, slots, and frames;
- measurement of radio characteristics including FER, SIR, interference power, etc., which are then reported to higher layers; and
- inner or closed-loop power control.

10.1.4.2 Basic Multiple Access Scheme and Physical Channels

The basic multiple access scheme employed in UTRA is direct-sequence code-division multiple access (DS-CDMA), with information spread over approximately 5 MHz of bandwidth, which is why this scheme is also referred to as wideband CDMA (WCDMA). Two duplex modes are supported, namely frequency-division duplex (FDD) and time-division duplex (TDD), the basic multiple access scheme of the latter also referred to as TD/CDMA. In both cases, a 10 ms *radio frame* is divided into 15 regular slots, at a chip-rate of 3.84 Mchip/s each slot measuring 2560 chips. The UTRA modulation scheme is quadrature phase shift keying (QPSK). In UTRA FDD, a double-length (i.e. 5120 chips) access slot format is also defined, with 15 access slots fitting into two radio frames.

The physical layer makes use of *physical channels* for the delivery of data over the air interface. In FDD mode, a physical channel is characterised by the code, the frequency and in the uplink also the relative phase, either I for *in-phase*, or Q for *quadrature-phase*. In TDD mode, in addition, the physical channel is also characterised by the time-slot.

UTRA supports variable Spreading Factors (SF):

- UTRA FDD from 256 to 4 on the uplink and from 512 to 4 on the downlink;
- UTRA TDD from 16 to 1 on either link.

Accordingly, the information rate of the channel is also variable.

Signals are first spread using channelisation codes, after which a scrambling code is applied at the same chip-rate as the channelisation code; hence scrambling does not alter the signal bandwidth. This is illustrated in Figure 10.3. *Channelisation codes* are used to separate channels from the *same source* (i.e. on the downlink different channels in one sector or cell, on the uplink different dedicated channels sent by one mobile terminal). *Scrambling codes* are used to separate signals from *different sources*.

The channelisation codes are based on the Orthogonal Variable Spreading Factor (OVSF) technique, which allows mutually orthogonal codes to be chosen from a *code-tree*, even when codes for different spreading factors are used simultaneously. It indeed makes sense to invest some effort in choosing orthogonal codes to separate channels from the same source. In 'benign' propagation conditions, in fact, this orthogonality is largely maintained at the receiving side. The number of codes available per tree is fairly limited though; it is equal to the spreading factor if all codes use the same spreading factor. An example with spreading factors from one (root of the tree) to eight is shown in Figure 10.4. The leaves of the tree at SF = 8 represent the available codes, if only SF = 8 is used in a cell. However, if a code at SF = 2 is assigned, then the tree is essentially pruned at that

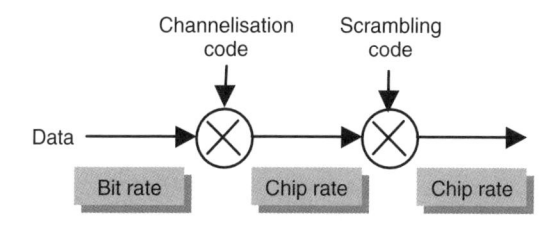

Figure 10.3 Spreading and scrambling in UTRA

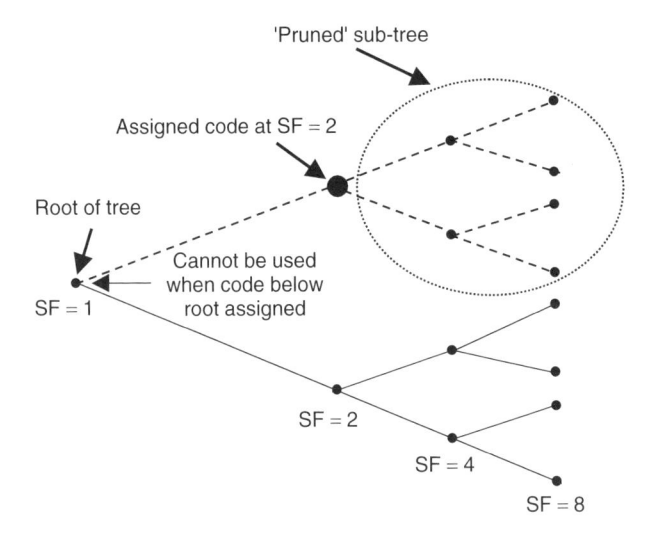

Figure 10.4 Example of an OVSF code tree

code, codes with higher spreading factors in the sub-tree below that specific code are not available anymore. More precisely, a code can be assigned to a mobile terminal if and only if no other code on the path from that code to the root of the tree or in the sub-tree below that code is assigned [279].

Without introducing special measures such as very tight synchronisation between different users, the orthogonality of signals sent by *different* sources would be lost at the receiving side even if orthogonal codes were selected at the transmitting side. This is why rather than orthogonality, other criteria such as the number of available codes and their auto-correlation properties were more important for the choice of suitable scrambling codes. In UTRA FDD, there are two types of scrambling codes, Gold codes with a 10 ms period (i.e. 38 400 chips) and so-called extended S(2) codes with a period of 256 chips, the latter optional and only applicable on the uplink. In UTRA TDD, the code-length of scrambling codes is 16.

For details on modulation and spreading, refer to TS 25.213 [280] (for FDD) and to TS 25.223 [281] (for TDD).

10.1.4.3 Transport Channels offered by the Physical Layer to the MAC

Various types of transport channels are offered by the PHY to the MAC. Transport channels are unidirectional channels. They can be classified into two groups, namely:

- *common transport channels*, where there is a need for inband identification of mobile terminals if a particular terminal is to be addressed; and

- *dedicated transport channels*, where, by virtue of a channel being dedicated to a particular communication, the terminal is identified by the physical channel it uses.

Common transport channels supported in R99 are:

- the Random Access CHannel (RACH) on the uplink;

- the Forward Access CHannel (FACH) on the downlink;
- the Downlink Shared CHannel (DSCH);
- the Common Packet CHannel (CPCH) on the uplink, only defined for UTRA FDD;
- the Uplink Shared CHannel (USCH), only defined for UTRA TDD;
- the Broadcast CHannel (BCH) on the downlink; and
- the Paging CHannel (PCH), also on the downlink.

There is only one type of dedicated transport channel defined in R99, namely the Dedicated CHannel (DCH).

10.1.4.4 Transport Channel Characteristics

The basic information unit delivered by the MAC on a transport channel to the physical layer is a *transport block*. Every so-called Transmission Time Interval (TTI), the MAC delivers either one or a set of transport blocks to the PHY for a given transport channel. Within a transport block set, all transport blocks are equally sized (but the block size can change from TTI to TTI). The TTI can assume integer multiples of the minimum interleaving period, which is 10 ms. More precisely, possible values are 10, 20, 40 or 80 ms. The TTI determines the interleaving depth, hence robustness against fading can be adjusted according to the delay constraints of the service to be supported.

The characteristics of a given transport channel are determined by its *transport format*, with attributes such as the transport block size, the number of transport blocks in a transport block set, the TTI, the error protection scheme to be applied (type and rate of channel coding), and the size of the CRC. Transport channel characteristics can be defined in terms of a *transport format set*. Some of the transport format attributes, such as those regarding the error protection scheme, must be the same within a transport format set. However, different transport block set sizes, optionally even different transport block sizes, can be chosen for transport formats within a transport format set. These two parameters affect the instantaneous bit-rate, and thus provide the means for a transport channel to support variable bit-rates. At every TTI, the MAC delivers the transport block set for a given transport channel to the PHY with the Transport Format Indicator (TFI) as a label, which indicates the transport format picked by the MAC from the transport format set.

Layer 1 can multiplex one or several transport channels onto a coded composite transport channel, each of them with its own transport format picked from its transport format set. However, not all possible permutations of these combinations are allowed. Rather, only a set of authorised Transport Format Combinations (TFC) may be used so that, for instance, the maximum instantaneous bit-rate of all transport channels added together can be limited. On the transmit side, the physical layer builds the Transport Format Combination Identifier (TFCI) from the individual TFIs, which is then appended to the physical control signalling. This is illustrated in Figure 10.9 provided in the next section on UTRA FDD. By decoding the TFCI on the physical control channel, the receiving side has all the parameters needed to decode the information on the physical data channels and deliver them to the MAC in the format of the appropriate transport channels. In UTRA FDD, if only a limited set of transport format combinations is used, then the receiving side may be in a position to perform blind detection, in which case TFCI signalling may be omitted.

More details on these matters including suitable illustrations are provided in the next few sections. Further information can also be found in TS 25.302 [282].

10.1.5 MAC Layer Basics

10.1.5.1 MAC Layer Functions

The MAC layer is specified in TS 25.321 [283]. Functions performed by the MAC include:

- the mapping between logical channels and transport channels;
- the selection of appropriate transport formats for each transport channel depending on the instantaneous source rate;
- various types of priority handling, be this between data flows from one terminal or from different terminals;
- the identification of mobile terminals on common transport channels; and
- the multiplexing of higher layer PDUs onto transport blocks to be delivered to the PHY on the transmitting side and demultiplexing of these PDUs from transport blocks delivered from the PHY on the receiving side.

Regarding the selection of appropriate transport formats (within the transport format sets defined for each transport channel), note that the assignment of transport format combination sets is done at layer 3. Therefore, the MAC has only a limited choice of transport formats, namely from the permitted combinations contained in the transport format combination set.

10.1.5.2 Logical Channels offered by the MAC to the RLC

Logical channels can be classified into two groups, namely *control channels* for the transfer of C-plane information, and *traffic channels* for the transfer of U-plane information. Except for the last one, the types of control channels defined for UMTS R99 will be familiar in name from GSM:

- the Broadcast Control CHannel (BCCH), a downlink channel used for broadcasting system control information;
- the Paging Control CHannel (PCCH), a downlink channel used to transfer paging information, when the network does not know the MS location at cell level or when the MS is in sleep mode;
- the Common Control CHannel (CCCH), a bi-directional channel used for transmitting control information;
- the Dedicated Control CHannel (DCCH), a point-to-point bi-directional channel used for the transmission of dedicated control information between an MS and the network; and
- the SHared Channel Control CHannel (SHCCH), a bi-directional channel defined for UTRA TDD only, which is used to transmit control information between MS and network relating to shared uplink or downlink transport channels.

10.1 UTRAN AND RADIO INTERFACE PROTOCOL ARCHITECTURE

Two types of traffic channels are distinguished:

- the Dedicated Traffic CHannel (DTCH), a point-to-point uplink or downlink channel dedicated to one MS for the transfer of user information; and

- the Common Traffic CHannel (CTCH), a point-to-*multipoint unidirectional* (downlink only) channel used for the transfer of dedicated user information for all or a group of specified mobile terminals.

It is important to note that the DTCH can be mapped onto dedicated *or common* transport channels. This is owing to the distinction mentioned earlier between the *type of information* transferred (as defined by the logical channel, here the DTCH) and *how* the information is transferred over the radio interface at the level of transport channels.

10.1.5.3 Types of MAC Entities and MAC Modes

Three different types of MAC entities are distinguished in TS 25.321, which handle different types of transport channels, namely:

- the *MAC-b* handling the BCH (hence b for broadcast), at the network side, it is situated at the node B;

- the *MAC-c/sh* handling all other common (or shared) transport channels, namely the DSCH, CPCH, FACH, PCH, RACH, and USCH; it is situated at the controlling RNC; and

- the *MAC-d* handling the only dedicated transport channel defined, namely the DCH. The MAC-d is situated at the serving RNC. When logical channels of dedicated type are mapped onto common transport channels, then the MAC-d, which provides these logical channels to the RLC, must interact with the MAC-c/sh, e.g. pass data to be transmitted through common transport channels on to the MAC-c/sh.

Obviously, the mobile terminal must support all different types of MAC entities.

Certain MAC features are not always required. For instance, inband identification of mobile terminals through a suitable identity contained in a MAC header are, with a few exceptions, only required when a dedicated logical channel is mapped onto a common transport channel. The case where no MAC header is required is referred to as *transparent* MAC transmission in TS 25.301.

10.1.6 RLC Layer Basics

The RLC provides three types of data transfer services to higher layers, namely transparent, unacknowledged, and acknowledged data transfer. In the case of *transparent data transfer*, higher layer PDUs are transmitted without adding any protocol information (e.g. RLC headers). In this transfer mode the 'RLC barely exists', although RLC segmentation and reassembly functionality may be used in transparent RLC mode. *Unacknowledged data transfer* means that higher layer PDUs are transmitted without guaranteeing delivery to the peer entity. However, the RLC performs error detection and delivers only SDUs free of transmission errors to higher layers. Finally, *acknowledged data transfer* implies error-free transmission (to the extent possible within specified delay limits, etc.). This is achieved by applying appropriate ARQ strategies.

Both RLC acknowledged mode and unacknowledged mode imply the addition of RLC headers to higher layer SDUs.

10.2 UTRA FDD Channels and Procedures

10.2.1 Mapping between Logical Channels and Transport Channels

All transport channels and logical channels listed in Subsections 10.1.4 and 10.1.5 respectively are defined for UTRA FDD, with the exception of the USCH and the SHCCH, which are only defined for UTRA TDD. The possible mapping between UTRA FDD logical channels and transport channels is depicted in Figure 10.5. As pointed out in the previous section, the DTCH can be mapped onto common or dedicated transport channels, hence onto the RACH, the CPCH, the DSCH, the FACH and the DCH (the first two obviously only in uplink direction, the DSCH and the FACH only in downlink direction). More than one DTCH can be mapped onto a single DCH, but different DTCHs can also be mapped onto different DCHs, depending on how the relevant radio bearers are configured.

10.2.2 Physical Channels in UTRA FDD

A UTRA FDD *physical channel* is characterised by the code, the frequency and in the uplink also the relative phase, either I for *in-phase*, or Q for *quadrature-phase*. More precisely, the uplink modulation is a dual-channel QPSK, which means separate BPSK modulation of different channels on I-channel and Q-channel. Downlink modulation is 'proper' QPSK (i.e. a single channel is modulated onto both in-phase and quadrature phase). It means that the symbol-rate of an up- and a downlink channel at a given spreading factor are the same, but that the downlink physical channel bit-rate is double that of the uplink physical channel, for example 30 kbit/s as compared to 15 kbit/s at a spreading factor of 256. As well as physical channels, there are also *physical signals*,

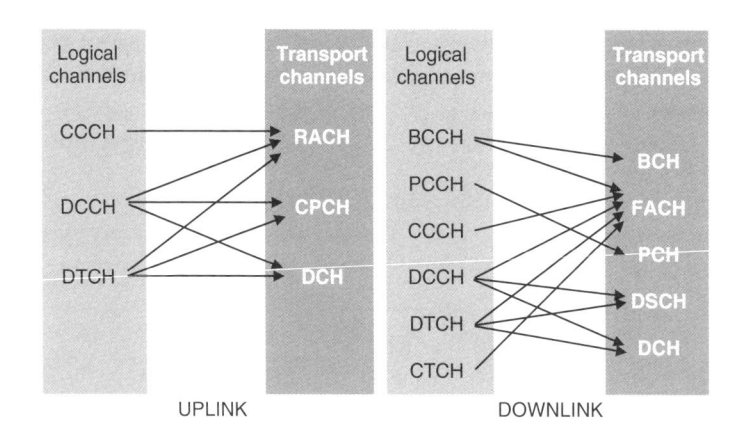

Figure 10.5 Mapping between logical and transport channels in UTRA FDD

10.2 UTRA FDD CHANNELS AND PROCEDURES

which do not have transport channels mapped to them. As usual, physical channels can be categorised as either dedicated or common physical channels.

10.2.2.1 Dedicated Physical Channels

All dedicated physical channels feature a radio frame length of 10 ms, with each frame subdivided into 15 slots. In the uplink direction, a Dedicated Physical Control CHannel (DPCCH) carrying layer 1 control information is code-multiplexed with the Dedicated Physical Data CHannel (DPDCH). In the downlink direction, there is effectively only one type of downlink Dedicated Physical Channel (DPCH), onto which data generated at layer 2 and above (i.e. the dedicated transport channel) is time-multiplexed with layer 1 control information. This is kind of similar to GSM bursts carrying layer 1 control information in the shape of training sequences and higher layer data in the payload portion of the burst format. With respect to the terminology used for the uplink, one could view the downlink DPCH as a time multiplex of a downlink DPDCH and a downlink DPCCH. The different slot formats in the uplink and downlink direction are illustrated in Figures 10.6 and 10.7 respectively.

The layer 1 control information consists of *pilot bits* (training sequences), TFCI bits discussed in Subsection 10.1.4, Transmit Power Control (TPC) bits, and (in the uplink

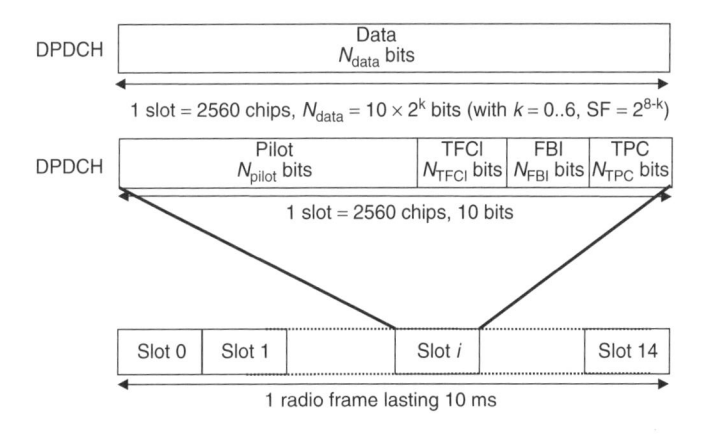

DPDCH: Data, N_{data} bits

1 slot = 2560 chips, $N_{data} = 10 \times 2^k$ bits (with $k = 0..6$, SF = 2^{8-k})

DPDCH: Pilot N_{pilot} bits | TFCI N_{TFCI} bits | FBI N_{FBI} bits | TPC N_{TPC} bits

1 slot = 2560 chips, 10 bits

Slot 0 | Slot 1 | ... | Slot i | ... | Slot 14

1 radio frame lasting 10 ms

Figure 10.6 Slot format and frame structure on the uplink DPDCH and DPCCH

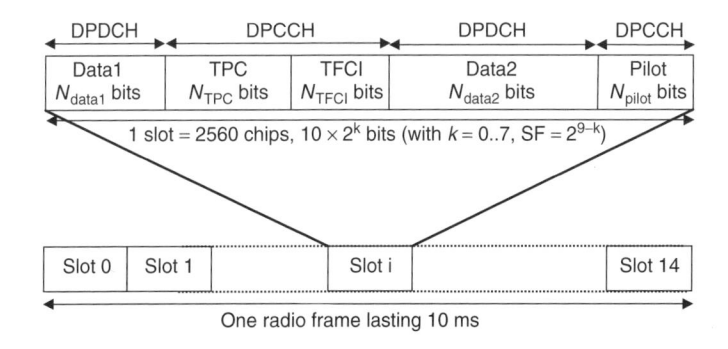

DPDCH: Data1 N_{data1} bits | DPCCH: TPC N_{TPC} bits | TFCI N_{TFCI} bits | DPDCH: Data2 N_{data2} bits | DPCCH: Pilot N_{pilot} bits

1 slot = 2560 chips, 10×2^k bits (with $k = 0..7$, SF = 2^{9-k})

Slot 0 | Slot 1 | ... | Slot i | ... | Slot 14

One radio frame lasting 10 ms

Figure 10.7 Slot format and frame structure for downlink DPCH

direction only) so-called feedback information bits used for closed-loop transmit diversity (see Reference [86] for details).

In the uplink direction, the DPDCH spreading factor is variable from 256 to 4, while that of the DPCCH is always 256, giving 10 bits per slot for physical layer overhead at a channel bit-rate of 15 kbit/s[1]. Different slot formats are defined, on the DPDCH specifying the spreading factor, on the DPCCH specifying how many bits are used as pilot, TPC, TFCI and feedback bits respectively (the last two types of bits are not always required). With single-code operation, the maximum gross data-rate (before error coding) on the uplink DPDCH is 960 kbit/s at $SF = 4$.

In the downlink direction, the DPCH spreading factor is variable from 512 to 4. The slot formats define the spreading factor, the number of DPDCH (i.e. data) bits, TPC, pilot and TFCI bits. Some slot formats specify zero TFCI bits. With single-code operation, the maximum gross data-rate at $SF = 4$ is 1920 kbit/s. Adjusted for the time-multiplexed TPC, pilot and TFCI bits, 1872 kbit/s remain. Certain mobile terminals may support multi-code operation (i.e. the use of multiple channelisation codes in parallel), allowing the data-rates to be further increased. Again, refer to Figures 10.6 and 10.7 for illustrations.

10.2.2.2 Uplink Common Physical Channels

The common physical channels defined on the uplink are the Physical Random Access CHannel (PRACH) and the Physical Common Packet CHannel (PCPCH). Not surprisingly they are used to carry the RACH and the CPCH respectively. They are both split into preamble parts and message parts. The *preambles* are 4096 chips long, fitting into *access slots* with a length of 5120 chips, i.e. double the length of a normal slot (hence there are 15 slots numbered from 0 to 14 every 20 ms or two radio frames). On the PRACH, there is only one type of preamble, while there are two mandatory preamble types on the PCPCH, namely the access preamble and the collision detection preamble. The *message part* on the PRACH is either one or two radio frames long, on the PCPCH one or several radio frames (up to 64). In both cases, the structure of the message part is very similar to that of the uplink dedicated physical channels, i.e. consisting of code-multiplexed data and control frames, the latter containing pilot and TFCI bits, in the case of the PCPCH also TPC and feedback bits. More details are provided in the next section.

10.2.2.3 Downlink Common Physical Channels

The following downlink common physical channels and signals are defined.

- The Common Pilot CHannel (CPICH), with exactly one mandatory Primary CPICH (P-CPICH) per cell and zero, one or several Secondary CPICH (S-CPICH). Both CPICH types are signals at a fixed rate of 30 kbit/s (i.e. $SF = 256$), which carry predefined bit sequences. The P-CPICH must be broadcast over the entire cell, whereas the S-CPICH may also be transmitted over only a part of the cell, e.g. as a result of the application of smart antennas.

---
[1] For comparison, on a GSM full-rate channel, if all except the encrypted symbols in the normal burst format shown in Figure 4.5 are taken to be physical layer overhead, an 'overhead bit-rate' of 8.72 kbit/s results. One might also add the SACCH overhead, since it includes power control and timing advance information (but not only!), resulting in a total rate of roughly 10 kbit/s. A fair comparison would also have to account for the fact that the DPCCH is transmitted at lower power than the DPDCH, e.g. 3 dB lower for voice according to Reference [86, Table 11.2], which halves the UTRA FDD overhead.

- The Primary Common Control Physical CHannel (P-CCPCH), which is a fixed rate channel (30 kbit/s, SF = 256) used to carry the BCH. This channel is not transmitted during the first 256 chips in each (regular) slot, which are instead used for the SCH.

- The Secondary Common Control Physical CHannel (S-CCPCH), a *variable* rate channel with spreading factors from 256 down to four used to carry the FACH and the PCH.

- The Synchronisation CHannel (SCH), a downlink signal used for cell search, which consists of two subchannels, namely the *primary* and the *secondary* SCH. They are transmitted during the first 256 chips in each (regular) slot, i.e. when the P-CCPCH is not transmitted.

- The Physical Downlink Shared CHannel (PDSCH) used to carry the DSCH. Unlike the DPCH, it does not carry any layer 1 control information. Instead, this information has to be carried by the DPCCH part of an associated downlink DPCH.

Also part of the downlink common physical channels are a number of indicator channels. Four of them provide fast downlink signalling required for the operation of the uplink common physical channels, i.e. the PRACH and the PCPCH. They are all fixed rate channels (SF = 256) making use of the double-length (i.e. 5120 chips or 1.33 ms) access slot format, the first three using only the first 4096 chips of each slot, the last one using the remaining 1024 chips, as follows.

- The Acquisition Indicator CHannel (AICH) used to carry Acquisition Indicators (AI) responding to PRACH preambles.

- The CPCH Access Preamble Acquisition Indicator CHannel (AP-AICH) carrying Access Preamble acquisition Indicators (API) responding to CPCH access preambles.

- The CPCH Collision Detection/Channel Assignment Indicator CHannel (CD/CA-ICH) carrying either Collision Detection Indicators (CDI), or, if channel assignment is used for the CPCH, Collision Detection Indicators/Collision Assignment Indicators (CDI/CAI) in response to CPCH collision detection preambles.

- The CPCH Status Indicator CHannel (CSICH) signalling the availability of CPCHs through Status Indicators (SI). This channel is always associated with a CPCH AP-AICH, the AP-AICH making use of the first 4096 chips per access slot, the CSICH of the remaining 1024 chips.

The fifth downlink indicator channel is the Paging Indicator CHannel (PICH), which is a fixed rate channel (SF = 256) like all other indicator channels. It carries Paging Indicators (PI), which are related to the PCH transport channel mapped onto an S-CCPCH.

10.2.2.4 Timing Relationships

On the downlink, CPICH, SCH/P-CCPCH and PDSCH have identical frame timings. Also, the 15 double-length downlink access slots carrying the various indicator channels used for RACH and CPCH operation are aligned in that slot 0 starts at the same time as an even-numbered P-CCPCH frame. All other channels are not aligned. Different rules apply for the timing offset of the different channels, with the constraint that the offset is in integer multiples of 256 chips. Details can be found in TS 25.211 [58].

On the uplink, the transmit timing at the mobile terminals depends always on the timing of the received signals at the terminals, which means that, unlike in GSM, there is no timing advance. For instance, the uplink DPCCH/DPDCH frame transmission is 1024 chips delayed with respect to the received DPCH frame. The timing relationship between PRACH and AICH, and between CPCH and the CPCH-related indicator channels are discussed in the next section, in the context of packet transmission on the RACH and the CPCH respectively.

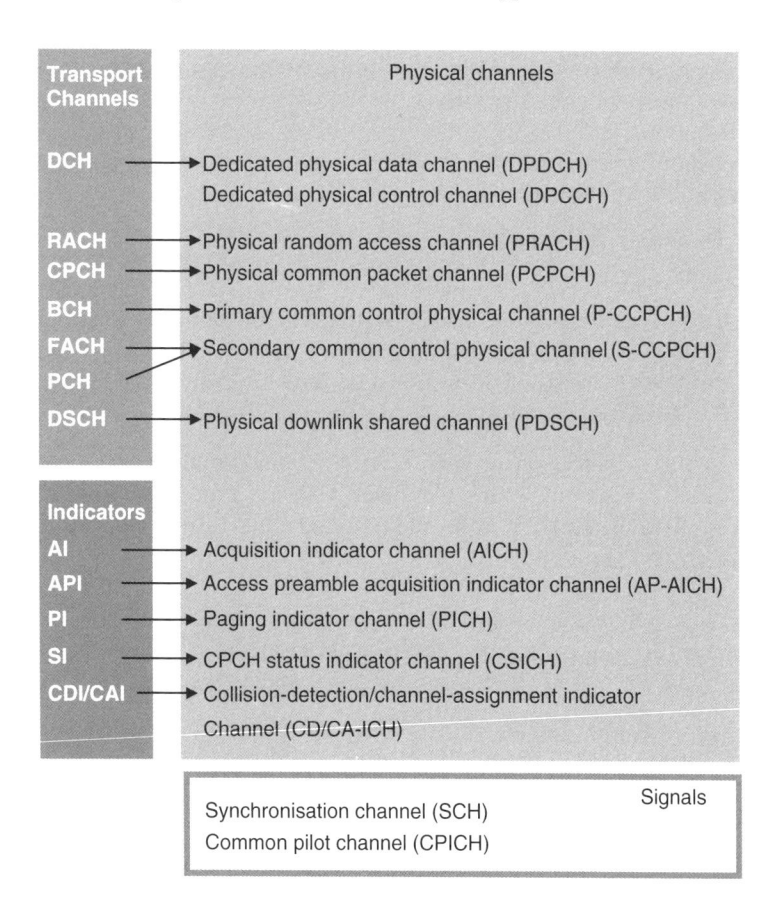

10.2.3 Mapping of Transport Channels and Indicators to Physical Channels

The physical layer offers transport channels as services to higher layers. It also offers *indicators*, which are fast low-level signalling entities that can be transmitted without relying on information blocks sent over transport channels. These indicators are either boolean (two-valued) or three-valued. The mapping of transport channels and indicators to physical channels is illustrated in Figure 10.8. This figure also shows physical signals, which do not have transport channels or indicators mapped to them.

Transport Channels	Physical channels
DCH	Dedicated physical data channel (DPDCH)
	Dedicated physical control channel (DPCCH)
RACH	Physical random access channel (PRACH)
CPCH	Physical common packet channel (PCPCH)
BCH	Primary common control physical channel (P-CCPCH)
FACH	Secondary common control physical channel (S-CCPCH)
PCH	
DSCH	Physical downlink shared channel (PDSCH)

Indicators	
AI	Acquisition indicator channel (AICH)
API	Access preamble acquisition indicator channel (AP-AICH)
PI	Paging indicator channel (PICH)
SI	CPCH status indicator channel (CSICH)
CDI/CAI	Collision-detection/channel-assignment indicator Channel (CD/CA-ICH)

Signals:
Synchronisation channel (SCH)
Common pilot channel (CPICH)

Figure 10.8 Mapping of transport channels and indicators to physical channels

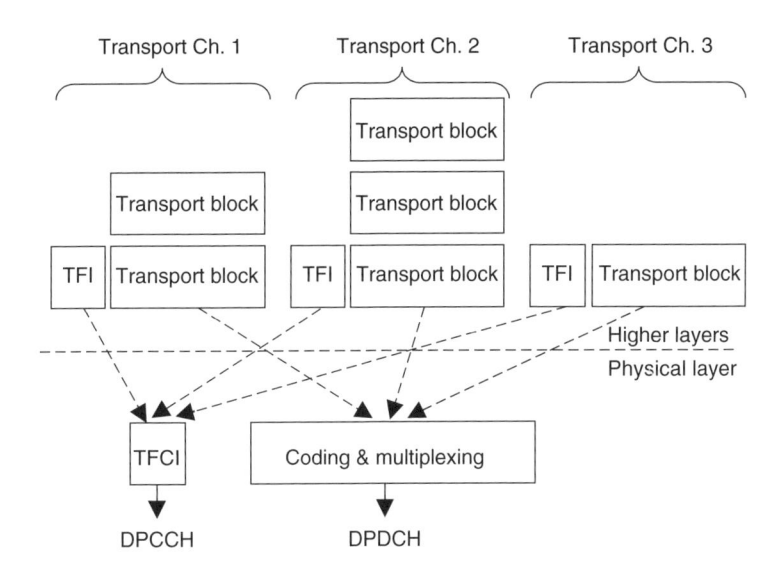

Figure 10.9 Multiplexing of transport channels onto physical channels

Figure 10.9 illustrates the multiplexing onto DPDCH and DPCCH of transport blocks and TFIs delivered by three transport channels at a given instant in time. Note that if, for example, the third transport channel has double the TTI of the other two, it will only deliver transport blocks to the PHY every second time the other two channels deliver them.

10.2.4 Power Control

It should be clear from earlier chapters that, because power is the shared resource in a CDMA system, transmit power control is a very important aspect of such a system. In fact, on the uplink, it is a vital feature to combat the near-far problem. Assuming homogeneous services (i.e. all users request the same bit-rates), one strategy is to control the transmit power in such a way that the received power levels at the base station are all equal. Other strategies can be thought of, such as SIR-based power control and differential power control in the case of heterogeneous services.

A rough means to control the uplink transmit power is *open-loop power control*. The terminal estimates the attenuation on the radio channel by listening to a pilot or beacon signal sent by the base station at a known power level and regulates its transmit power according to this estimate. The problem in an FDD system is that, due to frequency separation between the links, the uplink and downlink fast fading processes are pretty much independent. This method is therefore not very accurate and is only applied where closed-loop power control is not practicable, for instance on the RACH.

The solution used for instance on dedicated channels is fast *closed-loop power control*, where the base station measures received SIR levels, compares them with a target level and, based on the outcome of this comparison, orders the mobile terminals to either increase or decrease the transmit power level. In UTRA FDD, this happens once per slot, hence the power control rate is 1.5 kHz. This is fast enough to track pathloss and shadowing, and even fast fading of mobiles at low to moderate speeds. Closed-loop power control can be decomposed into outer-loop and inner-loop components. *Outer-loop* power control is a slow activity consisting of adjustments in the target SIR level based on the quality requirements and the current propagation conditions. *Inner-loop power control* is the fast ordering and adjustment process carried out once per slot to meet the target SIR.

On the downlink, since there is only a single signal source in a cell, power control is not needed to overcome the near-far problem. Instead, it is used to compensate for fading dips lasting longer than the interleaving period (which is mainly relevant for slow moving mobiles) and to aid mobiles at the cell edge suffering from increased intercell interference.

10.2.5 Soft Handover

As pointed out in the previous section, being connected to more than one base station during soft handover provides *macro-diversity*, which improves the transmission quality and helps, together with fast power control, to combat the near-far problem. The idea behind macro-diversity is as follows. If the same signal is transmitted via different propagation paths, which exhibit no or little correlation, then the probability that at least one of the paths delivers sufficient signal quality at the receiver at any given time is higher than when only a single path is relied upon.

In the uplink direction, during soft handover, a single signal transmitted by a mobile terminal is received by multiple base stations or node Bs. These base stations constitute the so-called *active set*. The different received signal copies are combined by the SRNC. From a terminal and interference perspective, nothing much changes with the exception that base stations in the active set try to execute closed-loop power control independently, which may result in the mobile terminal receiving conflicting power control commands. In this case, power-down commands have priority, since they imply that one base station in the active set receives the terminal's signal at sufficient quality (and this is exactly what is aimed for). Example results showing the benefit of soft handover for two base stations in the active set are provided in Reference [86, p. 203]. Under the propagation conditions considered in Reference [86], the maximum gain in terms of terminal transmit power reduction is close to 2 dB, when the pathloss from the terminal to each of the two base stations is equal. With increasing difference in the pathloss, the gain decreases. It disappears completely when the pathloss difference exceeds 5 dB.

In the downlink direction, soft handover has a quite fundamental implication in that signals directed to a single terminal in soft handover state are transmitted by multiple base stations (i.e. again those in the active set). This will obviously lead to increased downlink interference. Therefore, a clear trade-off exists between the positive effect of macro-diversity and the negative effect of increased interference. The maximum net gain at equal pathloss is 2.5 dB. At a pathloss difference of 6 dB, a net *loss* of 0.5 dB is incurred, at a difference of 10 dB a *loss* of even 2.5 dB. This is immediately intuitive: when the pathloss difference is large, the contribution of the 'weaker' base station to the received signal power at the terminal is negligible, hence the transmit power at the stronger base station cannot be reduced. At the same time, the signal is transmitted twice, which means that the total transmit power and thus the interference is increased by 3 dB (assuming equal transmit power levels). In simpler terms, the 'weaker' base station generates additional interference without providing any benefits. As a consequence, therefore, soft handover is only beneficial for a fraction of the terminal population. In practice, again according to Reference [86], the fraction of terminals in soft handover state will be below 30–40%.

10.2.6 Slotted or Compressed Mode

Unlike GSM terminals, which operate in half-duplex mode, UTRA FDD terminals transmit and receive simultaneously, thus need separate transmitter and receiver chains. Most handovers will be intra-frequency handovers, i.e. soft or softer handovers, but inter-frequency handovers can also be required, e.g. for traffic balancing between cell layers, or for a handover to UTRA TDD or GSM. To prepare inter-frequency handovers, measurements on possible target frequencies must be performed. However, this is not possible while a terminal transmits and receives continuously on a dedicated channel without an additional receiver chain.

The solution adopted in UTRA FDD to enable inter-frequency measurements without requiring an additional receiver chain is the so-called *slotted* or *compressed* mode. In slotted mode, transmission at the base station (and sometimes also at the terminal) is halted for a few milliseconds, during which the required measurements are performed. To be precise, the transmission gap length can be 3, 4, 7, 10 or 14 slots, either in a single frame or spread over two adjacent frames (the latter mandatory for gap lengths of 10 or 14 slots). Clearly, the fact that the link cannot be used during this gap for transmission of information needs to be compensated in some way. One solution is that higher layers are notified to reduce data-rates, enabling to keep the radio link parameters such as spreading factor, code-rate, etc., the same during slotted mode as during regular mode. Another solution is to halve the spreading factor, enabling the transmission of the same amount of information while making use of the link only during half of the time (which explains the term 'compressed mode'). A third solution listed in Reference [86] is that of puncturing at the physical layer, i.e. increasing the FEC code-rate. To achieve the same error performance as during regular mode, the last two solutions require more transmit power.

There are several downsides to compressed mode. The interleaving gain and the fast power control performance are reduced, requiring increased E_b/N_0 during compressed frames to maintain the same error performance, thus reducing system capacity. Where bit-rates cannot be reduced during compressed frames (e.g. for real-time services), and instead the spreading factor is halved, uplink coverage is reduced, as discussed in Reference [86]. There is an additional aspect not mentioned explicitly in Reference [86], which is strongly related to discussions in previous chapters of this book and the reason for discussing slotted mode here, namely that of instantaneous interference fluctuations. If terminals transmit intermittently at increased instantaneous power levels, then the interference variance increases compared to continuous transmission, which will also reduce capacity. Terminals in slotted mode should be scheduled in a staggered manner, such that transmission gaps do not occur simultaneously, to keep the aggregate instantaneous interference fluctuations as low as possible. In any case, slotted mode has capacity implications and should therefore only be applied with care, e.g. in circumstances, in which inter-frequency handovers are expected to be needed with high likelihood.

10.3 Packet Access in UTRA FDD Release 99

UTRA FDD provides considerable flexibility regarding the choice of suitable radio bearers for packet-data traffic. This is in part due to the conceptual split between logical channels (i.e. the DTCH for user data) and transport channels, several of which may be used for packet transfer. Small amounts of data can be sent on the RACH (in uplink direction) and on the FACH (in downlink direction). Small to medium amounts of data can be sent on the CPCH (in uplink direction). In the downlink direction, high-bit-rate non-real-time (NRT) packet-data traffic of bursty nature is best supported on the DSCH. For 'well-behaved' real-time (RT) traffic sources and for (more or less) continuous-stream packet-data (e.g. file transfers), the DCH is the most suitable vehicle of transmission, although it implies a certain overhead during idle phases which is not incurred on the CPCH for example. Each of these choices has advantages and disadvantages, as discussed in the following.

TS 25.302 describing the services offered by the physical layer states that a mobile terminal can only use one of the three possible uplink transport channels (i.e. RACH, CPCH and DCH) at any given time. On the downlink, certain terminals may support simultaneous use of DSCH and DCH, of FACH and DCH, and possibly even of all three together. However, while it is not possible to use, for example, RACH and DCH simultaneously on the uplink, it is possible to switch relatively fast between these two transport channels. This can be achieved for instance by negotiating at bearer setup a transport format combination set which contains some transport format combinations for common transport channels and other combinations, which could be used on a dedicated channel. It is then possible to switch between transport channels without having to perform a so-called transport channel reconfiguration. TS 25.303 [284] on 'interlayer procedures in connected mode' together with TR 25.922 [279] on 'radio resource management strategies' provide all the necessary details. The latter elaborates on radio bearer control through physical channel reconfiguration with or without transport channel switching, transport channel reconfiguration with or without transport channel switching, and radio bearer reconfiguration.

10.3.1 RACH Procedure and Packet Data on the RACH

The UTRA FDD random access transmission is based on a slotted ALOHA approach with fast acquisition indication combined with power ramping as a way of executing open-loop power control. The RACH transmission consists of two parts, namely preamble transmission and message part transmission. The preamble is 4096 chips (or roughly 1 ms) long, is transmitted with SF 256, uses one of 16 access signatures, and fits into one access slot. The message part can be transmitted with spreading factors from 32 to 256 and is 10 or 20 ms long, as determined by the TTI. The longer duration provides extended range [86]. The message part can be used for the transfer of signalling information or short user packets in uplink direction.

10.3.1.1 Random Access Prioritisation

Up to 16 different PRACHs can be offered in a cell, which may feature either 10 ms or 20 ms TTIs and a different choice of spreading factors for the message part. This choice is constrained by the minimum available spreading factor, which in turn determines the maximum possible data-rate for message transmission. The PRACH resources (i.e. access slots and signatures) can be partitioned flexibly. Two PRACHs may be distinguished by using different scrambling codes, or by assigning mutually exclusive access slots and signatures while using a common scrambling code [282]. Centralised probabilistic access control is performed individually for each PRACH through signalling of *dynamic persistence levels*. The update interval of these persistence levels can be chosen flexibly, its minimum value is 40 ms. Obviously, a lower value, which means more frequent updates, implies more downlink signalling overhead.

Within a single PRACH, a further partitioning of the resources between up to eight Access Service Classes (ASC) is possible, thereby providing a means of access prioritisation between ASCs by allocating more resources to high-priority classes than to low-priority classes. As an additional means for access prioritisation, in a very similar fashion to that discussed in Chapter 9, the dynamic persistence levels can be translated optionally into class-specific *persistence probability values* p_i, $i = 0, 1, \ldots, 7$ indicating the access service class. ASC 0 has highest priority, with p_0 always set to one, ASC 7 has lowest priority. ASC 0 is used for emergency calls or for reasons with equivalent priority. The probability of class 1, p_1, can be derived from the dynamic persistence level N as follows:

$$p_1 = 2^{-(N-1)}, \tag{10.1}$$

where N, which is signalled regularly, can assume integer values from 1 to 8 (i.e. providing a 3-bit resolution). Recall that we made a case for geometric quantisation of the permission probability values in Section 4.11. Coding of p_1 according to Equation (10.1) is indeed geometric, unlike the approach chosen in GPRS. A *persistence scaling factor* s_i relates p_i, $i = 2, \ldots, 7$ to p_1 through

$$p_i = s_i \cdot p_1, \tag{10.2}$$

where s_i can assume values between 0.2 and 0.9 in steps of 0.1 (i.e. at a 3-bit resolution). If the persistence scaling factors are not signalled, a default value of one is assumed. It is also possible to provide prioritisation with less than eight classes, by signalling values for less than six scaling factors. If, for instance, only values for s_2 and s_3 are signalled, then it is assumed that $s_i = s_3$ for $i = 4, \ldots, 7$.

The persistence probability value controls the timing of RACH transmissions at the level of radio frame intervals. When initiating a RACH transmission, after having received the necessary system information for the chosen PRACH and established the relevant p_i, the terminal draws a number r randomly between 0 and 1. If $r \leq p_i$, the physical layer PRACH transmission procedure is initiated. Otherwise, the initiation of the transmission procedure is deferred by 10 ms, then a new random experiment performed, and so on, until $r \leq p_i$. During this process, the terminal monitors downlink control channels for system information and takes updates of the RACH control parameters into account. This procedure is described in detail in TS 25.321.

To determine the relevant persistence value for initial access (e.g. at terminal power on), system information indicates the mapping to be applied from the access class stored in the SIM (similar to those in GSM and GPRS, see Chapter 4) to the ASC. For subsequent access, e.g. when the RACH is used for packet transmission, the ASC is linked to the MAC logical channel priority, which is assigned when the radio bearer is set up or reconfigured. Details are provided in TS 25.331 [285] specifying the radio resource control protocol, an enormous document (in terms of number of pages) with a very long 'document history' listing substantial changes well into the year 2001 (for release 1999!).

In summary, centralised access control can be applied in an extremely flexible manner, allowing load-based, backlog-based and hybrid access control schemes to be implemented, exactly as we proposed in Reference [286] presented at the second ETSI SMG2 UMTS workshop. At the same occasion, Cao also proposed to apply dynamic access control, more precisely backlog-based access control, to ensure system stability (see Reference [59]).

Independent control of multiple PRACHs is possible, allowing for instance access to high-rate PRACHs (with SF down to 32) to be controlled more restrictively than access to low-rate PRACHs. Additionally, prioritisation between up to eight ASCs within one PRACH can be performed through a parameterised solution similar to those discussed extensively in Chapter 9. For instance, it is possible to implement the non-proportional priority distribution algorithm proposed in Section 9.2.

10.3.1.2 The Physical PRACH Transmission Procedure

The UTRA FDD random access algorithm is a little bit more complex than 'just' simple slotted ALOHA with centralised probabilistic access control. Once a terminal obtains permission to access the PRACH at the MAC level (i.e. following a positive outcome of the random experiment, as described above), the physical layer PRACH transmission procedure is initiated. This entails open-loop power control through preamble power ramping and fast acquisition indication. The procedure is defined in TS 25.214 [287] entitled 'physical layer procedures (FDD)' and works according to the following steps.

(1) For the transmission of the first preamble, the terminal picks one access signature of those available for the given ASC (at most 16) and an initial preamble power level based on the received primary CPICH power level and some correction factors. To transmit this preamble, it picks randomly one slot out of the next set of access slots belonging to one of the PRACH subchannels associated with the relevant ASC. The concept of access slot sets is illustrated in Figure 10.10.

(2) The terminal then waits for the appropriate access indicator sent by the network on the downlink AICH access slot which is paired with the uplink access slot on which the preamble was sent. Sixteen tri-valued (+1, 0, −1) access indicators fit into an AICH access slot, one for each access signature.

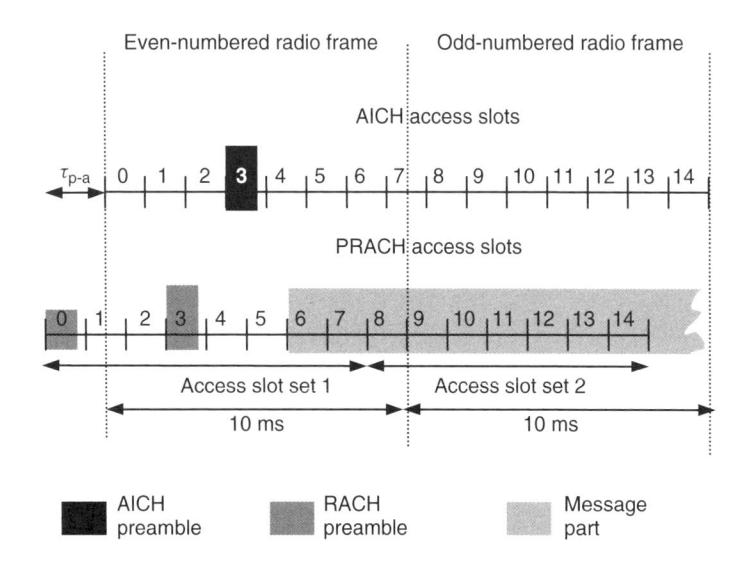

Figure 10.10 Timing relations and power ramping on the PRACH

- If the AI for the relevant signature signals a positive acknowledgement (+1), the terminal sends the message part after a predefined amount of time with a power level which is calculated from the level used to send the last preamble. Refer to TS 25.213 on spreading and modulation for the scrambling codes to be used for preambles and message part and for the channelisation code to be used for the message part.

- If the AI signals a negative acknowledgement (−1), the terminal stops with the transmission of preambles and hands control back to the MAC. After a backoff period, which is drawn from a uniform distribution between N_{min} and N_{max} radio frames (where $N_{min} = N_{max} = 0$ and $N_{min} = N_{max} \neq 0$ are permitted parameter choices), the terminal may regain access according to the MAC procedure based on persistence probabilities.

- If no acknowledgement is received, then this is taken as an indication that the network did not receive the preamble. If the maximum number of preambles that can be sent during a physical layer PRACH transmission procedure is not exceeded, the terminal sends another preamble with increased power, choosing an access slot in the same manner as for the first preamble transmission. It then continues by going to the beginning of step (2).

- If no acknowledgement is received and the maximum number of preambles that can be sent during a physical layer PRACH transmission procedure is exceeded, control is handed back to the MAC, where access can be gained again for a new power ramping cycle according to the persistence probabilities.

Similar to GSM and GPRS, it can happen that two terminals pick the same access signature for preamble transmission in the same access slot, and that the base station manages to decode one of the preambles and thus sends a positive acknowledgement. Both terminals will try to send their message part simultaneously and the base station will at best manage to receive one. Contention resolution is therefore also required in UTRA FDD. Higher layers handle it. If, for instance, the terminal does not receive a response to an RRC connection request message within a certain amount of time, it will simply have to try again.

10.3.1.3 RACH Subchannels and Timing Relations

The access slots are split between 12 RACH subchannels, hence every 12th access slot pertains to a specific subchannel (the exact mapping can be found in TS 25.214). The above-mentioned PRACH partitioning is defined in terms of subchannels and available signatures. Several (or all) subchannels may be associated with any one of up to 16 different PRACHs, similarly several or all subchannels associated with that PRACH may be used by a particular ASC. The partitioning is signalled through system information messages.

The 15 access slots are split into two access slot sets, the first eight slots are associated with set 1, the other seven with set 2, as illustrated in Figure 10.10. The downlink AICH access slots are aligned with the P-CCPCH frames, whereas the uplink RACH access slots are anticipated by τ_{p-a} chips (either 7680 or 12 800 chips, i.e. 1.5 and 2.5 access slots respectively, the former applying in Figure 10.10). The access indicators sent on an AICH slot relate to the access preambles sent on the PRACH slot with the same slot-number, leaving only $\tau_{p-a} - 4096$ chips for roundtrip propagation and processing delay. This is indeed quick and can justifiably be termed 'fast acquisition indication'. The *minimum* distance between two preambles (in case no acknowledgement is received on the relevant AICH slot) and the fixed distance between preamble and message part (in the case of a positive acknowledgement) measure both three access slots (15 360 chips) when $\tau_{p-a} = 7680$ chips, or four access slots (20 480 chips) when $\tau_{p-a} = 12 800$ chips. The distances are measured from the beginning of a preamble to the beginning of the subsequent preamble or message part. The *actual* distance between preambles depends also on the subchannels available for the given ASC. If only one is available, then the minimum distance between preambles measures 12 access slots.

Open-loop power control is realised on the RACH through power ramping between subsequent access preambles. The message part transmission power is determined by the transmission power of the last sent preamble. Clearly, this is a rough means of power control. Its accuracy is further compromised by the fact that there is a gap of at least 11 264 chips (or 3 ms) between preamble end and message start and, even worse, the power level is not adjusted during the message transmission. This is also why the message transmission is limited to 20 ms.

10.3.1.4 Packet Data Transmission on the RACH (and the FACH)

The RACH is used as an uplink transport channel to carry control information from the terminal, such as requests to set up a connection. It can also be used to send small amounts of data. For signalling transmission, typically a rather large spreading factor will be employed (e.g. SF = 256), corresponding to a channel bit-rate of 15 kbit/s. For data, if permitted by the network, the spreading factor can be as low as 32, providing a channel bit-rate of 120 kbit/s. While the 20 ms TTI may be used for signalling to extend the range of the RACH [86], it is unlikely to be used for the high rates, since the accuracy of the open-loop power control deteriorates with increasing message length. Assuming therefore a 10 ms TTI, at most 1200 channel bits can be transmitted in a RACH message, or around 600 net bits, if FEC coding at a code-rate of 0.5 is applied.

In summary, the RACH offers a means to convey short uplink data packets with, depending on the number of preambles that have to be sent, a very short access delay. However, it cannot be used for transmission of longer packets because of performance implications, such as inaccuracy of the open-loop power control, the fact that macro-diversity (or any other kind of handover) cannot be applied, and the collision risk during message part transmission.

The downlink counterpart to the RACH is the FACH mapped onto an S-CCPCH. It is used to respond to connection request messages and for other signalling purposes. Furthermore, like the RACH, it can be used for the delivery of short user packets.

10.3.2 The Common Packet Channel

The physical channels associated with CPCH transmission are the PCPCH on the uplink, the DPCCH, AP-AICH, CD/CA-ICH and the CSICH on the downlink. The PCPCH is split into preamble and message part. There are two different types of preambles, namely the *access preamble* and the *collision detection preamble*. The *message part* can be up to 64 radio frames long. It consists of code-multiplexed data and control frames, the latter containing pilot, TFCI, TPC and feedback bits, i.e. very similar to the uplink DCH shown in Figure 10.6. In various aspects, CPCH transmission resembles RACH transmission, e.g. concerning power ramping of the access preambles, and certain aspects of access control at the MAC. The major distinguishing features are the following:

- there is an additional collision detection mechanism realised through the collision detection (CD) preambles on the uplink, which are responded on the downlink by CDIs or CDI/CAIs sent on the CD/CA-AICH;
- fast closed-loop power control can be applied, requiring the CPCH (or rather the PCPCH) to be paired on the downlink with the DPCCH portion of a DPCH, which is a special DPCCH operating at a spreading factor of 512;
- the CPCH is suitable for short-to-medium length packets, the message part can extend to up to 64 radio frames and a spreading factor as low as 4 may be chosen for the message part.

These features are actually interrelated. On the PRACH, it would be risky to apply closed-loop power control during the message part, since there is a non-negligible risk of multiple users trying to transmit simultaneously (as collision resolution is dealt with by higher layers after message transmission). On the PCPCH, owing to the collision detection feature, which can be viewed as a partial contention-resolution mechanism, this risk is much lower. Also, closed-loop power control is a prerequisite for allowing the message part to extend beyond 20 ms.

In the following, we will highlight certain aspects of CPCH operation. To appreciate all the details related to the air interface, at least the following references need to be consulted:

- TS 25.211 for more details on the physical channels required for CPCH operation;
- TS 25.213 for the choice of PCPCH spreading codes;
- TS 25.214 for the CPCH physical access procedure;
- TS 25.321 for the control of the CPCH transmission at the MAC level; and finally
- TS 25.331 for the different parameters specifying the CPCHs available in a cell.

Other useful information may be found in TR 25.922 on radio resource management strategies.

10.3.2.1 The CPCH and Access Control

A cell can offer up to 16 CPCH sets, each containing up to 64 PCPCHs and up to 16 transport formats. Recall that the CPCH is an optional feature in release 1999, so cells may not offer any CPCH set at all and terminals need not support CPCH transmission either.

Access persistence can be controlled individually for each different transport format in each CPCH set, using the same eight persistence levels known from the RACH. It would not make sense to control persistence individually for each PCPCH, as little as it makes sense in MD PRMA to choose different permission probabilities for users requesting the same type of service who want to access code-slots in the same time-slot. However, it makes sense to control persistence on the basis of individual transport formats, as these determine the data-rate (up to 960 kbit/s at a spreading factor of four before FEC coding) and thus the level of interference generated. As a possible access control strategy, load-based access control could be used to react to the level of intercell interference and intracell interference, averaged over a suitable time window, generated by users transmitting on dedicated uplink channels. This strategy could also be applied on the RACH. Additionally, since the CPCH 'reservation phase' (i.e. the message part) can last up to 64 radio frames, load-based access control could in theory also be used to perform 'instantaneous load balancing' between all CPCH users within a cell and thus to reduce the variance of the intracell interference. This would be very similar to the type of load-based access control discussed extensively in Chapters 6 to 8 for MD PRMA. However, it would be 'inter-frame load balancing' as opposed to 'inter-slot load balancing' applied in MD PRMA and, as such, due to the longer time-scales involved, only suitable for non-real-time services (which is anyway what the CPCH is intended for). The condition for this to work is that the persistence-level update-interval is significantly shorter than the PCPCH message part duration. The minimum update interval is 40 ms, as in the case of the PRACH, so the maximum possible message part duration of 640 ms would likely have to be used for this type of 'short-term load-control' to provide a significant performance benefit.

10.3.2.2 The Physical CPCH Access Procedure

The first phase of the physical CPCH access procedure is very similar to that on the RACH. It consists of a power ramping of access preambles until reception of a positive access preamble acquisition indicator (or API) sent by the node B on the AP-AICH. As on the RACH, the power ramping procedure may be aborted, e.g. as a result of the reception of a negative API.

The second phase of the CPCH access procedure has no RACH equivalent. Following the reception of a positive API, the mobile terminal sends a collision detection preamble at the same power level as the last access preamble, picking one out of up to 16 collision detection signatures randomly. It then waits for a response on the CD/CA-ICH with the same signature. This provides a partial contention resolution mechanism, because two terminals would have to pick both the same access preamble signature in the same access slot and then also the same CD signature to risk transmitting the message part on the same PCPCH. This collision risk is much reduced compared to that on a PRACH. This is why one can afford to perform closed-loop power control and to extend the message part. It would not be a good idea to perform either of the two, if there were a significant collision risk, because of the increased interference this could cause.

Upon receiving a positive acknowledgement (i.e. a CDI or a CDI/CAI) on the CD/CA-ICH, the terminal sends its message on the appropriate PCPCH with its associated scrambling code. The message part can be preceded by a power control preamble lasting eight regular slots. The benefit of this preamble is that it allows the power control to converge before the message part transmission starts. This appears to make sense on the PCPCH because of the increased delay between the acknowledged access preamble determining the open-loop power level and the message part transmission, and because of the possibly high data-rates involved.

The approximate timing of the physical CPCH access procedure is illustrated in Figure 10.11. The access preamble timing and the node B response timing are the same as on the RACH. Both the intervals between the acknowledged access preamble and the CD

10.3 PACKET ACCESS IN UTRA FDD RELEASE 99

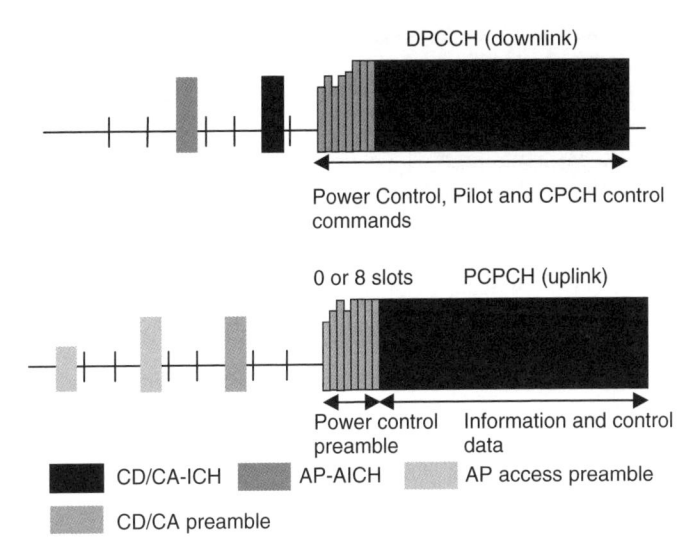

Figure 10.11 Timing of access and collision detection preambles and transmission of CPCH message part

preamble and between the CD preamble and the message part (or power control preamble, where used) are either three or four access slots. As on the RACH, this is measured from the beginning of one preamble to the beginning of the subsequent preamble or message part. Compared to RACH transmission, assuming the same number of signatures and no subchannel partitioning, the additional access delay amounts to the duration of either three or four access slots plus optionally that of eight regular slots, i.e. to roughly 10 ms.

We have ignored quite a bit of detail in the discussion of the CPCH access procedure, for which the reader is referred to the specifications listed above. We will briefly elaborate on one matter though, which was ignored so far, namely the fact that there are two variants to the physical CPCH access procedure, one with Channel Assignment (CA), and one without CA. For a start, this determines the format of the status indicators signalled on the CSICH. If CA is used, then the CSICH signals both the availability and the supported data-rate (i.e. the smallest available spreading factor) of each PCPCH offered in a cell. If CA is not used, only the availability is signalled. Before sending access preambles, the terminal listens to the CSICH to decide when and with what signature to send an access preamble.

Without CA, PCPCH selection, which is performed by the terminal, is a matter of picking a PCPCH with the right transport format supporting the requested data-rate. The combination of access preamble signature and access slot used determines the PCPCH. This is essentially *implicit resource assignment*, with the collision detection indicator (CDI) sent by the node B implicitly assigning the PCPCH for which the terminal contended. If this PCPCH is used by another terminal at the time the CD preamble is received, then the node B will have to respond with a negative collision detection indicator, and the terminal will have to try again on another PCPCH (starting from the MAC persistence check). Although there are various means for the terminal to avoid picking a busy PCPCH, in particular monitoring the status of the different PCPCHs as signalled by the CSICH, it may end up doing so all the same, e.g. because two terminals decided to go for the same PCPCH at roughly (but not exactly) the same time. In this respect, implicit resource assignment is not as foolproof on the CPCH as it is for instance with MD PRMA as defined in Chapter 6.

With CA, the terminal does not pick a specific PCPCH. Instead, the choice of access preamble signature and access slot indicates the required data-rate. Again, this choice is based on the CSICH signalling, which in this case does not only indicate PCPCH availability, but also the smallest available spreading factor. When the node B receives the collision detection preamble, it can assign *explicitly* one among possibly many PCPCHs supporting the requested data-rate, by sending a channel assignment indicator together with the collision detection indicator. This channel assignment scheme is called 'versatile channel assignment method'; it is defined in TS 25.331 and an example provided in TR 25.922.

In summary, access delay on the CPCH is somewhat bigger than on the RACH due to the collision resolution stage and the (optional) power control preamble. This increased access delay is the price to be paid for extending the message part from 10–20 ms to up to 640 ms, and for the smallest possible spreading factor to be reduced from 16 to 4.

10.3.3 Packet Data on Dedicated Channels

Dedicated channels with a *single* transport format are used for traffic with fixed or slow changing bit-rate. More interestingly for our purposes, they can also be used for fast changing bit-rates, by negotiating a *set* of suitable transport formats catering for all possible rates that may be requested by the traffic types or sources to be supported. While the network can execute some control regarding the aggregate data-rates to be delivered in a cell on the downlink, the question arises how interference control is possible on the uplink, if the terminal can select arbitrarily from a wide range of transport format sets. More precisely, the interference generated by an MS is determined by the data-rate of the CCTrCH, so what is relevant is the choice of TFC within the set of authorised TFCs.

There are several constraints regarding the choice of TFC, and accordingly mechanisms to control the interference.

Firstly, the physical channel may be configured with a minimum spreading factor, which may not be low enough to support all TFCs negotiated at radio bearer setup. One possible approach is therefore to negotiate a wide range of TFCs expected to be used during a call or session, but at the beginning to configure the physical channel with a medium spreading factor. When high transmission-rates are required, the terminal needs to request a lower spreading factor through physical channel reconfiguration. This may be triggered by a traffic measurement report sent from the terminal to the network.

A second mechanism is to specify a maximum allowed MS transmit power (below what the MS class would normally permit). If a given TFC requires more transmit power than allowed, it cannot be chosen for transmission. The subset of TFCs that are excluded because of this power limitation may vary with time, according to instantaneous channel conditions. Some TFCs may be permitted only during good conditions. For non-real-time traffic, this allows intercell interference to be minimised by transmitting with preference during good channel conditions.

A third mechanism is the TFC control procedure. This procedure can be used to exclude temporarily some TFCs in the negotiated set of TFCs from being selected by the terminal (and to re-allow their use at a later stage). It cannot be used to add new TFCs, though. It involves RRC signalling, and is thus rather slow.

A fourth approach, which is optionally supported by networks and terminals, is the so-called dynamic resource allocation control (DRAC). This is a probabilistic control scheme, which merits to be discussed in somewhat more detail below.

The most drastic approach for the network to control the uplink interference is to permit only a small set of TFCs at any given time, which covers only a limited subset of data-rates likely to be requested by the user during the course of a call or session. When the data-rate requirements increase, the MS needs to negotiate a new TFC set through a transport channel reconfiguration. This entails again RRC signalling, which may cause even more signalling overhead than for instance the TFC control procedure. It is therefore most likely too slow for real-time VBR traffic.

In summary, the DCH can support variable-bit-rate (VBR) packet-data traffic and provides a few features, which allow uplink interference fluctuations due to the variable bit-rates to be controlled to some extent. However, there are limits to their applicability, particularly when dealing with real-time traffic, which means that one may have to live without explicit means to control the uplink interference created by real-time traffic streams. In other words, when using DCHs for real-time VBR packet-data traffic, one will have to rely to a large extent on the inherent statistical multiplexing capability of CDMA. This is no problem with 'well-behaved' real-time traffic streams such as narrow-band on–off type voice, for which the inherent statistical multiplexing capability of a WCDMA system is near to perfect, as discussed in detail in Chapter 7. It is also possible to support a bursty NRT data-stream alongside RT voice on DCHs, because the QoS of the NRT stream should not suffer from delays incurred when having to renegotiate TFCs (assuming that the transport formats required for the voice transmission are available throughout a call). However, as TFC renegotiation is a slow process, it is not possible to react to fast interference fluctuations in this way. This problem can be overcome by applying DRAC, which can improve system capacity and QoS of the NRT stream, as discussed below.

DCH limitations are most likely to occur when dealing with not well behaved real-time VBR streams with medium-to-high average bit-rates, e.g. video streams exhibiting large data-rate fluctuations. In this case, we cannot expect the inherent statistical multiplexing capability of WCDMA to be near to perfect. This is due to the small number of users being multiplexed onto a common resource and, as a result, to the substantial variance of the aggregate intracell interference. A mechanism is therefore desirable, which allows a mobile station to access a wide range of data-rates without undue delay, while all the same providing some degree of interference control by the network. DRAC described below, although primarily targeted for NRT services, could be abused for this purpose.

One important feature distinguishing the DCH from RACH and CPCH, for example, is that soft handover can be applied, providing clear performance benefits in certain conditions, as outlined in the previous section. Fast closed-loop power control is also applied on the DCH. Since the DCH is assigned on a per-call or per-session basis, this implies a permanent overhead in the shape of the DPCCH, even during periods in which no user packets are transmitted. For instance, a total activity factor for a speech service of 0.67 is reported in Reference [86, p. 156], which is based on a voice activity factor of 0.5 plus DPCCH overhead during voice gaps. This means that the DPCCH contributes one third to the total DCH power during voice activity periods, which is consistent with a power difference between DPCCH and DPDCH of -3.0 dB reported elsewhere in Reference [86], provided that the same spreading factor is used for DPCCH and DPDCH.

In other words, the inherent near-perfect statistical multiplexing capability of CDMA systems for low-bit-rate users that is achieved on dedicated channels comes at the cost of permanent signalling overhead.

10.3.3.1 Dynamic Resource Allocation Control (DRAC) of Uplink DCH

In the early phases of UMTS standardisation in 3GPP, several proposals were made to introduce a new logical uplink channel for packet-data users, the Uplink Shared CHannel (USCH). The USCH was supposed to be paired with a new downlink channel, the Access Control CHannel (ACCH), which was intended to convey fast scheduling information, as required for a fully centralised MAC scheme proposed initially for NRT packet data. The drawback of such a scheme is the considerable overhead introduced on the downlink, since every individual packet would have to be scheduled explicitly. To alleviate the burden on the ACCH, one company proposed in Reference [60] to use load-based probabilistic access control (termed Dynamic Packet Admission Control, DPAC) instead of centralised scheduling, and to convey access control parameters, such as the permission probability, on the ACCH. This scheme is very much inspired by the approach to access control we proposed for joint CDMA/PRMA in Reference [30]. In fact, the simulation results presented in Reference [60] were obtained using the same type of channel access function with two linear segments we used in Reference [30]. These results illustrate how access and transfer delays can be reduced with this MAC-based access control scheme compared to RRC-based scheduling, the latter requiring transport-channel reconfiguration or similar signalling procedures. Furthermore, they show that if the packet-data traffic is generated by only a few high-bit-rate users, the mean supported throughput can be increased through access control (by more than 30% in the considered scenario). In other words, as pointed out above, the inherent statistical multiplexing capability of WCDMA for such a scenario is limited, thus multiplexing efficiency can be improved by resorting to explicit means.

It should be clear from previous sections that neither the USCH nor the ACCH made it into the final UTRA FDD R99 specifications (the USCH is featured in UTRA TDD). However, the DPAC scheme survived in a new incarnation, namely as Dynamic Resource Allocation Control (DRAC) on uplink DCH, as proposed in Reference [288] and, after further modifications, specified in TS 25.331. The network can use this procedure to dynamically control the allocation of resources on individual uplink DCHs. The required persistence parameters, namely a dynamic transmission probability p and a maximum bit-rate R_{max}, are contained in system information blocks signalled on the FACH. These parameters can be signalled individually for each of up to eight DRAC classes. R_{max} is coded in five bits representing data-rates from 16 kbit/s to 512 kbit/s in steps of 16 kbit/s, p in three bits from 0.125 to 1 in steps of 0.125. The update interval is an operator choice, the tightest interval being 40 ms, which is due to general constraints (which apply also to RACH and CPCH access parameters) regarding system information scheduling outlined in TS 25.331. As pointed out, DRAC is an optional capability in R99; it is optional for the network and needs only be supported by terminals that can receive simultaneously on one S-CCPCH and one DPCH, that is which can deal simultaneously with downlink DCH and downlink FACH.

We explain this procedure, which is designed to work also during soft handover, assuming that one DCH is controlled by DRAC (it could be more than one). The terminal listens periodically to the relevant DRAC system information for its class in each cell of its active set. If the system information is scheduled at the same time in multiple cells, then the terminal listens to the strongest one (determined by CPICH measurements). It stores the most stringent DRAC parameters among the last received values from each cell of its active set (by selecting the lowest product $p \times R_{max}$). R_{max} determines the allowable subset of the TFC set, only those TFCs being allowed, for which the sum of bit-rates on the DCH controlled by DRAC does not exceed R_{max}. Once equipped with all necessary parameters, at the start of the next TTI, the mobile terminal draws a random number r from a uniform distribution between 0 and 1. If $r < p$, the terminal transmits on the DCH controlled by DRAC during $T_{validity}$ frames using the allowed subset of TFCs. Once this 'reservation period' has expired, it performs again the random experiment based on updated DRAC parameters to obtain a new reservation. If it fails to obtain a reservation, it does not transmit on this DCH during T_{retry} frames, and thereafter performs again a random experiment based on the latest set of DRAC parameters to obtain a reservation. Transmission time validity, $T_{validity}$, and time duration before retry, T_{retry}, are indicated to the mobile terminal when the DCH controlled by DRAC is established, and can be changed through radio bearer or transport channel reconfiguration. Both parameters can assume values from 1 to 256 radio frames.

According to Reference [288], which recommends DRAC to be mandatory for all mobile terminals supporting high-bit-rate NRT packet services, the procedure 'aims at keeping a statistical control of the network load, considering the statistical nature of interference in a CDMA system, [and] the sharing of radio resources among cells and also with RT users'. Indeed, through appropriate choice of p, the load can be controlled based on the current level of intercell interference and the current load generated by 'long-term' users (i.e. real-time users on a DCH) within a cell with moderate 'statistical scheduling' overhead. The same is possible on the RACH and the CPCH. Additionally, by choosing $T_{validity}$ sufficiently long (in particular substantially longer than the update interval for the DRAC parameters), 'inter-frame' load balancing between NRT users is also possible, as already discussed for the CPCH. In fact, a DCH controlled by DRAC is better for this purpose than a CPCH since $T_{validity}$ can be up to 256 radio frames long compared to the maximum CPCH message length of 64 radio frames.

As pointed out above, this procedure could potentially be abused for real-time streams. The way in which this could be achieved would be to set $p = 1$ and control R_{max} dynamically. In the case of a VBR video stream, this may result in a picture 'freezing slightly' when a scene changes if R_{max} is smaller than the instantaneous bit-rate generated by the video codec during the scene change. It would therefore affect the QoS of a DRAC-controlled video user. However, it is better to put up with this impairment and avoid interference peaks rather than to let the video user create so much interference that it affects all other users in that cell.

10.3.4 Packet Data on the Downlink Shared Channel

On the downlink, OVSF code resources are precious since, whenever possible, a cell should only use one code tree operating under a single scrambling code to maintain orthogonality between downlink channels. On a DCH, as long as no reconfiguration takes place, codes are occupied on the code-tree according to the lowest possible spreading factor, hence code resources cannot be shared dynamically between DCHs assigned to different users according to the instantaneous (possibly higher) spreading factor. The DCH is not a suitable model of operation for packet-data streams with a high peak bit-rate, but a relatively low activity factor. The alternative is the DSCH, a downlink transport channel shared by multiple mobile terminals, which is designed exactly for high-rate NRT packet-data traffic. It allows precious downlink orthogonal code resources to be shared on a frame-by-frame basis between multiple users.

The PDSCH onto which the DSCH is mapped is characterised by a PDSCH root channelisation code. This root code determines the amount of code resources that are assigned semi-statically to the PDSCH, which can then be shared dynamically and flexibly between the multiple users to which data is directed on the respective DSCH. Assume that the root code is at SF = 4. In one frame, data may be directed to a single terminal using the root channelisation code at SF = 4, in the next frame to two different terminals at the two codes below the root code, i.e. at SF = 8. In the following frame, data could be directed to multiple users at different spreading factors, as long as the codes are all assigned on the DSCH sub-tree according to the rules explained in Subsection 10.1.4.

Each terminal to which data may be directed on a DSCH has also an associated (low-rate) downlink DCH assigned. The TFCI bits contained in its DPCH signal not only the TFC related to that DCH, but also both channelisation code to be used and TFC for the DSCH, if the latter carries data directed to that terminal. The DCH frames have to be anticipated with respect to DSCH frames, such that the terminal can first read the TFCI on the DCH, and then decide whether and how it has to decode data sent on the DSCH. Downlink power control can be applied, which is either slow power control or fast closed-loop power control, in the latter case requiring also an uplink DPCCH. Soft handover can be applied on the DCH associated with the DSCH, but not on the DSCH itself.

10.3.5 Time-Division Multiplexing vs Code-Division Multiplexing

Having discussed the characteristics of the various transport channels that can be used for packet-data traffic, let us step back and consider two generic approaches that can be adopted for the support of NRT services, namely Time-Division Multiplexing (TDM) and Code-Division Multiplexing (CDM). TDM implies time scheduling of users in such a manner that they transmit their packets sequentially one after the other. CDM means that multiple users transmit their packets simultaneously using different codes. In UTRA FDD, both these choices are available, time scheduling by using small spreading factors, code multiplexing by using large spreading factors. Obviously, also hybrid solutions between pure TDM and pure CDM are possible. But what is the best approach?

From the point of view of queuing theory, in terms of transfer delay performance, time scheduling is preferred. Without going into the theory, this can be illustrated easily using a simple example, which is depicted in Figure 10.12. Consider 10 users, each needing to transmit an equally sized packet that arrives at the same time $t = 0$ for each user. Assume that the total instantaneous bit-rate that can be supported is the same for pure time scheduling (i.e. 10 sequential packets) and pure code multiplexing (i.e. all 10 packets are transmitted simultaneously using different codes). If code multiplexing is used, the transfer delay, $D_{transfer}$, is the same for all users, say D_{CDM}. If time scheduling is used, the transfer delay of the user transmitting first is only $D_{CDM}/10$, of the second user $2 \times (D_{CDM}/10)$, etc. The transfer delay averaged over all users is only slightly more than half of that experienced with code multiplexing, namely $D_{TDM} = 0.55\, D_{CDM}$, so the choice should be clear?

10.3 PACKET ACCESS IN UTRA FDD RELEASE 99

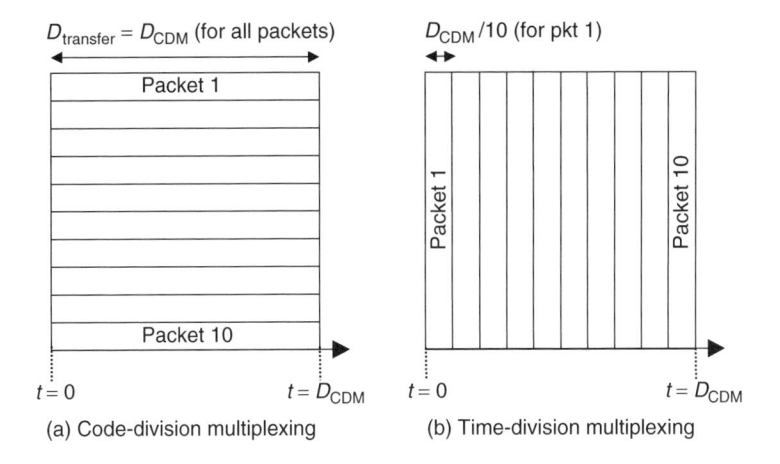

Figure 10.12 Code multiplexing vs time scheduling

Actually, this is not the whole story. First of all, while the network is in full control of the scheduling on the downlink, and can in theory do either time scheduling or code multiplexing as it pleases, one has to ask whether complexity and in particular scheduling overhead is the same on the uplink for both approaches. Code multiplexing can be performed by setting up DCHs with low data-rates. There have also been proposals to perform time scheduling on uplink DCHs [289]. A negative aspect of such an approach is that this implies RRC signalling to schedule individual packets, which is not expected to be very efficient. Both CDM and TDM could in theory be applied on CPCHs, by offering either many PCPCHs at low data-rates or only one or a few at a high data-rate. However, due to the non-deterministic access of users to the CPCH, a proper scheduling is not really possible, which is not so much an issue with CDM owing to the usual averaging effects, but it is one when trying to perform TDM, because a high-rate PCPCH cannot be 'fully packed'. This in turn leads to large interference fluctuations and wasted capacity. In this respect, it would appear that code multiplexing is the preferred option.

A similar concern with time scheduling is that neighbouring cells would suffer from undesired fast fluctuations of the intercell interference, even when 'proper' controlled scheduling can be performed by the network. This is because one user may be close to the cell edge towards neighbouring cell A, but far away from neighbouring cell B, while it is exactly the other way round for the user scheduled next, which will indeed lead to intercell interference fluctuations when dealing with high-bit-rate users. With code multiplexing, interference is averaged over the locations of multiple users, so that the fluctuations are reduced. Increased interference variance will result in reduced capacity, so from this point of view, TDM is clearly not a good idea on the uplink. It may, however, be possible to perform time scheduling in a co-ordinated manner across multiple cells so that, at least for NRT users, the negative impact of these interference fluctuations is alleviated[2]. On the other hand, even if this were possible, it would have considerable complexity implications. As a final remark on uplink multiplexing, our 'hobby', namely

---

[2] For RT users, the scope for improvements in the cells affected by fluctuating intercell interference would be more limited.

load-based access control, makes also much less sense with TDM than with CDM (on its own, evidently, this would not be a strong argument in favour of CDM).

The downlink is an entirely different matter. At least as long as we deal with omni-directional antennas[3], the same level of intercell interference is generated at any given time by a node B to any neighbouring node B, irrespective of whether time scheduling or code multiplexing is applied (e.g. on the DSCH). Considerations such as transfer delay performance become therefore more important on the downlink, tipping the balance more towards TDM.

10.4 Packet Access in UTRA TDD

10.4.1 Mapping between Logical and Transport Channels

Transport and logical channels supported by UTRA TDD are listed in Subsections 10.1.4 and 10.1.5 respectively. Note that the CPCH does not exist in UTRA TDD, instead, the Uplink Shared CHannel (USCH) serving similar purposes is defined. All logical channels defined for UTRA FDD are also supported by UTRA TDD. Additionally, the SHared Channel Control CHannel (SHCCH) is also defined, which is a bi-directional logical channel used for control information relating to shared channels. The possible mapping between UTRA TDD logical channels and transport channels is depicted in Figure 10.13. In UTRA TDD, the DTCH can be mapped onto the RACH, the USCH, the DSCH, the FACH and the DCH (the first two obviously only in uplink direction, the third and the fourth only in downlink direction).

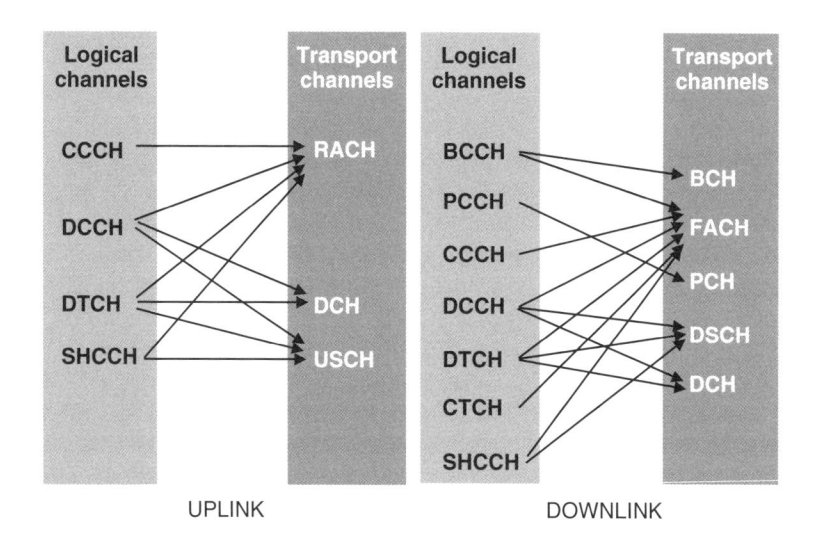

Figure 10.13 Mapping between logical and transport channels in UTRA TDD

---

[3] Matters can look differently when applying smart antennas, which allow the interference to be concentrated in space towards the direction of the target user(s).

10.4.2 Frame Structure and Physical Channels in UTRA TDD

The frame/slot timing of UTRA TDD is the same as that of UTRA FDD: a radio frame lasting 10 ms is divided into 15 time-slots. However, unlike in UTRA FDD, the time-slots are 'proper' slots providing a TDMA feature, hence users are not only separated in the code-domain, but also in the time-domain. Both approaches to TDD discussed in Section 6.3, namely alternating uplink and downlink slots (i.e. multiple link-switching-points in a frame) and a single link-switching-point are possible. However, due to universal frequency reuse, it is typically required that link-switching-points of cells covering a contiguous area are synchronised. When TDD is used in a cellular system (as opposed to uncoordinated use in unlicensed spectrum), at least two slots must be assigned to the downlink direction for SCH signalling and at least one slot for random access purposes to the uplink direction.

A physical channel is defined by the carrier frequency, the time-slot, the channelisation code, the burst type and the radio frame allocation. As in UTRA FDD, OVSF codes are used as channelisation codes, with a spreading factor of 1, 2, 4, 8 or 16 on the uplink, 1 or 16 on the downlink. A code-time-slot can be considered to represent the basic resource unit, in the same manner as for the system we investigated in detail in Chapters 8 and 9.

A similar list of common physical channels as in UTRA FDD is supported, namely primary and secondary common control physical channels (P-CCPCH and S-CCPCH respectively), the PRACH, the SCH, and the PDSCH. A physical channel exclusive to TDD is the Physical Uplink Shared CHannel (PUSCH), which is the channel onto which the USCH is mapped. The only indicator channel which exists in UTRA TDD is the PICH, the other indicator channels defined for UTRA FDD are not required, as there is no fast acquisition on the PRACH and no CPCH at all.

The UTRA TDD dedicated physical channel is the DPCH.

10.4.2.1 TDD Burst Types

The UTRA TDD burst structure is very similar to that of GSM described in Chapter 4. A burst is composed of two data symbol fields separated by a midamble or training sequence, and a guard period at the end of the burst. Unlike in GSM, timing advance is not always mandatory. It can be switched off in small cells.

Three burst types are defined. Types 1 and 3 share the same type of midamble spanning 512 chips (out of a total of 2560 chips per slot). They differ in that type 1 features a standard guard period of 96 chips (or 25 µs, which is slightly shorter than in GSM), while type 3 has a double-length guard period of 192 chips (or 50 µs). This extended guard period is useful for initial access or access to a new cell after handover. Burst type 2 is distinguished by a shorter midamble spanning only 256 chips, and has the same standard guard period of 96 chips as burst type 1. The length of the midamble determines how many different uplink channel impulse responses can be estimated. Burst type 2 is only applicable on the uplink if no more than four users share a time-slot (hence for high-bit-rate users), otherwise, burst type 1 (or 3) must be used, which allows up to 16 channel impulse responses to be estimated. This is consistent with the number of OVSF codes available, namely 16 if all users transmit at $SF = 16$. All bursts transmitted in the same time-slot must feature the same midamble type, hence types 1 and 3 can be mixed together, whereas type 2 bursts must be kept on separate time-slots.

Table 10.1 Number of data symbols per burst in UTRA TDD

SF	Burst Type 1	Burst Type 2	Burst Type 3
1	1952	2208	1856
2	976	1104	928
4	488	552	464
8	244	276	232
16	122	138	116

Table 10.1 shows the number of data symbols available per burst type for the different spreading factors (in the case of SF = 1, obviously, the number of data symbols corresponds to the number of chips). TFCI bits, where included, take resources away from the data symbol fields, and the same applies to TPC symbols. The latter are used for downlink power control, and are transmitted on the uplink at least once by each user per radio frame in which that user transmits. For a low-bit-rate user, which is assigned one code-time-slot or basic resource unit per frame, this equates to a power control rate of 100 Hz.

More details on bursts, types of midambles, etc., can be found in TS 25.221 [265] on TDD physical channels and mapping of transport channels onto them.

10.4.2.2 Dynamic Channel Allocation

Because universal frequency reuse is applied in UTRA TDD, this system is essentially interference limited, as is UTRA FDD. The interference limit may be reached before all code-time slots available in a cell are filled with bursts. Dynamic Channel Allocation (DCA) is used to determine which code-time-slots can be used in a cell without creating excessive interference to neighbouring cells. In TR 25.922, a distinction is made between *slow DCA*, which determines the resources that can be allocated to cells, and *fast DCA*, which entails allocation of individual resource units to bearer services within a cell.

With the limited UMTS spectrum available at 2 GHz in mind (see Chapter 2), choosing the best carrier frequency through slow DCA is typically not required, since operators deploying UTRA FDD in the paired band will only have one carrier available for UTRA TDD. Slow DCA will therefore mainly deal with the assignment of time-slots to cells, which may result in a clustering of time-slots in such a way that cells that would otherwise create strong mutual interference do not use the same time-slots. With fast DCA, the channels assigned through slow DCA to a given cell are allocated to individual bearer services or users according to their needs. Multiple resource units can either be assigned through code pooling, time-slot pooling or a combination thereof, preferably in a manner which maximises overall system capacity by taking for instance propagation characteristics and other environmental conditions into account.

10.4.3 Random Access Matters in UTRA TDD

Random access in UTRA TDD is based on a slotted ALOHA scheme and involves the transmission of individual bursts using burst type 3 in a time-slot designated for RACH use. There is a one-to-one correspondence between a RACH transport channel and the PRACH, its physical channel. Unlike UTRA FDD, there is no split in preamble and message part, and therefore no fast acquisition—the whole burst is either received by the node B or lost, e.g. due to a code collision.

As in UTRA FDD, multiple RACHs (up to 16) can be offered in a cell, each associated with a single time-slot. Up to eight different channelisation codes are available for RACH purposes on RACH time-slots, irrespective of the spreading factor to be used for this RACH, which can be either eight or 16. Splitting RACH resources into up to eight RACH subchannels is possible, so that an individual subchannel occurs only in a limited subset of radio frames. Different RACHs must either be assigned to different time-slots or, if they share a time-slot, must use non-overlapping partitions of that slot in terms of channelisation codes and RACH subchannels.

MAC-level access control mechanisms on the RACH are exactly the same as in UTRA FDD. Every RACH can have its own persistence level signalled, which may be translated into ASC-specific persistence probability values (with up to eight access service classes per RACH), exactly as outlined in Subsection 10.3.1.

Time-slots used for RACH purposes in one cell can also be used for RACH purposes in other cells. The specifications do not explicitly exclude a time-slot used for PRACH from being used to carry other channels in the same cell as well. Clearly, these other channels would have to use either burst type 1 or 3, since burst type 3 used for the PRACH cannot be mixed with burst type 2. However, because the mapping from midambles to channelisation codes on the PRACH described in TS 25.221 is different from that applied for other channels, such a time-slot sharing may not be possible in practice. Should this be the case, then UTRA TDD would not be suitable for in-slot protocols like MD PRMA (refer to the discussion in Section 3.6). Also, since the intracell interference on a time-slot used exclusively for RACH purposes has no slowly fluctuating component, load-based access control would only make (limited) sense in cases in which at least the intercell interference is more or less predictable. This excludes scenarios where the same time-slots are used for the PRACH across multiple cells. Prioritised backlog-based access control, on the other hand, is in any case suitable for UTRA TDD.

10.4.3.1 Packet Data and Signalling on the RACH

Unlike in GSM, where the access burst payload is very small, the RACH payload in UTRA TDD can be used to convey more than just a simple channel request containing very limited information. According to Table 10.1, 116 and 232 symbols fit into the data fields of a type 3 burst at a spreading factor of 16 and 8 respectively. Assuming rate 1/2 coding and given QPSK modulation (i.e. two bits per symbol), this happens also to be roughly the number of user bits which can be conveyed in a single burst (some overhead such as tail bits and CRC would have to be deducted). This is still rather little for user packets, hence the RACH in UTRA TDD is expected to be used mostly for signalling purposes, for instance in conjunction with the USCH and the DSCH, as discussed below.

Importantly, when using the RACH for signalling purposes, the payload is large enough to include short (i.e. 16-bit) unique identifiers assigned to mobile terminals while in 'connected mode' (as opposed to 'idle mode'), namely so-called Cell Radio Network Temporary Identities (C-RNTI). These can be used to identify mobile terminals unambiguously on common transport channels, which solves the contention resolution problem experienced in GSM and GPRS. Refer to TS 25.303 and TS 25.304 [290] for more information on connected mode and idle mode.

10.4.4 Packet Data on Dedicated Channels

In Subsection 10.3.3, we discussed the use of dedicated channels in UTRA FDD for packet traffic. It was outlined how they can be used for fast changing bit-rates, by negotiating a *set* of suitable transport formats catering for all possible rates that may be requested by the traffic types or sources to be supported. Depending on the type of traffic, WCDMA would then automatically provide statistical multiplexing of the various services owing to interference averaging on the fairly thick common resource or pipe. However, we also pointed at the need for explicit interference control mechanisms, particularly when dealing with not well-behaved real-time VBR streams with medium-to-high average bit-rates, e.g. video streams exhibiting large data-rate fluctuations.

With TD/CDMA, due to the temporal separation of users, 'automatic' interference averaging is less efficient, as discussed thoroughly in earlier chapters such as Chapter 7. It means that there is a more urgent need than in UTRA FDD for the network to have control over the instantaneous bit-rates at which uplink users transmit. When the data-rate requirements of a VBR source change, a channel reconfiguration is required, entailing RRC signalling, as already discussed for UTRA FDD. In the case of constant-bit-rate packet voice, the beneficial impact of interference reduction through voice activity detection is expected to be rather small, which is again due to limited interference averaging. One way to improve this situation could be time-hopping mentioned in TR 25.922, which could be used to provide better interference averaging. Time-hopping means swapping of time-slots between users from radio-frame to radio-frame, thus in a sense replicating the effect of slow frequency hopping in GSM, but now in the time-domain rather than in the frequency-domain. Time-hopping makes sense for voice and other low-to-medium-bit-rate services, but is of no benefit for high-bit-rate services, as they need to transmit during most time-slots anyway.

10.4.5 Packet Data on Shared Channels

Shared channel operation in UTRA TDD is in some respect a hybrid of GPRS and UTRA FDD shared channel operation. It means that a shared transport channel (USCH or DSCH) mapped onto one or several suitable physical channels (PUSCH or PDSCH) is shared in time between multiple users. Shared channels were designed to carry packet-data traffic in an efficient manner. The logical channel carrying user data, namely the dedicated traffic channel, is mapped onto the USCH in the uplink direction, and on the DSCH in the downlink direction. The logical channel defined for the signalling message exchange controlling shared channel operation is the SHared channel Control CHannel (SHCCH). It can be mapped onto USCH or RACH transport channels in the uplink, and onto DSCH or FACH in the downlink direction.

Unlike in UTRA FDD, certain classes of UTRA TDD mobile terminals may support on the uplink simultaneous use[4] of PRACH and DPCH, or of PRACH and PUSCH, or even all three together. This increases the flexibility for shared channel operation, particularly in terms of use of SHCCH. When simultaneous DPCH and PUSCH operation is supported, it is also possible to deal with real-time traffic on the DPCH, while concurrently using the PUSCH for NRT packet-data traffic.

---

[4] Simultaneous in this context means during the same radio frame, but not necessarily in the same time-slot.

10.4.5.1 Uplink Shared Channel Operation

USCH operation resembles very much the GPRS fixed allocation scheme. The selection of the mobile terminal allowed to transmit on the USCH for a certain period of time is performed by higher layer signalling. There are currently no means similar to the uplink state flag in GPRS which would allow terminal selection to be performed on a frame-by-frame basis (or rather, on a TTI basis). Terminals request USCH capacity by sending *capacity request messages* on the SHCCH, which is either mapped onto currently allocated USCH resources, thus corresponding to 'reservation mode' according to terminology used in earlier chapters, or onto the RACH, in which case these requests would be sent in 'contention mode'. If a capacity request is sent on the RACH, a 16-bit C-RNTI is included in the message, which is a unique MS entity in the relevant cell, so that contention resolution issues known from GPRS are avoided. In terms of the R-ALOHA protocol classification made in Chapter 3, sending capacity requests on the USCH would correspond to some kind of piggybacking scheme for capacity requests. If the network wants to allocate capacity on a suitable USCH, it sends a *physical shared channel allocation message* on the SHCCH, which is either mapped onto a DSCH, if one is already assigned to the requesting user, or on the FACH. This message specifies the physical resources to be used and the period of time (also referred to as *allocation period*) during which the terminal can use the USCH for data transfer. The allocation period can be selected from 1 to 256 radio frames.

A USCH is mapped onto one or several PUSCH. All spreading factors permitted on the uplink DPCH can also be applied to the PUSCH, i.e. $SF = 1, 2, 4, 8$ or 16. All three burst types can be used on the PUSCH. TFCI and TPC can be transmitted, where required. The USCH can be used with our without associated DCH. As in GPRS, mechanisms to maintain the timing advance, where required, are available.

10.4.5.2 Downlink Shared Channel Operation

Unlike in UTRA FDD, there is no need to associate a DSCH with DCHs in UTRA TDD. Individual users can be notified about DSCH reservations through higher layer signalling on the SHCCH, which can be mapped onto an already assigned DSCH or else onto a FACH. If an uplink acknowledgement of such signalling messages is required, it is either sent on the RACH or an already assigned USCH. This mode of operation is pretty much symmetric to USCH operation. Like on the USCH, this method is not suitable for switching between different users at every new TTI.

In the downlink direction, there are two additional, faster methods to signal to which terminal data on a shared channel is directed. One method makes use of mobile-terminal specific midambles instead of using the default midambles linked to the used scrambling and channelisation codes. Whenever bursts with a midamble assigned to one specific terminal are sent on the relevant downlink time-slot, the terminal in question knows that it has to decode the data. This method allows only one user per TTI to be addressed in a given time-slot, that is, no data may be directed to other terminals during this TTI on the relevant time-slot. The third method is similar to that used on the UTRA FDD DSCH, it is based on TFCI signalling. If a DCH is associated with the DSCH for a specific user, then TFCI signalling can be performed on the DCH as in UTRA FDD. Alternatively, if no DCH is associated, the TFCI signalling may also be performed on the DSCH itself.

A DSCH is mapped onto one or more PDSCH. As on the downlink DPCH, possible spreading factors are either $SF = 16$ or $SF = 1$. Burst types 1 or 2 can be used, and TFCI bits can be transmitted, if required (TPC bits are not required, since uplink power control in UTRA TDD is based on open-loop power control).

10.5 High-Speed Packet Access

Surely, if it is possible to squeeze a gross peak data-rate of 473.6 kbit/s out of a 200 kHz GSM carrier by using 8PSK modulation, then it must also be possible to increase peak data-rates significantly on a 5 MHz carrier. Very simplistically considering only the carrier bandwidth, the equivalent peak data-rate would exceed 10 Mbit/s.

During the specification of UMTS release 4, a feasibility study was conducted investigating techniques to increase data-rates on the downlink shared channels. These techniques are intended for use on a High Speed DSCH (HS-DSCH, to be introduced with release 5), and should be applicable to streaming, interactive and background services, but not necessarily to the fourth QoS class listed in TS 23.107, namely conversational (that is, real-time) services. The respective work item is, in a somewhat cumbersome fashion, referred to as High Speed Downlink Packet Access (HSDPA). Two technical reports were created by 3GPP on HSDPA for release 4, namely TR 25.848 [291] dealing with physical-layer aspects and TR 25.950 [292] considering mostly layer 2/3 aspects.

The techniques discussed in these two reports for HSDPA were the following:

- Adaptive Modulation and Coding (AMC), possibly combined with new scheduling techniques to exploit favourable channel conditions experienced by specific users;
- Hybrid ARQ (HARQ);
- Fast Cell Selection (FCS); and
- Multiple-Input Multiple-Output (MIMO) transmission.

These techniques are applicable to both UTRA FDD and UTRA TDD. In the following, where specific to a single mode, the discussion focuses on UTRA FDD.

10.5.1 Adaptive Modulation and Coding, Hybrid ARQ

The principle of matching the modulation and coding scheme to the average channel condition of each user has already been discussed extensively in Chapter 4 in the context of GPRS and EGPRS. GPRS provides adaptive FEC coding as a means of link adaptation, EGPRS, owing to the introduction of 8PSK on top of GMSK, also adaptive modulation. The AMC scheme investigated in TR 25.848 features seven different modulation and coding schemes (MCS), all using turbo coding with one of the following code-rates: 1/4, 1/2 or 3/4. The modulation schemes considered are QPSK (as per UTRA release 1999), 8PSK, 16QAM and 64QAM. The chip-rate is the same as in UMTS R99, i.e. 3.84 Mchip/s, and the spreading factor is fixed at SF = 32. Three binary symbols (i.e. bits) can be mapped onto an 8PSK symbol, four onto a 16QAM symbol, and six onto a 64QAM symbol. Multi-code transmission is considered using 20 codes in parallel. The MCS providing the highest bit-rate uses 64QAM modulation and a code-rate of 3/4, resulting in a peak bit-rate of 10.8 Mbit/s when using all 20 codes.

10.5 HIGH-SPEED PACKET ACCESS

When AMC is used in combination with time scheduling, then it is possible to take advantage of short-term variations in a mobile terminal's fading envelope so that a mobile terminal is always being served on a constructive fade. Obviously, there are limits: if the terminal is moving too fast, then the fading cannot be tracked anymore. One extreme scheduling strategy, which would maximise the capacity, would be to schedule always the user experiencing the highest instantaneous CIR on the HS-DSCH. The downside of this maximum CIR strategy is that it is extremely unfair to users experiencing unfavourable propagation conditions, hence it would have to be moderated to provide a minimum share of resources to such users as well. There are several prerequisites for a scheduling strategy attempting to exploit short-term channel fluctuations to work. First, the minimum TTI should be shorter than the release 1999 minimum of 10 ms (i.e. one radio frame). Some results presented in TR 25.848 are based on a TTI of 3.33 ms, i.e. five time-slots or a third of a frame, another option being considered for release 5 is that of three slots, i.e. 2 ms. Second, there must be a fast uplink signalling capability for instantaneous link quality measures to be available at the scheduler in a timely fashion. Third, scheduling should be performed by the node B instead of the RNC as in release 1999, because the RNC would be too slow to react to instantaneous channel conditions due to the signalling delays on the I_{ub} interface.

Hybrid ARQ schemes were already discussed in Section 4.12 in the context of EGPRS. In TR 25.950, an overview of the different types of ARQ schemes is provided, and a distinction is made between type II and type III HARQ schemes. In type III schemes, each retransmission is self-decodable, which is not the case with type II schemes. In GPRS and EGPRS, the RLC applies selective retransmission. In TR 25.950 it is argued that a selective retransmission scheme would be too demanding in terms of terminal memory requirements when combined with HARQ. Indeed, assuming constant acknowledgement delays, then the higher the data-rates, the tougher the memory requirements, so this is a matter of even more concern on the HS-DSCH than it is in EGPRS. A case is made in TR 25.950 for stop-and-wait ARQ, where the transmitter does not proceed with the sending of subsequent blocks until the current block is decoded successfully by the receiver (i.e. the transmitter needs indeed to stop and wait). This scheme requires very little overhead, a one-bit sequence number and one-bit acknowledgements for instance would, at least in theory, be sufficient. A stop-and-wait scheme reduces also the complexity encountered in EGPRS regarding sequence number encoding. Recall from Section 4.12 that the sequence number encoding must be very robust, because the receiver needs to decode the sequence number before being able to combine multiple block copies. The drawback of stop-and-wait ARQ is that the channel cannot be kept busy all the time. The transmitter has to wait until it receives an acknowledgement (or until the acknowledgement timer times out), before it can either retransmit the current block, potentially applying increased redundancy if an incremental redundancy strategy is adopted, or proceed with a new one. Fortunately, since we are dealing with a shared channel, other users can be scheduled during this 'dead time'. Furthermore, it is possible to extend the stop-and-wait approach to multiple parallel channels for a single user, each channel dealing with only one packet at a time, and therefore to alternate between these channels while waiting for acknowledgements.

In TR 25.950, it is reported that the use of AMC with HARQ allows average system throughputs close to 2.8 Mbits per second, sector and carrier to be achieved compared to 1.7 Mbits per second, sector and carrier when using only QPSK and rate 1/2 FEC coding.

The release 4 feasibility study conducted by 3GPP found that AMC, HARQ and scheduling at the node B were feasible and should be included in release 5 of the specifications. Indeed, this is the current working assumption for release 5.

10.5.2 Fast Cell Selection

The principle of fast cell selection is as follows. The mobile terminal indicates the best cell, which should serve it on the downlink, through uplink signalling. While multiple cells may be members of the active set, only one of them (i.e. the best) transmits at any one time, so that interference should be decreased and system capacity increased. FCS could be constrained to choosing between cells served by a single node B. Alternatively, inter-node B FCS may be applied. In the latter case, with scheduling at the node B, the issue of queue management arises, since there would be a need for distributed queues. Another issue is whether hybrid ARQ could be continued after a cell change, or whether it should be aborted. The benefits of FCS and the associated complexity need to be studied in more detail before a specific scheme can be adopted, possibly in the release 6 time-frame.

10.5.3 MIMO Processing

Multiple-input multiple-output processing employs multiple antennas at both the base station transmitter and the terminal receiver. This allows the peak throughput to be increased through a technique known as code re-use. With M transmit and M receive antennas, a single channelisation/scrambling code-pair allocated for HS-DSCH transmission can modulate up to M distinct data streams. Thus, in principle, the peak throughput with code re-use is M times the rate that can be achieved with a single transmit antenna. In practise, rather than maximising the peak bit-rate, more robust intermediate bit-rates might be sought through modulating M streams using for example 16QAM as opposed to having a single stream modulated with 64QAM, thereby reducing the required E_b/N_0 [291]. System level studies summarised in TR 25.848 would suggest that the average sector throughput using a MIMO system with four transmit and four receive antennas increases by a factor of up to 2.8 for maximum CIR scheduling compared to a conventional system with a single transmit and a single receive antenna. Under the considered conditions, peak data-rates in excess of 20 Mbit/s are possible. The throughput might be further increased by applying closed-loop MIMO techniques as opposed to the open-loop technique considered for the HSDPA feasibility study.

TR 25.848 concludes that MIMO represents a promising approach for both average throughput and peak bit-rates to be increased. However, the performance benefits depend strongly on the channel model assumed. Further work on channel models and more investigations on the relative performance benefits of the various MIMO techniques available are therefore needed before MIMO can be adopted for HSDPA.

10.5.4 Stand-alone DSCH

A stand-alone DSCH, which is a DSCH on a downlink carrier that is different from the WCDMA carrier carrying the companion DPCH, was also looked at in the HSDPA feasibility study. The drawback of the traffic segregation resulting from having a carrier dedicated to HS-DSCH is reduced trunking efficiency. From this perspective, we would prefer all types of traffic (i.e. RT traffic on DCHs and NRT traffic on DSCHs) to be multiplexed onto a common thick pipe. On the other hand, if the total traffic is downlink biased, then it makes sense to introduce downlink-only carriers. Since these are by nature not suitable for interactive RT traffic, using them only for HS-DSCH purposes is something one would simply have to accept. It may even be possible to further optimise such a stand-alone HS-DSCH (compared to a HS-DSCH on a 'normal' carrier), for example in terms of modulation schemes, since there would be no constraints regarding backwards compatibility with existing channel structures, modulations, etc.

This matter depends obviously also on spectrum availability. In the 2 GHz band, there is currently no suitable spectrum for 3G downlink-only carriers, but new 3G spectrum (e.g. between 2.5 and 2.7 GHz) could well contain chunks of downlink-only bands. A related question is that of mobile terminal complexity. Clearly, a mobile terminal would need an additional receiver chain to receive DPCH and DSCH on two different carriers, but the complexity associated with this varies with the spacing between the two carriers.

10.5.5 And What About Increased Data-rates on the Uplink?

It is generally assumed that it is more important to provide high data-rates on the downlink than it is on the uplink, which is why 3GPP considered initially only the downlink direction. However, if it were possible to increase the data-rates also on the uplink using similar techniques, then it would be desirable to do so. Indeed, some of the techniques proposed for HSDPA could also be applied to increase uplink data-rates, e.g. adaptive modulation and hybrid ARQ. However, because of the scheduling overhead and the scheduling delay involved on the uplink, optimisations that can be made on the downlink, such as maximum CIR scheduling, are more difficult to realise. Those techniques that are also suitable for the uplink may eventually (i.e. in the release 6 time-frame or even later) be implemented to provide what might be called High Speed Uplink Packet Access (HSUPA).

11

TOWARDS 'ALL IP' AND SOME CONCLUDING REMARKS

This concluding chapter provides first an introduction to some of the planned release 5 enhancements to UMTS and the GPRS/EDGE RAN (GERAN). These can be seen as the first step towards 'all IP'. The challenges when having to deliver real-time IP services over an air interface, in particular voice over IP services, are summarised and possible solutions to achieve spectrum efficiencies similar to those of optimised cellular voice services are outlined.

Unlike the UTRA modes, the GSM/GPRS air interface was not designed to handle real-time packet-data traffic. Further enhancements are required to support real-time IP bearers in GERAN. Possible alternatives are discussed and planned solutions are briefly described.

The last section provides summarising comments on multiplexing efficiency and access control, two key topics that kept reappearing throughout this book, for TDMA, hybrid CDMA/TDMA and CDMA systems.

11.1 Towards 'All IP': UMTS and GPRS/GERAN Release 5

In early 1999, a few operators and infrastructure manufacturers got together to form 3G.IP [92], an 'industrial lobby group' intended to influence 3GPP (for UMTS) and ETSI (for EDGE/GPRS) towards adoption of what was then termed an 'all IP' network architecture. This further evolved GPRS architecture, based on packet technologies and IP telephony, would function as a common core network to access networks based on both EDGE and WCDMA radio access technologies. The system would have to be able to deliver IP-based multimedia services efficiently, requiring also enhancements to the air interface. One of the main benefits provided by IP technology is the service flexibility, as already identified in Section 2.4. Another motivating factor for some operators is the wish to focus exclusively on the packet-switched infrastructure, once the technology is ready, to facilitate network management and possibly to save also on infrastructure costs. This would imply that existing services, including 'plain voice', would have to be replicated on the packet-switched infrastructure.

The top-level architecture devised by 3G.IP was adopted by 3GPP as a basis for enhancements to the packet-switched part of the 3GPP network architecture which, if further delays can be avoided, are to be incorporated in release 5 of the 3GPP

11 TOWARDS 'ALL IP' AND SOME CONCLUDING REMARKS

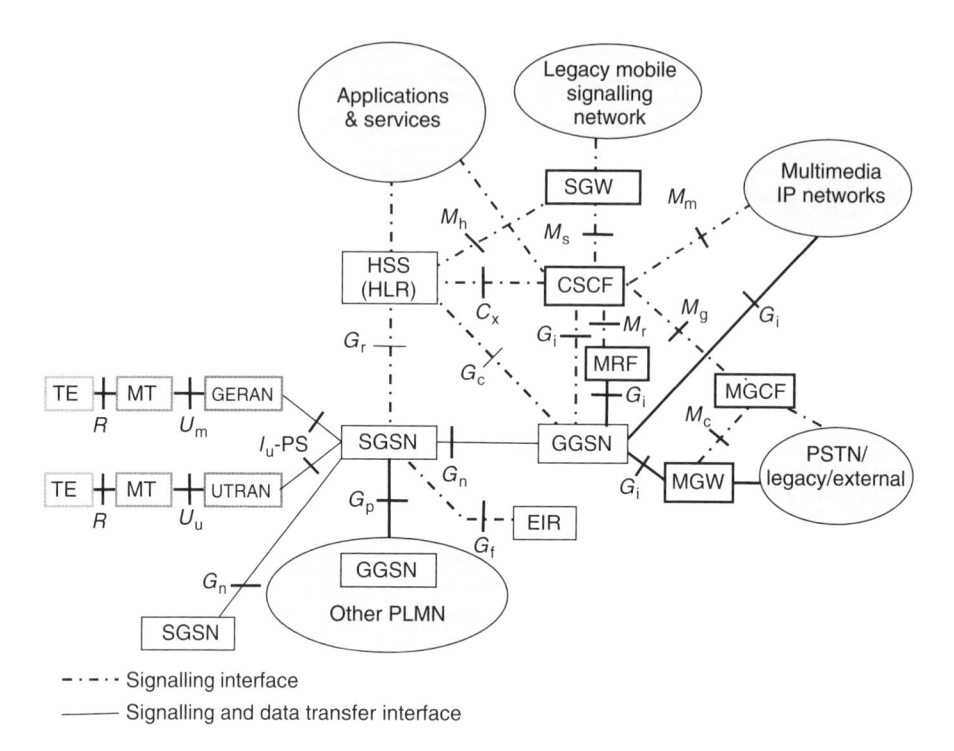

Figure 11.1 Simplified architecture for the support of IP-based multimedia services in 3GPP release 5

specifications. A few new functional entities are to be introduced, which form the IP Multimedia Subsystem (IMS), as shown in Figure 11.1. These are the Call State Control Function (CSCF), the Media GateWay (MGW), the Media Gateway Control Function (MGCF), the Media Resource Function (MRF), and Signalling GateWays (SGW). Additionally, the PS-domain core network of UMTS release 1999 composed of SGSNs and GGSNs (in itself an evolution from GPRS) has to be evolved to provide the necessary quality of service for real-time traffic. Finally, the concept of the HLR has evolved, it is to be substituted by a Home Subscriber Server (HSS).

One of the key components to provide IP-based multimedia services is the CSCF, which executes, among other things, the call control. To be precise, it was decided to base the required protocol on the IETF Session Initiation Protocol (SIP) [293]. 'Session' is in fact a more generic and appropriate term than 'call'. The latter is mostly associated with voice calls, while the aim is to provide all imaginable types of IP-based multimedia services, which may, but do not have to contain voice streams. A media gateway is required when providing an interconnection from the GGSN to legacy circuit-switched networks, such as the Public Switched Telephone Network (PSTN). The MGCF controls that gateway. The MRF performs multiparty call and multimedia conferencing functions. The signalling gateways perform signalling conversion as required.

Compared to the original 3G.IP reference architecture to be found in Reference [92], for the 3GPP network architecture shown in the release 5 version of TS 23.002 [294], some of the new elements were further decomposed. In particular, there are now different types of CSCF, for instance a proxy CSCF with a policy control function, the latter having a separate interface to the GGSN, namely G_o, on top of the G_i interface shown in Figure 11.1. Two types of signalling gateways were introduced, namely the Transport Signalling GateWay function (T-SGW) and the Roaming Signalling GateWay function (R-SGW). For details on the functionality of the individual components, see TS 23.002 [294] and TS 23.228 [295]. To provide a simple picture, Figure 11.1 shows the original 3G.IP reference architecture with a single type of CSCF and a single type of SGW, but with some of the terminology adapted to that now used in 3GPP.

The introduction of the IMS and the evolution of the PS-domain of the core network have a relatively moderate impact on UTRAN. In terms of support of real-time packet traffic over the air interface, the capabilities of the two UTRA modes, in particular the UTRA FDD mode, were discussed in some detail in Chapter 10 and it was shown that UTRA FDD provides considerable flexibility in this respect. One key concern relates to spectrum efficiency, mainly due to the overhead introduced by IP and higher layer headers, which has to be reduced or eliminated through appropriate means, as will be discussed in Section 11.2. Other than that, further improvements on top of what is available in release 1999 are being considered and may be introduced, if proven beneficial. These include both mechanisms to improve the radio link performance in general, and mechanisms specifically targeted for optimised wireless IP support, in particular bi-directional real-time and interactive IP-based applications. The latter could for instance consist of improved common downlink channels.

For the so-called GSM/EDGE RAN (GERAN), as the GSM radio network infrastructure is referred to from release 4 onwards, the situation looks different. Connecting GERAN infrastructure directly to the UMTS core network, as intended, means that I_u-CS and I_u-PS interfaces must be supported instead of the A and the G_b interface. According to Reference [296], the connection to the CS-domain via the I_u-CS interface is not so much different from that via the A interface. However, when comparing G_b to I_u-PS, which are the interfaces of relevance here, there are substantial differences. This has to do with the functional split between the core network and the radio access network, which are not the same in GSM and UMTS. It was decided to eliminate any radio-related functionality from the UMTS core network, so that different types of radio technologies could be connected to it (which is exactly what is happening here with UTRA and EGPRS). This resulted in functionality located in the GPRS core network in R97 (and also in EGPRS R99) to be pushed down to the RAN for UMTS R99. For instance, ciphering for the radio link, which used to be performed by the SGSN in GPRS R97, is performed by the RNC in UMTS. Accordingly, if a GERAN is to be connected via I_u-PS to a 3G SGSN, then the ciphering must be implemented in the GERAN. In terms of protocol stacks, LLC and SNDCP known from GPRS terminate in the core network, whereas in UMTS, where LLC and SNDCP were eliminated and PDCP introduced instead, the respective functionality is contained in the RAN. The reader is referred to Reference [296] for further information.

Another fundamental matter is that of the support of real-time packet traffic over the air interface. Essentially, neither GPRS R97 nor EGPRS R99 provide means to support real-time traffic over the air interface, so enhancements are necessary. Options and likely solutions will be discussed in more detail in Section 11.3, after having outlined some of the general challenges relating to the efficient support of voice over IP over air interfaces in Section 11.2. These are relevant for both UTRA and (E)GPRS.

To conclude this section, we would like to point out that 'all IP' means different things to different people. Evolving the GPRS core network, which makes use of some IP technologies, and adopting a few IETF protocols such as SIP for session control, while still keeping for instance cellular mobility management principles, is for a lot of people far from 'all IP'. For these people, release 5 provides only a first step towards 'all IP'. Possible evolutionary paths to 'real all IP' were briefly discussed in Section 2.5.

11.2 Challenges of Voice over IP over Radio

The Internet is working according to the end-to-end principle. It means that only the packet source and the packet destination have to be interested in the packet contents, while the network in between these two entities is assumed to be dumb. It does just one thing, namely sending packets from one place to another, in theory without discrimination. It is assumed that all packets are treated equally, and that their content is not tampered with. This is exactly how the often-quoted service flexibility is achieved: since no assumptions are made about the packets travelling across the network, there are no constraints on the uses to which they can be put. In practise, as multimedia traffic containing real-time streams is starting to be delivered over the Internet, means to provide appropriate QoS for these real-time streams have to be introduced. This implies often that packets are not treated equally anymore, but rather depending on the QoS requirements of the stream they belong to.

Regardless of QoS matters, the end-to-end principle is in direct conflict with what the cellular industry normally does, namely trying to optimise the use of precious spectrum resources depending on the nature of the data to be delivered. We have discussed this in detail for GSM voice in Chapter 4, where the importance of every single bit is known and it is treated accordingly. Efficiency is derived from the following means:

- low-bit-rate voice codecs optimised for wireless use, ideally adapting to the channel conditions;
- avoidance of header overheads since the application carried is known;
- Unequal Error Protection (UEP) according to the importance of the payload bits, so that FEC coding redundancy is only expended for bits for which it is worthwhile;
- Unequal Error Detection (UED) so that the frame erasure rate (FER) depends only on the residual error-rate of the most important bits (when used together with UEP, typically the same bits that enjoy the strongest error protection).

The same is not true for circuit-switched data in 2G systems, however, where, with the possible exception of payload compression (as an equivalent to low-bit-rate coding), none of these techniques is applied, nor is it for packet-switched data in 2.5G systems. So what are the concerns when dealing with real-time IP services?

Let us consider what is probably the most challenging service from an efficiency perspective, namely Voice over IP (VoIP). Assuming that it is desirable to carry voice over the PS-domain, then VoIP is the most obvious way in which this could be achieved. If a pure 'plain' voice service is to be offered (e.g. because it is required to replicate all current services on the packet-switched infrastructure), then this service has to compete against optimised circuit-switched voice in terms of efficiency.

There are two main reasons for which, without taking special measures, a VoIP service cannot compete, in terms of spectrum efficiency, with optimised circuit-switched voice:

- lacking payload optimisation, i.e. having to use equal error detection and protection (EED and EEP) instead of UED and UEP, if the end-to-end principle is respected (since the latter implies that there is no guarantee for the network to know the payload);
- the header overhead due to IP and higher layer headers.

It is important to note that these two factors are independent from whether voice is carried on dedicated or shared channels over the air interface. Choosing circuit-switched voice as a benchmark is simply due to the fact that this is the type of voice service which is typically supported in cellular communication systems, and which allows a high degree of optimisation. The same type of optimisation could also be performed if shared channels were used on the air interface, using for instance PRMA as a multiple access protocol.

11.2.1 Payload Optimisation

As regards UED and UEP, if the payload is known (e.g. both the type of codec applied and the ordering of the output bits according to their importance), these techniques can also be applied in conjunction with a VoIP service. According to Reference [86, p. 96], UEP performs around 1 dB better than EEP, in the conditions considered in Reference [297], the performance difference is 1.5 dB. Using UED and UEP violates somewhat the end-to-end principle, in so far as assumptions are made about the terminal behaviour and as the terminals would have to let the network know what they are doing on the bearers they are assigned. If a network-controlled 'plain voice service' is replicated on the packet-switched infrastructure, with exactly the same features as the original circuit-switched service, then the end-to-end principle is anyway *a priori* abandoned and the network should know what its bearers are used for.

Knowing the importance of the bits and being able to apply UED and UEP accordingly is only one important technique to improve the radio link performance, adapting the type of voice coding depending on the current radio conditions is another one gaining importance. For instance, with the recently standardised Adaptive Multi-Rate (AMR) codec, it is possible to trade off robustness against 'voice fidelity' depending on the current radio conditions. If they are bad, it is better to choose a lower rate, allowing more FEC redundancy to be added while keeping the channel rate constant, so that, even though fidelity is reduced, intelligibility can be improved. This requires either local control of the codec mode (e.g. by performing transcoding close to the radio link, that is converting a radio-independent non-AMR bit-stream into an AMR-coded stream according to the local conditions) or suitable end-to-end protocols to negotiate the rates. In the context of VoIP, the former is clearly not in tune with the end-to-end principle. The latter raises some challenges in terms of protocol architecture and information exchange. It would still leave the end terminals in charge of how they want to deal with media streams, so it could be considered end-to-end, but it would introduce dependence between the transport infrastructure and the applications running on top of it. It would therefore affect service flexibility.

11.2.2 VoIP Header Overhead

Real-time interactive traffic is often carried over IP using the Real Time Protocol (RTP) as an application protocol and the User Datagram Protocol (UDP) as a transport protocol. The header overhead specific to VoIP is therefore composed of IP, UDP and RTP headers. UDP headers are 8 octets long, RTP headers at least 12, and the length of the IP headers depends on the IP version applied. IPv4 headers are at least 20, IPv6 headers at least 40 octets long. Taken together, this means 40 octets or more with IPv4, and 60 octets or more with IPv6. Additional overhead results, for instance, if IP voice packets are encapsulated in an 'outer' IP packet, due for instance to the application of the mobile IP protocol [96] or the IP security protocol with encapsulating security payload [298].

The IP-related header overhead is particularly disturbing with low-bit-rate real-time services such as voice. Because of the packetisation delay, only a limited number of voice frames can be packed into an IP datagram or packet. Ideally, to provide a decent quality and keep the delay low, given a voice frame length of 20 ms typical for cellular communications, there should be a one-to-one relationship between voice frames and IP packets. Recall from Section 4.3 that a GSM full-rate voice frame lasting 20 ms measures 260 bits, hence roughly 33 octets before error coding, an enhanced full-rate frame 244 bits or 31 octets. In other words, with one frame per packet, the header overhead is bigger than the payload, and this even before adding lower-layer headers (e.g. at RLC and MAC), which are not required in an optimised voice solution, but may be required for VoIP.

Given that spectrum is the most precious resource for an operator of a mobile communications system, the key concern is inefficiency on the air interface. Hence the question is whether we do need to carry the headers over the air interface and, if we do, whether we can somehow compress them.

11.2.3 How to Reduce the Header Overhead

11.2.3.1 Header Removal

Clearly, the most drastic approach to remove the IP-related header overhead is to terminate the VoIP session in the RAN, i.e. at the BSC or the RNC, and to send conventional voice without any headers to the mobile terminal. This would imply that there is no VoIP client in the mobile terminal and the aspect of IP service flexibility would not be exploited. However, it would still allow reliance on the packet-switched core-network infrastructure for the delivery of the voice service.

11.2.3.2 Transparent Header Compression

The only approach that is compatible with the end-to-end principle and does not reduce service flexibility, is so-called *transparent header compression*. It means that IP/UDP/RTP headers are compressed, before a packet is sent over the air interface, and decompressed at the receiving end, before being handed over to the IP stack (e.g. in the terminal). This is shown in Figure 11.2 for the downlink direction. Transparency implies lossless compression, hence in the absence of transmission errors over the air interface, from an IP perspective, this process works as if no header compression were applied at all. Owing to the fact that these headers contain a lot of fields, which remain static over the

11.2 CHALLENGES OF VOICE OVER IP OVER RADIO

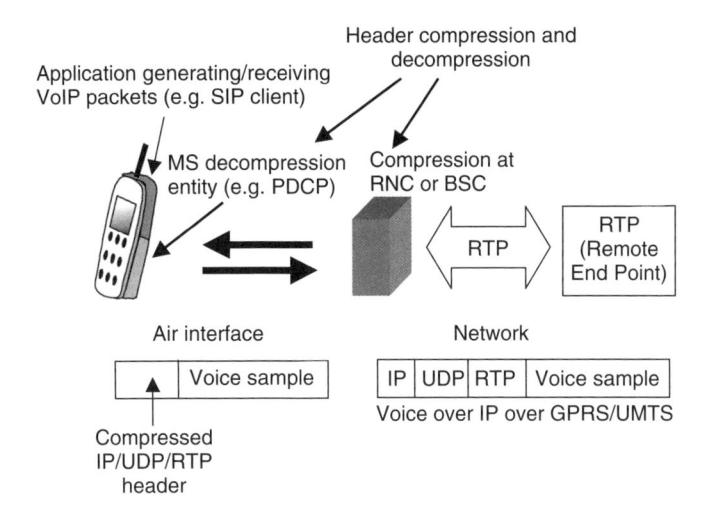

Figure 11.2 Header compression over the air interface

duration of a voice call (e.g. the source and destination IP addresses), or change in a predictable fashion (sequence numbers, RTP time stamps), very high compression ratios can be achieved.

'Early schemes' suitable for IP/UDP/RTP header compression, such as that specified in Reference [299], were not designed for radio links and are known not to be suitable for cellular communications [222,300]. Triggered by 3G.IP activities, a working group was set up in IETF to deal with so-called *robust* header compression schemes, which are at the same time very efficient and do not suffer unduly from errors experienced on the wireless transmission medium. This RObust Header Compression (ROHC) scheme was recently finalised and is specified in Reference [222]. It is supported by the UMTS PDCP protocol from release 4 onwards [301]. A short description of a preliminary scheme, which was fed into the ROHC standardisation process, can be found in Reference [302]. With ROHC, the average combined IP/UDP/RTP header size can be reduced to less than two octets for a conventional two-party voice call, hence the relative header overhead is reduced to a few percent, which appears to be acceptable. However, lossless compression comes at the price of variable sizes for the compressed headers, as unexpected or rare changes of certain header fields require longer compressed headers to be used. In the case of UTRA FDD, for example, owing to the inherent statistical multiplexing capability and the support of real-time VBR traffic, this can be tolerated (although it may, depending on the solution adopted, consume precious TFCI code points to signal what header size is currently being used). On the other hand, when trying to support packet-voice over the rather narrow-band GSM carriers, which are partitioned into even narrower basic physical channels, variable size headers are uncalled for.

Another problem in the end-to-end context is that the header compression entity can only guess what media streams are carried by the IP packets it is dealing with, through application of appropriate heuristics. To maximise the compression efficiency, it would be helpful to separate a voice stream running over IP/UDP/RTP from other IP/UDP/RTP streams to apply individually optimised ROHC profiles, and also from non-IP/UDP/RTP streams, for which other compression methods may be applied. This implies again that the mobile network needs to gain some knowledge on the services it is dealing with.

Finally, it is important to note that if end-to-end encryption is applied, then the redundancy in the header fields is eliminated and compression cannot be performed anymore. Compression would have to take place before encryption, but since compression is a hop-by-hop operation, here applied over the radio link, this is incompatible with end-to-end encryption. Additionally, when *block-ciphers* are applied for encryption, which work on a block of bits rather than single bits, then it is also not possible anymore to apply, for example, UEP, because there is no evident relation between the importance of a bit at a given position in a packet before and after ciphering. However, when *stream-ciphers* are applied, which work bit-by-bit (e.g. by performing an 'exclusive or' operation between a data bit to be encrypted and a key), then at least this problem does not arise.

11.2.3.3 Non-transparent Header Compression or Header Stripping

Transparent header compression exhibits drawbacks, namely that the remaining header overhead is still non-negligible and, in particular, variable in size. When dealing with a known service such as voice, which is delivered over a radio link from which timing and other information can be extracted from layers 1 and 2, one could be tempted to reduce the IP/UDP/RTP header overhead over the air interface to zero. The link information, together with information submitted at call set-up (e.g. IP source and destination addresses), should be sufficient to regenerate these headers at the other end, although it cannot be guaranteed that the regenerated header is bit-wise identical to the original header. This process is sometimes referred to as *header stripping* and regeneration. It is envisaged to be used with VoIP in cdma2000 systems and EDGE/GPRS release 5 systems[1]. There have been intense discussions in the IETF ROHC working group on whether such a header-stripping scheme should be dealt with by IETF at all, given that it can only be used in very specific circumstances, and violates fundamental Internet principles. Nobody can tell, for instance, what the impact on a remote VoIP client unaware of the applied compression scheme would be, when headers appearing to be semantically correct are not completely identical to the original headers. At the time of writing, this matter has not yet been resolved, see Reference [303] for up-to-date information.

Another approach, which solves the problem somewhat differently, is a gatewaying solution. It involves the setting up and interworking of two different VoIP sessions, one between the mobile terminal and the gateway (e.g. at the BSC or the RNC), and one between the gateway and the remote end (e.g. a VoIP client outside the domain of the mobile network). Header stripping applied over the air interface does in this case not affect the separate session between gateway and remote end. Also, both mobile terminal and gateway are aware of the header stripping method which is applied, and can therefore behave in such a manner that unexpected header field changes are avoided in the session between them and that the headers can be regenerated properly.

In both these cases, one would have to ask what the justification for a VoIP client in the mobile terminal is. If an optimised solution is sought for a specific service (i.e. plain voice) with little required IP service flexibility, why not use a header removal solution, in which the VoIP session is terminated at the BSC/RNC and interworked with a 'plain voice bearer' to the mobile terminal?

---

[1] For EDGE/GPRS, the process is referred to as header removal in TS 43.051, hence the use of terminology is somewhat different from that used here.

11.3 Real-time IP Bearers in GERAN

The suitability of UMTS radio bearers for real-time packet-date traffic has already been discussed in Chapter 10. The protocol architecture defined for UMTS release 1999 (as illustrated in Figure 10.2), which separates transport from logical channels, and enables various modes of operation for the MAC and the RLC layers, provides to a large extent the flexibility required for IP multimedia services to be supported. For VoIP to be carried efficiently over the air interface, header adaptation methods need to be introduced (e.g. ROHC header compression, which is supported from release 4 onwards) and means must be found to enable the application of UEP/UED. Whether all this will be available in the release 5 time-frame remains to be seen.

In the case of GERAN, the situation is somewhat different. In Section 4.9, we illustrated how the relative overhead created by the GPRS protocol stack can be detrimental when dealing with short IP packets such as those typical for VoIP. IP/UDP/RTP header compression or even header stripping help little in this case without introducing other system enhancements. In addition, neither GPRS R97 nor EGPRS R99 were designed for real-time traffic; further enhancements are therefore also required in this respect.

11.3.1 Adoption of UMTS Protocol Stacks for GERAN

Regarding the protocol stack, when a GERAN is connected to the GSM/UMTS core network in 'I_u-mode' (i.e. via the I_u-PS and I_u-CS interface), then the protocol model from UMTS is used, with UMTS MAC, RLC, and in particular the UMTS PDCP instead of the GPRS LLC and SNDCP. This reduces the overhead drastically, since RLC, MAC and PDCP can be used in modes which create minimal overhead (e.g. zero octets for MAC and RLC and one octet for PDCP). Additionally, the GERAN may also be connected to a GSM/UMTS core network via A and G_b interface, to support pre-release-4 GSM/GPRS terminals which do not support the new protocols, as shown in Figure 11.3. This figure also shows that an I_{ur}-like interface is envisaged both between GERAN base station systems and possibly even between a GERAN BSS and a UTRAN radio network subsystem. This has to do with a 3GPP work item on optimised radio resource management across different radio access technologies. The overall description of the GERAN is contained in 3GPP TS 43.051 [304].

11.3.2 Shared or Dedicated Channels?

In GPRS and EGPRS, non-real-time packet-data traffic is, with the exception of exclusive allocation in dual transfer mode, supported on shared channels. Voice and other real-time traffic, on the other hand, is only supported in the circuit-switched domain, using dedicated channels on the air interface. When wanting to support real-time traffic such as voice in the packet-switched domain, a very interesting question from a MAC perspective is whether it should be carried on dedicated channels or on shared channels. The latter would allow statistical multiplexing to be exploited by assigning physical channels to individual users only for the duration of their talk spurts. This matter was discussed in a contentious manner first in 3G.IP, then in ETSI and finally in 3GPP. It also ties into the discussion on interference-limited versus blocking-limited operation provided in Section 4.6.

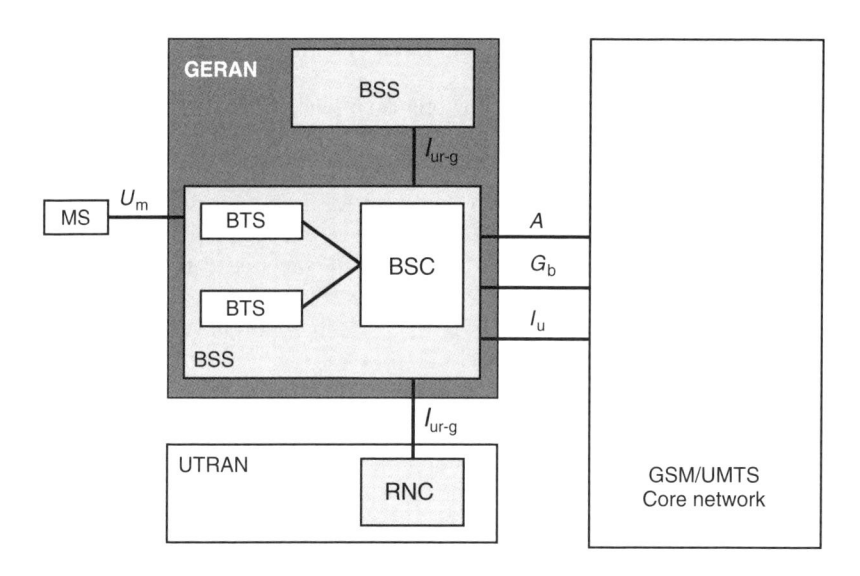

Figure 11.3 GERAN connected to GSM/UMTS core network

In Reference [305], a system capacity analysis for voice-only traffic with the same voice model we used in Chapter 7 is provided, considering 1/3, 3/9 and 4/12 frequency reuse patterns, and different values for the pathloss coefficient γ_{pl} and the standard deviation of the shadowing σ_s. 'Packet-switched operation' on shared channels is compared with 'circuit-switched operation' on dedicated channels. Only GMSK modulation is considered. The required carrier-to-interference ratio (CIR) is assumed to be 1 dB higher for calls carried on shared channels than for those on dedicated channels, due to additional header overhead that is needed (e.g. an RLC/MAC header identifying the user). A hard-blocking limit of 2 % is assumed and a CIR outage probability of 10 %. Furthermore, in the case of packet-switching, the dropping probability threshold is 1 %, with packets being dropped if they are delayed for more than 40 ms. Only the downlink is considered, because this is assumed to be the worst-case scenario from an interference perspective.

With dedicated channels, it is generally found that a tight 1/3 reuse pattern resulting in a soft-blocked capacity-limit delivers higher capacity than looser reuse patterns, the capacity being the higher, the higher γ_{pl} and the lower σ_s. However, when γ_{pl} is relatively low and σ_s sufficiently high, then 3/9 reuse provides higher (hard-blocked) capacity. For $\gamma_{pl} = 4$, this is the case when σ_s exceeds 8 dB, for $\gamma_{pl} = 3.5$ already at 7 dB. The capacity achieved with shared channels is in all cases higher than that with dedicated channels (in some cases by more than 50 %), and in most cases, the maximum shared-channel capacity, which is achieved either at a 3/9 or a 4/12 reuse, is hard-blocked. Judging from these results, shared-channel operation appears to be the preferred option. However, since the downlink is considered, capacity reductions in the reverse direction caused by the necessary uplink access mechanism are ignored. Furthermore, while interleaving on dedicated channels is typically diagonal over eight bursts, to make efficient use of resources and limit access and scheduling delay, realistically rectangular interleaving over four bursts will have to be applied on the shared channels in the same manner as in GPRS. According to Reference [297], this will increase the required CIR by 2 dB in a typical urban environment, assuming pedestrian speed and ideal frequency hopping (no values are provided for other channels). As a result, it is found in Reference [297] that interference-limited operation on dedicated channels provides higher capacity than blocking-limited operation with statistical multiplexing.

Another capacity comparison between shared and dedicated channels is provided in Reference [183]. When using GMSK modulation, the 'baseline performance' for voice with interference-limited operation on dedicated channels is found to be better than that with blocking-limited operation on shared channels. However, in the case of 8PSK modulation, when using only a half-rate physical channel per call, which is possible owing to the higher data-rates provided by this higher order modulation scheme, shared-channel operation outperforms that on dedicated channels in terms of capacity. On the other hand, when power control and dynamic channel assignment schemes are added, which can improve the performance in the interference-limited case, but not in the blocking-limited case, then the highest capacity is provided using dedicated channels in an interference-limited scenario.

11.3.3 Proposals for Shared Channels

In Reference [306], RLC/MAC design alternatives for the support of integrated services over EGPRS are discussed. It is noted that in-session access in GPRS R97 and EGPRS R99 is unnecessarily slow, because:

- the MS is not identified in the initial access request; and

- a fairly elaborate signalling exchange is taking place, which is not needed for an ongoing session.

We would add another reason, namely that the random access algorithm was not optimised for speed.

In Reference [306], additional control channels are proposed for GERAN, namely a fast packet access channel, a fast packet access grant channel and a fast packet polling channel. The fast packet access channel can either provide fast dedicated access, e.g. if a polling scheme is used, or fast random access, if an R-ALOHA scheme is used. In the R-ALOHA scheme, the random access is accelerated compared to the GPRS solution through various means. Unfortunately, assignments are still interleaved over 20 ms on the fast packet access grant channel, so the acknowledgement delay alone is 20 ms. Correspondingly, the access delay budged for voice is assumed to be 60 ms, which is rather on the generous side.

Similar proposals considering R-ALOHA-based schemes, polling schemes or hybrids thereof have been submitted to the relevant industry and standardisation fora (i.e. first 3G.IP, then ETSI SMG2 and later 3GPP). Polling schemes, for instance, have been proposed on the basis of the already existing USF in GPRS. However, because of the USF interleaving over 20 ms and the timing relationship between downlink USF signalling and uplink block allocation, USF-based polling schemes are too slow for voice unless the GPRS channel structure is modified to enhance their performance.

For us an interesting question is whether MD PRMA would be suitable for the support of real-time traffic on an EGPRS system. In References [48] and [49], we have investigated the performance of MD PRMA for some of the UMTS air interface proposals originally submitted to ETSI, taking into account those air-interface design parameters that are relevant for protocol operation, such as slots per frame, frame duration, etc. The GSM design parameters were also considered for these investigations. A rather aggressive design was assumed, where the delay threshold for packet dropping was set to the duration of a single TDMA frame (i.e. 4.615 ms in the case of GSM). Time-alignment for in-session random access was assumed, allowing normal bursts to be used for contention, so that enough signalling capacity is available for inclusion of an unambiguous temporary mobile identity. Interleaving was, if at all, only considered for the traffic channels, but assuming flexible block boundaries. This means that any TDMA frame can be the first one in a block, hence as soon as an acknowledgement is received, the mobile terminal can transmit on its assigned slot(s), without having to wait for the next block boundary. Most importantly, acknowledgement delays were ignored. The underlying assumption was that a resource assignment message for a voice terminal could be very short. If an implicit resource assignment strategy were adopted, for instance, it would be sufficient if it contained only the temporary mobile identity. Therefore, accepting some link performance loss, it should be possible to signal these assignments on a fast access grant channel without the need for interleaving over multiple TDMA frames, in a manner similar to that envisaged for VRRA, an early GPRS proposal (see Section 4.7).

Under the above assumptions, the multiplexing efficiency for conventional single-carrier PRMA was compared to that of MD PRMA with multiple carriers. Because only eight time-slots are available on a single carrier, the multiplexing efficiency achieved in this case is, at 0.67, rather low. On two carriers, it can be increased to 0.78, on four to 0.85. Pooling additional carriers together provides only a relatively moderate further improvement of the multiplexing efficiency, e.g. an increase to 0.9 when eight carriers are pooled together and to 0.92 with twelve carriers. It will be difficult to replicate these figures in a real system suffering from implementation constraints, resulting for instance in a non-negligible acknowledgement delay. However, a generally relevant conclusion can be drawn from these investigations, which will certainly not be such a big surprise for readers having studied Chapters 7 and 8 attentively. If shared channels are used for real-time traffic, then the number of basic physical channels being provided as shared resource units should be several tens to achieve a worthwhile multiplexing gain. In the particular case considered here, four carriers offering 32 resource units would appear to be the minimum required.

11.3.4 Likely GERAN Solutions

It is difficult to assess to the satisfaction of everybody whether shared-channel operation in a blocking-limited system or dedicated-channel operation in an interference-limited system will provide the highest capacity for real-time packet-data traffic. This matter depends on too many parameters, such as the voice activity factor assumed, the shadowing statistics, the pathloss coefficient, the accuracy of power control, assumptions on the radio link performance differences (which depend on the considered channel models) and finally, the impact of implementation constraints. In the uplink direction, in addition, the losses due to the multiple access protocol would have to be quantified, which depend on the choice of protocol and again on implementation constraints with regards to the chosen protocol.

In the end, rather than spending further effort to prove which solution was better, it was decided in 3GPP to focus first on dedicated channels for real-time traffic, because the standardisation effort associated with such a solution is comparatively limited. Further extensions can always be added later, if a clear need can be demonstrated.

In TS 43.051, which provides the overall description of GERAN, a distinction is made between physical channels (defined as a sequence of radio frequency channels and time-slots) and physical *subchannels*, which are defined as a physical channel or a part of a physical channel with an associated multiframe structure. These physical subchannels can either be dedicated (Dedicated Physical SubCHannel, DPSCH) or shared (Shared Physical SubCHannel, SPSCH). A DPSCH is for one user only and has an associated SACCH. A *dedicated MAC mode* is used on DPSCHs. An SPSCH is for one or more users and has an associated packet timing advance control channel. A *shared MAC mode* is used on SPSCHs. Both dedicated and shared subchannels can either be full-rate or half-rate channels.

A packet data traffic channel or PDTCH, in GPRS and EGPRS always a shared channel (with the exception of dual transfer mode with single-slot operation, see Section 4.8), can now also be mapped onto a DPSCH. In this case, it has its associated SACCH on top of the PACCH, in accordance with the above definition of a DPSCH.

Annex A in TS 43.051 shows the configurations which were adopted for the different Radio Access Bearers (RAB). A conversational RAB for real-time traffic makes always use of the dedicated MAC mode on DPSCHs. The traffic channel used for voice can be either a conventional voice TCH, a new 8PSK TCH designed for voice (featuring for instance interleaving parameters different from those on the E-TCH introduced in EGPRS R99), or a PDTCH (using either GMSK or 8PSK modulation). UEP is only applied on the TCH, whereas EEP is applied on the PDTCH. Other RAB types (i.e. streaming, interactive and background) can either use dedicated or shared MAC mode, and thus be mapped onto either dedicated or shared channels.

An optional fast random access scheme for shared channels may be introduced at some stage. This scheme is intended to feature *access request identifiers*, for inclusion in channel request messages. These are unambiguous temporary identities in the context in which they are used.

When exactly in terms of releases the different solutions will be introduced remains to be seen. Currently, it looks as if only basic VoIP capabilities will be introduced with release 5, while additional features providing performance enhancements will be introduced with later releases.

11.4 Summarising Comments on Multiplexing Efficiency and Access Control

Alongside with access control, a key topic in this book was that of statistical multiplexing and multiplexing efficiency. We have encountered this matter in different manifestations. Before reiterating them, let us define, for a system supporting a single service, statistical multiplexing efficiency η_{mux} as

$$\eta_{\text{mux}} = \frac{\alpha \cdot M}{U}. \tag{11.1}$$

This is a generalisation of Equation (6.1). M is the number of users that can be supported simultaneously while meeting the QoS requirements associated with the requested service, α the activity factor of these users (i.e. the fraction of time during which they have something to transmit), and U the number of resource units, which are shared between the users. In this single-service scenario, the definition of a resource unit is such that exactly one resource unit is required per user while transmitting. When $\alpha = 1$, therefore, $M \leq U$. Statistical multiplexing is said to be *perfect*, when $\eta_{\text{mux}} = 1$. For perfect statistical multiplexing to be possible, the instantaneous load would have to amount always to exactly $M \cdot \alpha$, hence the variance of the instantaneous channel load would have to be zero. Provided that the necessary scheduling capabilities exist, which may (particularly on the uplink) imply some overhead, this is in theory possible, but only if we are dealing with extremely delay-tolerant services. With real-time services, one will have to live with load-fluctuations and/or drop packets that exceed a certain delay threshold.

Strictly speaking, scheduling affects the statistics of the traffic sources, so some people might argue that this should not be called statistical multiplexing anymore. We refer to this as statistical multiplexing all the same, since its efficiency still depends on the statistical behaviour of the sources, even though it now depends also on other factors such as QoS requirements and the details of the scheduling or multiple access protocol employed. When scheduling or access control are required to improve the performance, these are said to be *explicit* means to provide statistical multiplexing.

So far, we have not specified whether M and U relate to a single cell or to multiple cells. For our results on MD PRMA presented in Chapters 7 to 9, our focus was on a single cell, so the resource units shared between a pool of users related to that specific cell. Where considered, we modelled cellular operation in a simplistic manner by accounting only for average intercell interference. It is possible to use dedicated resources within a cell, while sharing resources and performing statistical multiplexing between multiple cells. In this case, the definition of statistical multiplexing efficiency becomes fuzzier, particularly because such a sharing of resources between cells implies normally interference-limited operation, so that it can be tricky to assess the number of resource units U which are available. It is also possible to combine the two, i.e. share resources within a cell and between cells.

For completeness, we extend Equation (11.1) so that it can be used in a scenario with heterogeneous services of on–off nature. Define U now as the number of *basic* resource units and assume that a user requesting service i needs r_i such units while active. The activity factor for service i is α_i. With s different services supported by the system, the multiplexing efficiency can now be defined as

$$\eta_{\text{mux}} = \frac{1}{U} \sum_{i=1}^{s} \alpha_i r_i M_i. \qquad (11.2)$$

Both Equations (11.1) and (11.2) ignore call or session arrivals and departures, that is, M_i, the number of users requesting service i that can be supported simultaneously while meeting the QoS requirements of that service, is assumed to remain constant over the evaluation period.

11.4.1 TDMA Air Interfaces

First TDMA-based cellular communication systems were designed to support voice traffic. They provided resources for user data transfer exclusively in the shape of dedicated channels (control channels, e.g. for system information broadcast and random access, are a different matter) and were operated in a purely blocking-limited manner.

GSM, a TDMA-based system, provides a slow frequency-hopping feature. It is possible to operate a GSM system at tight frequency-reuse factors in an interference-limited manner. This implies fractional loading and soft-blocking, that is, not all time-slots that are in theory available in a cell can be used in practise, but there is no hard limit determining the number of time-slots that can be used, beyond which perfect quality suddenly turns into unacceptable quality. Interference is averaged between co-channel interferers across multiple cells. It means in essence that resources are shared between cells, and it can be viewed as an operation exploiting a form of statistical multiplexing, even though dedicated channels are used for data transfer.

Through introduction of suitable multiple access protocols, it is possible to share resources between users also within a cell and to perform 'proper' statistical multiplexing. For non-real-time traffic, this is the most appropriate solution on a TDMA air interface. In fact, it was adopted for GPRS, using an R-ALOHA-based protocol. Techniques such as adaptive modulation and coding and incremental redundancy discussed in Chapter 4 can be applied in this case to enhance the capacity.

When it was first studied how to support VoIP in GERAN, there was a debate on whether it would be better to use dedicated channels in interference-limited conditions or shared channels, the latter often, but not necessarily implying blocking-limited operation. A system using shared channels, even when not operating in an interference-limited fashion, can provide soft-blocking or soft-capacity, due to the fact that the service quality deteriorates gradually when the target load-limit is being exceeded. There is no absolute consensus on whether dedicated channels in interference-limited conditions or shared channels provide higher capacity for real-time traffic. This depends on many parameters and it is well possible that shared channels are beneficial in certain conditions and dedicated channels in others. It should be noted that some of the techniques that can be used for non-real-time services on shared channels, such as incremental redundancy, are not applicable for real-time services.

From an operator perspective, it would be best to have both possible options available, so that, depending on network deployment and frequency plan in operation, the more efficient of the two can be chosen. Because the standardisation effort required to define shared channels for use with real-time traffic is much bigger than that associated with reusing existing channel types, initially, the focus is on dedicated channels for real-time traffic. Shared channels may be considered for further GERAN evolution.

As regards the benefits of dynamic access control, this can improve the access delay performance, stabilise the random access protocol and be used for service prioritisation, as discussed in Section 4.11 for GPRS. For further benefits of access control, see also the discussion on blocking-limited systems in the next subsection.

11.4.2 Hybrid CDMA/TDMA Interfaces

Hybrid CDMA/TDMA interfaces can be operated both in a blocking-limited and in an interference-limited manner. Slow frequency hopping is in theory possible with such systems. In practise, however, given the substantial carrier bandwidths and the limited spectrum allocation to individual operators, it may often not be possible, and when it is, then it would only provide frequency and interference diversity, but not interference averaging. The feature enabling such a system to be operated in an interference-limited fashion is therefore the CDMA component, which is also providing a kind of interference averaging.

11.4.2.1 Blocking-limited Operation

In the blocking-limited case, the number of resource units U in Equation (11.1) available per cell 'is hard'. However, when suitable multiple access protocols for shared-channel operation such as MD PRMA are used, a soft-capacity feature can be obtained all the same. This is because the quality in terms of, for example, dropping probability, P_{drop}, deteriorates only gradually when the target load-level is exceeded.

We would argue that statistical multiplexing using shared channels is vital to achieve high capacity in blocking-limited scenarios, because only in this way can the resource utilisation be maximised — idle resources are wasted in a blocking-limited system. With carrier bandwidths of more than 1 MHz (e.g. 5 MHz for UTRA TDD), when dealing with low-bit-rate services, statistical multiplexing is not only vital, it is also efficient, owing to the large user population which can be multiplexed onto the shared resources available in a single cell.

MD PRMA in a blocking-limited scenario is investigated in Chapters 8 and 9. A purely blocking-limited scenario would imply that the bit-error-rates do not, or only to a limited extent, depend on the instantaneous load within the cell and the load in co-channel cells. The ideal assumption would be that of orthogonal code-time-slots as resource units, so that the channel can be modelled as a perfect-collision channel, as we did in these two chapters. In this case, dynamic access control, using a backlog-based schemes such as Bayesian broadcast control, is a nice feature to have, since it ensures low P_{drop} over a wide range of traffic levels and avoids stability problems experienced when applying static access control (i.e. fixed permission probabilities). However, unless there are very strict QoS requirements, Bayesian broadcast does not normally provide increased capacity compared to static access control. Dynamic access control is also useful to perform prioritisation at the random access. In particular, it is possible to trade off dropping performance of RT services (e.g. voice) against delay performance of NRT services in heterogeneous traffic scenarios. If one is prepared to put up with increased data access delay, which can be tolerated in the case of NRT services such as Web and particularly email traffic, it is possible to achieve better voice-dropping performance at a given traffic load than with homogeneous voice traffic.

Load-based access control does not make sense in the above-described scenario, where the impact of multiple access interference (MAI) is ignored. In reality, however, due to the CDMA component, MAI *will* affect the communications, namely in terms of the experienced packet erasure rate P_{pe}. Access control affects therefore both P_{pe} and P_{drop}. If access control is restrictive, P_{pe} can be kept low at the expense of possibly unnecessarily high P_{drop} and vice versa. The interesting and, given the complicated interdependencies between P_{pe} and P_{drop}, challenging aspect of access control in this case is to find the optimum trade-off between the two, such that the total rate of packet-loss P_{loss}, i.e. the sum of P_{pe} and P_{drop}, is minimised. This applies to both blocking-limited and interference-limited operation and is discussed in detail in Chapter 7 for the latter. Results for the blocking-limited case are presented in Section 8.4.

11.4.2.2 Interference-limited Operation

In the interference-limited case, the number of resource units U in Equation (11.1) available per cell 'is soft'. As usual, the question is whether it is better to support all traffic types, including real-time traffic, on shared channels, or whether dedicated channels should be used for real-time traffic. In Chapter 7, we compared the performance of MD PRMA with that of a 'circuit-switched benchmark' implying dedicated channels and found that, under the considered conditions, shared-channel operation provided significantly higher capacity. The reason for this is that, due to the low spreading factors on hybrid CDMA/TDMA air interfaces (typically not more than 16, we considered seven), the inherent multiplexing capability provided by the CDMA component is not good enough to compete with the explicit multiplexing mechanisms that are used on shared channels.

Suitable dynamic access control mechanisms can have a fundamental impact on system capacity in this case. With load-based access control, load balancing between time-slots can be performed to reduce instantaneous load fluctuations and increase the capacity. We used what we termed *channel access functions* to perform such load-based access control and found substantial capacity increases compared to both the circuit-switched benchmark and a random access protocol on shared channels operating without access control. It was also shown in Chapter 7 that the capacity gain, which can be obtained through dynamic access control, is not significantly affected by power control errors. This was demonstrated both through simulations assuming a power control error standard deviation σ_{pc} of 1 dB and theoretical results for various values of σ_{pc}.

11.4.3 CDMA Air Interfaces

A pure CDMA system is typically interference-limited. The different users served by a cell share a common power budget. The power budgets are also shared between cells, so that the load in neighbouring cells has a direct impact on capacity and quality in a specific cell, which is sometimes a cumbersome affair to manage. As a result of the large common pipe onto which the interference is multiplexed, wideband CDMA systems offer a near-perfect inherent statistical multiplexing capability, which can be obtained either on shared channels without any worry regarding access control (e.g. using the simple random access protocol), or on dedicated channels. To be precise, this inherent capability is only near-perfect as long as we deal with low-bit-rate traffic, so that the average user bit-rate R_{av} is much lower than the total bit-rate sustained in a cell R_{cell}, allowing a large population of users to be multiplexed onto the common resource.

From the point of view of radio link performance (i.e. required E_b/N_0 to achieve a given bit error rate), dedicated channels are preferred over shared channels, since they allow performance enhancing techniques such as fast power control and soft handover to be applied, as discussed in Chapter 10. The downside of dedicated channels is that they come at the price of permanent signalling overhead, hence they are not suitable for traffic of very bursty nature.

When dealing with low-bit-rate users, probabilistic access control provides very limited benefit, if any at all in a pure CDMA environment. This was demonstrated in Chapter 8, where the performance of reservation-code multiple access, a protocol using pure CDMA as basic multiple access scheme, was compared with that of PRMA and MD PRMA.

However, when high-bit-rate users have to be supported, that is, when $R_{av} \ll R_{cell}$ does not hold anymore, the inherent statistical multiplexing efficiency of a CDMA system is no longer near to perfect. Explicit means for statistical multiplexing must therefore be introduced to improve the performance. This may also include appropriate probabilistic access control schemes, such as load-based access control. Provided that dynamic access control is quick enough and applied to services with relaxed delay requirements, load-balancing in a manner similar to that in the hybrid CDMA/TDMA scenario can be achieved to reduce the instantaneous load fluctuations and, as a result, to increase the capacity of CDMA systems. The various available options for the support of packet traffic in UTRA FDD including the possible access control mechanisms are discussed in Chapter 10.

One issue specific to interference-limited systems, in particular systems with a CDMA element, is the nature of the multiple access interference, the spatial distribution of which might affect system operation significantly.

If only low-bit-rate services are provided, the large number of users, which can be served by each cell, should normally result in spatially more or less uniformly distributed and therefore 'benign' MAI. However, with high-bit-rate services, this is not the case any more. The location of a single user and its distance to the serving base station (or rather the attenuation, which depends on the distance, but also on fading) will heavily affect the interference it inflicts on neighbouring cells. For instance, a high-bit-rate user far away from its serving base station, thus needing to transmit at high power levels, may cause detrimental interference to the nearest neighbouring cell. Appropriate admission control algorithms will have to cater for such circumstances.

Irrespective of the chosen air-interface technology, admission control algorithms will also have to be complemented by resource reservation algorithms required to cater for terminal movements during a call. For a high-priority real-time service, at the time of call admission in a given cell, a check will have to be made of whether sufficient resources are available in the neighbouring cells to which the requesting user might move and, if so, these resources have to be reserved. Only in this case may the call be admitted. This type of resource reservation does not preclude temporary use of the reserved resources by other users, provided that their QoS requirements are such that they can be pre-empted at the time the high-priority user moves into their cell.

These last few considerations show that the support of high-bit-rate users in a mobile environment, using any type of wireless access technology, is a challenging affair, particularly when considering the limited spectrum typically available to individual operators of mobile communication networks. The fundamental problem is that of limited trunking and multiplexing efficiency, which in turn leads to a number of secondary problems. On top of that, as just outlined, in interference-limited systems, in particular CDMA systems, tight interference-control is a crucial matter when dealing with a heterogeneous service mix including high-bit-rate services.

BIBLIOGRAPHY

[1] Goodman DJ. Trends in cellular and cordless communications. *IEEE Commun. Mag.* 1991; 31–40.
[2] Lee WCY. *Mobile Cellular Telecommunications, Analog and Digital Systems.* New York: McGraw–Hill, 2nd edn, 1995.
[3] Steele R, Hanzo S (ed.). *Mobile Radio Communications, Second and Third Generation Cellular and WATM Systems.* Chichester: John Wiley & Sons, 2nd edn, 1999.
[4] Parsons JD. *The Mobile Radio Propagation Channel.* London: Pentech Press, 1991.
[5] Baier PW. CDMA or TDMA? CDMA for GSM? *Proc. PIMRC'94*, The Hague, The Netherlands, September 1994; 1280–1284.
[6] Lee WCY. Overview of cellular CDMA. *IEEE Trans. Veh. Technol.* 1991; 291–301.
[7] Goodman DJ. Cellular packet communications. *IEEE Trans. Commun.* 1990; 1272–1280.
[8] Goodman DJ, Valenzuela RA, Gayliard KT, Ramamurthi B. Packet reservation multiple access for local wireless communications. *IEEE Trans. Commun.* 1989; 885–890.
[9] Wilson ND, Ganesh R, Joseph K, Raychaudhuri D. Packet CDMA versus dynamic TDMA for multiple access in an integrated voice/data PCN. *IEEE J. Select. Areas Commun.* 1993; 870–884.
[10] ETSI, *ETSI GSM Technical Specifications (01–12 Series)*, European Telecommunications Standards Institute, Sophia Antipolis, France.
[11] *ATDMA System Definition.* Issue 4, RACE R2084, March 1996.
[12] Baier A, *et al*. Design study for a CDMA-based third-generation mobile radio system. *IEEE J. Select. Areas Commun.* 1994; 733–743.
[13] Blanz J, Klein A, Nasshan M, Steil A. Performance of a cellular hybrid C/TDMA mobile radio system applying joint detection and coherent receiver antenna diversity. *IEEE J. Select. Areas Commun.* 1994; 568–579.
[14] *Overview of the Omnipoint IS-661-based Composite CDMA/TDMA PCS System.* Omnipoint Corporation, Colorado Springs, 1995.
[15] Rasky PD, Chiasson GM, Borth DE, Peterson RL. Slow frequency-hop TDMA/CDMA for macrocellular personal communications. *IEEE Pers. Commun. Mag.* 1994; 26–35.
[16] *Code Time Division Multiple Access: CDMA and TDMA—A Marriage Made in Heaven?* Source Swisscom, presented at the 1st ETSI SMG2 UMTS Workshop, Sophia Antipolis, France, December 1996.

[17] Elhakeem AK, Di Girolamo R, Bdira IB, Talla M. Delay and throughput characteristics of TH, CDMA, TDMA, and hybrid networks for multipath faded data transmission channels. *IEEE J. Select. Areas Commun.* 1994; 622–637.

[18] Kautz R, Leon-Garcia A. Capacity and scheduling in hybrid CDMA/TDMA wireless ATM networks. *IEEE Globecom'96, Proc. Commun. Theory Mini Conf.*, London UK, November 1996; 147–152.

[19] Prasad R, Nijhof JAM, Cakil HI. Hybrid TDMA/CDMA multiple access protocols for multi-media communications. *Proc. IEEE Internat. Conf. Pers. Wireless Commun.*, New Delhi India, February 1996; 123–128.

[20] Leppänen PA, Pirinen PO. A hybrid TDMA/CDMA mobile cellular system using complementary code sets as multiple access codes. *Proc. IEEE Internat. Conf. Pers. Wireless Commun.*, India, Mumbai, December 1997; 419–423.

[21] Raychaudhuri D. Performance analysis of random access packet-switched code division multiple access systems. *IEEE Trans. Commun.* 1981; 895–901.

[22] Raychaudhuri D, Joseph K. Performance evaluation of unslotted random access packet CDMA channels. *J. Instit. Electron. and Telecommun. Eng.* 1990; 424–431.

[23] Tobagi FA, Storey JS. Improvements in throughput of a CDMA packet radio network due to a channel load sense access protocol. *Proc. 22nd Ann. Allerton Conf.*, 1984, 40–49.

[24] Abdelmonem AH, Saadawi TN. Performance analysis of spread spectrum packet radio network with channel load sensing. *IEEE J. Select. Areas Commun.* 1989; 161–166.

[25] Yin M, Li VOK. Unslotted CDMA with fixed packet lengths. *IEEE J. Select. Areas Commun.* 1990; 529–541.

[26] Prasad R. *CDMA for Wireless Personal Communications*. Boston, London: Artech House Publishers, 1996.

[27] Wieselthier JE, Ephremides A. A distributed reservation scheme for spread spectrum multiple access channels. *Proc. GLOBECOM'83*, San Diego, November/December 1983; 659–665.

[28] Brand AE, Aghvami AH. Joint CDMA/PRMA — A candidate for a third generation radio access protocol. *Proc. IEE Colloq. Mob. Commun. Towards the Next Millennium and Beyond*, May 1996; 9/1–9/6.

[29] Brand AE, Aghvami AH. Performance of the joint CDMA/PRMA protocol for voice transmission in a cellular environment. *Proc. ICC'96*, Dallas TX, June 1996; 616–620.

[30] Brand AE, Aghvami AH. Performance of a joint CDMA/PRMA protocol for mixed voice/data transmission for third generation mobile communication. *IEEE J. Select. Areas Commun.* 1996; 1698–1707.

[31] Brand AE. *A Joint CDMA/PRMA Protocol for Third Generation Mobile Communication*. Diploma Thesis, King's College London and Swiss Federal Institute of Technology (ETH) Zurich, March 1995.

[32] Toshimitsu K, Yamazato T, Katayama M, Ogawa A. A novel spread slotted Aloha system with channel load sensing protocol. *IEEE J. Select. Areas Commun.* 1994; 665–672.

[33] Dong X, Li L. A spread spectrum PRMA protocol with randomized arrival time for microcellular networks. *Proc. GLOBECOM'96*, London UK, November 1996; 1850–1854.

[34] Dong X, Li L. A minimum reservation capacity spread spectrum PRMA protocol for microcellular networks. *Proc. IEEE 47th Veh. Tech. Conf.*, Phoenix, AZ, May 1997; 1351–1355.

[35] Tan L, Zhang QT. A reservation random-access protocol for voice/data integrated spread-spectrum multiple access systems. *IEEE J. Select. Areas Commun.* 1996; 1717–1727.

[36] Mori K, Ogura K. An investigation of permission probability control in reserved/random CDMA packet radio communications. *Proc. PIMRC'97*, Helsinki Finland, September 1997.

[37] Mori K, Ogura K. An adaptive permission probability control method for integrated voice/data packet communications. *IEICE Trans. Fundam. Electron. Commun. Comput. Sci.* 1998; E81A: 1339–1348.

[38] Hoefel RPF, de Almeida C. Capacity loss of CDMA/PRMA systems with imperfect power control loop. *IEE Electron. Lett.* 1998; 1020–1022.

[39] Taaghol P, Taaghol P, Tafazolli R. Burst reservation CDMA protocol for mixed services S-PCS. *Proc. IEEE 48th Veh. Tech. Conf.*, Ottawa, Canada, May 1998; 91–95.

[40] Wang L, Wu J, Aghvami AH. Performance of CDMA/PRMA with adaptive permission probability in packet radio networks. *Proc. 49th IEEE Veh. Tech. Conf.*, Houston, TX, May 1999.

[41] Shin S, Lee L, Kim K. A modified joint CDMA PRMA protocol with an access channel for voice data services. *IEICE Trans. Fundam. Electron. Commun. Comput. Sci.* 1999; E82A: 1029–1031.

[42] Lee S, Oh C, Ahmad A, Lee JA, Kim K. A novel hybrid CDMA/TDMA protocol with a reservation request slot for wireless ATM networks. *IEICE Trans. Commun.* 1999; E82B: 1073–1076.

[43] Chang CJ, Chen BW, Liu TY, Ren FC. Fuzzy/neural congestion control for integrated voice and data DS-CDMA/FRMA cellular networks. *IEEE J. Select. Areas Commun.* 2000; 283–293.

[44] Liang Q. Explicit CDMA/PRMA in cellular environment. *Electron. Lett.* 2000; 2038–2040.

[45] Wen JH, Lai JK, Lai YW. Performance evaluation of a joint CDMA/NC-PRMA protocol for wireless multimedia communications. *IEEE J. Select. Areas Commun.* 2001; 95–106.

[46] Dunlop J, Irvine J, Robertson D, Cosimi P. Performance of a statistically multiplexed access mechanism for a TDMA radio interface. *IEEE Pers. Commun. Mag.* 1995; 56–64.

[47] Li Y, Andresen S. An extended packet reservation multiple access protocol for wireless multimedia communication. *Proc. PIMRC'94*, The Hague, The Netherlands, September 1994; 1254–1259.

[48] Brand AE. *Multidimensional PRMA (MD PRMA) — A Versatile Medium Access Strategy for the UMTS Mobile to Base Station Channel*. Presented at the 1st ETSI SMG2 UMTS Workshop, Sophia Antipolis, France, December 1996.

[49] Brand AE, Aghvami AH. Multidimensional PRMA (MD PRMA) — A versatile medium access strategy for the UMTS mobile to base station channel. *Proc. PIMRC'97*, Helsinki, Finland, September 1997; 524–528.

[50] Cunningham GA. Delay versus throughput comparisons for stabilized slotted ALOHA. *IEEE Trans. Commun.* 1990; 1932–1934.

[51] Rivest RL. Network Control by Bayesian Broadcast. *IEEE Trans. on Inform. Theory* 1987; 323–328.

[52] Brand AE, Aghvami AH. Multidimensional PRMA with prioritised Bayesian broadcast — A MAC strategy for multi-service traffic over UMTS. *IEEE Trans. Veh. Technol.* 1998; 1148–1161.

[53] Narasimhan P, Yates RD. A new protocol for the integration of voice and data over PRMA. *IEEE J. Select. Areas Commun.* 1996; 623–631.

[54] ETSI TS 101 350, *General Packet Radio Service (GPRS); Overall Description of the GPRS Radio Interface, Stage 2 (GSM 03.64 Ver. 6.4.0 Rel. 1997)*, November 1999.

[55] Fresco Diez C, Brand AE, Aghvami AH. Prioritised random access for GPRS with pseudo-Bayesian broadcast control, exponential backoff and stack based schemes. *Proc. ICT'98*, Chalkidiki, Greece, June 1998; 24–28.

[56] ETSI TR 101 112, *Selection Procedures for the Choice of Radio Transmission Technologies of the UMTS (UMTS 30.03 Ver. 3.1.0)*, November 1997.

[57] *Advanced Channel Structures for Future Personal Communications Systems*, a DTI/EPSRC LINK phase 2 research programme between King's College London, NEC Technologies UK, Plextek, Vodafone, June 1995 to May 1998.

[58] 3GPP TS 25.211, *Technical Specification Group Radio Access Network; Physical channels and mapping of transport channels onto physical channels (FDD) (Rel. 1999)*, Ver. 3.6.0, March 2001.

[59] Cao Q. Medium access control (MAC) for wide-band CDMA systems with optimal throughput. *Proc. IEEE 48th Veh. Tech. Conf.*, Ottawa, Canada, May 1998; 988–992.

[60] *MAC Multiplexing on Uplink for Packet Users*. Source Alcatel, submitted as Tdoc R2#2(99) 122 to 3GPP TSG RAN WG2, Stockholm, March 1999.

[61] Brand AE. *A PRMA-based Medium Access Control Protocol for the Uplink Channel of a Third Generation hybrid CDMA/TDMA Air Interface*. PhD Thesis, University of London, October 1999.

[62] Hoefel RPF, de Almeida C. Performance of CDMA/PRMA protocol for Nakagami-m frequency selective fading channel. *IEE Electron. Lett.* 1999; 28–29.

[63] Hoefel RPF, de Almeida C. The fading effects on the CDMA/PRMA network performance. *Proc. 49th IEEE Veh. Tech. Conf.*, Houston, TX, May 1999.

[64] Rappaport SS, Hu LR. Microcellular communication systems with hierarchical macrocell overlays. *Proc. IEEE*, September 1994; 1383–1397.

[65] Hu LR, Rappaport SS. Personal communication systems using multiple hierarchical cellular overlays. *IEEE J. Select. Areas Commun.* 1995; 406–415.

[66] Lee WCY. *Mobile Communications Engineering, Theory and Applications*. New York: McGraw–Hill, 2nd edn, 1997.

[67] Johnston W. Europe's future mobile telephony system. *IEEE Spectrum* 1998; 49–53.

[68] http://www.privateline.com/PCS/history.htm.

[69] Aghvami AH. *Digital Mobile and Personal Communication Systems*. Lecture Notes, vol. 2, King's College London, 1994.

[70] Website of the GSM Association at http://www.gsmworld.com.
[71] *An Overview of the Application of CDMA to Digital Cellular Systems and Personal Cellular Networks*. Qualcomm Incorporated, San Diego, May 1992.
[72] Uddenfeldt J. Evolution towards third generation mobile networks. *Proc. RACE Mobile Telecom Workshop*, Amsterdam, The Netherlands, May 1994; 1–2.
[73] Aghvami AH. Future CDMA cellular mobile systems supporting multi-service operation. *Proc. PIMRC'94*, The Hague, The Netherlands, September 1994; 1276–1279.
[74] ETSI, TR 101 111, *Requirements for the UMTS Terrestrial Radio Access System (UTRA) (UMTS 21.01 Ver. 3.0.1)*, October 1997.
[75] ETSI TS 101 625, *High Speed Circuit Switched Data (HSCSD)—Stage 1; (GSM 02.34 Ver. 6.0.0 Rel. 1997)*, January 1999.
[76] ETSI EN 301 344, *General Packet Radio Service (GPRS); Service Description; Stage 2 (GSM 03.60 Ver. 6.7.1 Rel. 1997)*, September. 2000.
[77] *EDGE Feasibility Study: Improved Data Rates through Optimised Modulation*. Source Ericsson, submitted as Tdoc SMG2#22 150/97 to ETSI SMG2, Munich, Germany, May 1997.
[78] ETSI EN 300 904, *Bearer Services (BS) supported by a GSM Public Land Mobile Network (PLMN); (GSM 02.02 Ver. 6.1.0 Rel. 1997)*, April 1999.
[79] Carneheim C *et al*. FH-GSM, frequency hopping GSM. *Proc. IEEE 44th Veh. Tech. Conf.*, Stockholm, July 1994; 1155–1159.
[80] Olofsson H, Näslund J, Sköld J. Interference diversity gain in frequency hopping GSM. *Proc. IEEE 45th Veh. Tech. Conf.*, Chicago, IL, July 1995; 102–106.
[81] Näslund J *et al*. An evolution of GSM. *Proc. IEEE 44th Veh. Tech. Conf.*, Stockholm, July 1994; 348–352.
[82] Chuang J, Timiri S, Whitehead J. Common control channel provisioning for wireless packet data system with compact frequency reuse and its impacts on EDGE system performance. *Proc. IEEE 51st Veh. Tech. Conf.*, Tokyo, Japan, May 2000.
[83] http://www.itu.int/imt/.
[84] ETSI, *Evaluation Report for ETSI UMTS Terrestrial Radio Access (UTRA) ITU-R RTT Candidate*. Submitted to ITU in October 1998, available at www.itu.int/imt/.
[85] *Specifications of Air-interface for a 3G Mobile System*. Japan, Association of radio industries and businesses (ARIB), Ver. 0, December 1997.
[86] Holma H, Toskala A (ed.). *WCDMA for UMTS*. Chichester: John Wiley & Sons, revised edn, 2001.
[87] Samukic A. UMTS Universal Mobile Telecommunications System: Development of standards for the third generation. *IEEE Trans. Veh. Technol*. 1998; 1099–1104.
[88] Concept Group Alpha, *Wideband Direct-Sequence CDMA, Evaluation Document*. Tdoc SMG2 270/97, ETSI SMG2, Bad Salzdetfurth, Germany, October 1997.
[89] Dahlman E, Beming P *et al*. WCDMA—The radio interface for future mobile multimedia communications. *IEEE Trans. Veh. Technol*. 1998; 1105–1118.
[90] Concept Group Delta, *WB TDMA/CDMA, System Description, Performance Evaluation, Ver. 2.0b*. Tdoc SMG 899/97, ETSI SMG, Madrid, December 1997.
[91] *Future Radio Wideband Multiple Access System—FRAMES, Basic Description of Multiple Access Scheme, Ver. 2.0*. ACTS AC090, November 1996.

[92] http://www.3gip.org.
[93] ETSI TR 101 683, *Broadband Radio Access Network (BRAN); HIPERLAN Type 2; System Overview*, Ver. 1.1.1, February 2000.
[94] Pereira JM. Fourth generation: now it is personal! *Proc. PIMRC'2000*, London, UK, September 2000.
[95] Nakajima N, Yamao Y. Development for 4th generation mobile communications. *Wirel. Commun. Mob. Comput.* 2001; 3–12.
[96] IETF IP Mobility Working Group, list of drafts and requests for comments at http://www.ietf.org/html.charters/mobileip-charter.html.
[97] IETF Seamoby Working Group, list of drafts and requests for comments at http://www.ietf.org/html.charters/seamoby-charter.html.
[98] http://www.mwif.org.
[99] http://www.ist-brain.org.
[100] Urban J, Wisely DR *et al*. BRAIN—an architecture for a broadband radio access network of the next generation. *Wirel. Commun. Mob. Comput.* 2001; 55–75.
[101] ETSI, *Broadband Radio Access Networks (BRAN); HIPERLAN Type 2; Requirements and Architectures for Interworking between HIPERLAN/2 and 3rd Generation Cellular Systems*, Ver. 0.e, April 2001.
[102] Rom R, Sidi M. *Multiple Access Protocols: Performance and Analysis*. New York: Springer, 1990.
[103] Schwartz M. *Telecommunication Networks: Protocols, Modeling and Analysis*. Reading, MA: Addison-Wesley, 1988.
[104] Bertsekas D, Gallager R. *Data Networks*. New Jersey: Prentice Hall, 2nd edn, 1992.
[105] ETSI TS 100 573, *Physical Layer on the Radio Path, General Description (GSM 05.01 Ver. 6.1.1 Rel. 1997)*, July 1998.
[106] Dell'Anna M, Mohebbi BB, Brand AE, Aghvami AH. A study of polarisation diversity for a microcellular environment. *Proc. PIMRC'97*, Helsinki, Finland, September 1997; 575–579.
[107] Special Issue on Space Division Multiple Access (SDMA), *Wireless Personal Communications*. Kluwer Academic Publishers, October 1999.
[108] Krishnamurthy SV, Acampora AS, Zorzi M. Polling based media access protocols for use with smart adaptive array antennas. *Proc. ICUPC'1998*, Florence, Italy, October 1998; 337–344.
[109] Baier PW. A critical review of CDMA. *Proc. IEEE 46th Veh. Tech. Conf.*, Atlanta, GA, April/May 1996; 6–10.
[110] Viterbi AJ. Spread spectrum communications—myths and realities. *IEEE Commun. Soc. Mag*. 1979; 11–18.
[111] Gilhousen KS *et al*. On the capacity of a cellular CDMA system. *IEEE Trans. Veh. Technol*. 1991; 303–312.
[112] Falciasecca G *et al*. Impact of non-uniform spatial traffic distribution on cellular CDMA performance. *Proc. PIMRC'94*, The Hague, The Netherlands, September 1994; 65–69.
[113] Sköld J, Gudmundson B, Färjh J. Performance and characteristics of GSM-based PCS. *Proc. IEEE 45th Veh. Tech. Conf.*, Chicago, IL, July 1995; 743–748.

[114] Mohr W. PCS 1900 and IS-95 CDMA system comparison for PCS applications in America. *Americas Telecom Technology Summit*, Rio de Janeiro, Brazil, June 1996; 153–159.

[115] Eriksson H, Gudmumdson B, Sköld J. Multiple access options for cellular based personal communications. *Proc. IEEE 43rd Veh. Tech. Conf.*, Secaucus, NJ, May 1993; 957–962.

[116] Reimers U. DVB-T: the COFDM-based system for terrestrial television. *IEE Electron. & Commun. Eng. J.* 1997; 28–32.

[117] Roberts LG. *ALOHA Packet System with and without Slots and Capture*. Stanford Research Institute, Advanced Research Projects Agency, Network Information Center, Stanford, CA, 1972.

[118] Zorzi M, Pupolin S. Slotted ALOHA for high-capacity voice cellular communications. *IEEE Trans. Veh. Technol.* 1994; 1011–1021.

[119] Dziong Z, Jia M, Mermelstein P. Adaptive traffic admission for integrated services in CDMA wireless-access networks. *IEEE J. Select. Areas Commun.* 1996; 1737–1747.

[120] Yao S, Geraniotis E. Optimal power control law for multi-media multi-rate CDMA systems. *Proc. IEEE 46th Veh. Tech. Conf.*, Atlanta, GA, April/May 1996; 392–396.

[121] Soroushnejad M, Geraniotis E. Multi-access strategies for integrated voice/data CDMA packet radio networks. *IEEE Trans. Commun.* 1995; 934–945.

[122] Sampath A, Holtzman JM. Access control of data in integrated voice/data CDMA systems: benefits and tradeoffs. *IEEE J. Select. Areas Commun.* 1997; 1511–1525.

[123] Bianchi G *et al*. A simulation study of cellular systems based on the capture-division packetized access (CDPA) technique. *Proc. ICC'95*, Seattle, WA, June 1995; 1399–1403.

[124] Abramson N. The ALOHA system—Another alternative for computer communications. *Proc. Fall Joint Comput. Conf., AFIPS Conf.* 1970; 37.

[125] Carleial AB, Hellman ME. Bistable behavior of ALOHA-type systems. *IEEE Trans. Commun.* 1975; 401–410.

[126] Raychaudhuri D, Joseph K. Performance evaluation of slotted ALOHA with generalized retransmission backoff. *IEEE Trans. Commun.* 1990; 117–122.

[127] Hu MC, Chang JF. Collision resolution algorithms for CDMA systems. *IEEE J. Select. Areas Commun.* 1990; 541–554.

[128] Amitay N, Nanda S. Resource auction multiple access (RAMA) for statistical multiplexing of speech in wireless PCS. *IEEE Trans. Veh. Technol.* 1994; 584–595.

[129] Zdunek KJ, Ucci DR, LoCicero JL. Packet radio performance of inhibit sense multiple access with capture. *IEEE Trans. Commun.* 1997; 164–167.

[130] Cidon I, Kodesh H, Sidi M. Erasure, capture, and random power level selection in multiple-access systems. *IEEE Trans. Commun.* 1988; 263–271.

[131] Robertson RC, Ha TT. A model for local/mobile radio communications with correct packet capture. *IEEE Trans. Commun.* 1992; 847–855.

[132] Brand AE. *Performance of Slotted Aloha with Capture and Various Retransmission Schemes*. Source Vodafone, Tdoc 72/95, ETSI SMG2 GPRS ad hoc meeting, Sophia Antipolis, France, October 1995.

[133] *Evaluation Criteria for the GPRS Radio Channel.* Source ETSI SMG2, Tdoc 29/96, ETSI SMG2 GPRS ad hoc meeting, Burnham, UK, February 1996.

[134] Cidon I, Sidi M. The effect of capture on collision–resolution algorithms. *IEEE Trans. Commun.* 1985; 317–324.

[135] Babich F. Analysis of frame-based reservation random access protocols for microcellular networks. *IEEE Trans. Veh. Technol.* 1997; 408–421.

[136] Brand AE, Schweickardt SJ. *Protokolle für DS/CDMA Funknetze.* Report of a semester project, Swiss Federal Institute of Technology (ETH) Zurich, July 1994.

[137] Morrow RK, Lehnert JS. Packet throughput in slotted ALOHA DS/SSMA radio systems with random signature sequences. *IEEE Trans. Commun.* 1992; 1223–1230.

[138] Kleinrock L, Tobagi F. Packet switching in radio channels: Part I — Carrier sense multiple-access modes and their throughput delay characteristics. *IEEE Trans. Commun.* 1975; 1400–1416.

[139] Tobagi F, Kleinrock L. Packet switching in radio channels: Part II — The hidden terminal problem in carrier sense multiaccess and the busy-tone solution. *IEEE Trans. Commun.* 1975; 1417–1433.

[140] Tobagi F, Kleinrock L. Packet switching in radio channels: Part IV — Stability considerations and dynamic control in carrier sense multiaccess. *IEEE Trans. Commun.* vol. COM-25, October 1997.

[141] Widipangestu I, 'T Jong AJ, Prasad R. Capture probability and throughput analysis of slotted ALOHA and unslotted np-ISMA in a Rician/Rayleigh environment. *IEEE Trans. Veh. Technol.* 1994; 457–465.

[142] Goodman DJ, Wei SX. Efficiency of packet reservation multiple access. *IEEE Trans. Veh. Technol.* 1991; 170–176.

[143] Nanda S, Goodman DJ, Timor U. Performance of PRMA: A packet voice protocol for cellular systems. *IEEE Trans. Veh. Technol.* 1991; 584–598.

[144] Nanda S. Analysis of PRMA voice data integration for wireless networks. *Proc. GLOBECOM'90*, San Diego, CA, December 1990; 1984–1988.

[145] Jalloul LMA, Nanda S, Goodman DJ. Packet reservation multiple access over slow and fast fading channels. *Proc. IEEE 40th Veh. Tech. Conf.*, Orlando, FL, May 1990; 354–359.

[146] Wong WC, Goodman DJ. A packet reservation multiple access protocol for integrated speech and data transmission. *IEE Proc.-I* 1992; 607–612.

[147] Frullone M *et al.* On the performance of packet reservation multiple access with fixed and dynamic channel allocation. *IEEE Trans. Veh. Technol.* 1993; 78–87.

[148] Frullone M, Riva G, Grazioso P, Missiroli M. Comparisons of multiple access schemes for personal communication systems in a mixed cellular environment. *IEEE Trans. Veh. Technol.* 1994; 99–108.

[149] Qi H, Wyrwas R. Markov analysis for PRMA performance study. *Proc. IEEE 44th Veh. Tech. Conf.*, Stockholm, Sweden, July 1994; 1184–1188.

[150] Wu G, Mukumoto K, Fukada A. Analysis of an integrated voice and data transmission system using packet reservation multiple access. *IEEE Trans. Veh. Technol.* 1994; 289–297.

[151] Hanzo L, Cheung JCS *et al*. A packet reservation multiple access assisted cordless telecommunication scheme. *IEEE Trans. Veh. Technol.* 1994; 234–244.
[152] Cheung JCS, Hanzo L, Webb WT, Steele R. Effects of PRMA on objective speech quality. *IEE Electron. Lett.* 1993; 152–153.
[153] Chua KC, Tan WM. Modified packet reservation multiple-access protocol. *IEE Electron. Lett.* 1993; 682–684.
[154] Mitrou M, Orinos TD, Protonotarios EN. A reservation multiple access protocol for microcellular mobile-communication systems. *IEEE Trans. Veh. Technol.* 1990; 340–350.
[155] Chua KC. Minislotted packet reservation multiple-access. *IEE Electron. Lett.* 1993; 1920–1922.
[156] Jang JS, Shin BC. Performance evaluation of fast PRMA protocol. *IEE Electron. Lett.* 1995; 347–349.
[157] Bianchi G *et al*. C-PRMA: A centralized packet reservation multiple access for local wireless communications. *IEEE Trans. Veh. Technol.* 1997; 422–436.
[158] Eastwood M, Hanzo L, Cheung JCS. Packet reservation multiple-access for wireless multimedia communications. *IEE Electron. Lett.* 1993; 1178–1180.
[159] Khan F, Zeghlache D. Analysis of aggressive reservation multiple access schemes for wireless PCS. *Proc. ICC'96*, Dallas, TX, June 1996; 1750–1755.
[160] Anastasi G, Grillo D, Lenzini L. An access protocol for speech/data/video integration in TDMA-based advanced mobile systems. *IEEE J. Select. Areas Commun.* 1997; 1498–1509.
[161] Delli Priscoli F. Adaptive parameter computation in a PRMA, TDD based medium access control for ATM wireless networks. *Proc. GLOBECOM'96*, London, UK, November 1996; 1779–1783.
[162] Jeong DG, Jeong WS. Performance of an exponential backoff scheme for slotted-ALOHA protocol in local wireless environment. *IEEE Trans. Veh. Technol.* 1995; 3: 470–479.
[163] Taaghol P, Tafazolli R, Evans BG. Performance optimisation of packet reservation multiple access through novel permission probability strategies. *Proc. IEE Colloq. Netw. Aspects of Radio Commun. Systems*, London, UK, 1996.
[164] Li JF, Cheng SX. A new improvement of PRMA's voice service. *Proc. ICUPC'1998*, Florence, Italy, October 1998; 191–194.
[165] Sindt JC. *Investigations on Physical and MAC Layers related to the Joint CDMA/PRMA Protocol for Third Generation Wireless Networks*. M.Sc. Thesis, King's College London and ISEM Toulon, September 1996.
[166] Akyildiz IF, Levine DA, Joe I. A slotted CDMA protocol with BER scheduling for wireless multimedia networks. *IEEE-ACM Trans. Netw.* 1999; 146–158.
[167] Wu J, Kohno R. A wireless multimedia CDMA system based on transmission power control. *IEEE J. Select. Areas Commun.* 1996; 683–691.
[168] Postel J. *Internet Protocol*. Request for Comments 791, Defense Advanced Research Projects Agency (DARPA), September 1981.
[169] Deering S, Hinden R. *Internet Protocol Ver. 6 (IPv6)*. Request for Comments 2460, the Internet Society, December 1998.

[170] Groupe Spécial Mobile (GSM), *Recommendation*, April 1988.
[171] ETSI TS 100 522, *Network architecture (GSM 03.02 Ver. 6.1.0 Rel. 1997)*, July 1998.
[172] ftp://ftp.3gpp.org/Specs/2000-09/<relevant phase/release>, for latest specifications from R97 and ftp.3gpp.org/Specs/<most recent date>/<release>.
[173] ETSI EN 300 924, *enhanced Multi-Level Precedence and Pre-emption service (eMLPP)—Stage 1 (GSM 02.67 Ver. 6.1.1 Rel. 1997)*, December 1999.
[174] ETSI EN 300 925, *Voice Group Call Service (VGCS)—Stage 1 (GSM 02.68 Ver. 6.0.0 Rel. 1997)*, April 1999.
[175] ETSI EN 300 926, *Voice Broadcast Service (VBS)—Stage 1 (GSM 02.69 Ver. 6.0.0 Rel. 1997)*, April 1999.
[176] Mouly M, Pautet M-B. *The GSM System for Mobile Communications*. Telecom Publishing, 1992.
[177] TS 100 936, *Layer 1; General requirements (GSM 04.04 Ver. 6.0.0 Rel. 1997)*, August 1998.
[178] ETSI EN 300 940, *Mobile radio interface layer 3 specification (GSM 04.08 Ver. 6.13.0 Rel. 1997)*, December 2000.
[179] ETSI EN 300 908, *Multiplexing and multiple access on the radio path (GSM 05.02 Ver. 6.9.0 Rel. 1997)*, May 2000.
[180] ETSI EN 300 959, *Modulation (GSM 05.04 Ver. 6.1.1 Rel. 1997)*, June 2000.
[181] ETSI TR 101 362, *Radio network planning aspects (GSM 03.30 Ver. 6.0.1 Rel. 1997)*, July 1998.
[182] Jakes WC. *Microwave Mobile Communications*. Piscataway, NJ: IEEE Press, 1993.
[183] Qiu X *et al*. Supporting voice over EGPRS: system design and capacity evaluation. *Proc. PIMRC'2000*, London, UK, September 2000.
[184] Ivanov K *et al*. Frequency hopping—spectral capacity enhancement of cellular networks. *Proc. ISSSTA 96*, Mainz, Germany, September 1996; 1267–1272.
[185] Rehfuess U, Ivanov K. Comparing frequency planning against 1×3 and 1×1 re-use in real frequency hopping networks. *Proc. IEEE 50th Veh. Tech. Conf.*, Amsterdam, The Netherlands, September 1999; 1845–1849.
[186] ETSI TS 100 552, *Mobile Station—Base Station System (MS—BSS) interface; Channel structures and access capabilities (GSM 04.03 Ver. 6.0.0 Rel. 1997)*, August 1998.
[187] ETSI TS 100 902, *Technical realization of Short Message Service Cell Broadcast (SMSCB) (GSM 03.41 Ver. 6.1.0 Rel. 1997)*, July 1998.
[188] ETSI/3GPP TS 04.18, *Mobile radio interface layer 3 specification, Radio Resource Control Protocol (GSM 04.18 Ver. 8.10.0 Rel. 1999)*, June 2001.
[189] ETSI EN 300 903, *Transmission planning aspects of the speech service in the GSM Public Land Mobile Network (PLMN) system (GSM 03.50 Ver. 6.2.0 Rel. 1997)*, July 2000.
[190] ETSI EN 300 963, *Full rate speech; Comfort noise aspect for full rate speech traffic channels (GSM 06.12 Ver. 6.0.1 Rel. 1997)*, June 1999.
[191] ETSI EN 300 964, *Full rate speech; Discontinuous Transmission (DTX) for full rate speech traffic channels (GSM 06.31 Ver. 6.0.1 Rel. 1997)*, June 1999.

[192] ETSI EN 300 965, *Full rate speech; Voice Activity Detector (VAD) for full rate speech traffic channels (GSM 06.32 Ver. 6.0.1 Rel. 1997)*, June 1999.
[193] Lüders C, Haferbeck R. The performance of the GSM random access procedure. *Proc. IEEE 44th Veh. Tech. Conf.*, Stockholm, Sweden, June 1994; 1165–1169.
[194] Kleinrock L, Lam S. Packet switching in a multiaccess broadcast channel: Performance evaluation. *IEEE Trans. Commun.* 1975; 410–423.
[195] ETSI EN 300 938, *Mobile Station–Base Station System (MS–BSS) interface; Data Link (DL) layer specification (GSM 04.06 Ver. 6.2.1 Rel. 1997)*, September 2000.
[196] ETSI TS 101 038, *High Speed Circuit Switched Data (HSCSD); Stage 2 (GSM 03.34 Ver. 6.0.0 Rel. 1997)*, April 1999.
[197] 3GPP TS 22.034, *Specification Group Services and System Aspects; High Speed Circuit Switched Data (HSCSD); Stage 1 (Rel. 1999)*, Ver. 3.2.1, April 2000.
[198] 3GPP TS 23.034, *Technical Specification Group Core Network; High Speed Circuit Switched Data (HSCSD)—Stage 2 (Rel. 1999)*, Ver. 3.3.0, December 2000.
[199] ETSI TS 100 945, *Rate adaption on the Mobile Station–Base Station System (MS–BSS) Interface (GSM 04.21 Ver. 6.1.0 Rel. 1997)*, August 2000.
[200] ETSI TS 100 528, *GSM Public Land Mobile Network (PLMN) connection types (GSM 03.10 Ver. 6.0.0 Rel. 1997)*, April 1999.
[201] ETSI TS 100 946, *Radio Link Protocol (RLP) for data and telematic services on the Mobile Station–Base Station System (MS–BSS) interface and the Base Station System–Mobile-services Switching Centre (BSS–MSC) interface (GSM 04.22 Ver. 6.2.0 Rel. 1997)*, January 2000.
[202] 3GPP TS 24.022, *Technical Specification Group Core Network; Radio Link Protocol (RLP) for circuit switched bearer and teleservices (Rel. 1999)*, Ver. 3.4.0, September 2000.
[203] ETSI TS 100 930, *Functions related to Mobile Station (MS) in idle mode and group receive mode (GSM 03.22 Ver. 6.2.0 Rel. 1997)*, July 1999.
[204] 3GPP TS 23.122, *Technical Specification Group Core Network; NAS Functions related to Mobile Station (MS) in idle mode (Rel. 1999)*, Ver. 3.5.0, December 2000.
[205] 3GPP TS 24.008, *Technical Specification Group Core Network; Mobile radio interface layer 3 specification; Core Network Protocols—Stage 3 (Rel. 1999)*, Ver. 3.7.0, March 2001.
[206] 3GPP TS 23.108, *Technical Specification Group Core Network (CN); Mobile radio interface layer 3 specification, Core Network Protocols—Stage 2 (Rel. 1999)*, Ver. 3.2.0, March 2000.
[207] ETSI TS 100 512, *Procedures for call progress indications (GSM 02.40 Ver. 6.0.0 Rel. 1997)*, April 1999.
[208] Rappaport SS. Blocking, hand-off and traffic performance for cellular communication systems with mixed platforms. *IEE Proc.-I* 1993; 389–401.
[209] Orlyk PV, Rappaport SS. A model for teletraffic performance and channel holding time characterisation in wireless communication with general session and dwell time distributions. *IEEE J. Select. Areas Commun.* 1998; 788–803.
[210] Ivanovich M *et al*. Performance analysis of circuit allocation schemes for half and full rate connections in GSM. *Proc. IEEE 46th Veh. Tech. Conf.* Atlanta, GA, April 1996; 502–506.

[211] Ivanovich M et al. Channel allocation methods for half and full rate connections in GSM. *Proc. ICC'96*, Dallas, TX, June 1996; 1756–1760.

[212] Kronestedt F, Frodigh M. Frequency planning strategies for frequency hopping GSM. *Proc. IEEE 47th Veh. Tech. Conf.*, Phoenix, AZ, May 1997; 1862–1866.

[213] ETSI EN 301 113, *General Packet Radio Service (GPRS); Service Description; Stage 1 (GSM 02.60 Ver. 6.3.0 Rel. 1997)*, October 1999.

[214] 3GPP TS 22.105, *Technical Specification Group Services and System Aspects; Services and Service Capabilities (Rel. 1999)*, Ver. 3.10.0, January 2001.

[215] 3GPP TS 23.107, *Technical Specification Group Services and System Aspects; QoS Concept and Architecture (Rel. 1999)*, Ver. 3.5.0, December 2000.

[216] Hämäläinen J et al. Multi-slot packet radio air interface to TDMA systems—variable rate reservation access. *Proc. PIMRC'95*, Toronto, Canada, September 1995; 366–371.

[217] ETSI TS 101 351, *General Packet Radio Service (GPRS); Mobile Station–Serving GPRS Support Node (MS-SGSN); Logical Link Control (LLC) layer specification (GSM 04.64, Ver. 6.8.0 Rel. 1997)*, December 2000.

[218] ETSI TS 101 297, *General Packet Radio Service (GPRS); Mobile Station (MS)–Serving GPRS Support Node (SGSN); Subnetwork Dependent Convergence Protocol (SNDCP) (GSM 04.65 Ver. 6.7.0 Rel. 1997)*, March 2000.

[219] ETSI TS 101 343, *General Packet Radio Service (GPRS); Base Station System (BSS)–Serving GPRS Support Node (SGSN); BSS GPRS Protocol (BSSGP) (GSM 08.18 Ver. 6.7.1 Rel. 1997)*, May 2000.

[220] ETSI TS 101 299, *General Packet Radio Service (GPRS); Base Station System (BSS)–Serving GPRS Support Node (SGSN) interface; Network Service (GSM 08.16 Ver. 6.3.0 Rel. 1997)*, July 1999.

[221] *Comparison between PDCP and SNDCP/LLC*. Source Ericsson, submitted as Tdoc 15/00 to ETSI SMG2 EDGE WS meeting, Stockholm, Sweden, February 2000.

[222] Bormann C et al. *Robust Header Compression (ROHC), Framework and four Profiles: RTP, UDP, ESP, and Uncompressed*. Request for Comments 3095, the Internet Society, July 2001.

[223] ETSI EN 300 910, *Radio transmission and reception (GSM 05.05 Ver. 6.7.0 Rel. 1997)*, April 2000.

[224] *Channel Coding Schemes for GPRS*. Source Ericsson, submitted as Tdoc 112/96 to ETSI SMG2 GPRS ad hoc meeting, Stockholm, Sweden, October 1996.

[225] Ameigeiras Gutiérrez PJ et al. Performance of link adaptation in GPRS networks. *Proc. IEEE 52nd Veh. Tech. Conf.*, Boston, MA, September 2000.

[226] ETSI EN 300 912, *Radio subsystem synchronization (GSM 05.10 Ver. 6.6.0 Rel. 1997)*, Nov. 1999.

[227] Malkamäki E, Deryck F, Mourot C. A method for combining radio link simulations and system simulations for a slow frequency hopped cellular system. *Proc. IEEE 44th Veh. Tech. Conf.*, Stockholm, Sweden, July 1994; 1145–1149.

[228] Olofsson H et al. Improved interface between link level and system level simulations applied to GSM. *Proc. ICUPC'1997*, San Diego, CA, October 1997; 79–83.

[229] Hämäläinen S et al. A novel interface between link and system level simulations. *ACTS Mobile Communications Summit*, Alborg, Denmark, 1997; 599–604.

[230] Fairhurst G, Wood L. *Advice to Link Designers on Link Automatic Repeat reQuest (ARQ)*. Work in progress, IETF WG on Performance Implications of Link Characteristics (see http://www.ietf.org/html.charters/pilc-charter.html).
[231] ETSI EN 300 911, *Radio Subsystem Link Control (GSM 05.08 Ver. 6.9.0 Rel. 1997)*, September 2000.
[232] ETSI EN 301 349, *General Packet Radio Service (GPRS); Mobile Station (MS)–Base Station System (BSS) interface; Radio Link Control/ Medium Access Control (RLC/MAC) protocol (GSM 04.60 Ver. 6.12.0 Rel. 1997)*, January 2001.
[233] *Fixed Allocation vs Dynamic Allocation for Non-Real Time Data Services in EGPRS Phase II: Performance Study*. Source Lucent, submitted as Tdoc 579/00 to ETSI SMG2 #35, Schaumburg, IL, April 2000.
[234] Brand AE. *Impact of Retransmission Probability Quantization on Delay Performance in Bayesian Retransmission Control*. Source Vodafone, submitted as Tdoc 10/96 to ETSI SMG2 GPRS ad hoc meeting, Tampere, Finland, January 1996.
[235] Jenq Y-C. Optimal retransmission control of slotted ALOHA systems. *IEEE Trans. Commun.* 1981; 891–895.
[236] Namislo C. Analysis of mobile radio slotted ALOHA networks. *IEEE J. Select. Areas Commun.* 1984; 583–588.
[237] Gallager RG. A Perspective on Multiaccess Channels. *IEEE Trans. on Inform. Theory*, 1985; 124–142.
[238] *IEEE Pers. Commun. Magazine*, 1999.
[239] *EDGE: Concept Proposal for Enhanced GPRS*. Source Ericsson, submitted as Tdoc 130/99 to ETSI SMG2 EDGE WS meeting, Paris, France, May 1999.
[240] *Concept Proposal for GPRS-136 HS EDGE*. Source UWCC, submitted as Tdoc 322/99 to ETSI SMG2 EDGE WS meeting, Paris, France, August 1999.
[241] Chuang J, Timiri S. *EDGE compact and EDGE classic packet data performance*. Available at http://www.uwcc.org/edge/papers/perf_abstract.html.
[242] Dell'Anna M, Aghvami AH. Performance of optimum and suboptimum combining at the antenna array of a W-CDMA system. *IEEE J. Select. Areas Commun.* 1999; 2123–2137.
[243] Dell'Anna M. *Analysis of the Performance of a W-CDMA Cellular Communications System with Smart Antennas*. PhD Thesis, University of London, October 2000.
[244] Roefs HFA, Pursley MB. Correlation parameters of random sequences and maximal length sequences for spread-spectrum multiple access communication. *Proc. 1976 IEEE Canadian Commun. and Power Conf.*, October 1976; 141–143.
[245] Pursley MB. Performance evaluation for phase-coded spread spectrum multiple-access communication—Part I: System analysis. *IEEE Trans. Commun.* 1977; 795–799.
[246] Morrow RK, Lehnert JS. Bit-to-bit error dependence in slotted DS/SSMA packet systems with random signature sequences. *IEEE Trans. Commun.* 1989; 1052–1061.
[247] Perle HC, Rechberger B. Throughput analysis of direct-sequence CDMA-ALOHA in a near/far environment. *Proc. PIMRC'94*, The Hague, The Netherlands, September 1994; 1040–1044.

[248] Morrow RK. Accurate CDMA BER calculations with low computational complexity. *IEEE Trans. Commun.* 1998; 1413–1417.

[249] Bronstein IN, Semendjajew KA. *Taschenbuch der Mathemathik*. Stuttgart & Leipzig: B. G. Teubner Verlagsgesellschaft. Moskau: Verlag Nauka, 25. Auflage, 1991.

[250] Liu Z, El Zarki M. Performance analysis of DS-CDMA with slotted ALOHA random access for packet PCN. *Proc. PIMRC'94*, The Hague, The Netherlands, September 1994; 1034–1039.

[251] Ganesh R, Joseph K, Wilson ND, Raychaudhuri D. Performance of cellular packet CDMA in an integrated voice/data network. *Internat. J. Wireless Information Netw.* 1994; **3**: 199–221.

[252] Lin S. *An Introduction to Error-Correcting Codes*. Englewood Cliffs, NJ: Prentice-Hall, 1970.

[253] Massey JL. *Applied Digital Information Theory II*. Lecture Notes, Swiss Federal Institute of Technology (ETH) Zurich, 1994.

[254] Viterbi AJ, Viterbi AM, Zehavi E. Other-cell interference in cellular power-controlled CDMA. *IEEE Trans. Commun.* 1994; 1501–1504.

[255] Newson P, Heath MR. The capacity of a spread spectrum CDMA system for cellular mobile radio with consideration of system imperfections. *IEEE J. Select. Areas Commun.* 1994; 673–684.

[256] Torrieri DJ. Performance of direct-sequence systems with long pseudonoise sequences. *IEEE J. Select. Areas Commun.* 1992; 770–781.

[257] Stern HP, Mahmoud SA, Wong K-K. A model for generating on–off patterns in conversational speech, including short silence gaps and the effects of interaction between parties. *IEEE Trans. Veh. Technol.* 1994; 1094–1100.

[258] Maglaris B *et al*. Performance models of statistical multiplexing in packet video communications. *IEEE Trans. Commun.* 1988; 834–844.

[259] Kleinrock L. *Queueing Systems, Vol. 2*. New York: Wiley, 1976.

[260] Stallings W. IPv6: The new internet protocol. *IEEE Commun. Mag.* 1996; 96–108.

[261] Anagnostou ME, Sanchez-P. J-A, Venieris IS. A multiservice user descriptive traffic source model. *IEEE Trans. Commun.* 1996; 1243–1246.

[262] Anderlind E, Zander J. A traffic model for non-real-time data users in a wireless radio network. *IEEE Commun. Lett.* 1997; 37–39.

[263] Uziel A, Tummala M. Modeling of low data rate services for mobile ATM. *Proc. PIMRC'97*, Helsinki, Finland, September 1997; 194–198.

[264] Taaghol P, Tafazolli R, Evans BG. Burst reservation multiple-access techniques for the GSM/DCS and DECT systems. *Proc. PIMRC'96*, Taipeh, Taiwan, October 1996; 412–416.

[265] 3GPP TS 25.221, *Technical Specification Group Radio Access Network; Physical Channels and Mapping of Transport Channels onto Physical Channels (TDD) (Rel. 1999)*, Ver. 3.6.0, March 2001.

[266] Gruber JG, Strawczynski L. Subjective effects of variable delay and speech clipping in dynamically managed voice systems. *IEEE Trans. Commun.* 1985; 801–807.

[267] Hoefel RPF, de Almeida C. Numerical analysis of joint CDMA/PRMA protocol based on equilibrium point analysis. *IEE Electron. Lett.* 1999; 2093–2095.

[268] Li Y, Andresen S, Feng B. On the performance analysis of EPRMA protocol with Markov chain model. *Proc. GLOBECOM'95*, Singapore, November 1995; 1502–1506.
[269] Lenzini L, Meini B, Mingozzi E. An efficient numerical method for performance analysis of contention MAC protocols: A case study (PRMA++). *IEEE J. Select. Areas Commun.* 1998; **5**: 653–667.
[270] Papantoni-Kazakos T, Likhanov NB, Tsybakov BS. A protocol for random multiple-access of packets with mixed priorities in wireless networks. *IEEE J. Select. Areas Commun.* 1995; **7**: 1324–1331.
[271] Frigon JF, Leung VCM. A pseudo-Bayesian ALOHA algorithm with mixed priorities for wireless ATM. *Proc. PIMRC'98*, Boston, MA, September 1998.
[272] *Priority Channel Access for GPRS Data Transfer.* Source Motorola, submitted as Tdoc 02/96 to ETSI SMG2 GPRS ad hoc, Tampere, Finland, January 1996.
[273] Buot TV. Priority scheme for mobile data access employing reservation. *Proc. 1995 Wireless Commun. Systems Symposium*, Long Island, NY, November 1995; 87–93.
[274] Buot TV, Watanabe F. Random access algorithm for users with multiple priorities. *IEICE Trans. Commun.* 1996; **3**(E79B): 237–243.
[275] Stavrakakis I, Kazakos D. A multiuser random-access communication-system for users with different priorities. *IEEE Trans. Commun.* 1991; **11**: 1538–1541.
[276] 3GPP TR 25.832, *Technical Specification Group Radio Access Network; Manifestations of Handover and SRNS Relocation (Rel. 1999)*, Ver. 3.0.0, October 1999.
[277] 3GPP TS 25.301, *Technical Specification Group Radio Access Network; Radio Interface Protocol Architecture (Rel. 1999)*, Ver. 3.8.0, June 2001.
[278] 3GPP TS 25.201, *Technical Specification Group Radio Access Network; Physical layer - General description (Rel. 1999)*, Ver. 3.1.0, June 2000.
[279] 3GPP TR 25.922, *Technical Specification Group Radio Access Network; Radio resource management strategies (Rel. 1999)*, Ver. 3.5.0, March 2001.
[280] 3GPP TS 25.213, *Technical Specification Group Radio Access Network; Spreading and modulation (FDD) (Rel. 1999)*, Ver. 3.5.0, March 2001.
[281] 3GPP TS 25.223, *Technical Specification Group Radio Access Network; Spreading and modulation (TDD) (Rel. 1999)*, Ver. 3.5.0, March 2001.
[282] 3GPP TS 25.302, *Technical Specification Group Radio Access Network; Services provided by the physical layer (Rel. 1999)*, Ver. 3.9.0, June 2001.
[283] 3GPP TS 25.321, *Technical Specification Group Radio Access Network; MAC protocol specification (Rel. 1999)*, Ver. 3.8.0, June 2001.
[284] 3GPP TS 25.303, *Technical Specification Group Radio Access Network; Interlayer Procedures in Connected Mode (Rel. 1999)*, Ver. 3.8.0, June 2001.
[285] 3GPP TS 25.331, *Technical Specification Group Radio Access Network; RRC Protocol Specification (Rel. 1999)*, Ver. 3.6.0, March 2001.
[286] Brand AE, Aghvami AH. MAC strategies for UMTS — random access strategies; review of reservation schemes; MD PRMA for TD/CDMA. Presented as SMG2 UMTS Tdoc 38/98 at *2nd ETSI SMG2 UMTS Workshop*, Sophia Antipolis, France, March 1998. Abstract (word) and slides (ppt) in SMG2/UMTS/UTRAN_WS_98_03 folder at ftp://docbox.etsi.org.
[287] 3GPP TS 25.214, *Technical Specification Group Radio Access Network; Physical layer procedures (FDD) (Rel. 1999)*, Ver. 3.6.0, March 2001.

[288] *Associated Control Channel and Soft Handover Issues related to 'Dynamic Resource Allocation Control (DRAC) of uplink DCH'*. Source Alcatel, submitted as Tdoc R2#3(99) 212 to 3GPP TSG RAN WG2, Yokohama, April 1999, available at ftp://ftp.3gpp.org/TSG_RAN/WG2_RL2/TSGR2_03/Docs/zips.

[289] Villier E, Legg P, Barrett S. Packet data transmission in a W-CDMA network — examples of uplink scheduling and performance. *Proc. IEEE 51st Veh. Tech. Conf.*, Tokyo, Japan, May 2000.

[290] 3GPP TS 25.304, *Technical Specification Group Radio Access Network; UE Procedures in Idle Mode and Procedures for Cell Reselection in Connected Mode (Rel. 1999)*, Ver. 3.7.0, June 2001.

[291] 3GPP TR 25.848, *Technical Specification Group Radio Access Network; Physical layer aspects of UTRA High Speed Downlink Packet Access (Rel. 4)*, Ver. 4.0.0, March 2001.

[292] 3GPP TR 25.950, *Technical Specification Group Radio Access Network; UTRA High Speed Downlink Packet Access (Rel. 4)*, Ver. 4.0.0, March 2001.

[293] IETF Session Initiation Protocol Working Group, list of drafts and requests for comments at www.ietf.org/html.charters/sip-charter.html.

[294] 3GPP TS 23.002, *Technical Specification Group Services and Systems Aspects; Network architecture (Rel. 5)*, Ver. 5.2.0, April 2001.

[295] 3GPP TS 23.228, *Technical Specification Group Services and System Aspects; IP Multimedia (IM) Subsystem — Stage 2 (Rel. 5)*, Ver. 5.1.0, June 2001.

[296] Turina D. Challenges of real-time IP support in GSM/EDGE radio access networks. *Proc. PIMRC'2000*, London, UK, September 2000.

[297] Eriksson M *et al*. The GSM/EDGE radio access network — GERAN; system overview and performance evaluation. *Proc. IEEE 51st Veh. Tech. Conf.*, Tokyo, Japan, May 2000.

[298] Kent S, Atkinson R. *IP Encapsulating Security Payload (ESP)*. Request for Comments 2406, the Internet Society, November 1998.

[299] Casner S, Jacobson V. *Compressing IP/UDP/RTP Headers for Low-Speed Serial Links*. Request for Comments 2508, the Internet Society, February 1999.

[300] Svanbro K, Wiorek J, Olin B. Voice-over-IP-over-wireless. *Proc. PIMRC'2000*, London, UK, September 2000.

[301] 3GPP TS 25.323, *Technical Specification Group Radio Access Network; Packet Data Convergence Protocol (PDCP) Specification (Rel. 4)*, Ver. 4.1.0, June 2001.

[302] Svanbro K, Hannu H, Jonsson L-E, Degermark M. Wireless real-time IP services enabled by header compression. *Proc. IEEE 51st Veh. Tech. Conf.*, Tokyo, Japan, May 2000.

[303] IETF Robust Header Compression Working Group, list of drafts and requests for comments at www.ietf.org/html.charters/rohc-charter.html.

[304] 3GPP TS 43.051, *Technical Specification Group GSM/EDGE Radio Access Network; Overall description - Stage 2 (Rel. 5)*, Ver. 5.2.0, June 2001.

[305] Samaras K, Demetrescu C, Luschi C, Yan R. Capacity calculation of a packet switched voice cellular network. *Proc. IEEE 51st Veh. Tech. Conf.*, Tokyo, Japan, May 2000.

[306] Qiu X *et al*. RLC/MAC design alternatives for supporting integrated services over EGPRS. *IEEE Pers. Commun. Mag.* 2000; 20–33.

[307] 3GPP TS 01.01, *Technical Specification Group Services and System Aspects; GSM Release 1999 Specifications (Rel. 1999)*, Ver. 8.1.0, December 2000.

[308] 3GPP TR 21.900, *Technical Specification Group Services and System Aspects; Technical Specification Group working methods (Rel. 1999)*, Ver. 3.6.0, March 2001.

Appendix

GSM AND UMTS STANDARD DOCUMENTS

As a service to the reader, this appendix attempts to shed some light onto the somewhat confusing matters of ETSI and 3GPP specification numbering and of GSM and UMTS release numbering. It also provides pointers to places, from where the relevant documents can be obtained. We appreciate that some of this information is highly volatile and risks being outdated soon. On the other hand, it is hoped that it will provide sufficient clues for the reader to catch up easily with latest policy changes.

GSM Phases and Releases

The standard documents specifying phase 1 of GSM are referred to as *recommendations*. Ignoring the preamble, these recommendations consist of 12 series of documents, each dealing with different aspects of the GSM system. First versions of these documents were published in 1988. All phase 1 recommendations carry version numbers 3.x.y (0, 1 and 2 were used for draft specifications in early stages of the standardisation process). ETSI members may find the last published version of each of these documents in /tech-org/smg/Document/smg/specs/PH1 on the password-protected ETSI FTP server at ftp://docbox.etsi.org. At the time of writing, as a result of the transfer of GSM-related work from ETSI to 3GPP that happened in the year 2000, they were also publicly accessible on ftp://ftp.3gpp.org, namely in the folder /Specs/2000-09/Ph1.

Phase 2 *specifications* (note the change in terminology), first published around September 1994, are identified by version numbers 4.x.y, and their final versions are also available on the two servers, either in /tech-org/smg/Document/smg/specs/PH2, or in /Specs/2000-09/Ph2.

What was initially subsumed under the heading *phase 2+* was later split into yearly releases. Specifications carrying version numbers 5.x.y pertain to release 1996 (R96). They can be found in the ETSI folder /tech-org/smg/Document/smg/specs/Ph2pl-v5 or in the 3GPP folder /Specs/2000-09/R1996. Every new yearly release up to release 1999 (R99) results in an increment of the first digit of the version number by one, that is, release 1997 (R97) carries version numbers 6.x.y, R99 version number 8.x.y. These can also be found on both servers.

Each release must feature a consistent set of functionality, which can be implemented without dependence on a subsequent release. When working on a new release, additions

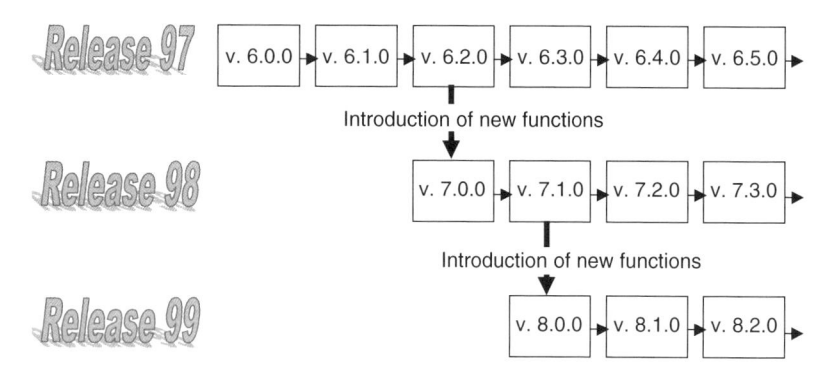

Figure A1 GSM Release Maintenance – thin arrows represent version changes as a result of corrections

to specifications due to new functionality should not affect old releases. To avoid interoperability problems, new functions should not be introduced in old releases. However, it is possible that errors are discovered relating to existing functionality while working on new releases. In this case, also the old releases need to be corrected and updated. It may also happen that errors are discovered when implementing functionality of a release long after the respective specifications were first published, one notorious example being GPRS in R97. This demonstrates that 'old' releases need continued maintenance. Therefore, when looking for the latest versions of R97 or later specifications, the reader is advised not to constrain his search to /Specs/2000-09/<relevant release>, but also to look in /Specs/<date>/<relevant release>, with a <date> after September 2000. The process of maintaining old and introducing new releases is illustrated in Figure A1.

In terms of document numbering, the evolution from phase 1 recommendations to release 1998 specifications was a relatively smooth one. Where new documents had to be added, they fitted into the existing document numbering scheme (i.e. they complemented one of the existing 12 series of documents), an example being GSM 04.60 specifying the GPRS MAC, which was added for release 1997. Some of the existing documents needed to be enhanced significantly to accommodate new functionality (e.g. GSM 04.08 specifying the radio interface layer 3 protocols). By looking at the second version digit, one can easily identify those documents that were affected most by functionality added in a specific release and, as a result, experienced a troublesome document evolution history. As an example, the latest release 1997 version of GSM 05.02, which was significantly affected by GPRS, is (currently!) version 6.9.0. GSM 05.01, by contrast, was much less affected, as indicated by its current version number 6.1.1.

Release 1999 of GSM and UMTS

UMTS release 1999 specifications consist of 15 series of documents, series 21 to 35. The 21-series covers the same topics as the 01-series in GSM, namely requirements, the 22-series the same as the 02-series, and so on. For version numbering of UMTS technical specifications, an approach very similar to that of ETSI for GSM was adopted. Accordingly, specifications of the first UMTS release, release 1999, are identified by version numbers 3.x.y.

The fact that UMTS built in many aspects on GSM had a quite profound impact on some GSM release 1999 specifications. GSM-only release 1999 specifications (e.g. the 05-series on the air interface) continued to evolve in the same manner as before, that is, new versions of documents carrying version number 8.x.y were created, which kept the same document number as the R98 documents they evolved from. However, in areas where UMTS built on GSM specifications, 3GPP effectively took over responsibility for standardisation from ETSI from the very start (i.e. before the remainder of the GSM work was transferred to 3GPP), and release 1999 versions of the respective specifications adopted 3GPP document and version numbers. For instance, there is no release 1999 version of GSM 03.60, the stage 2 GPRS service description, instead there is 3GPP TS 23.060 version 3.x.y. Note that UMTS document numbers contain three digits after the series numbers, so the equivalent of GSM 0x.yz is 3GPP TS 2x.0yz. More information on this topic can be found in GSM 01.01 [307].

Certain specifications have a quite complicated history, GSM 04.08 in particular. Up to R98, it contained the complete specification of the mobile radio interface layer 3 (RIL3). The release 99 version of this document, however, is only an empty shell pointing to three other specifications, into which the original document was split. The RR part of RIL3, which is GSM radio specific, is contained in GSM 04.18 [188], while MM and CM (the latter entailing both circuit-switched call-control and session management for packet-data services) are contained in 3GPP TS 24.008 [205], as they are common to GSM and UMTS. Finally, TS 23.108 [206] provides examples of structured procedures, such as the complete message exchange required when performing a location updating procedure.

The GSM-only release 99 specifications can be found on the ETSI server in the folder /tech-org/smg/Document/smg/specs/Ph2pl-v8. However, for up-to-date versions, it is best to look into /Specs/<newest date>/R1999 on the 3GPP server, where the GSM series of documents can be found alongside the UMTS series of documents.

After Release 1999: Releases 4, 5 and ?

And now to an additional twist: the release following R99 was initially referred to as release 2000, which was intended to contain the 'all-IP' architecture discussed in Chapters 2 and 11. Release 2000 was then split into two releases for various reasons, and the concept of yearly releases was abandoned. Instead, it was decided that releases would be referred to according to the document version number, hence the two releases after R99 are currently known as releases 4 and 5. Release 6 should come thereafter, if the release system is not changed again before this release is due.

With the further evolution of GSM specifications being handed over to 3GPP, a version number mismatch resulted, because the GSM equivalent of release 4 would have been release 9. The adopted solution was to introduce new series of documents in the 40s and 50s range with the same version numbers as the 20 and 30-series. The release 4 equivalent to GSM 01.xy is TS 41.0xy, that to GSM 12.xy TS 52.0xy. This is all summarised in TS 21.900 [308] on 3GPP working group methods.

As a final remark, when work starts at a new specification (i.e. one not existing in previous releases), first draft documents are identified by version numbers 0.x.y. They are then, according to progress, raised to version 1.x.y and possibly to 2.x.y. Once they are sufficiently stable, they are raised from 2.x.y to the version number matching the release, i.e. a release 5 specification may jump from 2.1.0 to 5.0.0 in one go, for example.

INDEX

The reader is advised first to consult major entries for keywords. If not found, he or she may consult subentries for GPRS, GSM, MD PRMA, slotted ALOHA, UMTS, UTRA FDD and UTRA TDD, as appropriate.

8-phase shift keying, 8PSK, 108, 135, 139, 386
   EGPRS, 212, 213, 218
   GERAN, 401, 403
access control, 13, 61, 90
   backlog-based (*see also* MD PRMA) 18–20, 367, 383
   centralised, 94, 95–96, 204, 366
   class-specific/prioritised, 91, 330, 367, 383, 406
   decentralised, 94
   dynamic, 405, 406, 407
   load-based (*see also* MD PRMA), 16–18, 85, 253, 407
   in UTRA, 367, 372, 376, 383
   with power control errors, 303–307, 407
   probabilistic, 292, 293
   for (W)CDMA, 366, 407–408
access delay, 89, 405
activity factor, 10, 404
adaptive modulation and coding, AMC, 386–387, 388, 405
admission control, 60, 302, 408
Advanced Mobile Phone System, AMPS, 25
   Digital AMPS, 26, 211, 212
'all IP', 40, 44, 391, 394
allocation cycle or period, 93, 95, 385
ALOHA, 51, 63, 64
always on, 164
automatic repeat request, ARQ, 52, 59–60, 90, 136
   hybrid ARQ (*see also* HSDPA), 181, 212
   type I-III, 181, 212, 387
   stop-and-wait ARQ, 387

backlog (estimation), 19, 292
backlogged mode, 67, 198–200
backward error control, 52, 181
Bayes' rule, 203, 270
Bayesian broadcast (control), *see also* pseudo-Bayesian broadcast, 204, 270
   for GPRS S-ALOHA, 198, 205
BCH code, 229, 230
Bernoulli experiment, 64, 78, 199, 260, 272
bit error rate, BER, 7
block coding, 229, 230
blocking-limited system, 11, 55, 87, 144, 145, 224
   GERAN, 399, 401, 402, 405
   GPRS, 159
   hybrid CDMA/TDMA, 405, 406
   MD PRMA, 302, 311, 406
blocking probability, 144–145, 159
broadcast channels, 58, 116, 258

C-slot, contention slot, 16, 76, 92, 95, 182
   in MD PRMA, *see* MD PRMA
call arrivals, 246
call blocking and dropping, 143, 144
capacity-limited system, 8
capture (effect), 59, 70–72, 81, 225, 241
   packet capture in GSM, 131, 132
capture-division packetized access, CDPA, 63, 83
capture probability, 200
capture ratio, 70, 203
carrier, frequency carrier, 4
carrier sense multiple access, CSMA, 47, 72–74

carrier-to-interference ratio, 8
CDMA, 4, 51, 54
  direct-sequence, DS-CDMA, 4, 353
  frequency-hopping, 4
  wideband (*see also* WCDMA), 14, 32
cdma2000, 32, 39, 41, 42, 398
cdmaOne, 26, 55
CDMA/NC-PRMA, 85, 291, 308
CDMA/PRMA, 15–21
CDMA/PRMA++, 85
CDMA/TDMA, 12–14, 32, 56–57, 84–86, 405–407
cell clusters, 6
cell dwell time, 147
cell planning, 37
cellular communication systems, 1
  1G & 2G, 3, 25
  3G, 8–9
  4G, 44–47, 56
Central Limit Theorem, CLT, 225, 226
channel access control, *see* access control
channel access function, CAF (*see also* MD PRMA), 16–18, 376, 407
channel feedback, 203, 204, 270, 271
channel (load) sensing, 63, 74–75, 269
chip, 5, 227
circuit-switch(ed), 10, 258
circuit-switched domain, CS-domain, 33, 393
circuit-switched voice, 11, 395
circuit-switching benchmark, CSB, 291, 298, 407
circuit-switching in CDMA, 310
code-division multiplexing, CDM, 378–380
code-rate, 229, 231
code-time-slot, 18, 84, 224, 238, 257, 381
  orthogonal, 406
collision resolution period, 68, 69
collision resolution protocols, 68, 71
comfort noise (parameters), 123–124, 243, 253
common channels, 10, 58, 59, 62, 258
  in GSM, 58, 118, 143
common resource, 13, 21, 58, 287
conflict-free protocol, 62
contention (mode), 50, 77, 253
contention-based protocol, 62
contention burst/packet, 94, 96, 256
contention resolution
  in GPRS, 180, 185, 189, 198
  in GSM, 131, 132
  in UTRA FDD, 369, 371, 372
  in UTRA TDD, 383, 385

convolutional coding, 229
core network, CN, 33, 349, 393
coverage-limited system, 8
cyclic redundancy check, CRC, 121, 229

data link control layer, DLC, 52
datagram (*see also* IP), 247, 248
DCS 1800, 107
dedicated channels, 10, 58, 258, 291, 395
  in CDMA, 310, 407
  for GERAN, 399, 402
deep fades, 261, 262
Digital Audio/Video Broadcast, 46
direct sequences, 227
diversity, 111, 112, 223
downlink channel, 2
dynamic channel assignment or allocation, DCA, 30, 382
dynamic resource allocation control, 20, 375, 376–377
dynamic resource assignment, DRA, 30

EDGE COMPACT, 30, 103, 211, 218–220
email message, 247, 250, 259, 264
encryption, 25, 398
end-to-end principle, 394, 395, 396
Enhanced Data Rates for Global Evolution, EDGE, 29, 40, 103, 139, 218, 391
Enhanced GPRS, EGPRS, 33, 103, 108, 211–220, 387
equal error detection and protection, 395, 403
Erlang, 145, 246, 302
Erlang B formula, 145, 302
ETSI, 26, 32, 100, 391, 399
explicit resource assignment, 83, 93, 263

fading, 3, 364
  fast, 3, 223, 239, 363
FDMA, 3, 54, 56
FEC coding, 112, 229
first-come first-serve policy, FCFS, 69
forward error control, 52
forward error correction (FEC) coding, 52, 90
FPLMTS, 31
frame-based protocols, 335
frame erasure rate, FER, 7
frame reservation multiple access, FRMA, 19, 83, 266, 335
  multidimensional, *see* multidimensional FRMA

INDEX

frequency diversity, 39, 54, 55, 57, 406
  in GSM, 111–114
frequency-division duplex(ing), FDD, 5, 32
  in UMTS, *see* UTRA FDD
frequency hopping, slow, SFH, 4, 54, 55, 57, 111
  baseband and synthesiser hopping, 155
  cyclic hopping (in GSM), 155
  perfect frequency hopping, 112
  random hopping (in GSM), 114, 152, 154, 155
frequency planning, 37
frequency reuse (*see also* reuse factor), 6
  universal, 7, 381
front-end clipping, 78, 263

Gateway GPRS Support Node, GGSN, 100, 160, 161, 393
Gaussian minimum shift keying, GMSK, 99, 108, 135, 401, 403
General Packet Radio Service, *see* GPRS
GERAN, 391, 393, 399–403, 405
  dedicated/shared MAC mode, 403
  real-time IP bearers, 399–403
Gilbert-Varshamov-bound, 229, 230, 280
global probabilistic control schemes, 68, 200, 270
Global System for Mobile Communications, *see* GSM
GPRS, 9, 19, 29, 405
  access burst (format), 182, 185, 197, 209
  ARQ, 162, 195, 218
  block check sequence, BCS, 171, 181, 191
  block error rate, BLER, 175
  block sequence number, BSN, 181, 185, 215, 217
  capacity on demand principle, 158, 169, 212
  cell (re)selection, 156, 169, 179, 211
  class A to C mobiles, 163, 170
  COMPACT, *see* EDGE COMPACT
  contention level, 208
  CS-1 to 4 (coding schemes), 172, 175–177
  downlink interference, 180
  downlink packet transfer, 194–196
  dual transfer mode, DTM, 169–170, 184, 217, 399
  dynamic allocation, 158, 168, 183–184, 190–191
  and power control, 180
  EGPRS RLC/MAC block, 213
  exclusive allocation, 184, 399

extended channel request message, 198
extended dynamic allocation, 184, 190
fixed allocation, 158, 168, 183–184, 191–193
  with EGPRS, 218
  for half-duplex operation, 196–197
  and power control, 180
frequency hopping, 214
GPRS attach procedure, 164
GPRS half-duplex mode or operation, 163, 179, 193, 196–197
GPRS mobility management, GMM, 162, 163
half-rate SPDCH, 170
header check sequence, HCS, 213, 215
incremental redundancy, 211, 212–213, 215
link adaptation, 176, 181, 211, 212–215
link quality control, LQC, 211, 212, 214, 216
LLC PDU, 171, 173, 174
logical link control, LLC, 53, 162, 173, 351, 399
Master PDCH, MPDCH, 158, 165, 166, 168
master-slave principle, 158, 169
MAX_RETRANS, 208, 209, 210
MM states (idle, ready, standby), 163, 169, 194
modulation and coding scheme, MCS, 213, 214
multiplexing, 180, 183, 190
one-phase access, 188–190, 197
PACCH, 165, 167, 403
packet access procedure, 209–211
packet channel request message, 178, 185, 189, 209, 217
packet control unit, 162, 216
packet data channel, PDCH, 158
packet data logical channels, 164–168
packet data protocol, PDP, 163
packet idle mode, 169, 170, 179, 182
packet paging, 194
packet queuing notification, 178, 186
packet system information (PSI) message, 181, 187
packet transfer mode, 169, 170, 179, 182
PAGCH, 165, 178
PBCCH, 165, 166, 167, 208
PCCCH, 165
PDP active state, 164
PDP context (activation), 163, 164

GPRS (continued)
  PDP inactive state, 163
  PDTCH, 158, 165, 167, 403
  persistence level, 208, 209, 210
  physical link layer, 171
  physical RF layer, 170
  polling (mechanism), 185, 195–196
  power control, 179, 180, 190
  PPCH, 165, 167
  PRACH control parameters, 207
  PRACH, 165, 168, 178, 198
  pre-emption, 190, 193
  (radio) priority classes, 207
  PTCCH, 165, 166, 168, 178–179
  radio block, 166, 167, 171, 172, 173, 184
    in EGPRS, 213
  radio resource operating mode, 169
  random access (procedure), 169, 198
  resource utilisation, 157, 158
  RLC data block, 171, 173, 184
  RLC/MAC block, 166, 171, 173–174, 183, 184
  RLC/MAC control block, 184
  S-parameter, 208, 210
  selective bitmap, 181, 186
  session management, 163
  Slave PDCH, SPDCH, 158, 165, 166
  SNDCP, 162, 173, 174
    elimination in UMTS, 351, 393, 399
  Temporary Block Flow, TBF, 167, 181–182, 183, 190
    close- and open-ended, 191, 192
    downlink TBF, 186, 193, 195, 197
    uplink TBF, 185–186, 193, 197
  Temporary Flow Identity, TFI, 167, 182, 183, 191, 195
  Temporary Logical Link Identity, TLLI, 185, 189
  time reuse, 219
  timing advance, TA, 165, 168, 177–179, 195
  two-phase access, 188–189, 197
  TX_INTEGER, 208, 210
  uplink packet transfer, 189–194
  Uplink State Flag, USF, 92, 167–168, 190, 191
    USF bits, 171, 172
    downlink packet transfer, 195
    in EGPRS, 214, 218
    for GERAN, 401
    and power control, 180
    for PRACH/PDTCH multiplexing, 158, 167
    for pre-emption, 93
    for uplink multiplexing, 183
    USF values, 167, 185
GPRS tunnelling protocol, GTP, 160
GPRS/EDGE RAN, see GERAN
Grade of Service, GoS, 143
GSM, 26, 29–31, 393
  access burst, 110–111, 143
  ACCH, 118
  adaptive multi-rate voice codec, AMR, 103, 116, 121, 395
  AGCH, 58, 118, 140
  BCCH frequencies, 118, 119, 126, 153, 155
  BCCH, 117, 124, 125, 140
  BSIC (decoding), 117, 120, 126, 179
  CBCH, 116, 125
  CCCH, 58, 118, 125
  cell allocation, CA, 114
  channel request message, 126, 127, 131
  data link layer, DLL, 105, 132
  dedicated control channels, 118
  dedicated mode, 140, 142
  discontinuous reception, DRX, 125, 167
  discontinuous transmission, DTX, 113, 123–124, 151, 152
  E-FACCH, E-IACCH, 139
  Enhanced Circuit-Switched Data, ECSD, 103, 108, 139
  enhanced full-rate voice codec, 116, 121
  enhanced traffic channel, E-TCH, 116, 135, 139, 403
  establishment cause, 128, 131, 140, 141, 198
  FACCH, 118, 121, 142–143
  FCCH, 117, 125, 166
  fractional loading, 153, 405
  frequency correction burst, FB, 109, 117, 120
  full-rate channel, 115
  full-rate voice (codec), 121, 396
  half-duplex (operation), 108, 111, 138, 163, 365
  half-rate channel, 115
  half-rate voice codec, 102
  handover, 142–143, 169
  hopping sequence, 114
  idle mode, 140
  immediate assignment procedure, 126, 128
  IMSI attach procedure, 126, 140

INDEX

International Mobile Subscriber Identity, IMSI, 125, 132
location updating (procedure), 126, 132, 133, 140–141
MAX_RETRANS, 128, 129
mobile allocation, MA, 114
mobility management, MM, 105, 163
multiframes, 115, 120, 123, 124–126, 166–168
multi-slot classes, 138, 163, 179, 182, 193
normal burst, NB, 109, 143
paging groups, 125
paging (response), 141, 161
PCH, 58, 118
phase 1 (recommendations), 101–102, 427
phase 2 (specifications), 101–102, 427
phase 2+, 102, 427
physical layer, PL, 105
power control, 118, 124, 126, 139
RACH, 58, 118, 126–134, 140, 157
radio interface layer 3, RIL3, 105, 140, 429
radio link protocol, RLP, 105, 136
radio resource management, RR, 105, 351, 429
release 4, 393
releases 1996 to 1999, 102–103, 427
resource utilisation, 143, 144
  heterogeneous traffic, 148–151
  homogeneous traffic, 146–148
S-parameter, 128, 198, 206, 262
SACCH, 118–119, 120, 124, 403
SCH, 117, 125, 166
SDCCH, 118, 125, 140
stealing flags or bits, 121, 213
Subscriber Identity Module, SIM, 128, 350
synchronisation burst, SB, 109, 117, 120
synchronisation sequence, 110–111
system information messages, 117–118, 125, 166
TDMA frame number, FN, 114, 115, 219
time-slot number, TN, 108
timing advance, TA, 108–111, 124, 129, 142, 143
traffic channel, TCH, 115–116, 120, 122, 403
transceiver, TRX, 153, 155
trunking efficiency, 135, 146–147, 151
TX_INTEGER, 128, 129–130, 199, 201

handover, 2
  in GSM/HSCSD, 136, 142, 148
  mobile assisted, MAHO, 109
  seamless, 28, 37, 88–89
  soft (*see also* UMTS), 37, 57, 88, 407
  vertical, 45
hard-blocking (level), 152, 302, 400
header compression, 41, 174, 396, 398
header removal, 396, 398
header stripping, 398
heterogeneous traffic/service mix, 148–151, 408
hidden terminal problem, 74
hierarchical cell layers, 38
hierarchical cellular structure, HCS, 24
High Speed Circuit-Switched Data, *see* HSCSD
High-Speed Downlink Packet-Data Access, *see* HSDPA
High-Speed Uplink Packet-Data Access, HSUPA, 389
HIPERLAN type 2, 43, 45, 46, 47, 56
homogeneous traffic, 148
HSCSD, 29, 103, 134–138
  (multi-slot) configuration, 116, 136, 138, 139
  half-duplex operation, 137, 138
  resource utilisation, 147, 149, 150
HSDPA, 42, 386–389
  fast cell selection, FCS, 386, 388
  hybrid ARQ, 386–387, 388, 389

i-mode, 27, 59
I-slot, information slot, 16, 76, 92, 182, 258
IEEE 802.11 standards, 47, 56, 74
immediate acknowledgement, 205
implicit resource assignment, 79, 83, 91, 93
  with MD PRMA, 260, 263
  on the UTRA FDD CPCH, 373–374
improved Gaussian approximation, IGA, 226, 228
IMT-2000, 31
in-slot protocol, 82, 84, 92, 240, 258
  with UTRA TDD, 85, 383
incremental redundancy, 211, 212–213, 387, 405
initial access, 62, 381, 401
interference averaging, 113, 152, 384, 406
  perfect, 154
interference diversity, 54, 111–114, 154, 406
interference, 379
  adjacent channel, ACI, 6, 36, 155
  co-channel, 6, 112, 143
  intercell, 7, 231, 237, 242
  intracell, 7, 224, 235

INDEX

interference-limited (system), 11, 87, 404, 408
  CDMA, 11, 54, 408
  hybrid CDMA/TDMA, 405, 407
  GERAN, 399, 401, 402, 405
  GSM, 144, 152
  MD PRMA, 279, 280, 312
interleaving, 111, 214, 239, 261
  diagonal, 121, 122
  rectangular, 124, 239, 261
Internet Engineering Task Force, IETF, 41, 397, 398
IP-based multimedia services, 40, 391
IP datagram, 256, 259, 264, 336, 396
IP Multimedia Subsystem, IMS, 392
IP packet, 174, 248
IPv4/v6, 174, 396
IP/UDP/RTP header, 396, 397, 398, 399
IS-136, 26
IS-95, 26, 39
ISMA, 15, 74, 75

Jakes' model, 112
Joint CDMA/PRMA (*see also* CDMA/PRMA), 257, 309, 376
joint detection, JD, 14
  for hybrid CDMA/TDMA, 57, 85, 223
  for TD/CDMA, 223, 224, 240, 242, 276

load balancing, 13, 291, 309, 407, 408
  in UTRA FDD, 372, 377
location area, 58, 133, 160

macro-diversity, 350, 352, 364
macrocell (layer), 24, 36, 37
Markov chain/model, 245
MD PRMA, 18–20, 84, 257–347
  access control, 262, 309, 310
    backlog-based, 270–276, 406
    combined load and backlog-based, 324, 325–326
    load-based, 267–269, 276, 279–310
  access permission probability, 259, 262
  acknowledgement delays, 262
  allocation cycle (length), 262, 264, 336, 342, 347
  C-slot, 258, 259, 347
  channel access function, CAF, 267, 276, 293–298
    heuristic, 293–298, 303
    semi-empirical, SECAF, 268, 294–298, 303, 326

  contention mode (or state), 260, 264, 292, 342
  contention procedure, 259, 260
  dedicated request burst, 261, 264, 320, 321
  delay (performance), 264, 338
  delay threshold, 260, 280
  email traffic, 341–346
  equilibrium point (analysis), 301, 313, 314
  explicit resource assignment, 263
  frame dropping ratio, 261
  frequency-division duplexing, 259
  for GERAN, 401
  idle mode, 260
  implicit resource assignment, 260, 263
  instability, 294
  interleaving, 320, 321
  known-backlog-based access control, KBAC, 291–293, 298, 318
  load balancing, 309
  MAI, 259, 260, 267, 276, 323
  Markov analysis, 313
  multiplexing efficiency, 263, 289, 300, 309, 327
  normalised throughput, 337
  NRT (data) services, 260, 261, 262
  optimized-a-posteriori-expectation-access-scheme, 292
  pseudo-Bayesian broadcast, 272–273, 312, 314, 406
    priority distribution, 336, 339, 345
    prioritisation parameter, 339, 347
  power control errors, 303–308, 309
  random coding, 260, 279
  reservation mode or phase, 260, 292, 336
  session interarrival time, 337
  soft capacity, 302
  time-division duplexing, TDD, 83, 259, 264–266
  timing advance, 264
  for UTRA TDD, 383
  voice activity factor, 280, 313
  Web traffic, 339–341, 344–346
medium access control layer, MAC, 3, 50
microcell (layer), 24, 36, 37
midamble, 110, 381, 385
MIMO processing/transmission, 386, 388
mini-slots, 81, 85, 94, 308
minimum-variance benchmark, MVB, 288–291, 298, 300
  with dropping, MVBwd, 289–290, 301
  with power control errors, 303, 305–307

Mobile Broadband System, MBS, 37, 45
mobile communication systems, 1
mobile IP, 44, 45, 396
mobile originated (call), MO, 58, 141
mobile terminated (call), MT, 58, 141
Mobile Wireless Internet Forum, MWIF, 44
mode (of a distribution), 244, 251
multi-access channel, 57, 58, 126
multidimensional FRMA, MD FRMA, 19, 266, 320–322
multidimensional PRMA, see MD PRMA
multiple access interference, MAI, 5, 16, 60, 223, 225–228
   packet retention, 241
   standard deviation, 284
multiple access protocols, 3, 50, 52
multiple access schemes, basic, 3, 50, 52
multiplexing efficiency (see also MD PRMA), 147, 263, 403–408
   MD FRMA, 327
   'random access protocol', 284–287, 305–307
   UTRA FDD, 376

near-far effect or problem, 54, 70, 223, 350, 363, 364
network layer, NWL, 256, 336, 350
Node B, 33, 349, 350, 357, 364
non-real-time (NRT) data services, 78, 89, 156
Nordic Mobile Telephone System, NMT, 25

opportunity driven multiple access, ODMA, 32, 49
origination mode, 65, 198–200
orthogonal frequency-division multiplexing, OFDM, 32, 47, 56
OSI layer, 50
out-slot protocol, 82, 85, 92

packet acquisition, 241
packet call, 247, 249
packet CDMA, 231
packet dropping ratio, 78, 260
packet-loss ratio, 16, 85, 263
packet reservation multiple access, PRMA, 8, 12–13, 395
   aggressive, 83, 95
   centralised, C-PRMA, 63, 83, 92
   equilibrium point analysis, 77
   integrated, IPRMA, 77, 83
   multidimensional, see MD PRMA
   soft-blocking, soft-capacity, 312
   spread spectrum, SSPRMA, 15
   TDD, 84, 335
packet retention, 241
packet spurt, 77, 248, 259
packet success probability, 228–229
packet-switch(ing), 8, 10, 258
packet-switched domain, PS-domain, 33, 392, 393, 394
packet-switched voice, 11
paired band, 6, 35
Pareto distribution, 247–249, 251, 264
pathloss, 3, 363, 364
pathloss coefficient, 232, 233
perfect collision channel, 64, 203, 224, 317, 331, 406
perfect collision slots, 238
perfect scheduling, 288
permission probability, 78
Personal Communications Network, PCN, 26, 107
Personal Communications System, PCS, 26, 107
Personal Digital Cellular, PDC, 27
Personal Handyphone System, PHS, 27
physical layer, PHY, 3, 50, 51, 52, 221–242
picocell, 24, 36
piggybacking, 93, 262
pilot channels, 258
Poisson distribution, 270, 275, 282
Poisson process, 65, 246
   splitting property, 272, 283
polarisation-division multiple access, PDMA, 51, 53
polling (protocol), 50, 62, 401
power control, 54, 60, 223, 363, 364
   closed-loop, 61, 363, 364
      UTRA FDD, 371, 372, 375
   errors, 21, 235, 236, 303–308, 407
   fast, 350, 407
   open-loop, 363, 366, 370, 386
power grouping, 308
pre-emption (mechanism), 95, 408
prioritisation at random access, 198, 206–207, 330, 367
PRMA++, 82, 85, 157, 254
protocol data unit, PDU, 162, 352
protocol efficiency (for multiple access protocols), 301, 309

pseudo-Bayesian broadcast (control), 19
  with delayed acknowledgements, 273
  impact of MAI, 275
  for MD FRMA, 274
  for MD PRMA, *see* MD PRMA
  prioritised, 330, 331–334, 368
  for PRMA and RCMA, 312
  for S-ALOHA, 203, 270–272
Public Land Mobile Network, PLMN, 100
Public Switched Telephone Network, PSTN, 100, 392

QPSK modulation, 353, 358, 386
Quality of Service, QoS, 89–92, 143, 303, 394, 403

radio link control, RLC, 53
Radio Network Controller, RNC, 33, 349–350, 387, 396, 398
  Controlling RNC, CRNC, 349, 357
  Drift RNC, DRNC, 350
  Serving RNC, SRNC, 350, 357, 364
radio resource control, RRC, 53, 91
RAKE receiver, 54, 223
random access, 50, 72
  fast random access, 403
  'random access CDMA', 281
random access protocols or schemes, 62, 63, 64–68
'random access protocol', RAP, 17, 281, 290, 298, 303
  power control errors, 305–307
random coding, 227, 241, 257
Rayleigh fading, 112, 113 (Fig.)
real-time (services), 89, 156, 391, 393
real-time protocol, RTP, 8, 396
'release 2000', 40, 429
release 5 (GERAN and UMTS), 391
reservation ALOHA, R-ALOHA, 47, 58, 75–76, 92–94
  for GERAN, 401
  in GPRS, 157, 169, 182, 405
  in-slot/out-slot protocol, 335
  resource utilisation, 87
  in UTRA TDD, 385
reservation mode, 77
reservation-code multiple access, RCMA, 20, 311, 312, 316, 407
resource allocation delay, 89
resource auction multiple access, RAMA, 69–70

resource reservation protocol, 408
resource unit, basic, 258, 404
resource utilisation, normalised, 145
resource utilisation in GSM/HSCSD, 140–155
reuse factor, 6, 25, 143
RLC frame, 239, 248, 256, 261
RLC protocol data unit, PDU, 239, 256, 261
robust header compression, ROHC, 397
routing area, 160

scheduling, maximum CIR, 387, 388
segmentation, 252
selective retransmission, 387
self-interference, 223, 230
sensing delay (normalised), 73
service
  bearer and tele-service, 30
  non-transparent, 136, 137, 139
  transparent, 136
service data unit, SDU, 352
service flexibility, 41, 391, 394, 395, 396
Serving GPRS Support Node, SGSN, 100, 160, 161, 349, 393
session arrival, 247, 249, 250
Session Initiation Protocol, SIP, 392, 394
shadowing, 3, 232, 233, 234, 363
shared channel(s), 10, 57, 59, 62, 258, 395
  for GERAN, 399, 402
  in CDMA, 407
Short Message Service, SMS, 102, 132, 133, 161
signalling channels, 258
signal-to-interference ratio, SIR, 7
signal-to-noise ratio, SNR, 7
simplified IGA, SIGA, 226
single-user detection, 223, 276
SINR, 8
sliding window protocol, 181
slotted ALOHA, S-ALOHA, 58, 59, 64–68, 74
  arrival rate, 199, 203, 270, 274
  bi-stable behaviour, 66, 130, 201
  deferred first transmission, DFT, 270
  drift, 67
  equilibrium points, 67
  exponential backoff, 198, 200, 201, 204, 207
  in GSM/GPRS, 128–130, 182, 198–211
  optimum (re)transmission scheme, 200, 203, 204
  prioritised pseudo-Bayesian broadcast, 331

INDEX

stabilisation, 68, 270
system state, 66, 199, 200
throughput, 65, 70–71
  in UTRA, 366, 368, 382
soft-blocking (level), 11, 152, 302, 400, 405
soft-capacity, 11, 152, 302, 406
soft-collision, 72
space-division multiple access, SDMA, 51, 53
spectral or spectrum efficiency, 225, 316, 393, 395
splitting algorithms, 68, 69, 71
spread spectrum ALOHA, 15
spread spectrum multiple access, 4, 54
spread spectrum PRMA, SSPRMA, 15
spreading factor, 5
stabilisation (of random access), 68, 198, 270, 405
stack (splitting) algorithm, 69, 207
stack-based algorithm, prioritised, 330, 331
stalling (of RLC protocol), 181, 216
standard Gaussian approximation, SGA, 223, 225, 228
statistical multiplexing, 10, 12–13, 254, 284, 403–408
  efficiency, *see* multiplexing efficiency
  explicit means, 404, 408
  inherent to CDMA, 60, 375, 397, 407
  near perfect, 307, 310, 375, 376, 408
  perfect, 12, 263, 288, 289, 404

talk gap, 243
talk spurt, 243, 244, 259, 336
TD/CDMA, 34, 237, 311, 353
TD/SCDMA, 42
TDMA, 4, 32, 54, 404–405
  TDMA frame, 4, 238, 258
TDMA/CDMA, *see* CDMA/TDMA
Third Generation Partnership Project, 3GPP, 26, 32, 100, 391, 399
  3GPP TR/TS, 352
3GPP2, 32
3G spectrum, 35
3G.IP, 40, 41, 44, 391–392
time-division duplex(ing), TDD, 5, 32, 87, 265, 266
  in CDMA and CDMA/TDMA, 57
  duplex interval, 265
  frame-based protocols, 335
  narrowband TDD, nTDD, 43
  in UMTS, *see* UTRA TDD
  wideband TDD, wTDD, 43
time-division multiplexing, TDM, 378–380

time-hopping, 384
time scheduling, 378, 387
time-slot, 4, 258
time-slot aggregation, 253
timing advance (*see also* GPRS, GSM, MD PRMA, UTRA FDD and TDD), 94, 240
Total Access Communications System, TACS, 25
traffic asymmetry, 38, 252–253
traffic heterogeneity, 149–150
training sequence, 110, 381
transmission backoff, 68
transmit diversity, 360
transport control protocol, TCP, 8
tree algorithm, 69
trunking efficiency (*see also* GSM), 37, 287, 322, 389, 408

UMTS, 9, 27, 100, 349, 393
  Access Service Class, ASC, 367, 368, 383
  broadcast channel, BCH, 355, 361
  channelisation code, 353, 381
  coded composite transport channel, CCTrCH, 352, 355, 374
  code-tree, 353–354
  control channel, 356
  control-plane, C-plane, 351, 356
  data link layer, DLL, 350
  dedicated channel, DCH, 355
  DSCH, 355, 357
  DTCH, 357, 365, 384
  FACH, 355, 357
  handover, 365
    soft, 349–350, 352, 364, 375
    softer, 349, 364, 365
  High Speed DSCH, HS-DSCH, 386, 387, 389
  HSDPA, *see* major entry
  interleaving depth/period, 355
  logical channel, 351, 356, 357
  MAC, 351, 356, 357, 399
    transparent mode, 352, 357
  Orthogonal Variable Spreading Factor (codes), OVSF, 353–354, 377, 381
  P-CCPCH, 361, 369, 381
  packet data convergence protocol, PDCP, 351, 393, 397, 399
  PCCH, 356
  PCH, 355, 361
  persistence level, 366, 367, 371, 383
  persistence probability value, 367, 383
  physical layer, PHY, 350

UMTS (continued)
  RACH, 354, 357
  radio bearer, 351, 366
  radio frame, 353, 359, 381
  radio resource control, RRC, 351, 367
  release 1999, 41, 349, 365
  releases 4 and 5, 41, 42, 391
  RLC, 351, 357–358, 399
  S-CCPCH, 361, 381
  scrambling code, 353, 354
  slot (format), 353, 360
  spreading factor, variable, 353
  stand-alone DSCH, 388–389
  traffic channel, 356
  Transmission Time Interval, TTI, 355, 363
  transport block, 355, 356, 363
  transport channel, 351, 354–355, 356
    common transport channel, 354, 357, 366
    dedicated transport channel, 354, 357
  transport format, 355, 356, 371, 374
  Transport Format Combination, TFC, 355, 356, 366, 374, 377
  Transport Format Combination Indicator, TFCI, 355, 397
  Universal Subscriber Identity Module, USIM, 350
  user-plane, U-plane, 351, 356
UMTS Terrestrial Radio Access, UTRA, 13, 27
unequal error detection and protection, 394, 395, 398, 399, 403
Universal Mobile Telecommunications System, see UMTS
unpaired band, 6, 32, 35
update interval, 204, 366
uplink channel, 2
user datagram protocol, UDP, 8, 396
UTRA FDD, 32–34, 353, 358–380
  access signature, 366, 368
  access slot (format), 353, 360, 361, 368, 369
  acquisition indication, fast, 366, 370
  compressed mode, 365
  CPCH, 13, 61, 355, 360, 370–374
    channel assignment, CA, 361, 373
    collision detection mechanism, 371
  dedicated channel, DCH, 355, 366, 374–377
  DPCCH, 359, 360, 370, 375
  DPDCH, 359, 360
  DSCH, 366, 377–378
  dynamic resource allocation control, DRAC, 375, 376–377
  FACH, 361, 366, 370, 376
  in-phase, I, 353, 358
  indicator channels, 361, 362, 368–370, 373
  indicators (AI, CAI, etc.), 361, 362, 368, 372–374
  logical channel, 358, 365
  message part, 360, 366
  packet-data traffic, 365–380
    non-real-time, 366, 377, 378
    real-time (VBR), 366, 375
  PCPCH, 360, 370, 371, 373, 374
  PDSCH, 361, 378
  persistence parameters for DRAC, 376
  physical channel, 358
    common, 360–361
    dedicated, DPCH, 359–360
  physical signal, 358
  pilot bits, 359, 360, 370
  power ramping, 366, 370, 371, 372
  PRACH, 360, 366–369
  preambles, 372
    collision detection, 360, 370, 371, 372
    CPCH access, 360, 361, 370
    PRACH, 360, 361, 366, 368, 369
  quadrature-phase, Q, 353, 358
  RACH, 360, 366–370
  SCH, 361
  slotted mode, 365
  spreading factor, 358
  TFC control procedure, 374
  TFCI bits, 359, 360, 370
  timing advance, 362
  transmission probability, dynamic, 376
  Transmit Power Control (bits), TPC, 359, 360, 370
  transport channel, 358, 362, 363, 365, 366
  transport channel reconfiguration, 375, 376
UTRA TDD, 32–34, 57, 237, 353, 380–386
  burst type, 381, 382
  dedicated channel, DCH, 380, 384, 385
  DPCH, 381, 384
  DSCH, 380, 384
  FACH, 380, 384
  indicator channel (i.e. PICH), 381
  packet-data traffic (RT and NRT), 383, 384
  PDSCH, 381, 384, 385
  physical channel, 381
  PRACH, 381, 382, 384

PUSCH, 381, 384, 385
RACH, 380, 381, 382–384
SCH, 381
shared (transport) channels, 380, 384
SHCCH, 356, 380, 384, 385
timing advance, 381, 385
USCH, 355, 357, 380, 384, 385
UTRAN, 33, 349, 351

variable-bit-rate (traffic), VBR, 89, 254, 355
variable-rate reservation-access, VRRA, 157–158
video traffic, 253–255
Viterbi decoder, 229

voice activity detection, VAD, 10, 77, 243, 301, 375
  in GSM, 123–124
voice activity factor, 243, 245, 375
Voice over IP, VoIP, 8, 11, 174, 394–399, 405
voice traffic model, 242–246
vulnerable period, 64, 65 (Fig.)

WCDMA, 34, 40, 61, 88, 353, 391
Web browsing session, 247
wireless local area network, WLAN, 43, 45, 46–47, 74

zero-variance benchmark, ZVB, 306